Bio-Based Packaging

Material, Environmental and Economic Aspects

Wiley Series in Renewable Resources

Series Editor:

Christian V. Stevens, Faculty of Bioscience Engineering, Ghent University, Belgium

Titles in the Series:

Wood Modification: Chemical, Thermal and Other Processes
Callum A. S. Hill

Renewables-Based Technology: Sustainability Assessment
Jo Dewulf, Herman Van Langenhove

Biofuels
Wim Soetaert, Erik Vandamme

Handbook of Natural Colorants
Thomas Bechtold, Rita Mussak

Surfactants from Renewable Resources
Mikael Kjellin, Ingegärd Johansson

Industrial Applications of Natural Fibres: Structure, Properties and Technical Applications
Jörg Müssig

Thermochemical Processing of Biomass: Conversion into Fuels, Chemicals and Power
Robert C. Brown

Biorefinery Co-Products: Phytochemicals, Primary Metabolites and Value-Added Biomass Processing
Chantal Bergeron, Danielle Julie Carrier, Shri Ramaswamy

Aqueous Pretreatment of Plant Biomass for Biological and Chemical Conversion to Fuels and Chemicals
Charles E. Wyman

Bio-Based Plastics: Materials and Applications
Stephan Kabasci

Introduction to Wood and Natural Fiber Composites
Douglas D. Stokke, Qinglin Wu, Guangping Han

Cellulosic Energy Cropping Systems
Douglas L. Karlen

Introduction to Chemicals from Biomass, 2nd Edition
James H. Clark, Fabien Deswarte

Lignin and Lignans as Renewable Raw Materials: Chemistry, Technology and Applications
Francisco G. Calvo-Flores, Jose A. Dobado, Joaquín Isac-García, Francisco J. Martin-Martínez

Sustainability Assessment of Renewables-Based Products: Methods and Case Studies
Jo Dewulf, Steven De Meester, Rodrigo A. F. Alvarenga

Cellulose Nanocrystals: Properties, Production and Applications
Wadood Hamad

Fuels, Chemicals and Materials from the Oceans and Aquatic Sources
Francesca M. Kerton, Ning Yan

Bio-Based Solvents
François Jérôme and Rafael Luque

Nanoporous Catalysts for Biomass Conversion
Feng-Shou Xiao and Liang Wang

Thermochemical Processing of Biomass: Conversion into Fuels, Chemicals and Power, 2nd Edition
Robert C. Brown

Chitin and Chitosan: Properties and Applications
Lambertus A.M. van den Broek and Carmen G. Boeriu

The Chemical Biology of Plant Biostimulants
Danny Geelen, Lin Xu

Biorefinery of Inorganics: Recovering Mineral Nutrients from Biomass and Organic Waste
Erik Meers, Evi Michels, René Rietra, Gerard Velthof

Process Systems Engineering for Biofuels Development
Adrián Bonilla-Petriciolet, Gade P. Rangaiah

Waste Valorisation: Waste Streams in a Circular Economy
Carol Sze Ki Lin, Chong Li, Guneet Kaur, Xiaofeng Yang

Forthcoming Titles:

High-Performance Materials from Bio-based Feedstocks
Andrew J. Hunt, Nontipa Supanchaiyamat, Kaewta Jetsrisuparb, Jesper T.N. Knijnenburg

Bio-Based Packaging

Material, Environmental and Economic Aspects

Edited by

S.M. SAPUAN
Universiti Putra Malaysia, Serdang, Selangor, Malaysia

R.A. ILYAS
Universiti Teknologi Malaysia, Johor, Malaysia

WILEY

This edition first published 2021

Registered Offices
John Wiley & Sons, Inc., 111 River Street, Hoboken, NJ 07030, USA
John Wiley & Sons Ltd, The Atrium, Southern Gate, Chichester, West Sussex, PO19 8SQ, UK

Editorial Office
The Atrium, Southern Gate, Chichester, West Sussex, PO19 8SQ, UK

For details of our global editorial offices, customer services, and more information about Wiley products visit us at www.wiley.com.

Wiley also publishes its books in a variety of electronic formats and by print-on-demand. Some content that appears in standard print versions of this book may not be available in other formats.

Library of Congress Cataloging-in-Publication Data

Names: Sapuan, S.M., editor. | R.A. Ilyas, editor.
Title: Bio-based packaging : material, environmental and economic aspects / edited by S.M. Sapuan, Universiti Putra Malaysia, Serdang, Selangor, Malaysia, R.A. Ilyas, Universiti Teknologi Malaysia, Johor, Malaysia.
Description: First edition. | Hoboken, NJ : Wiley, 2021. | Series: Wiley series in renewable resources | Includes index. | Includes bibliographical references and index.
Identifiers: LCCN 2020024015 (print) | LCCN 2020024016 (ebook) | ISBN 9781119381075 (cloth) | ISBN 9781119381044 (adobe pdf) | ISBN 9781119381051 (epub)
Subjects: LCSH: Packaging–Materials. | Packaging–Environmental aspects. | Packaging–Economic aspects. | Food–Packaging. | Biodegradable products.
Classification: LCC TS198.2 .B56 2021 (print) | LCC TS198.2 (ebook) | DDC 664/.09–dc23
LC record available at https://lccn.loc.gov/2020024015
LC ebook record available at https://lccn.loc.gov/2020024016

Cover Design: Wiley
Cover Images: Trees growing on coins © wk1003mike/Shutterstock, Turtle in ocean © Krofoto/Shutterstock, Eco friendly dishware © jannoon028/Shutterstock

Set in 10/12pt TimesLTStd by SPi Global, Chennai, India
Printed and bound by CPI Group (UK) Ltd, Croydon, CR0 4YY

C9781119381075_170321

Contents

List of Contributors

Sharmiza Adnan Pulp and Paper Laboratory, Biomass Technology Programme, Forest Products Division, Forest Research Institute Malaysia, Kepong, Selangor, 52109, Malaysia

Z.M.A. Ainun Program of Pulp and Paper and Pollution Control, Laboratory of Biopolymer and Derivatives, Institute of Tropical Forestry and Forest Products, Universiti Putra Malaysia, Serdang, Selangor, 43400, Malaysia

H.A. Aisyah Institute of Tropical Forest and Forest Products (INTROP), Universiti Putra Malaysia, Serdang, Selangor, 43400, Malaysia

Dogan Arslan Metallurgical and Materials Engineering Department, Faculty of Chemical and Metallurgical Engineering, Istanbul Technical University, Maslak, Istanbul, 34469, Turkey

M.R.M. Asyraf Department of Aerospace Engineering, Faculty of Engineering, Universiti Putra Malaysia, Serdang, Selangor, 43400, Malaysia

M.S.N. Atikah Department of Chemical and Environmental Engineering, Universiti Putra Malaysia, Serdang, Selangor, 43400, Malaysia

Min Min Aung Higher Education Centre of Excellence (HiCoE), Institute of Tropical Forestry and Forest Products, University Putra Malaysia

Department of Chemistry, Faculty of Science, University Putra Malaysia

Chemistry Division, Centre of Foundation Studies for Agricultural Science, University Putra Malaysia, Serdang, Selangor, 43400, Malaysia

Rafiqah S. Ayu Laboratory of Biocomposite Technology, Institute of Tropical Forestry and Forest Products (INTROP), Universiti Putra Malaysia, Serdang, Selangor, 43400, Malaysia

T.R.S. Cadaval Jr School of Chemistry and Food, Federal University of Rio Grande–FURG, km 8 Italia Avenue, 96203–900 Rio Grande, RS, Brazil

M. Chandrasekar School of Aeronautical Sciences, Hindustan Institute of Technology and Science, Padur, Kelambakkam, Chennai 603103, Tamilnadu, India

Ying Chen Centre for Polymer from Renewable Resources, South China University of Technology, Wushan, Guangzhou, 510640, China

Aamir Hussain Dar Department of Food Technology, Islamic University of Science and Technology, Awantipora, Jammu and Kashmir, India

B.S. Farias School of Chemistry and Food, Federal University of Rio Grande–FURG, km 8 Italia Avenue, 96203–900 Rio Grande, RS, Brazil

Gaiping Guo College of Materials Science and Engineering, Beijing Institute of Petrochemical Technology, Beijing 102617, China

Mohd Idham Hakimi Department of Bioprocess Technology, Faculty of Biotechnology and Biomolecular Sciences, Universiti Putra Malaysia, Serdang, Selangor, 43400, Malaysia

A.S. Harmaen Laboratory of Biocomposite Technology, Institute of Tropical Forestry and Forest Products (INTROP), Universiti Putra Malaysia, Serdang, Selangor, 43400, Malaysia

Carsten Herbes Nuertingen-Geislingen University, Germany

M.R.M. Huzaifah Institute of Tropical Forest and Forest Products (INTROP), Universiti Putra Malaysia, Serdang, Selangor, 43400, Malaysia

Rushdan bin Ibrahim Pulp and Paper Laboratory, Forest Research Institute Malaysia, Kepong, Selangor, 52109, Malaysia

R.A. Ilyas School of Chemical and Energy, Faculty of Engineering, Universiti Teknologi Malaysia, 81310 UTM Johor Bahru, Johor, Malaysia

Centre for Advanced Composite Materials, Universiti Teknologi Malaysia, 81310 UTM Johor Bahru, Johor, Malaysia

M.R. Ishak Department of Aerospace Engineering, Universiti Putra Malaysia, Serdang, Selangor, 43400, Malaysia

Ewelina Jamróz Institute of Chemistry, Faculty of Food Technology, University of Agriculture in Cracow, ul. Balicka Street 122, 30–149 Cracow, Poland

Jissy Jacob PG & Research Department of Chemistry, St. Thomas College, Pala, Kerala-686574, India

School of Chemical Sciences, Mahatma Gandhi University, Kottayam, Kerala-686560, India

Latifah Jasmani Pulp and Paper Laboratory, Biomass Technology Programme, Forest Products Division, Forest Research Institute Malaysia, Kepong, Selangor, 52109, Malaysia

Mohd Azwan Jenol Department of Bioprocess Technology, Faculty of Biotechnology and Biomolecular Sciences, Universiti Putra Malaysia, Serdang, Selangor, 43400, Malaysia

A. Khalina Institute of Tropical Forestry and Forest Products (INTROP), Universiti Putra Malaysia

Engineering Faculty, Universiti Putra Malaysia, Serdang, Selangor, 43400, Malaysia

Moe Tin Khaing Department of Chemistry, Yadanabon University, Mandalay, Myanmar

Shafat Khan Department of Food Technology, Islamic University of Science and Technology, Awantipora, Jammu and Kashmir, India

Piotr Kulawik Department of Animal Products Technology, Faculty of Food Technology, University of Agriculture in Cracow, Balicka Street 122, 30–149 Cracow, Poland

Rakesh Kumar Department of Biotechnology, Central University of South Bihar, Gaya 824236, India

T. Senthil Muthu Kumar Department of Mechanical Engineering, Kalasalingam Academy of Research and Education, Krishnankoil–626126, Tamil Nadu, India

Department of Mechanical and Process Engineering, The Sirindhorn International Thai German Graduate School of Engineering (TGGS), King Mongkut's University of Technology North Bangkok, 1518 Wongsawang Road, Bangsue, Bangkok 10800, Thailand

Centre for Composite Materials, International Research Centre, Kalasalingam Academy of Research and Education, Krishnankoil–626126, Tamil Nadu, India

Usman Lawal Electrochemical Power Source Division, CSIR- Central Electro Chemical Research Institute (CECRI), Karaikudi-630003, Tamil Nadu, India

Academy of Science and Innovative Research (AcSIR), Ghaziabad, 201002, India

C.H. Lee Institute of Tropical Forest and Forest Products (INTROP), Universiti Putra Malaysia

Faculty of Engineering, Universiti Putra Malaysia, Serdang, Selangor, 43400, Malaysia

Hongsheng Liu Centre for Polymer from Renewable Resources, South China University of Technology, Wushan, Guangzhou, 510640, China

Sino-Singapore International Joint Research Institute, Guangzhou Knowledge City, China

Sravanthi Loganathan Processing Engineering Division, CSIR- Central Electrochemical Research Institute (CECRI), Karaikudi, Tamil Nadu-630003, India

Kai Lu Centre for Polymer from Renewable Resources, South China University of Technology, Wushan, Guangzhou, 510640, China

Lu Lu Taung Mai Higher Education Centre of Excellence (HiCoE), Institute of Tropical Forestry and Forest Products, University Putra Malaysia, Serdang, Selangor, 43400, Malaysia

Ishrat Majid Department of Food Technology, Islamic University of Science and Technology, Awantipora, Jammu and Kashmir, India

P.C. Martins Federal University of Rio Grande, Rio Grande do Sul, Brazil

M.T. Mastura Faculty of Mechanical and Manufacturing Engineering Technology, Universiti Teknikal Malaysia Melaka (UTeM), Hang Tuah Jaya, Durian Tunggal, Melaka, 76100, Malaysia

V.G. Martins Federal University of Rio Grande, Rio Grande do Sul, Brazil

L.N. Megashah Department of Bioprocess Technology, Faculty of Biotechnology and Biomolecular Sciences, Universiti Putra Malaysia, Serdang, Selangor, 43400, Malaysia

J.M. Moura School of Chemistry and Food, Federal University of Rio Grande–FURG, km 8 Italia Avenue, 96203–900 Rio Grande, RS, Brazil

Syed Umar Faruq Syed Najmuddin Laboratory of Vaccines and Immunotherapeutics (LIVES), Institute of Bioscience, Universiti Putra Malaysia, Serdang, Selangor, 43400, Malaysia

Mohammadreza Nofar Metallurgical and Materials Engineering Department, Faculty of Chemical and Metallurgical Engineering, Istanbul Technical University Maslak, Istanbul, 34469, Turkey

Polymer Science and Technology Program, Istanbul Technical University, Maslak, Istanbul, 34469, Turkey

D. Nogueira Federal University of Rio Grande, Rio Grande do Sul, Brazil

K. Norfaryanti Institute of Tropical Forestry and Forest Products (INTROP), Universiti Putra Malaysia, Serdang, Selangor, 43400, Malaysia

Mohd Nor Faiz Norrahim Research Centre for Chemical Defence (CHEMDEF), Universiti Pertahanan Nasional Malaysia, Kem Perdana Sungai Besi, Kuala Lumpur, 57000, Malaysia

N. Mohd Nurrazi Laboratory of Biocomposite Technology, Institute of Tropical Forestry and Forest Products (INTROP), Universiti Putra Malaysia, Serdang, Selangor, 43400, Malaysia

Fatih Özogul Department of Seafood Processing Technology, Faculty of Fisheries, Cukurova University, Adana, Turkey

L.A.A. Pinto School of Chemistry and Food, Federal University of Rio Grande–FURG, Rio Grande, RS Brazil

Nur Farisha Abd Rahim Level 3, Putra Business School, Serdang, Selangor, 43400, Malaysia

Priya Rani Department of Biotechnology, Central University of South Bihar, Gaya 824236, India

Marwah Rayung Higher Education Centre of Excellence (HiCoE), Institute of Tropical Forestry and Forest Products, University Putra Malaysia, Serdang, Selangor, 43400, Malaysia

Nur Amira Mamat Razali Department of Physics, Center for Defence Foundation Studies, Universiti Pertahanan Nasional Malaysia, Kem Perdana Sungai Besi, Kuala Lumpur, 57000, Malaysia

M.H. Abdel Rehim Packaging Materials Department, National Research Centre, Egypt

V.P. Romani Federal University of Rio Grande, Rio Grande do Sul, Brazil

F.A. Sabaruddin Institute of Tropical Forest and Forest Products (INTROP), Universiti Putra Malaysia, Serdang, Selangor, 43400, Malaysia

S.O.A. SaifulAzry Institute of Tropical Forest and Forest Products (INTROP), Universiti Putra Malaysia, Serdang, Selangor, 43400, Malaysia

H.N. Salwa Institute of Tropical Forest and Forest Products (INTROP), Universiti Putra Malaysia, Serdang, Selangor, 43400, Malaysia

S.M. Sapuan Advanced Engineering Materials and Composites Research Centre (AEMC), Department of Mechanical and Manufacturing Engineering, Universiti Putra Malaysia, Serdang, Selangor, 43400, Malaysia

N.H. Sari Department of mechanical Engineering, University of Mataram, West Nusa Tenggara, Indonesia

K. Senthilkumar Department of Mechanical Engineering, Kalasalingam Academy of Research and Education, Krishnankoil–626126, Tamil Nadu, India

Department of Mechanical and Process Engineering, The Sirindhorn International Thai German Graduate School of Engineering (TGGS), King Mongkut's University of Technology North Bangkok, 1518 Wongsawang Road, Bangsue, Bangkok 10800, Thailand

Centre for Composite Materials, International Research Centre, Kalasalingam Academy of Research and Education, Krishnankoil–626126, Tamil Nadu, India

R.M. Shahroze Department of Aerospace Engineering, Universiti Putra Malaysia, Serdang, Selangor, 43400, Malaysia

Rafeeya Shams Sher-e-Kashmir University of Agricultural Sciences and Technology, Jammu, India

Suchart Siengchin Department of Mechanical and Process Engineering, The Sirindhorn International Thai German Graduate School of Engineering (TGGS), King Mongkut's University of Technology North Bangkok, 1518 Wongsawang Road, Bangsue, Bangkok 10800, Thailand

S. Suteja Department of Mechanical Engineering, University of Mataram, West Nusa Tenggara, Indonesia

I.S.M.A. Tawakkal Engineering Faculty, Universiti Putra Malaysia, Serdang, Selangor, 43400, Malaysia

Sabu Thomas School of Chemical Sciences, Mahatma Gandhi University

International and Inter University Center for Nanoscience and Nanotechnology, Mahatma Gandhi University

School of Energy Materials, Mahatma Gandhi University, Kottayam, Kerala-686560, India

Huafeng Tian Beijing Key Laboratory of Quality Evaluation Technology for Hygiene and Safety of Plastics, School of Material and Mechanical Engineering, Beijing Technology and Business University, Beijing 100048, China

Hiroshi Uyama Department of Applied Chemistry, Graduate School of Engineering, Osaka University, Suita 565-0871, Japan

Ravi Babu Valapa Electrochemical Power Source Division, CSIR- Central Electro Chemical Research Institute (CECRI), Karaikudi-630003, Tamil Nadu, India

Emre Vatansever Polymer Science and Technology Program, Istanbul Technical University, Maslak, Istanbul, 34469, Turkey

M. Vishnuvarthanan Department of Printing Technology, College of Engineering Guindy, Anna University, Chennai – 600 025, Tamil Nadu, India

Yunxuan Weng Beijing Key Laboratory of Quality Evaluation Technology for Hygiene and Safety of Plastics, School of Material and Mechanical Engineering, Beijing Technology and Business University, Beijing 100048, China

Tengku Arisyah Tengku Yasim-Anuar Department of Bioprocess Technology, Faculty of Biotechnology and Biomolecular Sciences, Universiti Putra Malaysia, Serdang, Selangor, 43400, Malaysia

Long Yu Centre for Polymer from Renewable Resources, South China University of Technology, Wushan, Guangzhou, 510640, China

Sino-Singapore International Joint Research Institute, Guangzhou Knowledge City, China

S. Zaiton Institute of Tropical Forestry and Forest Products (INTROP), Universiti Putra Malaysia, Faculty of Economics and Management, Universiti Putra Malaysia, Serdang, Selangor, 43400, Malaysia

Khairul Zaman Polycomposite Sdn Bhd, Taman Mutiara Galla, 70300 Seremban, Negeri Sembilan

M.Y.M. Zuhri Department of Mechanical and Manufacturing Engineering, Faculty of Engineering, Universiti Putra Malaysia, Serdang, Selangor, 43400, Malaysia

Series Preface

Renewable resources, their use and modification, are involved in a multitude of important processes with a major influence on our everyday lives. Applications can be found in the energy sector, paints and coatings, and the chemical, pharmaceutical, and textile industries, to name but a few.

The area interconnects several scientific disciplines (agriculture, biochemistry, chemistry, technology, environmental sciences, forestry, etc.), which makes it very difficult to have an expert view on the complicated interactions. Therefore, the idea to create a series of scientific books, focusing on specific topics concerning renewable resources, has been very opportune and can help to clarify some of the underlying connections in this area.

In a very fast-changing world, trends are not only characteristic of fashion and political standpoints; science too is not free from hypes and buzzwords. The use of renewable resources is again more important nowadays; however, it is not part of a hype or a fashion. As the lively discussions among scientists continue about how many years we will still be able to use fossil fuels – opinions ranging from 50 to 500 years – they do agree that the reserve is limited and that it is essential not only to search for new energy carriers but also for new material sources.

In this respect, the field of renewable resources is a crucial area in the search for alternatives for fossil-based raw materials and energy. In the field of energy supply, biomass- and renewables-based resources will be part of the solution alongside other alternatives such as solar energy, wind energy, hydraulic power, hydrogen technology, and nuclear energy. In the field of material sciences, the impact of renewable resources will probably be even bigger. Integral utilization of crops and the use of waste streams in certain industries will grow in importance, leading to a more sustainable way of producing materials. Although our society was much more (almost exclusively) based on renewable resources centuries ago, this disappeared in the Western world in the nineteenth century. Now it is time to focus again on this field of research. However, it should not mean a "retour à la nature", but should be a multidisciplinary effort on a highly technological level to perform research towards new opportunities, and to develop new crops and products from renewable resources. This will be essential to guarantee an acceptable level of comfort for the growing number of people living on our planet. It is "the" challenge for the coming generations of scientists to develop more sustainable ways to create prosperity and to fight poverty and hunger in the world. A global approach is certainly favored.

This challenge can only be dealt with if scientists are attracted to this area and are recognized for their efforts in this interdisciplinary field. It is, therefore, also essential that consumers recognize the fate of renewable resources in a number of products. Furthermore, scientists do need to communicate and discuss the relevance of their work. The use and modification of renewable resources may not follow the path of the genetic engineering concept in view of consumer acceptance in Europe. Related to this aspect, the series will certainly help to increase the visibility of the importance of renewable resources. Being convinced of the value of the renewables approach for the industrial world, as well as for developing countries, I was myself delighted to collaborate on this series of books focusing on the different aspects of renewable resources. I hope that readers become aware of the complexity, the interaction, and interconnections, and the challenges of this field, and that they will help to communicate on the importance of renewable resources.

I certainly want to thank the people of Wiley's Chichester office, especially David Hughes, Jenny Cossham, and Lyn Roberts, in seeing the need for such a series of books on renewable resources, for initiating and supporting it, and for helping to carry the project to the end.

Last, but not least, I want to thank my family, especially my wife Hilde and children Paulien and Pieter-Jan, for their patience, and for giving me the time to work on the series when other activities seemed to be more inviting.

Christian V. Stevens
Faculty of Bioscience Engineering, Ghent University, Belgium
Series Editor, "Renewable Resources"
June 2005

Preface

The plastic industry was initiated in the early 1900s when the first synthetic plastic was produced by Leo Hendrik Baekeland in the United States of America. Since the industry began, annual global plastic production has exploded from some 1.5 million metric tons in 1950 to 348 billion metric tons in 2017. Cumulatively, plastic production has reached 8.3 billion metric tons worldwide, with a dramatic increase in the amount of plastics anticipated to be produced globally by the year 2050. Of the amount produced, most plastics are single use and afterwards are immediately disposed of, usually into landfills. Currently, the packaging industry relies strongly on the use of petroleum-derived plastic materials, which raise some concerns from both an environmental and economic perspective. Furthermore, because of their lack of biodegradability, these products can pose significant waste disposal problems in some areas. Growing environmental concerns regarding usage of these toxic pollutants containing materials that are harmful to people, animals, and plants are motivating the use of bio-based materials as alternatives.

The global plastic sustainability issue has grown significantly in the last few years, encouraging academia as well as industry players to develop sustainable alternatives for preserving resources for future generations. Development of bio-based packaging is one of the important factors for sustainable growth of the packaging industry. Recent trends in the consumer market have moved towards greener packaging. In 2019, the global production capacity of biodegradable bioplastics was 941 000 metric tons. The global biodegradable plastic packaging market was valued at US$4.65 billion in 2019 and is expected to reach a market value of US$12.06 billion by 2025, registering a compound annual growth rate (CAGR) of 17.04% during the forecast period of 2020–2025. The successful use of renewable bio-based packaging for the production of packaging materials will overcome the environmental problems caused by petroleum-derived plastics. Bio-based packaging materials have attracted huge considerable research and development interest for a significant length of time, and in recent years these materials have reached the market. Eco-concerns played a major role in encouraging the development of biopolymers in packaging applications. This occurred directly via consumer demand for environmentally friendly products as well as indirectly via the political and ensuing environmental legislation and regulation. Furthermore, technological advances, such as nanotechnology, are forecasted to continue improving biopolymer properties and increasing the number of potential applications for such bio-based materials in packaging.

This book looks at how biopolymers might be used in packaging as a potential green solution. It addresses commentary from leading industrial and academic experts in the field who present cutting-edge research on bio-based materials for the packaging industry. It includes new potential materials to be used for food packaging applications, such as starch, protein, pectin, chitin, chitosan, natural fibre, cellulose, nanocellulose, polylactic acid (PLA), polyhydroxyalkanoates (PHA), etc., as well as their manufacturing processes and legislative demands for food contact packaging materials. It also gives an overview of the most recent advances and emerging new aspects of nanotechnology for development of composites for environmentally compatible packaging materials. Furthermore, this book covers a life cycle assessment of bio-based packaging products and the market for bio-based packaging, on how consumer perceptions and preferences regard bio-based packaging, as well as an assessment of supply chains, business strategies, and markets in biodegradable food packaging. This book covers economic issues and environmental issues of bio-based packaging, including the socioeconomic impact of bio-based packaging, the future of sustainable packaging, renewable sources for packaging material, environmental advantages and challenges of bio-based packaging materials, a life cycle assessment (LCA) of bio-based packaging products, as well as the reuse and recycle of bio-based packaging products. Lastly, it also elaborates the opportunities for biopolymers in key end-use sectors and penetration of biopolymer-based concepts in the packaging market, as well as barriers to widespread commercialization.

1

Starch-Based Packaging Materials

Ying Chen[1], Kai Lu[1], Hongsheng Liu[1,2], and Long Yu[1,2]

[1] *School of Food Science and Engineering, South China University of Technology, Wushan, Guangzhou, 510640, China*

[2] *Sino-Singapore International Joint Research Institute, Guangzhou Knowledge City, Guangzhou, 510663, China*

1.1 Introduction

Packaging has played an important role in world pollution in recent decades as multimillions tons of packaging waste have resulted in environmental issues [1, 2]. Due to the environmental challenges many researchers have formulated biodegradable and ecofriendly composite materials to replace conventional packaging materials [3]. In fact, more and more countries have recently passed regulations or laws to ban the application of disposable plastic packages, which enhanced the work in this area. The high cost of producing bioplastics has hindered its further development, so alternative low-cost and renewable substrates have been proposed using agricultural waste [4]. Starch-based materials are the promised alternative to synthetic ones in food packaging and handling in many aspects [5].

Starch films have been widely used in food and medicine packaging [6–10], in which the film should be edible in many cases, such as applications in candy wrappers and medicine capsules, etc. [11, 12]. Such films maintain the shelf life of foods for distant marketing, either by acting as a barrier for gases and volatile compounds or by controlling water permeability [13–15]. Improvement of mechanical properties and reducing moisture sensitivity are two ongoing challenges for starch-based materials. To achieve these two goals, various blending and compositing techniques have been developed, such as blending with other polymers or reinforcing with particle or fiber-fillers [16–19]. However, incorporation of

Bio-Based Packaging: Material, Environmental and Economic Aspects, First Edition. Edited by S.M. Sapuan and R.A. Ilyas

any additive is sensitive when developing food contactable or edible packaging films, due to safety issues.

On the other hand, the unique microstructures of different starches and their multiphase transitions during thermal processing can be used as an outstanding model system to illustrate our conceptual approach to understanding the structure–processing–property relationships in polymers [20–27].

This chapter introduces the development of starch-based materials, in particular their application in packaging, including both fundamental and application researches. It starts from fundamental issues of starch microstructures and phase transition to application techniques of extrusion processing, and then moves to show how to solve the well-recognized weaknesses of starch-based materials.

1.2 Macrostructures and Phase Transitions of Starch

1.2.1 Microstructures of Starch Granules

Synthetic polymers have been developed to the point where microstructures can be designed and molecular weight and molecular weight distribution can be controlled. However, the mesoscopic structure within a starch granule has evolved to suit a plant's own needs and is therefore much more complex. Starch is a polysaccharide produced by mostly higher-order plants as a means of storing energy. It is stored intracellularly in the form of spherical granules 2–100 μm in diameter. Most commercially available starches are isolated from grains such as corn, rice, and wheat or from tubers such as potato and cassava (tapioca).

Chemically, starch is a polymeric carbohydrate consisting of anhydroglucose units linked together primarily through α-d-($1 \rightarrow 4$) glucosidic bonds [22, 26–38]. Although the detailed microstructures of different starches are still being clarified, it has generally been established that starch is a heterogeneous material containing two types of microstructures – linear (amylose) and branched (amylopectin). Amylose is essentially a linear structure of α-1,4 linked glucose units, while amylopectin is a highly branched structure of short α-1,4 chains linked by α-1,6 bonds. Figure 1.1 shows the chemical structure and a schematic representation of amylose and amylopectin starches. The linear structure of amylose makes its behavior more closely resemble that of conventional synthetic polymers. Depending on its source and the processing conditions employed during its extraction, the molecular weight of amylose is about ×106, which is 10 times higher than conventional synthetic polymers. Amylopectin, on the other hand, is a branched polymer and its molecular weight is much greater than amylose, with light-scattering measurements indicating a molecular weight in the millions. The high molecular weight and branched structure of amylopectin reduce the mobility of the polymer chains and interfere with any tendency for them to become oriented closely enough to permit significant levels of hydrogen bonding. Between the linear amylose and short-branched amylopectin, a long-branched structure has been detected, such as that found in tapioca starch.

Physically, most native starches are semi-crystalline, having a crystallinity of about 20–45% [22, 39–41]. Amylose and the branching points of amylopectin form amorphous regions. The short-branched chains in the amylopectin are the main crystalline components in granular starch. Crystalline regions are present in the form of double helices with a length of ~5 nm. The amylopectin segments in the crystalline regions are all parallel to

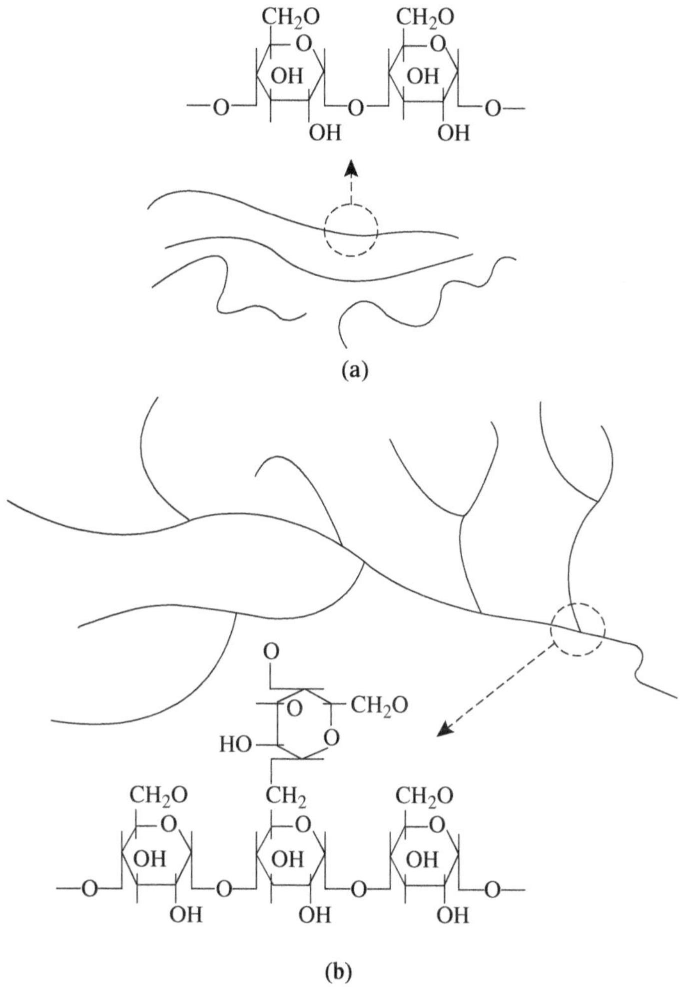

Figure 1.1 *Chemical structures and physical schematic representation of (a) amylose starch and (b) amylopectin starch.*

the axis of the large helix. The molecular weight of amylopectin is about 100 times higher than that of amylose. The ratio of amylose to amylopectin depends upon the source and age of the starch and it can also be controlled by the extraction process employed. Starch granules also contain small amounts of lipids and proteins.

Figure 1.2 shows the chemical structures and physical schematic representation of (a) amylose starch and (b) amylopectin starch.

1.2.2 Phase Transition During Thermal Processing

The thermal processing of starch-based polymers involves multiple chemical and physical reactions, e.g. water diffusion, granule expansion, gelatinization, decomposition, melting,

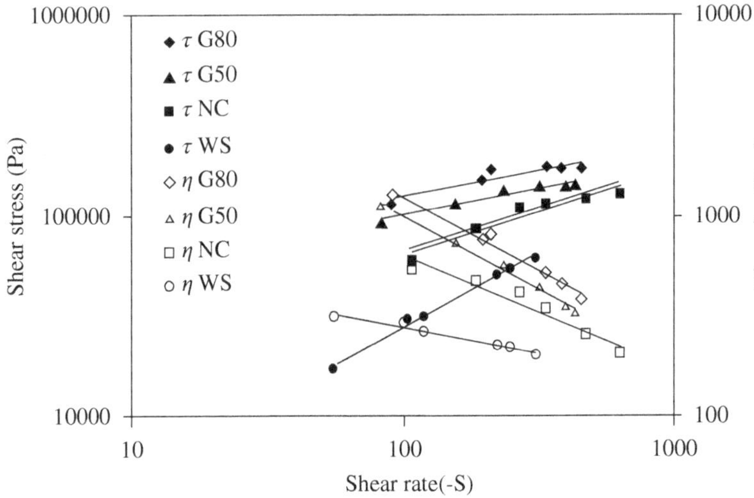

Figure 1.2 *Effect of amylose content on shear stress and melt viscosity of various starches (moisture content 23%; extrusion temperature 130 °C). Source: Reproduced from [59] with permission.*

and crystallization. Among the various phase transitions, gelatinization is particularly important because it is closely related to the others and it is the basis for the conversion of starch to a thermoplastic. Furthermore, the decomposition temperature of starch is higher than its melting temperature before gelatinization. The well-accepted concept of "gelatinization" refers to the destruction of the crystalline structure in starch granules, which is an irreversible process that includes, in a broad sense and in time sequence, granular swelling, native crystalline melting (loss of birefringence), and molecular solubilization [26, 32–43]. Starch gelatinization is a multistage process that can be detected by differential scanning calorimetry (DSC) [44–50].

Previous researches on starch suspensions (starch/water: 30w/70w) have shown that amylose and amylopectin are partly separated during gelatinization [51, 52] because they are incompatible [25, 53, 54] and that less water in a thermoplastic starch system may hinder this separation. On the other hand, shear stress during extrusion may enhance separation, whereby amylose may be partly leached out of the amylopectin. However, most of the amylose still remains in the amylopectin even in a 10% starch system [20]. Gelatinized amylopectin initially remains in an amorphous state, even though there is still a small amount of ungelatinized amylopectin. Although the double-helical crystalline structure formed by the short-branched chains in amylopectin is torn apart during gelatinization, the chains remain in a regular pattern and keep a certain "memory." It has been considered that these short-branched chains form gel-balls [55], which mainly contain chains from the same sub-main chain. One amylopectin molecule may form a relatively separated super-globe. The molecular entanglements between gel-balls and super-globes are much less than those between linear polymer chains, due to the size and length of the chains (only 4–6 glucose). V-type single helix crystals have been found to form directly after extrusion, which initially results in higher modulus and yield stress for amylose-rich materials. The crystallinity of the V-type crystals increases with time.

1.3 Extrusion Processing for Starch

1.3.1 Phase Transition During Extrusion

Extrusion processing is one of the most popular techniques used for polymer processing. Unlike most conventional polymers, starch granules undergo unique and complicated phase changes during extrusion (normally an aqueous environment), including starch swelling, loss of birefringence, melting, and solubilization [20, 25, 53–64]. Without shear stress, the process of gelatinization depends mainly on water content and temperature conditions. During the initial stage of thermal processing as the temperature is increased from 20 to 60 °C, water is thought to be reversibly complexed with starch molecules, thus decreasing its mobility [65]. At this time, the starch granular structure remains stable, with the constituent molecules held together in spherulites by van der Waals forces or hydrogen bonds, and the granules still exhibit a typical birefringence under polarized light [66] due to their largely intact crystalline components. The temperature at this stage is usually referred to as the "initial gelatinization temperature." When an aqueous starch suspension is heated beyond its gelatinization temperature, hydrogen bonding is disrupted and water molecules become attached to the hydroxyl groups in the starch molecules, resulting in greater swelling and the dissolution of crystallites [66]. During this time, all amylopectin double helices become dissociated, although the swollen granular structure will be retained until a higher temperature is applied to disrupt it [67].

Because extrusion processing involves high-shear and high-pressure conditions, gelatinization is typically achieved at a low moisture content (MC), since the shear forces physically tear the starch granules apart, allowing faster transfer of water into the interior molecules [68]. Therefore, during extrusion, the loss of crystallinity is not caused by water penetration but by the mechanical disruption of molecular bonds due to the intense shear fields within the extruder [69]. In fact, during extrusion at a low moisture content (MC), small amounts of gelatinized and melted starch, as well as starch fragmentation (also degradation or decomposition), exist simultaneously [70]. A previous study [69] has shown that shear stress can result in fragmentation of starch granules during extrusion. Indeed, both the mechanical and thermal energy transferred to starch dough during extrusion will affect the breakdown of the main and secondary valence bonds and the hydrogen bonds between neighboring starch polymers in a starch structure [69]. These structural changes increase the susceptibility of starch to enzyme action, reduce the hydrogen bonds, and increase the free hydroxyl groups. Extrusion can also affect amylopectin much more than amylose [70]. Colonna et al. [71], for example, found that the average molecular weight of amylose and amylopectin decreased by factors of 1.5 and 15, respectively, during extrusion.

The main objectives of most starch processing techniques are melting and mixing, which are adjusted to minimize chain degradation. However, during processing under shear, starch fragmentation is unavoidable, with the extent of fragmentation during extrusion depending on operating conditions such as screw speed, temperature, and moisture content, as well as the type of starch. Davidson et al. [72] found that the extent of amylopectin fragmentation decreased with decreasing screw speed and increasing temperature (121–177 °C) or MC (20–25%). Similar results were observed by Wen et al. [70] in a study of degerminated yellow corn meal with 20–30% MC at 100–200 °C. In another study of waxy maize starch initially extruded using a co-rotating twin-screw extruder and then re-extruded in a single-screw extruder, Willet et al. [73] found that molecular weight reductions during the second pass increased with increasing specific mechanical energy, but they were less drastic than in the initial extrusion. A more recent study [74] showed that, under intensive mixing, an increase in glycerol content reduced starch degradation, whereas an increase in fiber content led to increased degradation. Controlling the fragmentation (degradation) of starch-based materials during processing is of prime importance, especially when multiple extrusion runs are needed, since fragmentation not only impacts on processing properties such as viscosity, but it can also affect the performance of final products.

Due to the multiphase transitions of starch, the microstructures and mechanical properties of starch-based materials strongly depend on the processing techniques and conditions. For example, Baltá Calleja et al. [75] studied the microstructural differences in potato starch after different processing techniques and found that compression molding of the starch granules led to sintered, relatively brittle materials, since the amylopectin crystals in the native powder remained largely preserved. Preparation of dry films from aqueous gels resulted in the disintegration of the structure of the native starch granules, and the formation of a new semi-crystalline structure comprised of crystallized amylose molecules. Injection molding of native starch increased the amorphous phase, leading to superior mechanical properties.

1.3.2 Rheological Behaviors of Starch-Based Materials

The melt rheology of starch-based materials has both scientific and industrial importance and thus it has been widely investigated. For example, Llo, Tomschik, Berghofer, and Mundigler [76] studied the effects of extrusion conditions on the apparent viscosity of maize grits, including the effects of melt temperature, shear rate, material composition, and extrusion processing history. Willett, Millardt, and Jasberg [73] reported that waxy corn starch exhibited shear thinning behavior, becoming more Newtonian as either the temperature or moisture content was increased. González, Torres, de Greef, and Guadalupe [77] investigated the effects of several factors including screw speed, die restriction (l/r), and moisture content, together with corn endosperm hardness and rice amylose content, on apparent melt viscosity using a Brabender single-screw extruder. Thuwall, Boldizar, and Rigdahl [78] studied the effects of amylose content, moisture content, and the starch/glycerol ratio on the apparent viscosity of potato starch.

Xie et al. [64] have investigated the rheological properties of corn starches with different amylose/amylopectin ratios (80/20, 50/50, 23/77, and 0/100) systematically using Haake rheometry. The starches were initially pre-compounded with water to designated moisture content levels using a twin-screw extruder. A single-screw extruder with a slit capillary die was then used to characterize the shear stress and melt viscosity characteristics of sample pellets, as a function of both moisture content and extrusion temperature. The melts exhibited a shear thinning behavior under all conditions, with the power law index increasing with increasing temperature and moisture content in the majority of cases. Viscosity was higher for the amylose-rich starches, while amylopectin-rich starches exhibited increased Newtonian behavior. The higher the amylose content is, the higher is the viscosity, which is opposite to the sequence of molecular weight. These rheological behaviors are attributed to the higher gelatinization temperature of amylose-rich starches, and in particular the multiphase transitions that occur in these starches at higher temperatures, and the gel-ball structure of gelatinized amylopectin.

Figure 1.2 shows the effect of amylose content on the shear stress and melt viscosity of various samples at a moisture content of 23% and extrusion temperature of 130 °C, as a function of the shear rate. It was found that the higher amylose content resulted in a higher apparent viscosity over the same shear rate range.

Actually, extrusion processing has also been widely used to modify starch, such as grafting and pre-gelatinization [79–85]. The high efficiency and accelerated rates at which modification reactions were carried out in the extruder, compared to batch processes, is particularly noteworthy.

1.4 Improving Mechanical Properties by Reinforcement

Improving performance and decreasing cost are two never-stop strategies for all kinds of materials. Like most polymers from petroleum, polymers from renewable resources are rarely used by themselves. In fact, the history of composites from renewable resources is far longer than conventional polymers. In the biblical Book of Exodus, Moses' mother built the ark from rushes, pitch, and slime: a kind of fiber-reinforced composite, according to the modern classification of material. During the opium war in the 1870s, the Chinese built their castles to defend against invaders using a kind of mineral particle-reinforced composite made from gluten rice, sugar, calcium carbonate, and sand.

Polymer blending and compositing are well-used techniques whenever modification of properties is required, because it uses conventional technology at low cost. The usual objective for preparing a novel blend of two or more polymers is not to change the properties of the components drastically, but to capitalize on the maximum possible performance of the blend. In principle, the properties of starch-based materials can be significantly improved by blending with synthetic polymers. In the 1970s and 1980s, numerous blends of starch with various polyolefins were developed. However, these blends were not biodegradable and thus the advantage of using a biodegradable polysaccharide was lost. In this paper, polymer blends and composites only from natural raw materials are discussed.

1.4.1 Reinforcement by Natural Fillers

Fiber-reinforced plastics have successfully proven their value in various applications because of their excellent specific properties [86, 87], e.g. high strength and stiffness, and low weight. Fiber-reinforced composites have previously been discussed and reviewed by many authors and many books are available. In that respect, natural fibers are of basic interest since they not only have the functional capability to substitute the widely used glass fibers but they also have advantages from the point of view of weight and fiber–matrix adhesion, specifically with polar matrix materials. They have good potential for use in waste management due to their biodegradability and their much lower production of ash during incineration.

Starch reinforced cellulose is a typical example of natural polymer composites. These materials have the advantages of being renewable, biodegradable, abundantly available, and inexpensive, and as such they have attracted great attention in the last two decades. Like many other fiber-reinforced polymer composites, improving mechanical properties is one of the major driving forces to reinforce starch-based materials using cellulose fibers. Wollerdorfer and Bader [88] reported that reinforced thermoplastic wheat starch was four times better (37 N/mm^2) than without fibers. Enhanced properties were obtained by using various natural cellulosic fibers, such as sisal, cotton, bamboo, jute, straw, ginger, kenaf, and wood. Compared to inorganic fillers, composites based on nature fibers offer a number of benefits. These include a renewable nature, a wide variety of fillers available throughout the world, low energy consumption, low cost, low density, high specific strength and modulus (desirable fiber aspect ratio), high sound attenuation, comparatively easy processability (due to their flexibility and non-abrasive nature, which allow high filling levels, resulting in significant cost savings), and a relatively reactive surface (which can be used for grafting specific groups) [89]. Various techniques and products have been developed and reviewed in detail [20].

Recently, Ali et al. reported starch-based composite films reinforced by corn/wheat hulls [90] and apricot and walnut shells [91], which are by-products of rigid particles from natural food resources. Figure 1.3 shows the photos of wheat and corn hulls used in this work and their images under scanning electron microscopy (SEM). Both hulls are yellow-brown colored particles. Wheat hull looks more like cellulose particles with an irregular shape, while corn hull is flat in shape. Figure 1.4 shows the granulated wheat and corn hull particles observed under an optical microscope with normal and polarized lights. Both wheat and corn hull particles are irregular in shape and the crystalline particles can be clearly observed under polarized lights. Results showed that, as expected, the materials became stronger after addition of these rigid fillers. The gas barrier properties of the starch films were also improved by adding the hulls with higher density and higher crystallinity, especially the corn hull because of its flat structure. SEM observations indicated that the

Figure 1.3 Photos of wheat and corn husks, and their images under SEM. Source: Reproduced from [88] with permission.

Figure 1.4 *Granulated wheat (A,a) and corn (B,b) husk particles observed under optical microscope with normal and polarized lights. Source: Reproduced from [88] with permission.*

compatibility between starch and the two hulls were quite good, which was expected since all the components are hydrophilic and exhibit polar behavior.

Chen et al. [92] have improved the gas permeability of starch film by adding laver. When the laver was added to a cold starch suspension, the laver flack (fiber) simply acted as a reinforcing agent to improve the mechanical properties of the starch matrix, which resulted in increasing the modulus and tensile strength. When the laver was added to a hot starch suspension and mixed under shear stress, the protein contained in the laver is released and mixes with the starch, which results in starch–protein–laver fiber hybrid composites. In the hybrids system, fiber improves the mechanical properties of the starch-based film by about 25%, while the protein reduces the moisture sensitivity and gas permeability by about 45%. The scanning electron microscopic observation indicated good compatibility between the starch matrix and the laver. Since all the components used in this work are from food sources, the prepared films are safe for food packaging and application as edible films.

Figure 1.5 shows the time dependence of water vapor permeability (WVP) of the starch-based film containing different laver contents and prepared by different methods. It is seen that WVP of the starch-based films was decreased significantly after adding laver, and further decreased with increasing laver content. The decrease in the WVP by incorporation of laver is due to the formation of a twisted pathway in the film matrix to allow the water vapor molecules to pass through. Actually, the fillers with a thin flat shape, such as delaminated talc, clay, graphene, etc., have been widely used to decrease the gas permeability in polymeric materials. It has been noticed that WVP of the films prepared using the hot method was lower than those prepared using the cold method with the same concentrations. This phenomenon could be attributed to the formation of cross-links between the starch and protein [93]. The good bonding between starch and protein provides less free volume and moisture diffusion through it (Figure 1.6).

1.4.2 Starch-Based Nanocomposites

The starch-based nanocomposites have continuously attracted attention and have been successfully developed and commercialized. One of the challenges using nanoclay is to exfoliate it, due to the strong boundary between the layers by iron. Another issue is to improve the hydrophilic surface of clay using the hydrophobic surface of most conventional polymers, especially polyolefin. These two challenges can be simply solved in starch-based materials. Firstly, hydrophilic water is a very efficient chemical to delaminate clay with enhancement using ultrasonic methods [94–101]. Secondly, starch itself is a hydrophilic polymer that

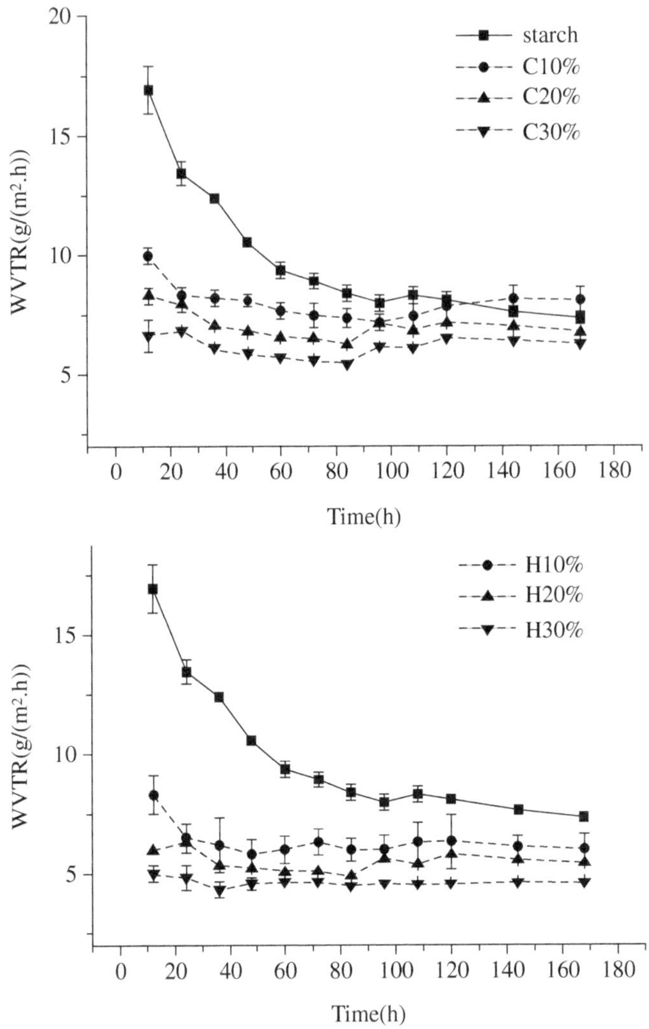

Figure 1.5 *Effect of laver on water vapor transmission rate of starch films prepared by: cool (top) and hot (bottom). Source: Reproduced from [90] with permission.*

can be very compatible with clay and water. Actually, water is a plasticizer for starch-based materials.

Recently, polysaccharide-based crystals have attracted huge attention to optimize the mechanical properties of starch-based biodegradable composites since they have a naturally perfect interface and are edible [102–105]. It was found that both cellulose nanocrystals and starch nanocrystals can improve tensile strength as well as Young's modulus [106–108]. Li et al. [110] found that starch-based nanocomposite films filled with cellulose nanofibrils could increase the creep resistance and reduce the creep recovery rate. Panaitescu et al. [111] showed that starch films filled with 7% cellulose nanocrystals had strength and stiffness

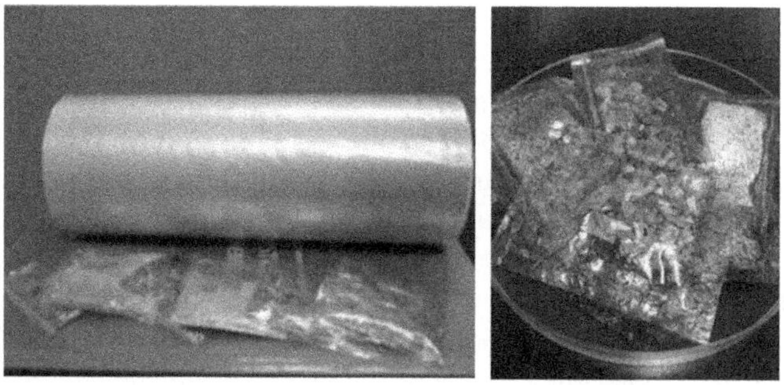

(a) (b)

Figure 1.6 Photo of starch-based film with developed flavor bags for easy muddle (a) and laver reinforced the products (b). Source: Reproduced from [90] with permission.

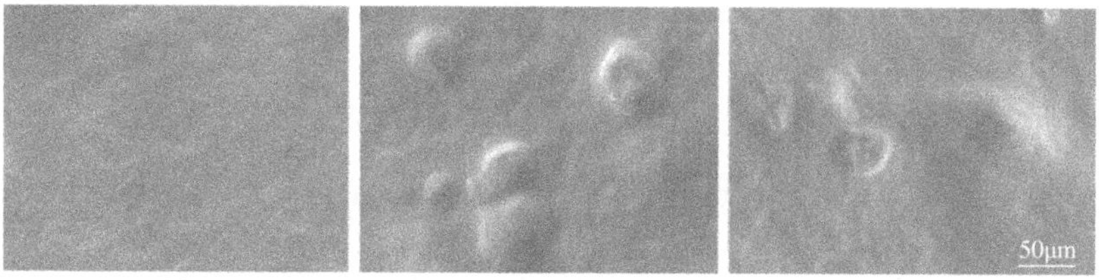

(a) (b) (c)

Figure 1.7 Surface SEM images of starch film (a: pure starch film; b: containing starch crystals; c: containing cellulous crystals). Source: Reproduced from [110] with permission.

close to that of polyolefins and were regarded as a low-cost "green" substitute for application in food packaging and conservation.

More recently, Ali et al. [112] reported the starch-based edible films, reinforced by two types of polysaccharide-based crystals: starch and cellulous crystals. The films were transparent and their surface was smooth in appearance. Addition of crystals increased the Young's modulus and tensile strength of starch-based materials and decreased elongation at break. Scanning electron microscopic observations (see Figure 1.7) indicated good compatibility between the starch matrix and the reinforcing fillers due to the same chemical unit (glucose). Cellulose crystals have a higher thermal stability than that of starch crystals; this provides better processability and superior mechanical properties to starch films filled with cellulose crystals. On the other hand, starch films filled with starch crystals demonstrated higher protection against ultraviolet (UV) radiation. Since all the components used in this work belong to food sources, the prepared films are biodegradable, safe for food packaging, and can also be used to develop edible films and medicinal capsules.

1.4.3 Self-Reinforced Composites

Self-reinforced composites (SRCs), or single-polymer composites, in which a polymer matrix is reinforced with oriented fibers and tapes or particles of the same polymer, are an interesting concept. The advantages of SRC systems include the ability to achieve perfect interfaces between composite components, their pure chemical functionality, and their higher value as recyclable products due to their relative homogeneity compared to composites composed of different classes of components [113–118]. This homogeneity is particularly important in biomaterial applications, since any additives composed of different chemicals could affect biocompatibility and biodegradability.

Typical starch-based SCR was reported by Lan et al. [119]. A unique technique of designing and preparing self-reinforced starch-based films was developed using a cornstarch that was chemically modified in different ways. Hydroxypropylation was used to decrease the gelatinization temperature and to improve processability. Lowering the processing temperature by reducing the gelatinization temperature was due mainly to the presence of the substituted group, which weakened or strained the internal bond structure holding the granules together [120]. The reinforcing component consisted of cross-linked starch granules, where the cross-linking increased granular thermal stability and moisture resistance.

Many starch materials reinforced by polysaccharide-based crystals can also be classified as a kind of SRC since they basically have similar chemical components [102–112].

1.4.4 Blending with Other Natural Polymers

Starch has also been used to improve processibility and reduce the cost of other natural polymers. For example, the blended films of polysaccharides and proteins have shown unique properties, such as a better gas barrier than any pure film [10, 124, 125]. Previous researches have shown that gelatin and starch are immiscible; however, their morphologies and compatibility are affected by various factors, such as processing time, temperature, pH value, and solid concentration.

Zhang et al. [120–124] have discussed the edible films developed by blending starch with gelatin. The well-recognized complex issue of compatibility between starch and gelatin was investigated based on their interface and phase composition using a synchrotron Fourier transform infrared spectroscopy (FTIR) microspectroscope. Investigation of the composition and interface of the blend using FTIR microspectroscopy enables a unique insight into the interface and morphologies of the system since it is based on chemical contrast between constituents. Figure 1.8 shows a typical count contour map for a starch–gelatin 40 : 60 blend. The image was generated from the data of a single measurement set by allocating a color to each pixel based on the ratio of Bands-1 (starch) and Bands-2 (gelatin), which was used to characterize the starch–gelatin content. The red peak in the scale denotes a high value of starch while the blue bottom line denotes a high value of gelatin. It was observed that the starch phase (Bands-1) distribution was a dispersed phase, while gelatin (Bands-2) was a continuous phase. However, the sharp peak, not the column, of Bands-1 indicates that there are certain chain diffusions on the interface of starch domains, which means that the complete de-mixing of starch and gelatin domains did not occur.

Another example is the blends of hydroxypropyl methylcellulose (HPMC) and hydroxypropyl starch (HPS) [126–128], which are receiving increasing attention due to their unique thermal/cooling gel systems and wide applicability. The research on this system has both scientific and commercial importance. Scientifically, the blend system provides a good scientific model to investigate the thermal/cooling gel system, based on the rheological characteristics, phase transitions, morphologies, compatibility, etc.

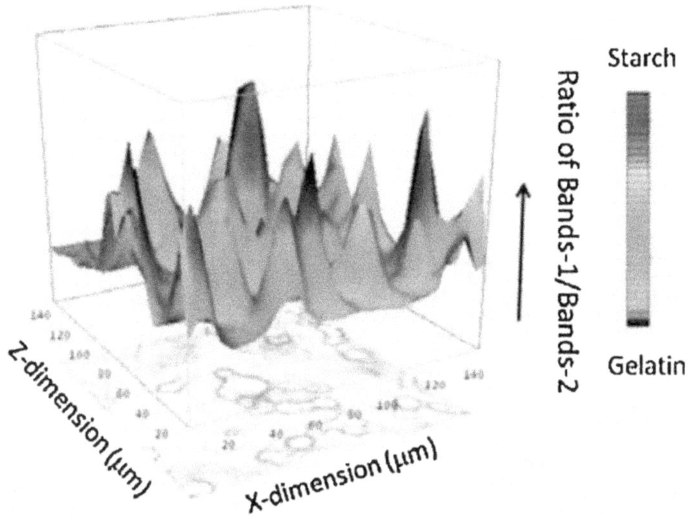

Figure 1.8 *Variation of the ratio of the area of the Bands-1 (starch)/Bands-2 (gelatin) plotted as 2D and3D contour maps for starch–gelatin (60 : 40) blend. Source: Reproduced from [120] with permission.*

Commercially the blends have been widely used from food packaging and medicine capsules to cartilage tissue engineering applications. The relative advantageous properties of the two components were partially combined in such blends, which presented quite good processability, mechanical properties, and barrier properties to oxygen and water vapor.

Figure 1.9 shows the viscosities of the solutions with different hydroxypropyl methylcellulose (HPMC) and hydroxypropyl starch (HPS) ratios under same shear rate and at different temperatures [129]. It is seen that viscosity of the starch HPS is much higher that of HPMC and the viscosity is increased gradually with increasing HPS content during both increasing and decreasing temperatures. The viscosity of pure HPS decreased with increasing temperature, as expected. The viscosity of pure HPMC decreased with increasing temperature at a lower temperature (<50 °C) and then increased in the range between 50 and 75 °C as in many previous reports. The acceptable mechanism of thermal gelation of HPMC is that the sufficient energy results in outer layers of water molecules, which are driven from the cellulose ether chains; the chains then lock and the solution is transformed into a gel. The viscosity peak of HPMC is shafted to a shorter time with an increasing HPS content. It has been noted that the effect of HPS on the viscosity of HPMC is not linear. When the HPS content is lower than 70%, the viscosity peak of HPMC still exists and the peak temperatures are almost same, which provide an operation window for making a capsule under similar conditions as pure HPMC. When the HPS is higher than 70%, no viscosity peak was observed and only a shoulder was detected. The XRD

Figure 1.10 shows the photos of capsules made from the solutions with different HPMC/HPS solutions. It can be seen that the transparencies of both pure HPMC and HPS capsules are reasonably good. The transparency of the capsules decreased with increasing HPS content, which indicates that there is phase separation for the blends and the results correspond with the measurements of film transparencies. The capsule wall thickness

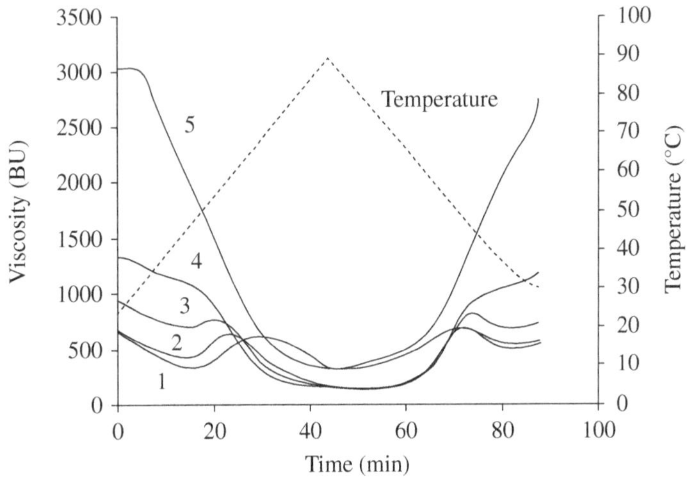

Figure 1.9 Effect of temperature on the viscosity of HPMC/HPS blends: (1) 100/0; (2) 70/30; (3) 50/50; (4) 30/70; and (5) 0/100. Source: Reproduced from [127] with permission.

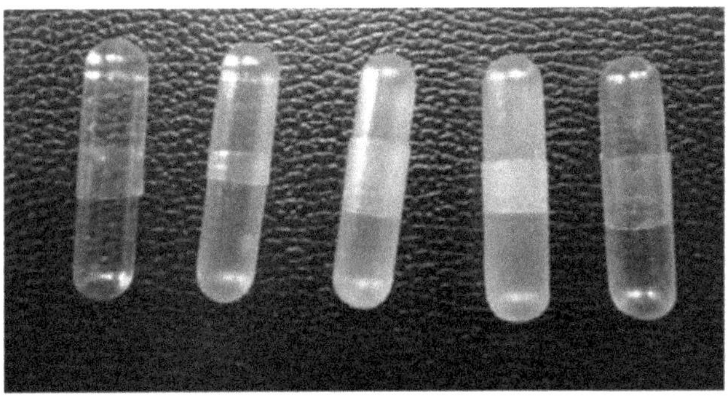

Figure 1.10 Photos of capsules made from solutions with different HPMC/HPS contents: (a) 100 : 0; (b) 70 : 30; (c) 50 : 50; (d) 30 : 70; and (e) 0 : 100. Source: Reproduced from [125] with permission.

increased slightly with increasing HPS content under the same dipping conditions, since the HPS increased the viscosity of the solutions.

1.4.5 Functionalized Composites

In order to add some functions and meet market requirements, some functionalized starch-based materials have been developed. For example, Ali et al. [130] have reported that pomegranate peel (PGP) as an antimicrobial agent as well as a reinforcing agent was utilized in developing starch-based films. The demand for safe and ready-to-serve food prepared by generally recognized as safe (GRAS) preservatives to control microbial contamination is increasing. Previous studies have shown that PGP extract demonstrated positive results against microbes [131, 134]. However, there had been no previous report about the incorporation of pomegranate peel as a filler in starch-based films to develop

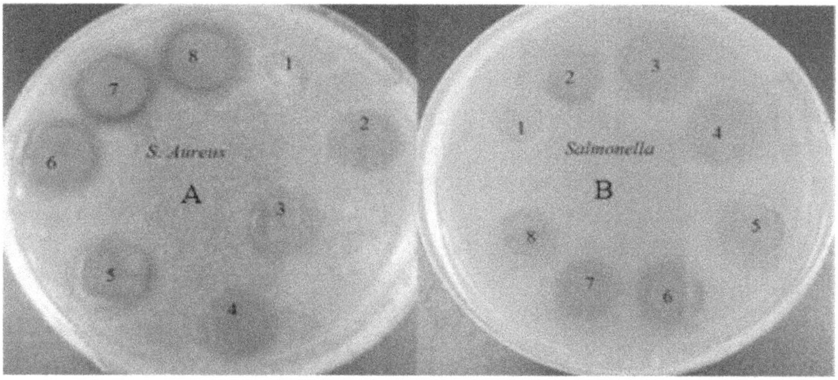

Figure 1.11 Inhibition zones of starch films filled with PGP against S. aureus (a) and Salmonella (b), PGP content (% wt) in sample-1, 2, 3, 4, 5, 6, 7 and 8 is 0, 2, 4, 6, 8, 10 and 12 respectively. Source: Reproduced from [125] with permission.

antimicrobial materials. Results clearly demonstrated that PGP inhibited the growth of both gram-positive (*Staphylococcus aureus*) and gram-negative (*Salmonella*) bacteria. Meanwhile, PGP also increased Young's modulus, tensile strength, and stiffness of the starch-based films. The XRD Both the optical microscope (OM) and scanning electron microscope (SEM) observations indicated reasonably good compatibility between starch and PGP particles. The X-ray diffraction (XRD) results indicated that PGP retained its semi-crystalline structure in the film, which can be used to explain the mechanism of mechanical reinforcement. Since all the components are food ingredients, it is expected that the developed material can be used as an edible film and food grade packaging material.

Figure 1.11 shows the antimicrobial activity of film samples against *S. aureus* and *Salmonella*. Starch-based films filled with PGP demonstrated larger inhibition zones against *S. aureus* than those of *Salmonella*. With the increase in concentration of PGP in the starch matrix, the diameter of the inhibition zone was also increased. The highest value of zone inhibition was shown by 12–14% PGP reinforced films against both microorganisms. The results are in agreement with previous reports whereby the efficacy of pomegranate peel extract was reported against *S. aureus, Escherichia coli*, and *Salmonella* strains. Similar to PGP extract, the possible mechanism of action of PGP particles against microorganisms can be related to the phenolic toxicity that interacts with sulfhydryl groups of proteins in micro-organisms. It can be concluded that the starch matrix released the PGP particles, which in turn demonstrated the antimicrobial properties.

1.5 Reducing Moisture Sensitivity by Coating

The well-recognized weakness of starch-based materials is their moisture sensitivity. The hydrophilic nature of starch makes it susceptible to moisture uptake, resulting in inadequate performance, in particular WVP and tensile strength. To improve the moisture resistance and mechanical properties of starch-based materials, various blending and compositing techniques have been developed [133–138]. However, a blend containing starch cannot become hydrophobic until the starch content is low enough to form a separate domain phase in a hydrophobic polymer matrix, in which case starch only acts as a filler. The advantages of surface coating and cross-linking include ease of processing, lower cost, flexibility to control thickness, maximizing natural properties of matrix, etc. In fact, cross-linking is a common approach to improve the performance of starch for various applications.

Starch and starch products have been cross-linked with various cross-linking agents, such as phosphorus oxychloride, sodium trimetaphosphate, sodium tripolyphosphate, epichlorohydrin, 1,2,3,4-diepoxybutane, and citric acid to improve the mechanical and moisture stability properties [139–144]. Most of the cross-linking agents have to be added to a starch solution during film casting. However, this method is not suitable for industrial manufacture, in which a film or sheet is mainly produced by extrusion casting. Surface coating and surface cross-linking are more suitable for post-treatment. Recently, Ge et al. [145] reported that an acrylated epoxidized soybean oil (AESO)-based coating was developed to reduce the moisture sensitivity and improve the gas permeability of a starch-based film. Similar to starch, soybean oil also comes from renewable resources. The coating was applied on starch-based films by dipping the samples on AESO-based coating solutions, followed by cross-linking with ultraviolet (UV) light. The results showed that the moisture sensitivity of the starch-based sheets was reduced significantly since the cross-linked AESO acted as a hydrophobic layer. Moisture permeability was decreased more than 10 times after AESO treatment. It was found that the cross-linking density acted as one of the key factors, so even a very thin layer of AESO could achieve good water resistance.

In order to improve the surface adhesion between starch and AESO coating and further improve the moisture sensitivity, Meng et al. [146] used polyethylenimine (PEI) to improve the bonding between hydrophilic starch and hydrophobic coatings of AESO. Analysis including nuclear magnetic resonance (NMR), Fourier transform infrared (FTIR), and X-ray photoelectron spectroscopy (XPS) revealed that PEI interacted effectively with starch films via hydrogen bonding and bonded chemically with AESO through the Michael addition reaction. Shear adhesion testing demonstrated that the interfacial adhesion between the starch and AESO coating was significantly improved following surface treatment with PEI. Starch films with PEI treatment and AESO coating maintained original mechanical strength after water soaking for more than three hours. These results demonstrated the potential of developing new waterproof materials based on renewable and environmentally friendly starch and AESO using PEI as a coupling agent.

1.6 Applications in Packaging

So far various different starch-based packaging products, such as films, capsules, sheets, trays, and foam, etc., have been developed and commercialized. Starch-based materials have showed great potential, since they offer a very attractive low-cost base for new biodegradable polymers due to their low material cost and ability to be processed with conventional plastic processing equipment. A special group of starch-based packaging and edible packaging materials has also showed increased attractions. Figure 1.12 shows some commercialized starch-based products, which are not only fully biodegradable but also can be used to feed animals or are even edible.

Figure 1.12 Photos of commercialized starch-based products.

Recently, applications of starch foam have increased due to new regulations banning disposable plastics announced by more and more countries in the world. Starch-based foam can simply replace many traditional plastic foams, such as polystyrene (PS), polyethylene (PE), polyurethane (PU), etc. Meng et al. [147] have studied the relationship between phase transition and foaming behavior of starch-based material by investigating the effects of water content on melting temperature (T_m), crystallinity, foaming process, cell structure, as well as mechanical properties. It was found that there was a critical point of water content (between 16% and 18%) at which the expanding ratio changed significantly since the cell structure changed from open to closed. The lower the water content, the higher the T_m of starch-based materials, which resulted in lower melting strength and an open cell structure [147, 148]. From an application-oriented point of view, in order to decrease the cost of transportation, the foam was prepared in two steps in this work [145, 146]: (i) compounding and (ii) foaming. This is because the foamed materials consist of more than 90% gas, so the transportation cost could be higher than the cost of the material itself.

Figure 1.13 shows the photos of various starch foams and their applications. It is seen that various foams from ball to sheet as well as many post-processed products have been developed.

1.7 Summary and Future Work

Various starch-based packaging products, such as films, capsules, sheets, thermal forming, and foam, have been developed and commercialized. Starch-based materials have showed great potential, especially when more and more countries have passed the low or regulation of banning disposable plastics. Like all other materials, improving performance and decreasing cost are two on-going strategies. In principle, the properties of starch-based materials can be significantly improved by blending with synthetic oil-based polymers. However, these blends are not biodegradable and thus the advantage of using a biodegradable polysaccharide will be lost. The polymer blends and composites made only from natural raw materials were discussed in this chapter.

Starch reinforced by cellulose is a typical example of natural polymer composites. These materials have the advantages of being renewable, biodegradable, abundantly available, and inexpensive. Recently, starch films reinforced by both cellulose and starch-based crystals have been developed. There is good compatibility between the starch matrix and the reinforcing fillers due to the same chemical unit (glucose). Since all the components used belong to food sources, the prepared films are biodegradable, safe for food packaging, and can also be used to develop edible films and medicinal capsules. The new technique of self-reinforced composites has also been used in developing starch-based materials.

Figure 1.13 Photos of starch foams: from foam ball to form sheet and post-processed products.

The starch-based nanocomposites have continuously attracted attention and have been successfully developed and commercialized. One of the challenges using nanoclay in conventional polymers is to exfoliate it because of the strong boundary between the layers by iron. Another issue is to improve the hydrophilic surface of clay with a hydrophobic surface of most conventional polymers, especially polyolefin. These two challenges can be simply solved in starch-based materials. Firstly, the hydrophilic water is a very efficient chemical to delaminate clay with the enhancement by ultrasonics. Secondly, starch itself is a hydrophilic polymer that can be compatible with clay and water. In fact, water is a plasticizer for starch-based materials.

In order to add some functions and to meet market requirements, some functionalized starch-based materials have been developed. For example, pomegranate peel has been used as an antimicrobial agent as well as a reinforcing agent for starch-based films. Starch-based films filled with PGP demonstrated larger inhibition zones against both *S. aureus* and *Salmonella*. It has been concluded that the starch matrix released the PGP particles, which in turn demonstrated the antimicrobial properties.

The well-recognized weakness of starch-based materials is their moisture sensitivity. To improve the moisture resistance and mechanical properties of starch-based materials, various blending and compositing techniques have been developed. However, the most efficient and easy way is coasting. An AESO-based coating was developed to reduce the moisture sensitivity and improve gas permeability of a starch-based film. Similar to starch, soybean oil also comes from a renewable resource. In order to improve the surface adhesion between starch and the AESO coating and further improve the moisture sensitivity, PEI was used to improve the bonding between hydrophilic starch and hydrophobic coatings of AESO.

From an application-oriented point of view, in order to decrease the cost of transportation, two foaming steps have been developed: (i) compounding and (ii) foaming. This is because the foamed materials consist of more than 90% gas, so the transportation cost could be higher than the cost of the material itself. It was found that there was a critical point of water content (between 16% and 18%) at which the expendable ratio changed significantly as the cell structure changed from open to closed.

Based on above discussion, it is seen that both challenges and opportunities are exciting for starch-based materials. Future work will continuously focus on improving performance and decreasing cost, in particular improving the well recognized weakness of starch-based materials: moisture sensitivity.

Acknowledgments

The authors from SCUT, China, would like to acknowledge the research fund National Key R&D Program of China (2018YFD04007002), NSFC (31571789), and 111 Project (B17018).

References

1. J. Song, R. Murphy, R. Narayan, G. Davies, Biodegradable and compostable alternatives to conventional plastics, *Philos. Trans. R. Soc. B.* **364**(1526) (2009) 2127–2139.
2. L. Lebreton, A. Andrady, Future scenarios of global plastic waste generation and disposal, *Palgrave Commun.* **5**(1) (2019) 6.
3. M.M. Reddy, S. Vivekanandhan, M. Misra, S.K. Bhatia, A.K. Mohanty, Biobased plastics and bionanocomposites: current status and future opportunities. *Prog. Polym. Sci.* **38**(10–11) (2013) 1653–1689.
4. R. Jain, A. Tiwari, Biosynthesis of planet friendly bioplastics using renewable carbon source, *J. Environ. Health Sci. Eng.* **13** (2015) 11.
5. T.J. Gutiérrez, V.A. Alvarez, Cellulosic materials as natural fillers in starch-containing matrix-based films: a review, *Polym. Bull.* **74**(6) (2017) 2401–2430.
6. U.V. Deore, H.S. Mahajan, Isolation and characterization of natural polysaccharide from Cassia Obtustifolia seed mucilage as film forming material for drug delivery, *Int. J. Biol. Macromol.* **15** (2018) 1071–1078.
7. P. Devi, V. Bajala, V. Garg, S. Mor, K. Ravindra, Heavy metal content in various types of candies and their daily dietary intake by children, *Environ. Monit. Assess.* **188** (2) (2016) 86.
8. A. Jiménez, M.J. Fabra, P. Talens, A. Chiralt, Edible and biodegradable starch films: a review, *Food Bioprocess Technol.* **5**(6) (2012) 2058–2076.
9. M. Chaichi, M. Hashemi, F. Badii, A. Mohammadi, Preparation and characterization of a novel bionanocomposite edible film based on pectin and crystalline nanocellulose, *Carbohydr. Polym.* **157** (2017) 167–175.

10. I. Arvanitoyannis, A. Nakayama, S.Aiba, Edible films made from hydroxypropyl starch and gelatin and plasticized by polyols and water, *Carbohydr. Polym.* **36** (2–3) (1998) 105–119.
11. I. Arvanitoyannis, Totally and partially biodegradable polymer blends based on natural and synthetic macromolecules: preparation, physical properties, and potential as food packaging materials, *J. Macromol. Sci. Polym. Rev.* **39**(2) (1999) 205–271.
12. V.P. Romani, C. Prentice-Hernández, V.G. Martins, Active and sustainable materials from rice starch, fish protein and oregano essential oil for food packaging, *Ind. Crops Prod.* **97** (2017) 268–274.
13. L.C. Garcia, L.M. Pereira, C.I.G. de Luca Sarantópoulos, M.D. Hubinger, Effect of antimicrobial starch edible coating on shelf-life of fresh strawberries, *Packag. Technol. Sci.* **25**(7) (2012) 413–425.
14. M.R. Sucaldito, D.H. Camacho, Characteristics of unique HBr-hydrolyzed cellulose nanocrystals from freshwater green algae (*Cladophora rupestris*) and its reinforcement in starch-based film, *Carbohydr. Polym.* **169** (2017) 315–323.
15. R.Q. Assis, C.H. Pagno, T.M.H. Costa, S.H. Flôres, A.d.O. Rios, Synthesis of biodegradable films based on cassava starch containing free and nanoencapsulated β-carotene, *Packag. Technol. Sci.* **31**(3) (2018) 157–166.
16. R. Jumaidin, S.M. Sapuan, M. Jawaid, M.R. Ishak, J. Sahari, Thermal, mechanical, and physical properties of seaweed/sugar palm fibre reinforced thermoplastic sugar palm starch/agar hybrid composites, *Int. J. Biol. Macromol.* **97** (2017) 606–615.
17. M. Shahbazi, G. Rajabzadeh, S. Sotoodeh, Functional characteristics, wettability properties and cytotoxic effect of starch film incorporated with multi-walled and hydroxylated multi-walled carbon nanotubes, *Int. J. Biol. Macromol.* **104**(Pt A) (2017) 597–605.
18. F.X. Espinach, M. Delgado-Aguilar, J. Puig, F. Julian, S. Boufi, P. Mutjé, Flexural properties of fully biodegradable alpha-grass fibers reinforced starch-based thermoplastics, *Composites Part B* **81** (2015) 98–106.
19. A.T. Le, A. Gacoin, A. Li, T.H. Mai, N. El Wakil, Influence of various starch/hemp mixtures on mechanical and acoustical behavior of starch-hemp composite materials, *Composites Part B* **75** (2015) 201–211.
20. L. Yu, G. Christie, Microstructure and mechanical properties of orientated thermoplastic starches, *J. Mater. Sci.* **40**(1) (2005) 111–116.
21. L. Yu, K. Dean, L. Li, Polymer blends and composites from renewable resources, *Prog. Polym. Sci.* **31**(6) (2006) 576–602.
22. H. Liu, F. Xie, L. Yu, L. Chen, L. Li, Thermal processing of starch-based polymers, *Prog. Polym. Sci.* **34**(12) (2009) 1348–1368.
23. F. Xie, P.J. Halley, L. Avérous, Rheology to understand and optimize processibility, structures and properties of starch polymeric materials, *Prog. Polym. Sci.***37** (4) (2012) 595–623.
24. L. Liao, H. Liu, X. Liu, L. Chen, L. Yu, P. Chen,Microstructures and phase transitions of starch,*Acta. Polym. Sin.***6** (2014)761–773.
25. K. Svegmark, A.M. Hermansson, Distribution of amylose and amylopectin in potato starch pastes: effects of heating and shearing, *Food Struct.* **10** (2) (1991) 117–129.
26. P. Chen, L. Yu, L. Chen, X. Li, Morphology and microstructure of maize starches with different amylose/amylopectin content, *Starch-Stärke.***58** (12) (2006) 611–615.
27. P. Liu, L. Chen, P.A. Corrigan, L. Yu, Z. Liu, Application of atomic force microscopy on studying micro-and nano-structures of starch, *Int. J. Food Eng.***4** (7) (2008) 1–23.
28. P. Chen, L. Yu, G. Simon, E. Petinakis, K. Dean, L. Chen, Morphologies and microstructures of cornstarches with different amylose–amylopectin ratios studied by confocal laser scanning microscope, *J. Cereal Sci.* **50**(2) (2009) 241–247.
29. H. Liu, L. Yu, G. Simon, X. Zhang, K. Dean, L. Chen, Effect of annealing and pressure on microstructure of cornstarches with different amylose/amylopectin ratios, *Carbohydr. Res.* **344**(3) (2009) 350–354.

30. H. Liu, L. Yu, G. Simon, K. Dean, L. Chen, Effects of annealing on gelatinization and microstructures of corn starches with different amylose/amylopectin ratios, *Carbohydr. Polym.* **77**(3) (2009) 662–669.
31. P. Chen, L. Yu, G.P. Simon, X. Liu, K. Dean, L. Chen, Internal structures and phase-transitions of starch granules during gelatinization, *Carbohydr. Polym.* **83**(4) (2011) 1975–1983.
32. W.A. Atwell, The terminology and methodology associated with basic starch phenomena, *Cereal Foods World* **33** (1988) 306–311.
33. J. Lelievre, Starch gelatinization, *J. Appl. Polym. Sci.***18** (1) (1974) 293–296.
34. H. Liu, L. Yu, F. Xie, L. Chen, Gelatinization of cornstarch with different amylose/amylopectin content, *Carbohydr. Polym.* **65**(3) (2006) 357–363.
35. P. Chen, L. Yu, T. Kealy, L. Chen, L. Li, Phase transition of starch granules observed by microscope under shearless and shear conditions, *Carbohydr. Polym.***68**(3) (2007) 495–501.
36. X. Liu, L. Yu, H. Liu, L. Chen, L. Li, In situ thermal decomposition of starch with constant moisture in a sealed system, *Polym. Degrad. Stab.* **93**(1) (2008) 260–262.
37. X. Liu, L. Yu, H. Liu, L. Chen, L. Li, Thermal decomposition of corn starch with different amylose/amylopectin ratios in open and sealed systems, *Cereal Chem.* **86** (4) (2009) 383–385.
38. P. Liu, L. Yu, X. Wang, D. Li, L. Chen, X. Li, Glass transition temperature of starches with different amylose/amylopectin ratios, *J. Cereal Sci.***51**(3) (2010) 388–391.
39. X. Liu, L. Yu, F. Xie, M. Li, L. Chen, X. Li, Kinetics and mechanism of thermal decomposition of cornstarches with different amylose/amylopectin ratios, *Starch-Stärke* **62**(3–4) (2010) 139–146.
40. H. Liu, M. Li, P. Chen, L. Yu, L. Chen, Z. Tong, Morphologies and thermal properties of hydroxypropylated high-amylose corn starch, *Cereal Chem.* **87**(2) (2010) 144–149.
41. C. Lan, H. Liu, P. Chen, L. Yu, L. Chen, X. Li, X. Zhang, Gelatinization and retrogradation of hydroxypropylated cornstarch, *Int. J. Food Eng.* **6**(4) (2010) 1–12.
42. D. Qiao, L. Yu, H. Liu, W. Zou, F. Xie, G. Simon, E. Petinakis, Z. Shen, L. Chen, Insights into the hierarchical structure and digestion rate of alkali-modulated starches with different amylose contents, *Carbohydr. Polym.* **144** (2016) 271–281.
43. J.W. Sullivan, J.A. Johnson, Measurement of starch gelatinization by enzyme susceptibility, *Cereal Chem.* **41**(2) (1964)73–77.
44. L. Yu, G. Christie, Measurement of starch thermal transitions using differential scanning calorimetry, *Carbohydr. Polym.* **46** (2) (2001) 179–184.
45. H. Liu, F. Xie, L. Chen, L. Yu, K. Dean, S. Bateman, Thermal behaviour of high amylose cornstarch studied by DSC, *Int. J. Food Eng.* **1**(1) (2005) 1–6.
46. H. Liu, L. Yu, L. Chen, L. Li, Retrogradation of corn starch after thermal treatment at different temperatures, *Carbohydr. Polym.* **69**(4) (2007) 756–762.
47. P. Liu, L. Yu, H. Liu, L. Chen, L. Li, Glass transition temperature of starch studied by a high-speed DSC, *Carbohydr. Polym.* **77**(2) (2009) 250–253.
48. H. Liu, L. Yu, K. Dean, G. Simon, E. Petinakis, L. Chen, Starch gelatinization under pressure studied by high pressure DSC, *Carbohydr. Polym.* **75** (3) (2009) 395–400.
49. F. Xie, W.C. Liu, P. Liu, J. Wang, P.J. Halley, L. Yu, Starch thermal transitions comparatively studied by DSC and MTDSC, *Starch-Stärke* **62** (7) (2010) 350–357.
50. H. Liu, L. Yu, Z. Tong, L. Chen, Retrogradation of waxy cornstarch studied by DSC, *Starch-Stärke* **62**(10) (2010) 524–529.
51. C. Keetels, G. Oostergetel, T. Van Vliet, Recrystallization of amylopectin in concentrated starch gels, *Carbohydr. Polym.* **30** (1) (1996) 61–64.
52. C. Keetels, T. Van Vliet, P. Walstra, Gelation and retrogradation of concentrated starch systems: 2. Retrogradation, *Food Hydrocoll.* **10** (3) (1996) 355–362.
53. M.T. Kalichevsky, S.G. Ring, Incompatibility of amylose and amylopectin in aqueous solution, *Carbohydr. Res.* **162** (2) (1987) 323–328.

54. J. Doublier, G. Llamas, A rheological description of amylose-amylopectine mixtures. In *Food Colloids and Polymers: Stability and Mechanical Properties*; E. Dickinson, P. Walstra, Eds.; Royal Society of Chemistry: Cambridge, U.K., (1993) 138–146.
55. C. Biliaderis, C. Page, L. Slade, R. Sirett, Thermal behavior of amylose-lipid complexes, *Carbohydr. Polym.* **5** (5) (1985) 367–389.
56. F. Xie, H. Liu, P. Chen, T. Xue, L. Chen, L. Yu, P. Corrigan, Starch gelatinization under shearless and shear conditions, *Int. J. Food Eng.* **2**(5) (2007)1–29.
57. F. Xie, L. Yu, L. Chen, L. Li, A new study of starch gelatinization under shear stress using dynamic mechanical analysis, *Carbohydr. Polym.* **72**(2) (2008) 229–234.
58. T. Xue, L. Yu, F. Xie, L. Chen, L. Li, Rheological properties and phase transition of starch under shear stress, *Food Hydrocoll.***22**(6)(2008) 973–978.
59. B. Su, F. Xie, M. Li, P.A. Corrigan, L. Yu, X. Li, L. Chen, Extrusion processing of starch film, *Int. J. Food Eng.* **5**(1) (2009) 1–12.
60. F. Xie, L. Yu, B. Su, P. Liu, J. Wang, H. Liu, L. Chen, Rheological properties of starches with different amylose/amylopectin ratios, *J. Cereal Sci.* **49**(3) (2009) 371–377.
61. J. Wang, L. Yu, F. Xie, L. Chen, X. Li, H. Liu, Rheological properties and phase transition of cornstarches with different amylose/amylopectin ratios under shear stress, *Starch-Stärke* **62**(12) (2010) 667–675.
62. M. Li, P. Liu, W. Zou, L. Yu, F. Xie, H. Pu, H. Liu, L. Chen, Extrusion processing and characterization of edible starch films with different amylose contents, *J. Food Eng.* **106**(1) (2011) 95–101.
63. X. Liu, Y. Wang, L. Yu, Z. Tong, L. Chen, H. Liu, X. Li, Thermal degradation and stability of starch under different processing conditions, *Starch-Stärke* **65**(1–2) (2013) 48–60.
64. X. Liu, H. Ma, L. Yu, L. Chen, Z. Tong, P. Chen, Thermal-oxidative degradation of high-amylose corn starch, *J. Therm. Anal. Calorim.* **115**(1) (2013) 659–665.
65. J. Olkku, C. Rha, Gelatinisation of starch and wheat flour starch–a review, *Food Chem.* **3** (4) (1978) 293–317.
66. M.H. Lim, H. Wu, D.S. Reid, The effect of starch gelatinization and solute concentrations on T g' of starch model system, *J. Sci. Food Agric.* **80** (12) (2000) 1757–1762.
67. R.F. Tester, S.J. Debon, Annealing of starch–heavy metal content in various types of candies and their a review, *Int. J. Biol. Macromol.* **27** (1) (2000) 1–12.
68. B.C. Burros, L.A. Young, P.A. Carroad, Kinetics of corn meal gelatinization at high temperature and low moisture, *J. Food Sci.* **52**(5) (1987) 1372–1376.
69. L. Wen, P. Rodis, B. Wasserman, Starch fragmentation and protein insolubilization during twin-screw extrusion of corn meal, *Cereal Chem.* **67** (1990) 268–275.
70. L. Lai, J. Kokini, Physicochemical changes and rheological properties of starch during extrusion (a review), *Biotechnol. Progr.* **7** (3) (1991) 251–266.
71. P. Colonna, J. Doublier, J. Melcion, F. De Monredon, C. Mercier, Extrusion cooking and drum drying of wheat starch, *Cereal Chem.* **61** (1984) 538–554.
72. V. Davidson, D. Paton, L. Diosady, G. Larocque, Degradation of wheat starch in a single screw extruder: characteristics of extruded starch polymers, *J. Food Sci.* **49** (2) (1984) 453–458.
73. J. Willett, M. Millard, B. Jasberg, Extrusion of waxy maize starch: melt rheology and molecular weight degradation of amylopectin, *Polymer* **38** (24) (1997) 5983–5989.
74. A.J.F. Carvalho, M.D. Zambon, A.A.S. Curvelo, A. Gandini, Size exclusion chromatography characterization of thermoplastic starch composites 1. Influence of plasticizer and fibre content, *Polym. Degrad. Stab.* **79** (1) (2003) 133–138.
75. F.J. Baltá Calleja, D.R. Rueda, T. Secall, R.K. Bayer, M. Schlimmer, Influence of processing methods on starch properties, *J. Macromol. Sci B.* **38**(4) (2006) 461–469.
76. S. Ilo, U. Tomschik, E. Berghofer, N. Mundigler, The effect of extrusion operating conditions on the apparent viscosity and the properties of extrudates in twin-screw extrusion cooking of maize grits, *LWT-Food Sci. Technol.* **29** (7) (1996) 593–598.

77. R.J. González, R.L. Torres, D.M. De Greef, B.A. Guadalupe, Effects of extrusion conditions and structural characteristics on melt viscosity of starchy materials, *J. Food Eng.* **74** (1) (2006) 96–107.
78. M. Thuwall, A. Boldizar, M. Rigdahl, Extrusion processing of high amylose potato starch materials, *Carbohydr. Polym.* **65** (4) (2006) 441–446.
79. F. Xie, L. Yu, H. Liu, L. Chen, Starch modification using reactive extrusion, *Starch-Stärke* **58**(3–4) (2006) 131–139.
80. G. Moad, Chemical modification of starch by reactive extrusion, *Prog. Polym. Sci.* **36** (2) (2011) 218–237.
81. D. Qiao, W. Zou, X. Liu, L. Yu, L. Chen, H. Liu, N. Zhang, Starch modification using a twin-roll mixer as a reactor, *Starch-Stärke* **64**(10) (2012) 821–825.
82. D. Qiao, H. Ma, L. Yu, H. Liu, W. Zou, L. Chen, P. Chen, Synthesis and characteristics of graft copolymerization of starch-G-PAM using a twin-roll mixer as reactor for cornstarch with different amylose/amylopectin ratios, *Int. Polym. Proc.* **29** (2) (2014) 252–259.
83. D. Qiao, X. Bao, H. Liu, X. Liu, L. Chen, L. Yu, X. Zhang, P. Chen, Preparation of cassava starch-based superabsorbent polymer using a twin-roll mixer as reactor, *Chin. J. Polym. Sci.* **32**(10) (2014) 1348–1356.
84. X. Bao, L. Yu, S. Shen, G.P. Simon, H. Liu, L. Chen, How rheological behaviors of concentrated starch affect graft copolymerization of acrylamide and resultant hydrogel, *Carbohydr. Polym.* **219** (2019) 395–404.
85. X. Bao, L. Yu, G.P. Simon, S. Shen, F. Xie, H. Liu, L. Chen, L. Zhong, Rheokinetics of graft copolymerization of acrylamide in concentrated starch and rheological behaviors and microstructures of reaction products, *Carbohydr. Polym.* **192** (2018) 1–9.
86. H. Abral, J. Ariksa, M. Mahardika, D. Handayani, I. Aminah, N. Sandrawati,A.B. Pratama, N. Fajri, S. M. Sapuan, R. A. Ilyas, Transparent and antimicrobial cellulose film from ginger nanofiber, *Food Hydrocoll.* **98** (2020),105226.
87. H. Abral, J. Ariksa, M. Mahardika, D. Handayani, I. Aminah, N. Sandrawati, N. Fajri, S. M. Sapuan, R. A. Ilyas, Highly transparent and antimicrobial PVA based bionanocomposites reinforced by ginger nanofiber, *Polym. Test.* (2019), 106186.
88. M. Wollerdorfer, H. Bader, Influence of natural fibres on the mechanical properties of biodegradable polymers, *Ind. Crop. Prod.* **8**(2) (1998) 105–112.
89. A. Dufresne, D. Dupeyre, M.R. Vignon, Cellulose microfibrils from potato tuber cells: processing and characterization of starch–cellulose microfibril composites, *J. Appl. Polym. Sci.***76** (14) (2000) 2080–2092.
90. A. Ali, L. Yu, H. Liu, S. Khalid, L. Meng, L. Chen, Preparation and characterization of starch-based composite films reinforced by corn and wheat hulls, *J. Appl. Polym. Sci.* **134**(32) (2017)45159–45165.
91. A. Ali, S. Ali, L. Yu, H. Liu, S. Khalid, A. Hussain, M.M.N. Qayum, C. Ying, Preparation and characterization of starch-based composite films reinforced by apricot and walnut shells, *J. Appl. Polym. Sci.* (2019) 10.1002/app.47978.
92. Y. Chen, L. Yu, X. Ge, H. Liu, A. Ali, Y. Wang, L. Chen, Preparation and characterization of edible starch film reinforced by laver,*Int. J. Biol. Macromol.* **129** (2019) 944–951.
93. S. Sukhija, S. Singh, C.S. Riar, Analyzing the effect of whey protein concentrate and psyllium husk on various characteristics of biodegradable film from lotus (Nelumbo nucifera) rhizome starch, *Food Hydrocoll.* **60** (2016) 128–137.
94. F. Xie, E. Pollet, P.J. Halley, L. Avérous, Starch-based nano-biocomposites, *Prog. Polym. Sci.* **38**(10–11) (2013) 1590–1628.
95. F. Chivrac, H. Angellier-Coussy, V. Guillard, E. Pollet, L. Avérous, How does water diffuse in starch/montmorillonite nano-biocomposite materials?, *Carbohydr. Polym.* **82**(1) (2010) 128–135.
96. F. Chivrac, O. Gueguen, E. Pollet, S. Ahzi, A. Makradi, L. Avérous, Micromechanical modeling and characterization of the effective properties in starch-based nano-biocomposites, *Acta Biomater.* **4**(6) (2008) 1707–1714.

97. F. Chivrac, E. Pollet, L. Avérous, Progress in nano-biocomposites based on polysaccharides and nanoclays,*Mater. Sci. Eng. R.* **67**(1) (2009) 1–17.
98. K. Dean, L. Yu, Biodegradable protein–nanoparticle composite. In *Biodegradable Polymers*; R. Smith, Ed.; Woodhead: Cambridge, U.K., **11** (2005) 289–309.
99. K. Dean, L. Yu, D.Y. Wu, Preparation and characterization of melt-extruded thermoplastic starch/clay nanocomposites, *Compos. Sci. Technol.* **67**(3–4) (2007) 413–421.
100. K. Dean, M. Do, E. Petinakis, L. Yu, Key interactions in biodegradable thermoplasticstarch/PVOH /montmorillonite micro-and nanocomposites, *Compos. Sci. Technol.* **68**(6) (2008) 1453–1462.
101. K. Dean, E. Petinakis, L. Goodall, T. Miller, L. Yu, N. Wright, Nanostabilization of thermally processed high amylose hydroxylpropylated starch films, *Carbohydr. Polym.* **86**(2) (2011) 652–658.
102. E.d.M. Teixeira, D. Pasquini, A.A.S. Curvelo, E. Corradini, M.N. Belgacem, A. Dufresne, Cassava bagasse cellulose nanofibrils reinforced thermoplastic cassava starch, *Carbohydr. Polym.* **78** (3) (2009) 422–431.
103. E. Robles, A.M. Salaberria, R. Herrera, S.C. Fernandes, J. Labidi, Self-bonded composite films based on cellulose nanofibers and chitin nanocrystals as antifungal materials, *Carbohydr. Polym.***144** (2016) 41–49.
104. D. Liu, T. Zhong, P.R. Chang, K. Li, Q. Wu, Starch composites reinforced by bamboo cellulosic crystals, *Bioresour. Technol.* **101**(7) (2010) 2529–2536.
105. M. Roohani, Y. Habibi, N.M. Belgacem, G. Ebrahim, A.N. Karimi, A. Dufresne, Cellulose whiskers reinforced polyvinyl alcohol copolymers nanocomposites, *Eur. Polym. J.* **44** (8) (2008) 2489–2498.
106. N. Soykeabkaew, N. Laosat, A. Ngaokla, N. Yodsuwan, T. Tunkasiri, Reinforcing potential of micro- and nano-sized fibers in the starch-based biocomposites, *Compos. Sci. Technol.* **72**(7) (2012) 845–852.
107. J.S. Alves, K.C. dos Reis, E.G. Menezes, F.V. Pereira, J. Pereira, Effect of cellulose nanocrystals and gelatin in corn starch plasticized films, *Carbohydr. Polym.* **115** (2015) 215–222.
108. N. El Miri, K. Abdelouahdi, A. Barakat, M. Zahouily, A. Fihri, A. Solhy, M. El Achaby, Bio-nanocomposite films reinforced with cellulose nanocrystals: rheology of film-forming solutions, transparency, water vapor barrier and tensile properties of films, *Carbohydr. Polym.* **129** (2015) 156–167.
109. X. Li, C. Qiu, N. Ji, C. Sun, L. Xiong, Q. Sun, Mechanical, barrier and morphological properties of starch nanocrystals-reinforced pea starch films, *Carbohydr. Polym.* **121** (2015) 155–162.
110. M. Li, D. Li, L. Wang, B. Adhikari, Creep behavior of starch-based nanocomposite films with cellulose nanofibrils, *Carbohydr. Polym.* **117** (2015) 957–963.
111. D.M. Panaitescu, A.N. Frone, M. Ghiurea, I. Chiulan, Influence of storage conditions on starch/PVA films containing cellulose nanofibers, *Ind. Crops Prod.* **70** (2015) 170–177.
112. A. Ali, F. Xie, L. Yu, H. Liu, L. Meng, S. Khalid, L. Chen, Preparation and characterization of starch-based composite films reinfoced by polysaccharide-based crystals, *Composites Part B* **133** (2018) 122–128.
113. M.S. Amer, S. Ganapathiraju, Effects of processing parameters on axial stiffness of self-reinforced polyethylene composites, *J. Appl. Polym. Sci.* **81** (5) (2001) 1136–1141.
114. P.J. Hine, R.H. Olley, I.M. Ward, The use of interleaved films for optimising the production and properties of hot compacted, self reinforced polymer composites, *Compos. Sci. Technol.* **68**(6) (2008) 1413–1421.
115. M. Manninen, U. Päivärinta, H. Pätiälä, P. Rokkanen, R. Taurio, M. Tamminmäki, P. Törmälä, Shear strength of cancellous bone after osteotomy fixed with absorbable self-reinforced polyglycolic acid and poly-L-lactic acid rods, *J. Mater. Sci. Mater. Med.* **3** (4) (1992) 245–251.
116. R. Suuronen, L. Wessman, M. Mero, P. Törmälä, J. Vasenius, E. Partio, K. Vihtonen, S. Vainionpää, Comparison of shear strength of osteotomies fixed with absorbable self-reinforced poly-L-lactide and metallic screws, *J. Mater. Sci. Mater. Med.* **3**(4) (1992) 288–292.
117. C. Gao, L. Yu, H. Liu, L. Chen, Development of self-reinforced polymer composites, *Prog. Polym. Sci.* **37**(6) (2012) 767–780.

118. C. Gao, L. Meng, L. Yu, G.P. Simon, H. Liu, L. Chen, S. Petinakis, Preparation and characterization of uniaxial poly(lactic acid)-based self-reinforced composites, *Compos. Sci. Technol.* **117** (2015) 392–397.
119. C. Lan, L. Yu, P. Chen, L. Chen, G. Simon, X. Zhang, Designing and preparing starch-based self-reinforced composites, *Macromol. Mater. Eng.* **295**(11) (2010) 1025–1030.
120. M. Stading, A.M. Hermansson, P. Gatenholm, Structure, mechanical and barrier properties of amylose and amylopectin films, *Carbohydr. Polym.***36** (2–3) (1998) 217–224.
121. N. Zhang, H. Liu, L. Yu, X. Liu, L. Zhang, L. Chen, R. Shanks, Developing gelatin–starch blends for use as capsule materials, *Carbohydr. Polym.* **92** (1) (2013) 455–461.
122. N. Zhang, X. Liu, L. Yu, R. Shanks, E. Petinaks, H. Liu, Phase composition and interface of starch-gelatin blends studied by synchrotron FTIR micro-spectroscopy, *Carbohydr. Polym.* **95**(2) (2013) 649–653.
123. N.Z. Zhang, R. Shanks, X.X. Liu, L. Yu, Phase composition of starch-Gelatin blends studied by FTIR, *Adv. Mater. Res.* **875-877** (2014) 106–109.
124. I. Arvanitoyannis, M. Kalichevsky, J.M. Blanshard, E. Psomiadou, Study of diffusion and permeation of gases in undrawn and uniaxially drawn films made from potato and rice starch conditioned at different relative humidities, *Carbohydr. Polym.***24** (1) (1994) 1–15.
125. S. Esteghlal, M. Niakosari, S.M. Hosseini, G.R. Mesbahi, G.H. Yousefi, Gelatin-hydroxypropyl methylcellulose water-in-water emulsions as a new bio-based packaging material, *Int. J. Biol. Macromol.* **86** (2016) 242–9.
126. L. Zhang, Y. Wang, H.. Liu, N. Zhang, X. Liu, L. Chen, L. Yu, Development of capsules from natural plant polymers,*Acta Polym. Sin.* **13** (1) (2013) 1–10.
127. L. Zhang, Y. Wang, H. Liu, L. Yu, X. Liu, L. Chen, N. Zhang, Developing hydroxypropyl methylcellulose/hydroxypropyl starch blends for use as capsule materials, *Carbohydr. Polym.***98** (1) (2013) 73–79.
128. X. Liu, Y. Wang, N. Zhang, R.A. Shanks, H. Liu, Z. Tong, L. Chen, L. Yu, Morphology and phase composition of gelatin-starch blends, *Chin. J. Polym. Sci.* **32** (1) (2014) 108–114.
129. L. Zhang, Y. Wang, L. Yu, H. Liu, G. Simon, N. Zhang, L. Chen, Rheological and gel properties of hydroxypropyl methylcellulose/hydroxypropyl starch blends, *Colloid. Polym. Sci.* **293**(1) (2014) 229–237.
130. A. Ali, Y. Chen, H. Liu, L. Yu, Z. Baloch, S. Khalid, J. Zhu, L. Chen, Starch-based antimicrobial films functionalized by pomegranate peel, *Int. J. Biol. Macromol.***129** (2019) 1120–1126.
131. H. Hayrapetyan, W.C. Hazeleger, R.R. Beumer, Inhibition of listeria monocytogenes by pomegranate (Punica granatum) peel extract in meat paté at different temperatures,*Food Control* **23** (1) (2012) 66–72.
132. S. Khalid, L. Yu, M. Feng, L. Meng, Y. Bai, A. Ali, H. Liu, L. Chen, Development and characterization of biodegradable antimicrobial packaging films based on polycaprolactone, starch and pomegranate rind hybrids, *Food. Packag. Shelf.* **18** (2018) 71–79.
133. F. Xie, L. Yu, H. Liu, K. Dean, Effect of compatibilizer distribution on thermal and rheological properties of gelatinized starch/biodegradable polyesters blends, *Int. Polym. Proc.* **21** (4) (2006) 379–385.
134. L. Yu, K. Dean, Q. Yuan, L. Chen, X. Zhang, Effect of compatibilizer distribution on the blends of starch/biodegradable polyesters, *J. Appl. Polym. Sci.* **103**(2) (2007) 812–818.
135. K. Dean, L. Yu, S. Bateman, D.Y. Wu, Gelatinized starch/biodegradable polyester blends: processing, morphology, and properties, *J. Appl. Polym. Sci.* **103**(2) (2007) 802–811.
136. E. Petinakis, X. Liu, L. Yu, C. Way, P. Sangwan, K. Dean, S. Bateman, G. Edward, Biodegradation and thermal decomposition of poly (lactic acid)-based materials reinforced by hydrophilic fillers, *Polym. Degrad. Stab.* **95** (9) (2010) 1704–1707.
137. X. Liu, S. Khor, E. Petinakis, L. Yu, G. Simon, K. Dean, S. Bateman, Effects of hydrophilic fillers on the thermal degradation of poly(lactic acid), *Thermochim. Acta* **509**(1–2) (2010) 147–151.

138. L. Yu, E. Petinakis, K. Dean, H. Liu, Q. Yuan, Enhancing compatibilizer function by controlled distribution in hydrophobic polylactic acid/hydrophilic starch blends, *J. Appl. Polym. Sci.* **119**(4) (2011) 2189–2195.
139. E. Olsson, C. Menzel, C. Johansson, R. Andersson, K. Koch, L. Jarnstrom, The effect of pH on hydrolysis, cross-linking and barrier properties of starch barriers containing citric acid, *Carbohydr. Polym.* **98** (2) (2013) 1505–1513.
140. N. Reddy, Y. Yang, Citric acid cross-linking of starch films, *Food Chem.* **118** (3) (2010) 702–711.
141. B. Sreedhar, D.K. Chattopadhyay, M.S.H. Karunakar, A.R.K. Sastry, Thermal and surface characterization of plasticized starch polyvinyl alcohol blends crosslinked with epichlorohydrin, *J. Appl. Polym. Sci.* **101**(1) (2006) 25–34.
142. I. Šimkovic, M. Hricovι´ ni, R. Mendichi, J.J. van Soest, Cross-linking of starch with 1, 2, 3, 4-diepoxybutane or 1, 2, 7, 8-diepoxyoctane, *Carbohydr. Polym.***55**(3) (2004) 299–305.
143. J.B. Hirsch, J.L. Kokini, Understanding the mechanism of cross-linking agents (POCl3, STMP, and EPI) through swelling behavior and pasting properties of cross-linked waxy maize starches, *Cereal Chem.* **79** (1) (2002) 102–107.
144. J. Zhou, J. Zhang, Y. Ma, J. Tong, Surface photo-crosslinking of corn starch sheets, *Carbohydr. Polym.* **74**(3) (2008) 405–410.
145. X. Ge, L. Yu, Z. Liu, H. Liu, Y. Chen, L. Chen, Developing acrylated epoxidized soybean oil coating for improving moisture sensitivity and permeability of starch-based film, *Int. J. Biol. Macromol.* **125** (2019) 370–375.
146. L. Meng, S. Li, W. Yang, R. Simons, L. Yu, H. Liu, L. Chen, Improvement of interfacial interaction between hydrophilic starch film and hydrophobic biodegradable coating, *ACS Sustainable Chem. Eng.* **7**(10) (2019) 9506–9514.
147. L. Meng, H. Liu, L. Yu, Q. Duan, L. Chen, F. Liu, Z. Shao, K. Shi, X. Lin, How water acting as both blowing agent and plasticizer affect on starch-based foam, *Ind. Crops Prod.* **134** (2019) 43–49.
148. Q. Duan, L. Meng, H. Liu, L. Yu, K. Lu, S. Khalid, L. Chen, One-step extrusion to minimize thermal decomposition for processing PLA-based composites, *J. Polym. Environ.* **27**(1) (2018) 158–164.
149. Ilyas RA, Sapuan SM, Atikah MSN, Asyraf MRM, Ayu Rafiqah S, Aisyah HA, Mohd Nurazzi N, Norrahim MNF. Effect of Hydrolysis Time on the Morphological, Physical, Chemical, and Thermal Behavior of Sugar Palm Nanocrystalline Cellulose (Arenga Pinnata (Wurmb.) Merr). *Textile Research Journal*. 2020. In print. 10.1177/0040517520932393

2

Protein-Based Materials for Packaging Applications

V. G. Martins, V. P. Romani, P. C. Martins, and D. Nogueira

School of Chemistry and Food, Federal University of Rio Grande, Avanida Italy km 8, Rio Grande, Rio Grande do Sul, 96203-900, Brazil

2.1 Introduction

Plastic production increased significantly, reaching about 350 million tons each year [1]. Durability, resistance, lightness, flexibility, and low cost are some of the characteristics that makes the plastic ubiquitous and indispensable in daily life [2, 3]. However, the qualities that make the plastic a good material for consumer products can also make it extremely resistant to biodegradation and can remain in the environment for decades [4, 5]. The world produces about 300 million tons of plastic waste each year and only 9% of this waste has been recycled [6]. Each year at least 8 million metric tons of plastic go to sea and oceans, often decomposing into small microplastics that end up stopping in our food chain [7].

Biodegradable polymers are one of the potential solutions to the problems associated with discarded wastes due to their fast degradation by the action of naturally occurring microorganisms in the environment [8]. The most widely used molecules for the development of these polymers include proteins, polysaccharides, and lipids, which are well known for their ability to form three-dimensional polymer networks [9]. Basically two processes are used to produce biodegradable films, the casting methodology and the extruder. The casting process is used more on the laboratory scale and is a simple process used to produce protein-based films. The extrusion techniques are widely used in industry due to their higher productivity, but it is still limited in some aspects to produce biodegradable films, mainly because of the temperatures used.

Bio-Based Packaging: Material, Environmental and Economic Aspects, First Edition. Edited by S.M. Sapuan and R.A. Ilyas

Polymers produced from proteins are one of the most studied research topics in this field, since the mechanical properties obtained are better when compared to polysaccharide and lipid films [10]. Many proteins have been used as potential film-forming agents such as corn zein, wheat gluten, whey, egg white, collagen, gelatin, fish proteins, among others. Some of these films have good oxygen barrier and flavor properties, but in general have a hydrophilic behavior and poor mechanical properties that are responsible for the low performance of the developed films compared to the synthetic polymers. This characteristic affects the ability of protein materials to perform the basic functions of protection of the packaged products [11, 12]. Based on that, strategies can be used to improve the performance of protein-based films, such as mixing of different raw materials to combine their properties, incorporation of reinforcing agents, use of chemical and enzymatic methods for modification of polymer chains, as well as physical methods, including cold plasma and ultraviolet (UV) radiation, that alter the properties of the polymeric network.

Another factor that drives the need to develop new protein-based materials for food packaging is the consumer demand for natural products, combined with the need to increase the shelf life of foods that reduce waste. Natural compounds, such as essential oils and plant extracts, are being extensively studied because of their potential to replace chemical additives for food preservation [13]. The incorporation of natural active compounds into packaging materials, which is one of the active packaging mechanisms, is a trend in the development of new technologies for food preservation due to the possibility to reduce chemical additives in their composition and/or increase their shelf life.

This chapter is focused on protein-based films, including production processes that are used in film development, raw materials used for protein extraction, discussions about the properties measured in the development of new materials for food packaging, existing and potential uses of these protein polymers, and challenges and future trends in the field.

2.2 Proteins

Proteins are the most abundant biological macromolecules in cells, occurring in a wide variety of species and ranging in size from relatively small peptides to polymers with high molar mass. These molecules exhibit diverse biological functions [14], providing structure or biological activity in plants and animals. Proteins are distinguished from other macromolecules due to their structure, which is based on approximately 20 amino acid monomers, rather than just a few, or even one monomer, such as glucose in the case of cellulose and starch. Most proteins contain 100–500 amino acid residues [15].

The functional diversity of proteins essentially arises from their chemical structure. Depending on the sequential order of the amino acids, the protein will assume different structures along the polymer chain, based on Van der Waals, hydrogen bonding, electrostatic, hydrophobic, and disulfide cross-link interactions among the amino acid units [15]. Structurally the proteins are organized into primary, secondary, tertiary, and quaternary structures, as shown in Figure 2.1. Billions of proteins with unique properties can be synthesized by changing the amino acid sequence, the type and ratio of amino acids, and the chain length of polypeptides. Traditionally, milk, meats (including fish and poultry), eggs, cereals, vegetables, and oilseeds have been the major sources of food proteins. The functional properties of proteins in foods are related to their structural and

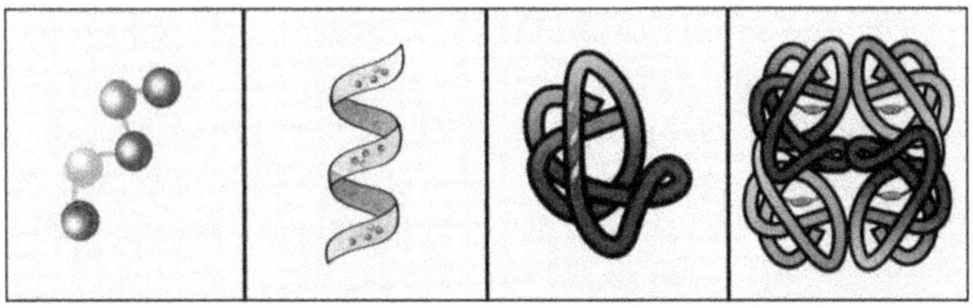

Figure 2.1 Structural protein organization scheme.

other physicochemical characteristics [16]. The structures of proteins can be modified by various physical and chemical agents, including heat, mechanical treatment, pressure, irradiation, lipid interfaces, acids and alkalis, and metal ions [15]. Such agents have the capacity to change the structures of proteins and affects their functional properties. In film development, these modifications are often used in the formation process to optimize protein configurations and interactions, resulting in better film properties.

Protein polymers generally have good mechanical strength and low permeability, gaining greater attention from researchers regarding their development and application [17]. Protein films are stabilized through electrostatic interactions, hydrogen bonds, Van der Waals forces, and covalent and disulfide bonds [18]. These are formed by the partial denaturation of polypeptide chains by solvent addition, pH change, and heat application [19]. In the film production process, initially the heating phase is responsible when protein structure is disrupted, when some disulfide bridges are broken, and when hydrophobic and sulfidic groups are exposed. Then, in the drying phase, new hydrophobic interactions and disulfide and hydrogen bonds are formed [20]. Proteins from animal sources include collagen, gelatin, fish myofibrillar protein, keratin, egg white protein, casein, whey protein, among others, while some examples of proteins from plant sources are corn zein, wheat gluten, soy, peanut, ora pro nobis, cottonseed, gelatin, whey protein, among others. Examples of protein sources that have been used in the development of films are discussed in this chapter.

The study of the physical, chemical, and functional properties of proteins are fundamental to understanding the performance of these molecules in the production of biomaterials. Furthermore, significant changes in the properties of the proteins are induced by environmental factors and processing conditions, which will also affect the properties of the films produced.

2.3 Protein Films for Food Packaging

Films, generally applied as primary packaging in food industry, are defined as thin layers and can be produced from different polymeric matrices [21]. Films have been widely used with the intention to improve the sensory properties of products and also as a barrier against gases and humidity, as well as supporting the incorporation of active components to extend the shelf life of packaged products [22]. Several biopolymers can be used in the formation process of films such as proteins, polysaccharides, lipids, and polymer blends [23, 24]. Protein-based films are interesting for use in the food industry as they show distinct characteristics from other films, such as transparency, optimum carbon dioxide and lipid barrier properties, and high water sensitivity [22].

During film production processes, additives in general are used to improve the properties of polymers, such as plasticizers (e.g. sorbitol, glycerol, and polyethylene glycol), to facilitate the manipulation and malleability of the packaging [25, 26]. There are many processes used in the production of films, such as pH modification, salt addition, heating, enzymatic modification, drying, solvent use, and mechanical forces. Many chemical reactions are involved in the film development, to form the three-dimensional polymer network that results in the packaging, such as covalent bonds, inter/intramolecular forces, electrostatic interactions, and ionic forces [27]. The characteristics of the native protein, such as the forces that compose the structure and the initial amino acid sequence, and the methods used for their extraction, are important aspects that affect the final characteristics of the formed film [28]. The final characteristics of the packaging will be formed according to the structure of the protein used and the method used to obtain the film.

A wide variety of proteins from animal/vegetable sources can be used to produce films, as shown in Table 2.1 and Figure 2.2. In addition to the use of proteins for film production,

Table 2.1 *Some research using different protein bases in the preparation of films combining strategies to improve the properties of films.*

Protein type	Strategies	Effect observed	Reference
	Animal Protein		
Argentine anchovy (*Engraulis anchoita*)	Sorbic or benzoic acids	Antimicrobial	[10]
Chicken feet	–	–	[30]
Gelatin	Longan seed extract	Antioxidant	[31]
Fish gelatin	Chitosan nanoparticles	Structure reinforce	[32]
Porcine meat and bone meal	Coriander oil	Antimicrobial	[33]
Shrimp (*Litopenaeus vannamei*) muscle	Cinnamaldehyde/thermal treatment	Crosslinking	[34]
Whey protein	Gum tragacanth	Structure reinforce	[35]
Whey protein	Ultraviolet treatment	Structure reinforce	[36]
Whey protein isolate	*Lactobacillus rhamnosus*	Antimicrobial	[37]
Whitemouth croacker muscle	Pink pepper phenolic	Browning reduction	[38]
	Vegetable Protein		
Bitter vetch (*Vicia ervilia*) seed/ corn zein	Bilayer	Improve structure	[39]
Amaranth	Native waxy and maize starch nanocrystals	Structure reinforce	[40]
Canola	–	–	[41]
Pea	Lysozyme	Antimicrobial	[42]
Sesame meal	–	–	[43]
Soy	Chestnut (*Castanea mollissima*) bur extracts	Antioxidant	[44]
Soy/agar	Blend/Extrusion	Structure reinforce	[45]
Soy protein isolate	Peanut protein nanoparticles	Structure reinforce	[46]
Sunflower protein	Clove essential oil	Structure reinforce and improve shelf-life	[47]

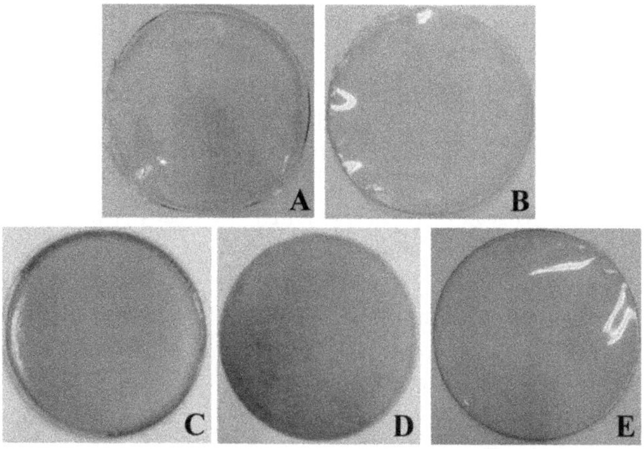

A: hake protein isolate film; B: wheat gluten film; C: corn zein film;
D:zein/wheat gluten blend; E: hake protien isolate/wheat gluten blend.

Figure 2.2 *Biodegradable films produced from different proteins and their blends [48].*

researchers are focusing on the study of some strategies to improve their performance and to provide bioactive properties. In the group of animal proteins, the most used are caseins, whey protein, collagen, gelatin, myofibrillar proteins, and egg proteins, and among the vegetable proteins, the most used are soy protein, gluten, and zein [25, 29].

Animal proteins, such as milk proteins (casein and whey protein), have been widely explored, mainly due to their high nutritional value and excellent functional properties. Films formed from milk proteins are generally cohesive, tasteless, edible, transparent or translucent, and flexible [28]. Other widely used groups of proteins are collagen and gelatin. Collagen can be obtained from skin, tendons, and various connective tissues, representing up to 30% of the body. Collagen films have high surface homogeneity, are flexible, have low mechanical strength, and show low oxygen permeability and humidity. Gelatin is used in films because it gives them packaging characteristics such as transparency, resistance, flexibility, and impermeability [10, 49]. Other examples are the myofibrillar proteins that are obtained from fish muscle and form films with high tensile strength, high water vapor permeability (WVP), and intermediate solubility [49].

As mentioned in Table 2.1, there are several proteins of vegetable origin used to produce food packaging, such as wheat gluten, corn zein, and soy protein. Wheat gluten is subdivided into two important proteins that are responsible for producing gluten, gliadins, and glutenins. The use of wheat gluten in the industry is promising, mainly as coatings, films, and adhesives. For the production of wheat gluten films, it is necessary to apply solvents such as alcohol due to partial solubility of the protein in water. Films produced from this protein have great potential for use as packaging with selective gas permeability and can be easily applied to control respiratory exchange in fruit and vegetable conservation [21]. Zein is a protein present in corn and the amino acid sequences in its composition make it hydrophobic and consequently insoluble in water. Due to the zein structure characteristics, the films from this protein have low WVP and water solubility. Soy protein is another type of protein largely studied for packaging production. Different from gluten and zein, soy protein can be solubilized in water using heat and pH adjustment [25]. The use of soy protein as an alternative for packaging has many advantages, since soy has low cost and it is completely biodegradable. Films made from isolated soy protein are partially oxygen-soluble at low humidity and have moderate mechanical properties [50].

Although there are advantages and great potential for the use of protein biopolymers, the physical properties of films are still inferior to some synthetic polymers used in the food industry, such as low density polyethylene (LDPE), high density polyethylene (HDPE), or cellophane. Thus, there are several strategies used to overcome these limitations and make the characteristics of natural polymers equivalent to the properties of synthetic polymers. It is possible to observe in Table 2.1 that most studies used strategies to improve the properties of films or to provide bioactivity to the packaging. Some of them are: the incorporation of composites [46, 51], incorporation of lipids, [52], cross-linking [53], pH changes, use of phenolic extracts [44], polymer blends [45], application of plasma [54], ultraviolet radiation treatment [55], among others.

In the current context, although the search for biodegradable polymers with suitable properties for industrial applications is extremely broad, limited packaging options are commercially available. There are many types of packaging produced with biodegradable characteristics, such as cup, bags, straws, lids, gloves, cutlery, bottles, plates, and many others. However, the composition of these materials is generally protected by patents and most of them do not use exclusively biopolymers from natural and renewable sources in the production process. Most films use natural polymers as an additive in the packaging formulation process or incorporate more degradable polymers from nonrenewable sources, such as poly(vinyl alcohol) (PVA) [56]. In commercial uses, carbohydrates are the most explored natural polymers, once they have wide applicability and low cost. Although proteins have a great utilization potential, it is still necessary to explore other ways to enhance their packaging properties and production processes.

2.4 Film Production Processes

Protein film preparation techniques are classified into two main categories, namely solution-based (wet and most common) and mass-based (dry and less common) [57, 58]. The casting method (wet) first involves dispersing or solubilizing the protein in a solvent and then the film suspension or solution is placed in a suitable mold, where the solvent is evaporated under controlled conditions [59, 60]. During this process, primarily polymer–polymer interactions are formed, resulting in a stable three-dimensional network [61]. The result of the casting process depends mainly on the mixing conditions: temperature, time, type, and concentration of solvent, plasticizer, pH, relative humidity, and drying temperature. The dissolution of the component is generally temperature-assisted in order to promote disruption of native protein interactions. This temperature will strongly depend on the denaturation temperature of the protein fractions. The pH of the solution should also be adjusted and then protein folding and denaturation occurs, exposing reactive polar and non-polar groups. In addition, disulfide interactions are promoted at alkaline pH.

Another important aspect to be considered in film production is the concentration of the plasticizer used. Protein films produced without the addition of any plasticizer are fragile, although high levels of plasticizer may promote low mechanical strength or migration to the surface of the film [62]. Therefore, plasticizers are necessary to improve the processing capacity and fluidity of proteins, and the choice of plasticizer as well as the concentration to be used must be studied. Most research on protein-based films use the casting method because it is simple, reproducible in most laboratories, and useful as a first approximation to protein film formation. However, the drying step makes this method slow and discontinuous, making it unfeasible for an industrial scale [57, 63, 64].

Dry processing of polymers typically includes hot pressing or compression molding, as well as melting and extrusion techniques [63]. It is based on the application of thermomechanical processes to promote denaturation and mobility of protein chains, which is essential for film formation, and the addition of plasticizers can improve processability, favoring increased free volume and chain mobility [65]. Although thermomechanical processing has received less attention, these technologies may use pre-existing thermoplastic equipment with minimal modification [62], favoring the industrial production of polymers. When compared to the solution-based technique, it has a lower energy expenditure and is more practical on an industrial scale. This technique includes extrusion, hot pressing, and injection molding [58].

Extrusion is one of the main processes used in the conventional plastic packaging industry [57]. During the extrusion, the polymeric solid material is processed by one or two screws, which progressively increases pressure and mixes the ingredients needed to make the continuous polymer pass through the die [57, 58]. Twin-screw extruders have considerably more heat exchange capacity and higher production efficiency than single-screw extruders. The length/diameter ratio, temperature, and screw speed are the main parameters that influence the dispersion effect [58].

Compared to the traditional solution casting method, extrusion is an attractive alternative method for producing protein films due to their processing speed, fewer processing steps, and lower power consumption, due to the fact that more film solution concentrates can be fed into the extruder. However, this technique involves the use of different parameters to successfully control film production and slight variations in processing conditions can have a considerable effect on film characteristics. The manufacture of extruded biodegradable materials is limited to the research scale due to a lack of understanding of the behavior of these materials in the extruder [57]. Studies on the use of extrusion to produce biodegradable films are limited, especially those derived from agriculture sources, because the behavior and chemical interaction of these components during extrusion is difficult to understand and many processing parameters need to be considered [57, 58].

Compression molding, which is assisted by a prior intensive blending, is generally studied on a laboratory scale as a precursor to continuous extrusion in order to determine the appropriate processing conditions [61]. During thermomechanical processing, proteins disaggregate, denature, and dissociate, resulting in a complete restructuring of protein molecules and allowing them to recombine, cross-link, and aggregate through specific bonds. Parameters such as temperature, pressure, time, and plasticizers determine the degree of conformational changes, protein aggregation, and chemical cross-linking that occur during processing. The mixing temperature used in the process should be sufficient to ensure the flowability of the mixture, but below 100 °C it is essential to prevent water loss at this stage as well as thermomechanical degradation of low molecular weight protein fragments. Mixing speed and time must also be studied in order to produce a homogeneous mixture without producing high degradation. Excessive temperatures or high mixing speeds may produce intense yellow–brown color mixtures due to protein degradation in addition to degradation of the carbohydrate fraction present in unpurified proteins [62]. Cross-linking via covalent interactions (sulfhydryl and disulfide groups) is highly temperature dependent; thus, the processing temperature will have significant effects on the final properties of the films [66, 67]. A higher degree of network cross-linking results in films with higher microstructure density, lower solubility, and better mechanical strength and barrier properties [62].

Regardless of the method chosen for the preparation of protein films, a thorough study of protein structures is required to define the process parameters (solvent, time, temperature, pH, plasticizer, among others) in order to obtain films with better properties.

2.5 Characterization of Films

The integrity of films used as packaging materials must be maintained during the processing, transportation, and handling of foods [68]. In general, the properties of the biodegradable films depend on the nature of the material used and their structure, which is strongly related to the ability of the polymer to form strong or numerous bonds among the polymer chains.

2.5.1 Mechanical Properties

Polymer structure plays an important role in the mechanical properties and, consequently, in the process used to obtain the final product (injection molding, sheet extrusion, blow molding, thermoforming, casting) [69]. The mechanical properties of biopolymer films, including tensile strength, elongation, deformability, and modulus of elasticity, are extremely important since packaging materials must have adequate mechanical strength to maintain their integrity during handling and storage [12].

The maximum tensile strength is the strength offered by the material at the point of failure. Elongation is the percentage ratio between the elongation of the material in the test and its initial length. These values are important to obtain mechanical information from biopolymer materials in order to compare them with nonbiodegradable materials [69]. The mechanical properties of biopolymer films depend on the composition and environmental conditions. For example, the incorporation of plasticizers increases the mobility of polymer chains [26], which leads to increased elongation and decreased tensile strength of plasticized films [70]. Incorporating other additives, such as lipids or cross-linking materials, may increase the strength and extensibility of films [71]. Relative humidity can also affect the mechanical properties of biopolymer films; in high humidity environments hydrophilic films quickly absorb moisture, thereby increasing the plasticizing effect of water, which subsequently reduces tensile strength and increases the extensibility of the films. In addition, the interaction between the packaged product and the packaging material may also affect the performance of a biopolymer [12].

Incorporating fillers, particularly nanoscale sized ones, is an effective way to improve the mechanical strength of protein films. The mechanical properties of a resulting nanocomposite are influenced by the intrinsic properties of nanoparticles, such as modulus and morphology (including shape, aspect ratio, etc.), the nature of the matrix, and the dispersibility of nanoparticles in the matrix, as well as the interaction (compatibility) between the filler and the matrix. For high surface energy, nanoparticles are prone to agglomeration, which impairs mechanical properties. Good dispersion effects of nanoparticles in polymeric matrices are the prerequisite to preparing high mechanical performance nanocomposite materials [58].

2.5.2 Barrier Properties

Biopolymeric packaging must have mass transfer properties suitable for its use. Significant losses in food quality may occur due to the transfer of moisture, gas, aroma, taste, or color to and from the surrounding environment [12]. Water vapor and oxygen are the main permeants studied in packaging applications as they can transfer from the internal or external environment through the polymer package wall, resulting in a continuous change in product quality and shelf life [69]. WVP indicates the amount of water that permeates per unit area and the time through the packaging material. This property is of great importance to extend the shelf life of food products, where physicochemical and microbiological spoilage is closely related to water activity [62]. For fresh food it is important to avoid dehydration, while for bakery products it is important to avoid water permeation [69].

The hydrophilic nature as well as the amount of plasticizer significantly affect the moisture barrier capacity of protein films. A higher plasticizer content (glycerol and sorbitol) promotes a higher moisture content. Both glycerol and water act as plasticizers with a synergistic effect, dominated by the plasticizing power of water. In addition, water plasticization reduces protein–protein interactions, favors polymer chain mobility, increases free volume, and, as a result, increases moisture diffusion increasing the WVP in hygroscopic or hydrophilic films [72, 73]. Although WVP should be a thickness-independent parameter because the thickness value is used in the equation, experimental results for biodegradable films suggest that WVP is thickness dependent [74]. A linear relation between WVP and the thickness of edible protein-based films has been established [75].

The oxygen barrier property of a fresh food package (fruits, salads, ready meals) plays an important role in preserving them. The oxygen barrier is quantified by oxygen permeability coefficients (OPC), which indicate the amount of oxygen that permeates per unit of area and time in a packaging material. Thus, when a protein film package has a low oxygen permeability coefficient, the oxygen pressure inside the container drops to the point where oxidation is delayed, extending the shelf life of the product [69]. Generally, biodegradable polymers have OPC values of one or more orders of magnitude below the synthetic polymer used in the same field, such as low- and high-density polyethylene, polypropylene, and polystyrene [62, 69]. Most protein films have excellent oxygen barriers since their ordered hydrogen bonding network structure is well compacted [62].

As the oxygen and water vapor barrier properties, the carbon dioxide barrier property is also of particular importance in food packaging applications. Carbon dioxide barrier is quantified by the carbon dioxide permeability coefficients that indicate the amount of carbon dioxide that permeates per unit area and time in a packaging material [69]. Carbon dioxide is especially important for modified atmosphere packaging because it can potentially reduce the problems associated with fresh products, leading to significantly longer shelf life [58, 62, 69]. Improved mechanical properties, low WVP, and high oxygen barrier properties are the main characteristics of biodegradable films [58] and therefore are the most important analysis to be performed on protein films for their use in application as food packaging [69].

Although protein films generally have excellent gas barrier properties, they tend to have higher WVP, but they are mechanically weaker, having less elongation when compared to most synthetic polymers. Many studies have been conducted to improve these properties; however, it is common to improve one property while another reduces, which limits its application as packaging.

2.5.3 Structural Properties

The mechanical and barrier properties of biopolymer films are closely related to their microstructural characteristics, which are highly influenced by the film formulation and manufacturing process [61]. These properties can be analyzed by different techniques, such as scanning electron microscopy (SEM), X-ray diffraction (XRD), atomic force microscopy (AFM), Fourier transform infrared spectroscopy (FTIR), and solid-state nuclear magnetic resonance (NMR).

Structural information of the film, such as homogeneity or the presence of ruptures and failures, is verified by SEM. The presence of ruptures and failures can significantly affect the mechanical and barrier properties of the material [61]. A uniform and homogeneous microstructure indicates the homogeneity of the film. Evaluation of the cross-sections of protein films may show, for example, the removal of dissolved air from film-forming solutions by applying a controlled vacuum, which reduces the size and number of pores in the formed films. This is a desirable effect because the presence of large pores within the films results in high WVP levels [12].

The XRD technique allows morphological studies in materials to determine the percentage of crystallinity. Through this technique, it is possible to evaluate the functional groups of the films, as well as to determine possible interactions among the polymers when more than one is used [76]. The FTIR technique can be used to study molecular interaction of the compounds that formed the films. Through this technique, hydrogen bond formation can be observed by displacing characteristic amide peaks. The FTIR spectrum is important in film production as it determines the potential functional groups that perform specific functions in film properties, such as tensile strength and WVP [76, 77].

NMR and AFM techniques are best applied to investigate protein systems. Researchers employ the NMR instrument to determine protein structures, protein folding, as well as condensed microstructures and their interactions. These techniques can be used in studies of intermolecular interactions between plasticizers and matrices in plasticized protein systems. Investigation of nanoscale material structures is most effectively assessed through tools such as the transmission electron microscope (TEM) and AFM [58].

These structural evaluation techniques (FTIR, NMR, AFM, and TEM) are most applied when a molecular and nanometric characterization of the material structure is desired, being more sophisticated and expensive techniques. Therefore, XRD and SEM techniques are often used in the structural evaluation of protein films.

2.5.4 Thermal and Optical Properties

In order to predict the performance of protein-based films under different end-use conditions, the thermal transitions of the films should be investigated. Among the most important techniques in thermal analysis are thermogravimetric analysis (TGA) and differential exploratory calorimetry (DSC) [78]. TGA is based on the study of the mass variation of a sample resulting from a physical (sublimation, evaporation, and condensation) or chemical (degradation, decomposition, and oxidation) transformation as a function of time or temperature. This is a continuous process that measures the change in mass (loss or gain) of a substance or material as a function of temperature and/or time. DSC is a technique that measures the temperature and heat rate associated with material transitions as a function of time and temperature in a controlled atmosphere. In polymer samples, the technique is employed for measuring glass transition (Tg), melting (Tm), and crystallization temperatures [58, 77], which are important parameters to predict the type of food product to which the films can be applied. Those temperatures inform whether the packaged product can be frozen, refrigerated, or heated without degradation of the biopolymers.

The optical properties (color, transparency, and brightness) of films can have a major impact on the appearance of the food product and therefore on consumer acceptability [79], but it is not a limiting factor in the use of film as packaging. For a good visual presentation of the product, it is desirable that the packaging has high brightness and high transparency. However, light protection is often required (low or zero transparency) when products sensitive to light-catalyzed deterioration reactions are packaged [76].

2.5.5 Biodegradability of Polymers

Biodegradable food packaging is a demanding area of great interest for researchers because of the short time of degradation compared to classic petroleum-based polymers [80]. Renewable polymers such as polysaccharides and proteins can be decomposed into CO_2, water, and small organic molecules (humus) after disposal in bioactive media by microorganisms such as bacteria, fungi, and algae and by freshwater or seawater [58]. Biodegradability of biopolymers can be achieved by anaerobic or aerobic fermentation in the presence of extracellular enzyme-secreting microorganisms to fragment the polymeric chains in the low molecular weight packaging material [80]. In soil, microorganisms use films as a food source in the natural environment and accelerate biodegradation [81]. In freshwater, total biodegradability can be achieved quickly as well, due to the fact that microorganisms and algae present in freshwater can attack polymeric chains and cause chain cleavage. In salt water, the absence or reduction of these microorganisms can promote polymeric degradation more slowly [80].

The study of biodegradability in materials is performed by weight loss evaluation during composting, degraded surface evaluation by SEM, and structural changes that can be characterized by FTIR [58]. The physicochemical properties and the structural and biodegradation properties presented by the films are extremely dependent on the protein structure used, as well as the method chosen to produce the film.

2.6 Protein Films Application

Many studies in the literature reported the production and evaluation of films from proteins using the most diverse raw materials from agriculture, as mentioned previously in this chapter. Some studies also evaluated the capacity of protein films to keep the quality and safety of food products. However, in most of them, active compounds were incorporated in the protein film matrices. In Table 2.2, a summary was made of some of the studies where protein films were evaluated by their capacity to maintain and increase the shelf life of products, including the films incorporated with active compounds. Among the sources used, whey protein and gelatin were widely used, as demonstrated in Table 2.2.

Gelatin is a water-soluble protein and is obtained from collagen hydrolysis. Skin, bones, cartilage, and tendons are the major raw materials used to obtain gelatin. The yield and recovery of gelatin depend upon the type of collagen and animal age [82]. Films developed using gelatin can act as a barrier to water, gas, aroma, flavor, and/or oil, protecting foods from the external environment. The properties of gelatin films in general depend on the source of the gelatin, the film production conditions (including concentration, solvent, pH, and temperature), the presence of additives, and testing conditions. In general, gelatin films do not break during handling of food and have desirable sensory characteristics during storage [83]. On the other hand, whey protein is a mixture of globular proteins isolated from milk as a byproduct of cheese production that contains approximately 7% dry matter. Whey protein has been widely explored in packaging applications because of the excellent barrier properties, abundance, and good processing [84–86]. Besides these two raw materials, others have been explored, such as fish proteins, corn zein, soy proteins, etc., which showed promising packaging-related properties and consist of good matrices for incorporation of active compounds.

Most of the studies using proteins to produce polymeric films evaluated the packaging related properties, including mechanical and barrier properties, chemical structure, and morphology. However, few studies evaluated the performance of these materials in real foods and the ones that evaluated had in general incorporated functional substances to design active packaging in order to prevent reactions responsible for food degradation. The incorporation of these active substances has an important role for the development of new packaging technologies for the food industry, to keep food safety and security and to prevent food waste. Especially when natural substances are incorporated, the materials developed respond to consumer demand for clean label products. For example, the use of natural preservatives incorporated in the packaging material may enable the reduction of synthetic preservatives in food formulation, which are known to cause health problems [13, 101]. Promising results have been obtained incorporating a wide range of functional substances, such as essential oils, plant extracts, enzymes, acids, and others. Akcan, Estévez, Serdaroglu [90] incorporated phytochemicals from *Laurus nobilis L.* and *Salvia officinalis* in whey protein films and observed that oxidative changes were retarded in cooked meatballs. Wang et al. [94] developed gelatin films with anthocyanins and obtained the delay in olive oil oxidative deterioration. In these two examples of studies, the capacity of the films produced to keep the quality and prevent degradation reactions were attributed to the incorporation of functional substances in the protein matrices.

There are few studies that compared the protein materials developed with synthetic polymers used in the food industry. Brink, Šipailienė, Leskauskaitė [92] used cranberry

Table 2.2 Studies evaluating the use of protein films in foods shelf life.

Raw material	Active compound	Effect observed	Reference
Whey protein	*Lactobacillus sakei* cell-free supernatant	Antimicrobial films reduced *Listeria monocytogenes* and *Escherichia coli*	[87]
Catfish skin gelatin	Longan seed extract/butylated hydroxytoluene	Retardation of lipid oxidation in soybean oil	[88]
Fish protein isolate/fish skin gelatin	ZnO nanoparticles and basil leaf essential oil	Decrease in growth of psychrophilic bacteria, lactic acid bacteria and spoilage microorganisms in sea bass slices	[31]
Furcellaran-gelatin	Se-AgNPs	Increase of mini kiwi shelf life evaluated by visual changes, reduction of weight loss and fungal growth	[89]
Whey protein	Phytochemicals from *Laurus nobilis L.* and *Salvia officinalis*	Oxidative changes were retarded in cooked meatballs	[90]
Soy protein isolate	β-cyclodextrins	Sequester cholesterol of cholesterol rich food (such as milk)	[91]
Whey protein/chitosan	Cranberry/quince juice and organic acids (benzoic, ascorbic, and sorbate)	Stopped microbiological deterioration of turkey meat and the development of pathogenic microorganisms: *Salmonella typhimurium*, *Escherichia coli*, and *Campylobacter jejuni*	[92]
Chitosan/zein blend	–	Mushroom had lower weight loss rate, respiration rate, relative leakage rate, polyphenol oxidase and peroxidase activity, however, higher whiteness index.	[93]
Gelatin	Anthocyanins	Delayed olive oil oxidative deterioration	[94]
Bovine skin gelatin	Zinc oxide nanorods/clove essential oil	Films showed antibacterial activity against *Listeria monocytogenes* and *Salmonella Typhimurium* inoculated in shrimp during refrigerated storage	[95]
Whey protein isolate	Montmorillonite and citric acid	Reduction of enzymatic browning and loss of quality associated to the maintenance of acidity, soluble solids, water activity, reduction of polyphenol oxidase and peroxidase activity of the apple slices	[96]
Gelatin	Tannic acid	Decrease in the weight loss and browning index of cherry tomatoes and grapes, and antibacterial properties against *Escherichia coli* and *Staphylococcus aureus*	[97]
Puffer fish skin gelatin	*Moringa oleifera* Lam. leaf extract	Gouda cheese packaging with extract containing film inhibited the microbial growth and retarded the lipid oxidation	[98]
Whey protein	Lysozyme	Lower *Listeria innocua* counts in smoked salmon slices	[99]
Skate skin gelatin	Thyme essential oil	Antimicrobial activity against *Listeria monocytogenes* and *Escherichia coli* O157:H7 in chicken tenderloin	[100]

and quince juice and organic acids (benzoic, ascorbic, and sorbate) in the preparation of whey protein/chitosan films and evaluated the active films compared to control films (no additives) and polyethylene. These authors studied the microbiological deterioration of turkey meat and the development of pathogenic microorganisms (*Salmonella typhimurium*, *Escherichia coli*, and *Campylobacter jejuni*), and observed that films with additives retarded microbial growth. Arfat et al. [31] followed the same trend in the evaluation of active films compared to control films (without active compound) and polypropylene films. These authors incorporated ZnO nanoparticles and basil leaf essential oil in fish protein isolate/fish skin gelatin films and obtained a decrease in growth of psychrophilic bacteria, lactic acid bacteria, and spoilage microorganisms in sea bass slices.

Jamróz et al. [89] studied the shelf life of mini kiwi by wrapping with furcellaran-gelatin films containing nanoparticles (Se-AgNPs) and compared them to films without nanoparticles and low-density polyethylene films. The fruit's shelf life was studied through visual changes, reduction of weight loss, and fungal growth and the furcellaran-gelatin films presented a capacity to prevent kiwi degradation, especially the films incorporated with nanoparticles. Here it is important to mention how essential the knowledge is about the most appropriate packaging material to use for each type of food product. For example, high moisture products (e.g., meat and dairy products) require more hydrophobic materials, as solubility in water is a drawback. However, this property is advantageous in the case of solubilization of edible films to be consumed with food products, such as coffee/tea sachets. Regarding barrier performance, for snacks in general a very low barrier is important, in order to keep the product crunchiness, while fruits and vegetables need polymers with permeability, due to their respiration processes [102–104]. Thus, in the case of the mini kiwi study, the use of nanoparticles in the furcellaran-gelatin films decreased the WVP, but enough to keep changes with the external environment, which are necessary due to the metabolism of the fruits. With controlled permeability, the weight loss and other changes of the kiwis occurred at a lower rate, increasing their shelf life.

The significance of barrier performance was also studied in films using zein, the corn protein, blended with chitosan for mushroom packaging, reported by Zhang et al. [93]. They developed blends using different ratios of zein and chitosan and studied the effects on weight loss, respiration and relative leakage rates, polyphenol oxidase and peroxidase activities, and the whiteness index. The use of films sealed on top of plastic containers with mushroom samples inside resulted in a decrease in respiration and transpiration rates, consequently increasing the shelf life of the samples. The authors observed that superior ratios of zein protein resulted in a higher barrier that was responsible for better performance, even with some condensation phenomenon at the inner surface of the films. In this study, the authors compared the results of the use of the blends with samples without films.

On the other hand, there are also studies that evaluated only active films, without comparison with controls, such as that of Beristain-Bauza et al. [87] with the incorporation of *Lactobacillus sakei* cell-free supernatant in whey protein films with antimicrobial properties, and González and Igarzabal [91] with the development of soy protein isolate films with β-cyclodextrins to sequester cholesterol of foods. In this regard, it is worth noting that, for some applications, the comparison with synthetic films is important, while not that significant for others. That is why the knowledge about the product characteristics for which the packaging is being developed has a major importance.

Overall, the researchers evaluated the capacity of the active films to keep the quality of products in comparison to the control films (without active compounds). However, there is a lack of information regarding their performance compared to polymers used in the food industry. Thus, it is necessary that more investigations are performed concerning how these new materials would behave in real situations compared to synthetic polymers and/or what foods these new protein polymers are indicated for, based on their characteristics. These evaluations would contribute to identify how far we are from the desired properties to advance in the way for the large use of these sustainable materials. In addition, potential strategies to overcome the limitations of the performance of the materials for commercial use would be better driven.

2.7 Challenges and Future Perspectives

Promising properties of protein films for food packaging have been reported using a wide range of raw materials and active compounds. However, the total production of bio-based plastics, especially using proteins, is still small compared to the conventional polymers. Nevertheless, it is increasing because of the growing consumers awareness toward the environment pollution and the changes in their behavior for the use of ecoproducts. In many researches the development of these bio-based materials with the objective to enhance their performance as well as the effectiveness and efficiency in their production are being studied. However, although significant improvements have been obtained, the wide introduction of these new materials in the market still face challenges that need to be addressed, such as the development of the new technology, financial capacity, composting infrastructure, and regulation policies [105, 106].

Among these hurdles to the commercialization of sustainable/renewable material development, the technology transfer failure can occur in different stages between the food packaging innovation and end use application. For example, industries in general adopt and license unique raw materials for plastic production and then industry adjustment for production of the new polymers is needed. In addition, the academic research generally stops at the proof of concept stage and patented technologies are commonly incomplete in scope to be adopted by industry [107–109]. Another aspect is the process scale-up because the laboratory development of these protein films in general is not directly translatable or scalable to industrial production. Films are, in general, fabricated by solvent casting, microcompounding, and compression molding, and industrial polymers are in general produced by extrusion processes. Then, for commercialization, the proof of concept needs to be successfully fabricated by the industry. It is important that the pilot scale fabrication and characterization demonstration is viable, otherwise the manufacturers take a high risk in adopting the new technology and it is likely to be financially inviable. In the case of protein materials, industrial processes (extrusion and thermoplasticization) are challenging due to the molecular architecture and spatial arrangement of the molecules resulting in inefficient reproducibility. When regarding the hurdles, national and international regulatory requirements are still a critical aspect for the commercial translation of these new packaging technologies [110–112].

According to Caba et al. [106], for the development of new packaging, socioeconomic aspects such as resource efficiency, profitable value added, sustainable processes, etc., also need to be considered. In addition, study of the environmental benefits of the developed materials through a life cycle assessment (LCA) has an important role to play in measuring the impact of the films to confirm the better use of them in comparison to synthetic polymers. As reported by Conte et al. [113], LCA is a powerful tool to measure emissions throughout the production value chain of goods and services. In the case of food packaging, LCA is particularly important because some materials have less impact than others to be produced, but it might not protect the food properly and then higher food spoilage can result in a higher environmental impact at other points in the value chain. According to Babu, O'Connor, and Seeram [114], in order to make these new packaging economically viable, the logistic for biomass feedstocks and the new manufacturing routes to have high yields are fundamental, in agreement with the already mentioned aspects.

The potential of gelatin and whey proteins for the manufacture of polymer for food packaging has been widely described previously. Especially in the use of whey proteins, much progress has already been made and these materials are likely to be commercially used in the near future. Schmid and Müller [115] reported that whey protein films can replace synthetic polymers as an oxygen barrier, such as ethylene vinyl alcohol (EVOH). Whey protein films have excellent optical and barrier properties and were positively validated for storage of many food products. In this regard, the use of raw materials, such as whey protein, present combined advantages, firstly due to the environmentally friendly characteristics and also because it can be obtained from wastes/byproducts, which do not compete with food production. Regarding gelatin films, according to the most recent studies, many food products can be packed, such as fruits and vegetables, meat and poultry, fish and crustaceans, low moisture foods and oily foods. However, further studies are required to obtain the necessary performance for translation to commercial packaging [83].

Since many sources of proteins, besides those mentioned here, have been explored and important advances have already been made to develop these new materials for food packaging, there are opportunities for new research to deal with the challenges to the commercial translation of these technologies and the coming years are expected to solve these issues. University researchers could work to close the gap between the laboratory research and the commercialization by the development of partnerships with supply chain researchers, industries, and government. Strategic partnerships may help the researchers to identify market needs and modifications needed for material production and conversion, fabrication scale-up, and product applications of the finished package [108, 109, 116, 117]. As Licciardello [118] has mentioned, the collaboration among companies, decision makers, and packaging researchers could optimize packaging systems, ensuring that the changes in materials do not affect the food shelf life. In addition, the development of protein materials for food packaging is a very multidisciplinary, multiscale, and multifactorial field and it is important that professionals with different expertise become involved, considering the importance of the knowledge of food science needed coupled with materials performance and various aspects. Also, other initiatives together with scientists, including politicians, industries, and consumers are important, due to their relevance in the sustainability of the food chain and resources use. Finally, the research on protein polymers is very dynamic and a lot of progress can still be made.

2.8 Conclusions

There are advantages and a great potential to use protein biopolymers. However, the physical properties of films are still inferior to some synthetic polymers used in the food industry, such as LDPE, HDPE, or cellophane. Despite several strategies to overcome these limitations and make the characteristics of natural polymers equivalent to the properties of synthetic polymers in commercial uses, carbohydrates are the most explored natural polymers, once they have wide applicability and low cost. Thus, although proteins have a great utilization potential, it is still necessary to explore other ways to enhance their packaging properties and production processes. The casting technique is the most used process to produce protein films because it is simple and reproducible in most laboratories, but industries in general produce plastics through the extrusion process. Regarding the properties of protein films, they depend on the structure of the protein used, as well as the production method and the parameters chosen. Although protein films generally have excellent gas barrier properties, they tend to have high WVP, they are mechanically weaker, and have less elongation when compared to synthetic polymers in general. Among the most promising sources for protein film production are whey protein and gelatin. The potential of their use for film production is described in the literature and a lot of progress was already been made for their commercialization. Despite most of the studies using proteins to produce polymeric films evaluated the packaging-related properties, few studies evaluated the performance of these materials in real foods. The studies that evaluated this had in general the incorporation of functional substances to design active packaging in order to prevent reactions responsible for food degradation. In addition, there are few studies that compared the protein materials developed with synthetic polymers used in the food industry. Besides that, the wide introduction of these new materials in the market still faces challenges that need to be addressed, such as the development of the new technology, financial capacity, composting infrastructure, and regulation policies. Ways to overcome this are initiatives together with scientists, including politicians, industries, and consumers, which are important due to their relevance in the sustainability of the food chain and resources use.

References

1. Plastics Europe – Association of Plastics Manufacturers. PlasticsEurope [Internet]. An Analysis of European Plastics – The Facts 2018. 2018 [cited 2019 Aug 14]. p. 60. Available from: https://www.plasticseurope.org/application/files/6315/4510/9658/Plastics_the_facts_2018_AF_web.pdf
2. Heidbreder LM, Bablok I, Drews S, Menzel C. Tackling the plastic problem: a review on perceptions, behaviors, and interventions. *Sci. Total Environ.* 2019;**668**:1077–93.
3. Malizia A, Monmany-Garzia AC. Terrestrial ecologists should stop ignoring plastic pollution in the Anthropocene time. *Sci. Total Environ.* 2019;**668**:1025–9.
4. Horton AA, Walton A, Spurgeon DJ, Lahive E, Svendsen C. Microplastics in freshwater and terrestrial environments: evaluating the current understanding to identify the knowledge gaps and future research priorities. *Sci. Total Environ.* 2017;**586**:127–41.
5. Wan Y, Wu C, Xue Q, Hui X. Effects of plastic contamination on water evaporation and desiccation cracking in soil. *Sci. Total Environ.* 2019;**654**:576–82.

6. ONU. ONU Meio Ambiente aponta lacunas na reciclagem global de plástico [Internet]. 2019 [cited 2019 May 25]. Available from: https://nacoesunidas.org/onu-meio-ambiente-aponta-lacunas-na-reciclagem-global-de-plastico.
7. ONU. Fungos podem ajudar no combate à poluição por plásticos, diz ONU Meio Ambiente [Internet]. 2019 [cited 2019 May 25]. Available from: https://nacoesunidas.org/fungos-podem-ajudar-no-combate-a-poluicao-por-plasticos-diz-onu-meio-ambiente.
8. Sintim HY, Bandopadhyay S, English ME, Bary AI, DeBruyn JM, Schaeffer SM, et al. Impacts of biodegradable plastic mulches on soil health. *Agric. Ecosyst. Environ.* 2019;**273**:36–49.
9. Adilah ZAM, Jamilah B, Nur Hanani ZA. Functional and antioxidant properties of protein-based films incorporated with mango kernel extract for active packaging. *Food Hydrocoll.* 2018;**74**:207–18.
10. Rocha M, Loiko MR, Tondo EC, Prentice C. Physical, mechanical and antimicrobial properties of Argentine anchovy (*Engraulis anchoita*) protein films incorporated with organic acids. *Food Hydrocoll.* 2013;**37**:213–20.
11. Benbettaïeb N, Gay JP, Karbowiak T, Debeaufort F. Tuning the functional properties of polysaccharide–protein bio-based edible films by chemical, enzymatic, and physical cross-linking. *Compr. Rev. Food Sci. Food Saf.* 2016;**15**:739–52.
12. Wihodo M, Moraru CI. Physical and chemical methods used to enhance the structure and mechanical properties of protein films: a review. *J. Food Eng.* 2013;**114**:292–302.
13. Domínguez R, Barba FJ, Gómez B, Putnik P, Bursać Kovačević D, Pateiro M, et al. Active packaging films with natural antioxidants to be used in meat industry: a review. *Food Res. Int.* 2018;**113**:93–101.
14. Nelson DL, Cox MM. *Aminoácidos, Peptídeos e Proteínas*. 4 Sarvier; 2007. 74–113 p.
15. Cheftel JC, Cuq J-L, Llorient D. Amino acids, peptides, and proteins. In: O.R. Fennema, editor. *Food Chemistry*. New York: Marcel Dekker; 1985. pp. 245–370.
16. Damodaran S, Parkin KL, Fennema OR. *Química de Alimentos de Fennema*. 4°. Cláudia Bittencourt, editor. Artmed, São Paulo; 2010.
17. Gu L, Wang M, Zhou J. Effects of protein interactions on properties and microstructure of zein–gliadin composite films. *J. Food Eng.* 2013;**119**:288–98.
18. Rezvani E, Schleining G, Sümen G, Taherian AR. Assessment of physical and mechanical properties of sodium caseinate and stearic acid based film-forming emulsions and edible films. *J. Food Eng.* 2013;**116**:598–605.
19. Rhim JW, Ng PKW. Natural biopolymer-based nanocomposite films for packaging applications. *Crit. Rev. Food Sci. Nutr.* 2007;**47**:411–33.
20. Krochta JM. *Edible Protein Films and Coatings in Food Proteins and their Applications in Foods*. New York: Marcel Dekker; 1997.
21. Hassan B, Ali S, Chatha S, Hussain AI, Zia KM. Recent advances on polysaccharides , lipids and protein based edible films and coatings: a review. *Int. J. Biol. Macromol.* 2018;**109**:1095–107.
22. Chiralt A, González-Martínez C, Vargas M, Atarés L. Edible films and coatings from proteins. In: Rickey Y. Yada, editor. *Protein-Based Films and Coatings*. 2018. p. 477–500.
23. Flores-López ML, Cerqueira MA, de Rodríguez DJ, Vicente AA. Perspectives on utilization of edible coatings and nano-laminate coatings for extension of postharvest storage of fruits and vegetables. *Food Eng. Rev.* 2016;**8**:292–305.
24. Pagno CH, Costa TMH, De Menezes EW, Benvenutti E V., Hertz PF, Matte CR, et al. Development of active biofilms of quinoa (*Chenopodium quinoa* W.) starch containing gold nanoparticles and evaluation of antimicrobial activity. *Food Chem.* 2015;**173**:755–62.
25. Lacroix M, Vu KD. Edible coating and film materials. In: Jung H. Han, editor. *Innovations in Food Packaging*. 2014. p. 325–50.
26. Vieira MGA, Silva MA, Santos LO, Beppu MM. Natural-based plasticizers and biopolymer films: a review. *Eur. Polym. J.* 2011;**47**:254–63.
27. Han JH. Edible films and coatings: a review. In: Jung H. Han, editor. *Innovations in Food Packaging*. Texas; 2005. p. 213–55.

28. Schmid M, Merzbacher S, Müller K. Time-dependent crosslinking of whey protein based films during storage. *Mater. Lett.* 2018;**215**:8–10.
29. Cader Mhd Haniffa MA, Ching YC, Abdullah LC, Poh SC, Chuah CH. Review of bionanocomposite coating films and their applications. *Polymers* 2016;**8**:1–33.
30. Lee J, Lee J, Song KB. Development of a chicken feet protein film containing essential oils. *Food Hydrocoll.* 2015;**46**:208–15.
31. Arfat YA, Benjakul S, Vongkamjan K, Sumpavapol P, Yarnpakdee S. Shelf-life extension of refrigerated sea bass slices wrapped with fish protein isolate/fish skin gelatin-ZnO nanocomposite film incorporated with basil leaf essential oil. *J. Food Sci. Technol.* 2015;**52**:6182–93.
32. Hosseini SF, Rezaei M, Zandi M, Farahmandghavi F. Fabrication of bio-nanocomposite films based on fish gelatin reinforced with chitosan nanoparticles. *Food Hydrocoll.* 2015;**44**:172–82.
33. Lee J-H, Won M, Song K Bin Physical properties and antimicrobial activities of porcine meat and bone meal protein films containing coriander oil. *LWT-Food Sci. Technol.* 2015;**63**:700–5.
34. Gómez-Estaca J, Montero P, Gómez-Guillén MC. Shrimp (*Litopenaeus vannamei*) muscle proteins as source to develop edible films. *Food Hydrocoll.* 2014;**41**:86–94.
35. Tonyali B, Cikrikci S, Halil M. Physicochemical and microstructural characterization of gum tragacanth added whey protein based films. *Food Res. Int.* 2018;**105**:1–9.
36. Díaz O, Candia D. Whey protein film properties as affected by ultraviolet treatment under alkaline conditions. *Int. Dairy J.* 2017;**73**:84–91.
37. Beristain-Bauza SC, Mani-López E, Palou E, López-Malo A. Antimicrobial activity and physical properties of protein films added with cell-free supernatant of *Lactobacillus rhamnosus*. *Food Control* 2016;**62**:44–51.
38. Romani VP, Hernández CP, Martins VG. Pink pepper phenolic compounds incorporation in starch/protein blends and its potential to inhibit apple browning. *Food Packag. Shelf Life.*2018;**15**:151–8.
39. Arabestani A, Kadivar M, Amoresano A, Illiano A, Di Pierro P, Porta R. Bitter vetch (*Vicia ervilia*) seed protein concentrate as possible source for production of bilayered films and biodegradable containers. *Food Hydrocoll.* 2016;**60**:232–242.
40. Condés MC, Añón MC, Mauri AN, Dufresne A. Amaranth protein films reinforced with maize starch nanocrystals. *Food Hydrocoll.* 2015;**47**:146–57.
41. Shi W, Dumont MJ. Processing and physical properties of canola protein isolate-based films. *Ind. Crops Prod.* 2014;**52**:269–77.
42. Fabra MJ, Sánchez-gonzález L, Chiralt A. Lysozyme release from isolate pea protein and starch based films and their antimicrobial properties. *LWT-Food Sci. Technol.* 2014;**55**:22–6.
43. Sharma L, Singh C. Sesame protein based edible films: development and characterization. *Food Hydrocoll.* 2016;**61**:139–47.
44. Wang H, Hu D, Ma Q, Wang L. Physical and antioxidant properties of flexible soy protein isolate films by incorporating chestnut (*Castanea mollissima*) bur extracts. *LWT-Food Sci. Technol.* 2016;**71**:33–9.
45. Garrido T, Etxabide A, Guerrero P, Caba K. Characterization of agar/soy protein biocomposite films: effect of agar on the extruded pellets and compression moulded films. *Carbohydr. Polym.* 2016;**151**:408–16.
46. Xiaojing L, Na J, Chao Q, Mingtao X, Liu X, Qingjie S. The effect of peanut protein nanoparticles on characteristics of protein- and starch-based nanocomposite films: a comparative study. *Ind. Crops Prod.* 2015;**77**:565–74.
47. Salgado PR, López-Caballero ME, Gómez-Guillén MC, Mauri AN, Montero MP. Sunflower protein films incorporated with clove essential oil have potential application for the preservation of fish patties. *Food Hydrocolloids* 2013;**33**:74–84.
48. Nogueira D, Martins VG. Use of different proteins to produce biodegradable films and blends. *J. Polym. Environ.* 2019;**21**.

49. Lacroix M, Cooksey K. Edible films and coatings from animal-origin proteins. In: Jung H. Han, editor. *Innovations in Food Packaging*. 2005. p. 301–17.
50. Wang K, Wang W, Ye R, Liu A, Xiao J, Liu Y, et al. Mechanical properties and solubility in water of corn starch-collagen composite films: effect of starch type and concentrations. *Food Chem.* 2017;**216**:209–16.
51. Campos-Requena VH, Rivas BL, Pérez MA, Figueroa CR, Figueroa NE, Sanfuentes EA. Thermoplastic starch/clay nanocomposites loaded with essential oil constituents as packaging for strawberries À in vivo antimicrobial synergy over *Botrytis cinerea*. *Postharvest Biol. Technol.* 2017;**129**:29–36.
52. Atarés L, Hiralt A. Essential oils as additives in biodegradable films and coatings for active food packaging. *Trends Food Sci. Technol.* 2016;**48**:51–62.
53. Azeredo HMC, Waldron KW. Crosslinking in polysaccharide and protein films and coatings for food contact – a review. *Trends Food Sci. Technol.* 2016;**52**:109–22.
54. Romani VP, Vera A, Olsen BD, Martins VG,. Effects of pH modification in proteins from fish (*Whitemouth croaker*) and their application in food packaging films. *Food Hydrocoll.* 2018;**74**:307–14.
55. Díaz O, Candia D, Cobos Á. Effects of ultraviolet radiation on properties of films from whey protein concentrate treated before or after film formation. *Food Hydrocoll.* 2016;**55**:189–99.
56. Rydz J, Musioł M, Zawidlak-W B, Sikorska W. Polymers for food packaging applications. In: Alexandru M. Grumezescu, Alina M. Holban, editors. *Biopolymers for Food Design*. 2018. p. 431–67.
57. Hanani ZAN, Beatty E, Roos YH, Morris MA, Kerry JP. Manufacture and characterization of gelatin films derived from beef, pork and fish sources using twin screw extrusion. *J. Food Eng.* 2012;**113**:606–14.
58. Tian H, Guo G, Fu X, Yao Y, Yuan L, Xiang A. Fabrication, properties and applications of soy-protein-based materials: a review. *Int. J. Biol. Macromol.* 2018;**120**:475–90.
59. Mellinas C, García A, Ramos M, Burgos N, Garrigós M. Active edible films: current state and future trends. *J. Appl. Polym. Sci.* 2015;**133**.
60. Nilsuwan K, Benjakul S, Prodpran T. Physical/thermal properties and heat seal ability of bilayer films based on fish gelatin and poly(lactic acid). *Food Hydrocoll.* 2017;1–9.
61. Hernandez-Izquierdo VM, Krochta JM. Thermoplastic processing of proteins for film formation – a review. *J. Food Sci.* 2008;**73**:30–9.
62. Ciannamea EM, Stefani PM, Ruseckaite RA. Physical and mechanical properties of compression molded and solution casting soybean protein concentrate based films. *Food Hydrocoll.* 2014;**38**:477–500.
63. Guerrero P, Retegi A, Gabilondo N, Caba K. Mechanical and thermal properties of soy protein films processed by casting and compression. *J. Food Eng.* 2010;**100**:145–51.
64. Musso YS, Salgado PR, Mauri AN. Gelatin based fi lms capable of modifying its color against environmental pH changes. *Food Hydrocoll.* 2016;**61**:523–30.
65. Andreuccetti C, Carvalho RA, Galicia-garcía T, Martinez-bustos F, González-nuñez R, Grosso CRF. Functional properties of gelatin-based films containing *Yucca schidigera* extract produced via casting, extrusion and blown extrusion processes: a preliminary study. *J. Food Eng.* 2012;**113**:33–40.
66. Park JW, Whiteside SW, Cho SY. Mechanical and water vapor barrier properties of extruded and heat-pressed gelatin films. *LWT-Food Sci. Technol.* 2008;**41**:692–700.
67. Rhim JW, Gennadios A, Fu D, Weller CL, Hanna MA. Properties of ultraviolet irradiated protein films. *LWT-Food Sci. Technol.* 1999;**32**:129–33.
68. Mirmoghtadaie L, Shojaee Aliabadi S, Hosseini SM. Recent approaches in physical modification of protein functionality. *Food Chem.* 2016;**199**:619–27.
69. Siracusa V, Rocculi P, Romani S, Rosa MD. Biodegradable polymers for food packaging: a review. *Trends Food Sci. Technol.* 2008;**19**:634–43.
70. Al-Hassan AA, Norziah MH. Starch e gelatin edible films: water vapor permeability and mechanical properties as affected by plasticizers. *Food Hydrocoll.* 2012;**26**:108–17.

71. González A, Strumia MC, Igarzabal CIA. Cross-linked soy protein as material for biodegradable films: synthesis, characterization and biodegradation. *J. Food Eng.* 2011;**106**:331–8.
72. Cuq B, Gosntard N, Aymard C, Guilbert S. Relative humidity and temperature effects on mechanical and water vapor barrier properties of myofibrillar protein-based films. *Polymer Gels Netw.* 1997;**5**:1–15.
73. Kokoszka S, Debeaufort F, Lenart A, Voilley A. Water vapour permeability, thermal and wetting properties of whey protein isolate based edible films. *Int. Dairy J.* 2010;**20**:53–60.
74. Longares A, Monahan FJ, Riordan EDO, Sullivan MO. Physical properties and sensory evaluation of WPI films of varying thickness. *Swiss Soc. Food Sci. Technol.* 2004;**37**:545–50.
75. Park HJ, Chinnan MS. Gas and water vapor barrier properties of edible films from protein and cellulosic materials. *J. Food Eng.* 1995;**25**:497–507.
76. Suderman N, Isa MIN, Sarbon NM. The effect of plasticizers on the functional properties of biodegradable gelatin-based film: a review. *Food Biosci.* 2018;**24**:111–9.
77. Tian H, Guo G, Xiang A, Zhong WH. Intermolecular interactions and microstructure of glycerol-plasticized soy protein materials at molecular and nanometer levels. *Polym. Test.* 2018;**67**:197–204.
78. Ahmed I, Lin H, Zou L, Brody AL, Li Z, Qazi IM, et al. A comprehensive review on the application of active packaging technologies to muscle foods. *Food Control* 2017;**82**:163–78.
79. Shojaee-Aliabadi S, Hosseini H, Mohammadifar AM, Mohammadi A, Ghasemlou M, Ojagh SM, et al. Characterization of antioxidant-antimicrobial k-carrageenan films containing *Satureja hortensis* essential oil. *Int. J. Biol. Macromol.* 2013;**52**:116–24.
80. Moustafa H, Youssef AM, Darwish NA, Abou-Kandil AI. Eco-friendly polymer composites for green packaging: future vision andchallenges. *Composites Part B* 2019;**172**:16–25.
81. Maran JP, Sivakumar V, Thirugnanasambandham K, Sridhar R. Degradation behavior of biocomposites based on cassava starch buried under indoor soil conditions. *Carbohydr. Polym.* 2014;**101**:20–8.
82. Rawdkuen S, Thitipramote N, Benjakul S. Preparation and functional characterisation of fish skin gelatin and comparison with commercial gelatin. *Int. J. Food Sci. Technol.* 2013;**48**:1093–102.
83. Tongdeesoontorn W, Rawdkuen S. Gelatin-based films and coatings for food packaging applications. In: Jari Vartiainen, editor. *Reference Module in Food Science*. 2019. p. 41.
84. Azevedo VM, Borges SV, Marconcini JM, Yoshida MI, Neto ARS, Pereira TC, et al. Effect of replacement of corn starch by whey protein isolate in biodegradable film blends obtained by extrusion. *Carbohydr. Polym.* 2017;**157**:971–80.
85. Cinelli P, Schmid M, Bugnicourt E, Wildner J, Bazzichi A, Anguillesi I, et al. Whey protein layer applied on biodegradable packaging film to improve barrier properties while maintaining biodegradability. *Polym. Degrad. Stab.* 2014;**108**:151–7.
86. Sharma S, Luzinov I. Whey based binary bioplastics. *J. Food Eng.* 2013;**119**:404–10.
87. Beristain-Bauza S del C, Mani-López E, Palou E, López-Malo A. Antimicrobial activity of whey protein films supplemented with Lactobacillus sakei cell-free supernatant on fresh beef. *Food Microbiol.* 2017;**62**:207–11.
88. Sai-Ut S, Benjakul S, Rawdkuen S. Retardation of lipid oxidation using gelatin film incorporated with longan seed extract compared with BHT. *J. Food Sci. Technol.* 2015;**52**:5842–9.
89. Jamróz E, Kopel P, Juszczak L, Kawecka A, Bytesnikova Z, Milosavljevic V, et al. Development of furcellaran-gelatin films with Se-AgNPs as an active packaging system for extension of mini kiwi shelf life. *Food Packag. Shelf Life* 2019;**21**:1003–39.
90. Akcan T, Estévez M, Serdaroglu M. Antioxidant protection of cooked meatballs during frozen storage by whey protein edible films with phytochemicals from *Laurus nobilis* L. and *Salvia officinalis*. *LWT-Food Sci. Technol.* 2017;**77**:323–31.
91. González A, Alvarez Igarzabal CI. Nanocrystal-reinforced soy protein films and their application as active packaging. *Food Hydrocolloids* 2015;**43**:777–84.

92. Brink I, Šipailienė A, Leskauskaitė D. Antimicrobial properties of chitosan and whey protein films applied on fresh cut Turkey pieces. *Int. J. Biol. Macromol.* 2019;**130**:810–7.
93. Zhang L, Liu Z, Wang X, Dong S, Sun Y, Zhao Z. The properties of chitosan/zein blend film and effect of film on quality of mushroom (*Agaricus bisporus*). *Postharvest Biol. Technol.* 2019;**155**:47–56.
94. Wang S, Xia P, Wang S, Liang J, Sun Y, Yue P, et al. Packaging films formulated with gelatin and anthocyanins nanocomplexes: physical properties, antioxidant activity and its application for olive oil protection. *Food Hydrocoll.* 2019;**96**:617–24.
95. Ejaz M, Arfat YA, Mulla M, Ahmed J. Zinc oxide nanorods/clove essential oil incorporated type B gelatin composite films and its applicability for shrimp packaging. *Food Packag. Shelf Life* 2018;**15**:113–21.
96. Azevedo VM, Dias MV, Elias HHS, Fukushima KL, Silva EK, Carneiro JDS, et al. Effect of whey protein isolate films incorporated with montmorillonite and citric acid on the preservation of fresh-cut apples. *Food Res. Int.* 2018;**107**:306–13.
97. Halim ALA, Kamari A, Phillip E. Chitosan, gelatin and methylcellulose films incorporated with tannic acid for food packaging. *Int. J. Biol. Macromol.* 2018;**120**:1119–26.
98. Lee K-Y, Yang H-J, Song K Bin. Application of a puffer fish skin gelatin film containing *Moringa oleifera* Lam. Leaf extract to the packaging of Gouda cheese. *J. Food Sci. Technol.* 2016;**53**:3876–83.
99. Boyacı D, Korel F, Yemenicioğlu A. Development of activate-at-home-type edible antimicrobial films: an example pH-triggering mechanism formed for smoked salmon slices using lysozyme in whey protein films. *Food Hydrocoll.* 2016;**60**:170–8.
100. Lee K, Lee J, Yang H, Song K Bin. Original article production and characterisation of skate skin gelatin films incorporated with thyme essential oil and their application in chicken tenderloin packaging. *Int. J. Food Sci. Technol.* 2016;1465–72.
101. Shahidi F, Zhong Y. Novel antioxidants in food quality preservation and health promotion. *Eur. J. Lipid Sci. Technol.* 2010;**112**:930–40.
102. Guilbert S, Gontard N. Agro-polymers for edible and biodegradable films : polymeric materials , physical and mechanical characteristics. In: Jung H. Han, editor. *Innovations in Food Packaging.* 2005. p. 263–76.
103. Stuchell YM, Krochta JM. Enzymatic treatments and thermal effects on edible soy protein films. *J. Food Sci.* 1994;**59**:1332–7.
104. Yang L, Paulson AT, Nickerson MT. Mechanical and physical properties of calcium-treated gellan films. *Food Res. Int.* 2010;**43**:1439–43.
105. Rydz J, Musioł M, Zawidlak-Węgrzyńska B, Sikorska W. Present and future of biodegradable polymers for food packaging applications. In: Alexandru M. Grumezescu, Alina M. Holban, editors. *Biopolymers for Food Design.* Elsevier; 2018. p. 431–67.
106. Caba K, Guerrero P, Trung TS, Cruz-Romero M, Kerry JP, Fluhr J, et al. From seafood waste to active seafood packaging: an emerging opportunity of the circular economy. *J. Cleaner Prod.* 2019;**208**:86–98.
107. Inns GR. The packaging supply chain. In: Anne Emblem, editor. *Packaging Technology.* Elsevier; 2012. p. 10–23.
108. Nerkar A, Shane S. Determinants of invention commercialization: an empirical examination of academically sourced inventions. *Strateg. Manag. J.* 2007;**28**:1155–66.
109. Tolfree D, Jackson MJ. *Commercializing Micro-Nanotechnologies.* Boca Raton, FL: CRC Press; 2008.
110. Butler TI, Ealer GE. *Film Extrusion Manual: Process, Materials, Properties.* Atlanta: TAPPI Press; 2005. 616 pp.
111. Mensitieri G, Di Maio E, Buonocore GG, Nedi I, Oliviero M, Sansone L, et al. Processing and shelf life issues of selected food packaging materials and structures from renewable resources. *Trends Food Sci. Technol.* 2011;**22**:72–80.

112. Werner BG, Koontz JL, Goddard JM. Hurdles to commercial translation of next generation active food packaging technologies. *Curr. Opin. Food Sci.* 2017;**16**:40–8.
113. Conte A, Cappelletti GM, Nicoletti GM, Russo C, Del Nobile MA. Environmental implications of food loss probability in packaging design. *Food Res. Int.* 2015;**78**:11–7.
114. Babu RP, O'Connor K, Seeram R. Current progress on bio-based polymers and their future trends. *Prog. Biomater.* 2013;**2**:8.
115. Schmid M, Müller K. Whey protein-based packaging films and coatings. In: Hilton C. Deeth, Nidhi Bansal, editors. *Whey Proteins*. 2018. p. 407–37.
116. Conceição O, Fontes M, Calapez T. The commercialisation decisions of research-based spin-off: targeting the market for technologies. *Technovation* 2012;**32**:43–56.
117. Ilyas RA, Sapuan SM, Atikah MSN, Asyraf MRM, Ayu Rafiqah S, Aisyah HA, Mohd Nurazzi N, Norrahim MNF. Effect of hydrolysis time on the morphological, physical, chemical, and thermal behavior of sugar palm nanocrystalline cellulose (Arenga Pinnata (Wurmb.) Merr). *Textile Research Journal*. 2020. In print. 10.1177/0040517520932393
118. Licciardello F. Packaging, blessing in disguise. Review on its diverse contribution to food sustainability. *Trends Food Sci. Technol.* 2017;**65**:32–9.

3

Protein-Based Biodegradable Polymer: From Sources to Innovative Sustainable Materials for Packaging Applications

Huafeng Tian[1], Yunxuan Weng[1], Rakesh Kumar[2], Priya Rani[2], and Gaiping Guo[3]

[1] *Beijing Key Laboratory of Quality Evaluation Technology for Hygiene and Safety of Plastics, School of Material and Mechanical Engineering, Beijing Technology and Business University, Beijing, 100048, China*

[2] *Department of Biotechnology, Central University of South Bihar, Gaya, 824236, India*

[3] *College of Materials Science and Engineering, Beijing Institute of Petrochemical Technology, Beijing, 102617, China*

3.1 Introduction

Before the advent of synthetic plastics as packaging materials, the world civilizations were dependent on biobased packaging materials. In new forms, the biobased packaging material is resurfacing because of problems being created by nonbiodegradable-based synthetic packaging materials [1]. The edible materials, particularly fruits and vegetables, undergo progressive deterioration in the form of ripening. One way to increase shelf life and to improve the quality of edible materials is by modified atmosphere packaging (MAP) [2]. By reducing the level of O_2 and increasing the level of CO_2, shelf life can be increased [3]. Polymeric films semi-permeable to gases are good candidates to create MAP. Edible films

Bio-Based Packaging: Material, Environmental and Economic Aspects, First Edition. Edited by S.M. Sapuan and R.A. Ilyas

based on wheat gluten (WG), soy, zein, and casein proteins have good film-forming and gas permeability properties [4]. Barrier properties of the synthetic or natural films are affected by temperature and the hydrophilic films are affected by humidity.

Soy protein isolate (SPI) or soy protein concentrate (SPC) is a byproduct of the soy oil industry. Presently, most SPI or SPC is utilized in applications in animal feed and food supplements. The main commercial soy protein products obtained from the market are soy flour (SF), SPC, and SPI, with around 50%, 70%, and 90% protein content, respectively. Zein is a class of alcohol-soluble prolamine proteins and is more hydrophobic than soy proteins owing to a high percentage of nonpolar amino acid residues, including proline, leucine, alanine, and glutamine. Specifically, zein has more than 50% hydrophobic domains on the surface. It has been reported that the contact angle of water on zein film is 55.4°, which is much higher than that of a film of SPI film (26.5°) [5]. WG is primarily composed of high molecular weight glutenins and gliadins and low molecular weight albumins and globulins. Protein films formed from wheat gluten have good film-forming capabilities due to their unique cohesive and elastic properties [6]. The film prepared from WG is considered hydrophilic. S—S bonds are predominantly intermolecular in glutenin, thus cross-linking polypeptide chains. Casein proteins that constitute ~80% of the total proteins in milk exist as calcium and sodium caseinates (CaCas and NaCas) [7].

Proteins have great potential to produce edible films and these films can form a good barrier to oxygen and other nonpolar molecules such as hydroxyl and amino groups. It should also be noted that protein films tend to shrink during drying and can become brittle. Therefore, glycerol as a plasticizer has been widely used to prepare flexible protein films [8]. Every protein polymer has its specific material characteristics for packaging purposes. To increase specific material properties in these polymers, nanoparticles and additives are incorporated [8, 9]. The aim of this chapter is to discuss the nanoparticles or additives that have been incorporated to date in the protein polymers to make them a commercially acceptable packaging material.

3.2 Forms of Packaging Materials

Protein materials such as casein, soy protein, wheat gluten, and zein can be successfully casted into film. Most of the protein films are prepared at alkaline pH (9.0–9.5). During processing or fabrication, the proteinous film can be exploited to tailor some of the properties in various ways in order to broaden their range of potential commercial applications, such as edible wrappers or soluble pouches for single-serve foods, packaging films for cheese or oxygen-sensitive foods, coatings for nuts and cereals, or multilayered packaging films. WG as a proteinous film has attractive mechanical, oxygen barrier, and film-forming properties and renewability [10]. The proteinous film and film-forming properties of proteinous materials are strongly dependent on the pH of the dispersion, and they are normally toward higher alkaline side [9–11] or higher acidic side [2–4].

3.3 Commercially Available Proteinous Material for Packaging

There are four protein fractions in SPI that are classified according to their sedimentation properties, which are 2S, 7S, 11S, and 15S fractions. It has been reported that 2S, 7S,

11S, and 15S comprise 8%, 35%, 52%, and 5% of the total protein content, respectively [11]. SPI contains two major components: glycinin, or 11S globulin, and β-conglycinin, or 7S globulin. The 11S globulin consists of two hexagonal rings, each containing three hydrophobically associated subunits [12]. The glycinin subunits consist of two polypeptide components linked via disulfide bonds, one acidic (A) and the other with a basic (B) isoelectric point [13]. The 7S globulin is a trimeric glycoprotein that consists of three types of subunit: α, α', and β. In contrast to 11S, the 7S globulin does not contain disulfide bonds. The rest of the SPI consists of whey proteins, such as β-conglycinin, the basic 7S globulin, lipoxygenase, agglutinins, and β-amylases [14]. The major storage protein of soybean (*Glycine max*) is globulin.

Soy protein produces more flexible, smooth, and clear films compared to films from other plant protein sources such as whey, casein, and zein [15]. The formation of films from soy proteins or any proteins as described in this book chapter has been described as a process involving the denaturation of proteins followed by surface dehydration. Denaturation of protein involves its solubilization and unfolding. Denatured proteins form the film network through disulfide cross-linking and hydrophobic bonds [16]. Soy protein-based edible film is most commonly prepared from SPI rather than SPC because the nonprotein fraction in the latter adversely affects the film-forming abilities [17]. Casein films can be used in combination with other packaging materials to protect products prone to oxidation. Pure casein film has the potential to provide films with lower water vapor permeability (WVP). The greater film homogeneity of a protein's film gives better oxygen barrier properties. The zein biofilm is formed through development of hydrophobic, hydrogen, and limited disulfide bonds between zein chains. The WVP values of the protein films are normally two–four orders of magnitude higher than that of low density polyethylene (LDPE) [18]. Lactic acid, tannic acid, ionized calcium, or transglutaminase can be used to cross-link proteins, resulting in increased resistance to water vapor and gas transport. Preparation of protein films from aqueous emulsions of proteins and lipids decrease WVP.

3.4 Preparation Methods for Protein-Based Materials for Different Packaging Applications

Currently, the main techniques of preparing protein-based packaging materials include solution casting, solution lyophilization, melt processing, and electrospinning. The solution casting method is based on the dispersion or solubility of proteins in a solvent medium. For soy protein, wheat gluten, and casein, etc., water is the best solvent, while for some hydrophobic proteins, for example, zein, alcohol/water mixtures of a certain ratio, would be the best choice. After dissolving and evaporating the solvents, either water or alcohol protein-based films can be obtained. For some protein-based blends and nanocomposites, the solution casting method could be used, but the blending moiety is restricted to the solvent soluble/dispersed polymers, for example, soy protein/agar blends in water systems [19], soy protein/water-borne polyurethane blends in alkaline aqueous solution [20], zein/chitosan/nano-TiO_2 composite films in glacial acetic acid solution [21], etc. All the protein materials (6–8%), such as SPI, casein, and WG (except zein, which is dispersed in alcohol) powder, are dispersed in water solution in the presence of plasticizers and the mixture is homogenized by magnetic stirring at 200–500 rpm. All protein films are generally cast on silicone baking mats because of good compatibility and the ease of peeling off the dried films. Generally, the solution casting technique is a low efficiency method with a high energy consumption, and thus only viable in the laboratories.

In addition to casting, the mixed solution could also be dried by lyophilization or by adding the precipitant of soy protein. The protein and other blending moieties (including nanofillers and other polymers) can be dispersed and mixed in a solution, and then protein composite powders can be obtained by freeze-drying the mixture solution [22] or by adding a precipitant such as acetone [23, 24]. The protein-based films or sheets could be further prepared by hot-pressing or extrusion. A lyophilization-related method has a lower productivity with a high cost.

The melt processing technique is more efficient and practical for industrial-scale production. This technique includes extrusion, hot-pressing, and injection molding. Extrusion is one of the prime processes used in the plastic industry. However, for most protein materials, the flowing ability would be poor in the extruders, so plasticizers are added to improve the flowability. Plasticizers with polar groups, such as water, polyols, amide, hydroxylamine, etc., are generally incorporated and the resulting material possesses fine flexibility during the plasticizing effect [25]. Extrusion is an easy scale-up process and is used to produce casein (rennet casein, acid casein, and sodium caseinate)-based sheets in the presence of glycerol plasticizer with improved WVP [26]. During the extrusion process, if the raw material contains a mixture of various components such as impurities, phase separation, decomposition, and gas generation then it will interfere with the process.

Nowadays, the electrospinning technique has gained great interest due to its ability to fabricate nanosized fibers. This technique involves the stretching of a polymer solution in a high electric field to produce fibers with a diameter ranging from tens of nanometers to submicron size [5]. In food packaging, electrospun fibers have been proposed for stabilizing or controlling the release of the active compounds. The composite fibrous films are developed from zein and poly(lactic acid) (PLA) by incorporating carvacrol using electrospinning [27]. These carvacrol loaded electrospun fibers with antioxidant and antimicrobial properties can be used to extend the shelf life of fresh foods as a new approach of using electrospun fibers in food applications.

Protein samples can also be prepared by compression molding techniques. For the production of compression molded samples, protein powder can be mixed with plasticizers and can be subjected to compression at an appropriate temperature and pressure. Compression molded soy protein samples can be fabricated at 120–130 °C and 15–20 MPa pressure [8]. Extruded protein samples can be ground to a powder and then can be subjected for compression molding as reported for zein protein compression molded samples prepared at 138 °C and pressure of 17 MPa [28].

3.5 Properties of Protein-Based Packaging Materials

3.5.1 Mechanical Properties

The mechanical property is one of the key properties needed when protein-based materials are used as packaging materials. The characterization methods for protein-based packaging materials mainly include tensile tests for films and sheets, impact tests for sheets, as well as compression tests for foam materials.

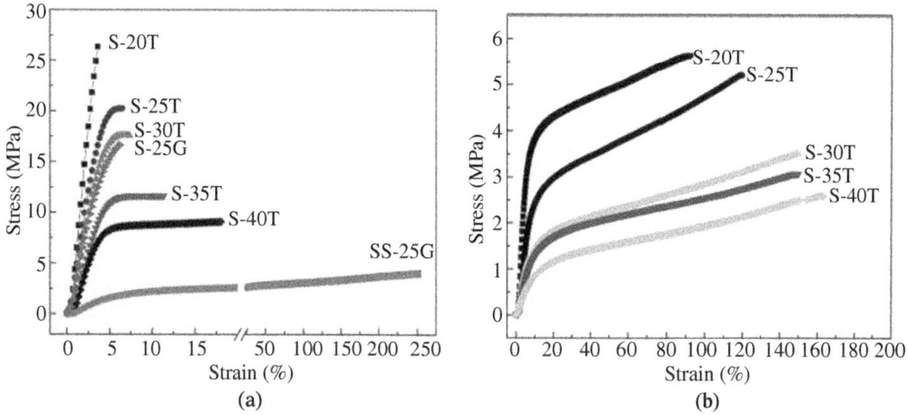

Figure 3.1 *Mechanical properties of soy protein films at (a) 57% and (b) 97% RH. Source: Reprinted with permission from Kumar, R., L. Wang, and L. Zhang. 2009.Structure and mechanical properties of soy protein materials plasticized by thiodiglycol. Journal of Applied Polymer Science 111 (2): 970–977. Copyright 2019, Wiley.*

Neat protein-based films usually exhibit low mechanical strength. Incorporation of fillers, in particular, those with a nanoscale size, constitutes an effective way to improve the strength of protein plastics. The mechanical properties of the resulting nanocomposite are impacted by parameters such as the intrinsic properties of nanofillers, such as the modulus and aspect ratio, the nature of the matrix, and the dispersibility of the nanofillers as well as compatibility between the filler and the matrix. Especially, compatibility is the key parameter to influence the mechanical performance, which would also impact the dispersing effect of the nanoparticles.

The synthetic polymers usually possess fine mechanical properties. Thus, blending proteins with some synthetic biodegradable polymers is another effective way to improve the mechanical properties of protein-based materials. The interaction and compatibility between the components also play a key role in influencing the mechanical properties. When blending with some polymers that have poor compatibility with proteins, some compatibilizing methods must be adopted to enhance the compatibility as well as the dispersing effect of the two moieties. The ternary components, called compatibilizers, are added to improve the interaction of the two matrix moieties, such as isocyanate, maleic anhydride-grafted polymers, etc. [25]. Kumar et al. reported in Figure 3.1 the mechanical properties of soy protein film in the presence of thiodiglycol plasticizer at relative humidity (RH) of 57% and 97% [29]. Tannic acid has been explored as an effective cross-linking agent for casein protein, so that the yielded films exhibit improved physicochemical properties. Casein film with improved mechanical properties can be employed for food packaging applications [30]. WG induced by the formation of disulfide bonds can be used to form anisotropic product structures with an increase in hardness and Young's modulus [31].

Fibers from banana stems were mixed with wheat gluten and films were prepared by solution casting followed by compression molding. Further, cross-linking of fiber and gluten was done with citric acid. Cross-linking with citric acid led to a substantial increase in the strength of the films from 3.5 MPa to 13 MPa [32]. Glycerol and a series of polyethers (PEs), including poly(ethylene glycol), poly(propylene glycol), and poly(tetramethylene glycol), were chosen as model plasticizers to soften zein films with a very high elongation at break that can be used as flexible packaging [33]. Zein film having glycerol and PE plasticizer in the ratio of 1 : 1 in mass showed excellent elongation at break of 245–350% at 0.45 g plasticizer/g zein [33].

3.5.2 Moisture Resistance

Most of the proteins possess an abundance of polar groups and therefore exhibit poor water resistance. Thus, moisture sensitivity of protein-based biomaterials is one of the main drawbacks that can prevent their application. The moisture sensitivity results in variable mechanical properties under different environment relative humidities (RHs). The high moisture contents are also prone to the bacterial growth for these biopolymer-based materials. Similar to the mechanical properties, incorporating nanofillers and blending with hydrophobic polymers could effectively improve the water resistance of protein-based materials to a certain extent.

Water resistance of soy protein-based materials has been considerably increased by the addition of benzilic acid and mandelic acid [34–36]. Soy protein films incorporated with mandelic and benzilic showed a water uptake of around 30 ± 5% as compared to neat soy protein having a water uptake of 120 ± 10%. Banana fiber-reinforced wheat gluten films without a citric acid cross-linker showed maximum water absorption values up to 500%, whereas cross-linking reduced water absorption drastically to about 200% [32].

3.5.3 Barrier Properties

Fine mechanical properties, low WVP, and high oxygen barrier properties are the key parameters for using as packaging films. The barrier of oxygen, carbon dioxide, and some odors are essential for food packaging materials. Especially, the oxygen barrier performance is one of the determining factors that guarantee the maximum shelf life of food products. The protein films usually exhibit effective barriers to lipid, oxygen, and carbon dioxide, but exhibit low moisture barrier properties. The small molecular weight plasticizers usually damage the barrier properties.

Incorporating nanoparticles could effectively improve the barrier properties. Nanofillers, especially some two-dimensional dimension particles, for example layered silicate clays, including montmorillonite (MMT), rectorite, etc., generate a tortuous pathway for the small molecule permeable elements, such as water and oxygen, to penetrate through the nanocomposite, therefore decreasing the WVP and oxygen permeability (OP) significantly [37].

Zein film without plasticizer form hydrophobic film and showed high WVP because of the formation of many void spaces and cavities in the films. WVP is influenced by the hydrophobic or hydrophilic nature of the material, by the presence of voids or cracks, by steric hindrance, and by tortuosity in the structure. Zein films plasticized by glycerol and PEs plasticizer in the ratio of 1 : 1 had a higher WVP because of their unique porous structure [33]. Casein-based sheets produced by the extrusion process showed improved WVP [26].

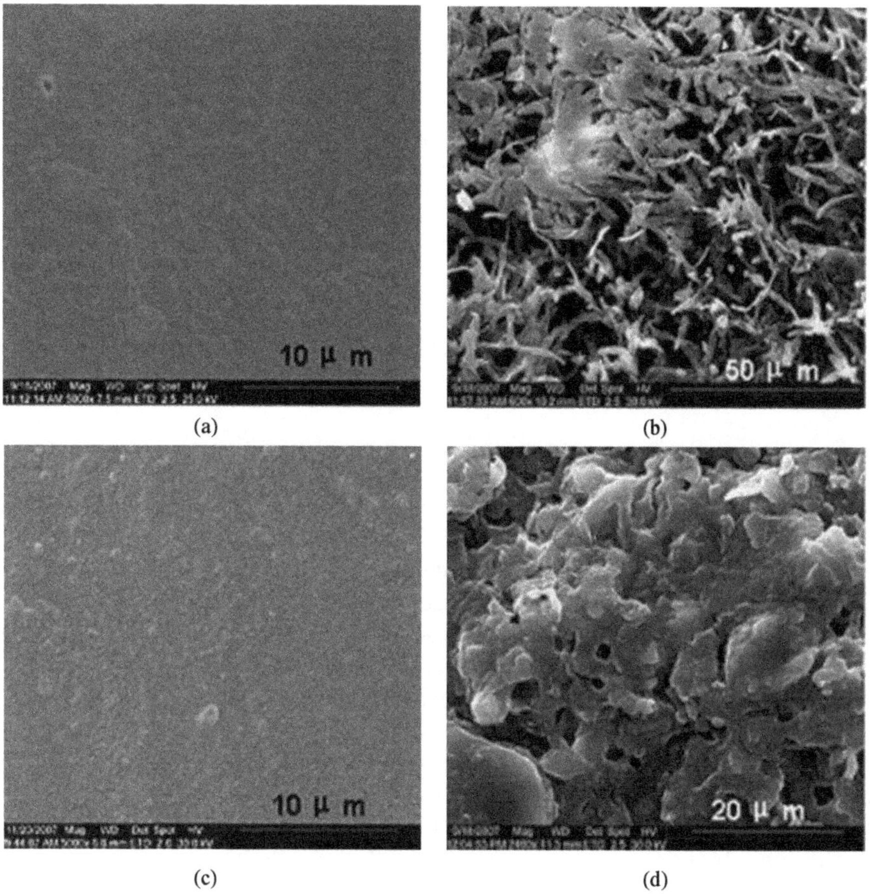

Figure 3.2 SEM image of neat (top) and benzilic acid (bottom) incorporated soy protein film taken out from buried soil after 0 days (left) and 7 days (right). Source: Reprinted with permission from Kumar, R., and L. Zhang. 2008. Water-induced hydrophobicity of soy protein materials containing 2, 2-diphenyl-2-hydroxyethanoic acid. Biomacromolecules 9 (9): 2430–2437. Copyright 2019, American Chemical Society.

3.5.4 Biodegradability

All the proteins discussed in this chapter are obtained from renewable resources and are the kind of renewable polymers that could be decomposed into CO_2, water, and organic small molecules by microorganisms during composting conditions. The biodegradability ratio could be studied by evaluating weight loss during composting. The degraded surface can be observed by scanning electron microscopy (SEM), as has been done for benzilic acid incorporated soy protein films, and is shown in Figure 3.2 [34] and the structural changes can be characterized by Fourier transformed infrared spectroscopy (FTIR) [38]. In general, natural protein possessed a higher biodegradation rate than synthetic polymers [39]. Therefore, incorporation of natural proteins into these polymers speeds up the biodegradation rate of the blends.

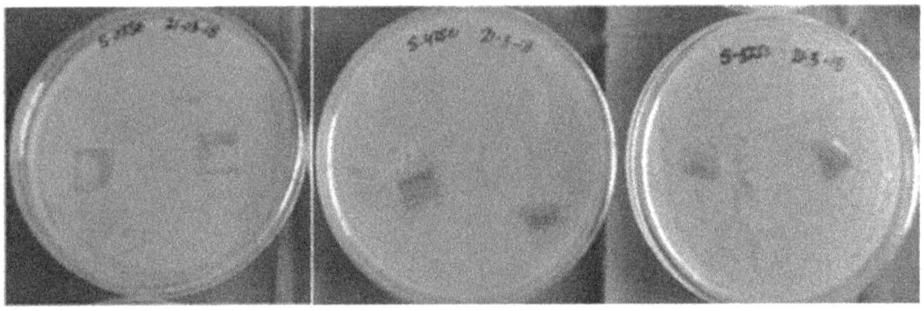

Figure 3.3 Antimicrobial properties of ZnSe incorporated SPI films. Source: Reprinted with permission from Kumar, R., R. Praveen, S. Rani, K. Sharma, K. P. Tiwary, and K. D. Kumar. 2019. ZnSe nanoparticles reinforced biopolymeric soy protein isolate film. Journal of Renewable Materials (Open Access) 7 (8): 749–761.

3.5.5 Antimicrobial Properties

The ability to protect products from microbial contamination is preferred in food packaging since microbial spoilage is responsible for a shortened shelf life and degraded food quality. In general, protein films show no antibacterial property. The gelatin-based electrospun fibrous membranes display remarkable antibacterial properties against *Escherichia coli* and *Staphylococcus aureus*. The electrospun fibrous membrane was produced with fish gelatin as the structural matrix and bovine lactoferrin as the active antimicrobial agent [40].

Encapsulation of metal oxide or sulfide nanoparticles with high antimicrobial activity and a specific surface area into polymers is a promising approach to enhance the antimicrobial activity of nanoparticles due to the improved stability and sustained release. Recently, Kumar et al. have incorporated ZnS and ZnSe nanoparticles in SPI to prepare antimicrobial soy protein films [41, 42]. Figure 3.3 shows the antimicrobial properties of ZnSe nanoparticles SPI films [42].

3.6 Nanomaterials Incorporated Protein-Based Packaging Materials

The combination of nanometric particles with natural proteins is a promising option for improving mechanical, barrier, water resistance properties, etc. In general, the methods to produce nanoparticles in a protein matrix include *in situ* synthesis of nanoparticles by the precursor and compounding the existing nanoparticles with the protein matrix. The synthesis of nanoparticles from the protein solution by the *in situ* method is not easily controlled, leading to a nonuniformed size and structure, although the compatibility between the fillers and matrix as well as the dispersion effects is usually better [43]. Therefore, most of the nanocomposites are fabricated by incorporating existing nanoparticles. A few kinds of protein nanocomposites could be fabricated by both the methods mentioned above, for example, soy protein/silica nanocomposites [23, 44, 45].

3.6.1 Protein/Inorganic Filler Nanocomposites

Inorganic fillers usually exhibit much better thermal stability as compared with organic fillers. Incorporation of original or surface modified nanoparticles, such as polyhedral oligomericsilsesquioxanes (POSS) [46], hydroxyapatite [47], eggshell nanopowder [48], and carbon nanotubes [24] have been reported by several authors. The mechanical properties, water resistance, thermal stability, etc., of the resulting protein-based composite films was dramatically enhanced by the incorporation of these nanoparticles.

Nanoclays are widely used as fillers to reduce flammability and permeability to liquids and gases, as well as reinforcement in a wide range of plastics. Nanoclays are commercially exploited for the improvement in barrier properties that will have a wide application in food packaging areas. The U.S. Food and Drug Administration has already approved clay-based composites for use as foods, medicines, beverages, and biomedical devices [49]. Intercalation and exfoliation can be viewed as possible arrangements of layered clays dispersed in a polymer matrix, including phase-separated or immiscible microcomposites. These two technical terms, i.e. intercalation and exfoliation, are highly responsible for the high dispersion and disordered degree of the nanolayers. The dispersion effect depends on the processing methods, the nature of both the polymer and the clay, and finally the interaction between them. MMT is the most widely used layered silicate clay. Soy protein/MMT nanocomposites could be fabricated by both solution intercalation [37, 50] and melt intercalation [51]. The structure and properties of the nanocomposites were strongly dependent on the MMT content [50]. Other kinds of layered silicate, such as rectorite, was also reported to prepare soy protein nanocomposites [52]. The reduction of gas permeability would be the most dramatic besides the improvement of mechanical performance, due to the prolonged passing-through route of gas molecules in the matrix by the highly dispersed nanolayers.

The antimicrobial property is one of the characteristics concerned in food packaging fields, especially for these biopolymer-based materials, which are more prone to attack by bacteria and fungus compared with synthetic polymers. By incorporating certain antibacterial nanoparticles, such as titanium dioxide (TiO_2), silver, etc., these protein films could prolong their shelf life and endow their applications. Antibactericidal efficiency against *E. coli* and *S. aureus* have been observed in SPI/titanium dioxide composite films [53]. The bactericidal activity of films containing TiO_2 nanoparticles (2.00 g/150 ml) against gram-negative *E. coli* and gram-positive *S. aureus* can reach 71.01% and 88.49%, respectively.

Silver is a widely used antimicrobial agent used against many kinds of bacteria, fungi, and viruses. Antimicrobial silver nanoparticles (AgNPs)/SPI films were prepared *in situ* and synthesized from SPI/$AgNO_3$ solution by the reducibility of tyrosine residue in SPI [54]. The resulting SPI nanocomposite films exhibited excellent antimicrobial properties for both Gram-positive bacteria (*S. aureus*) and Gram-negative bacteria (*E. coli*).

Casein-based ZnO nanocomposites were prepared via double *in situ* polymerization with γ-(2,3-epoxypropoxy) propytrimethoxysilane (KH560) as the silane coupling agent. The average size of casein-based ZnO nanocomposite particles was approximately 259 nm. At 3% content of ZnO precursor, the water resistance of the composite film was maximum. Also, the 9.0 mm zone of inhibition with *E. coli* was observed for casein-based ZnO nanocomposite film [55].

Whey protein-based bionanocomposite films (WPI/CNFs) containing whey protein isolate (WPI), cellulose nanofibers (CNFs with amounts of 2.5%, 5%, 7.5%, and 10%), titanium dioxide (TiO_2 with amounts of 0.5%, 1%, and 1.5%) nanoparticles, and rosemary essential oil (REO with amounts of 1.5% and2%) have been reported [56]. Water resistance of composite films increased significantly with the addition of TiO_2 nanoparticles and REO. At the content of 7.5% CNFs and 1% TiO_2, the WPI/CNFs films showed a significant increase in tensile strength and Young's modulus. Also, these films showed strong antibacterial activity against pathogenic and spoilage bacteria, especially gram-positive bacteria [56].

Zein films are generally prepared through casting in aqueous ethanol solutions to a contact surface. However, the use of organic solvents in many food applications is restricted. Aqueous dispersions of zein nanoparticles were prepared by antisolvent precipitation from 90% v/v of aqueous ethanol zein solutions followed by casting the dispersions in silicone molds and air-drying at 50°C to prepare zein films. This type of zein film showed grainy morphology and higher surface roughness [57]. A series of TiO_2-modified zein/chitosan (CS) films prepared through a process involving blending, solution casting, and evaporation have been reported [21]. The tensile strength of the zein/CS/TiO_2 films was reported as 28.33 ± 1.53 MPa at 0.15 wt % TiO_2 nanoparticles [21]. As well as *S. aureus, E. coli,* and *Salmonella Enteritidis*, the film showed an antibacterial effect [21]. Zein protein films embedded with core-and-shell nanoparticles, with titanium dioxide as the core and silica as the shell, have been also prepared by a solution-casting method [58]. *In vitro* antibacterial, antibiofilm, and mosquito larvicidal effects of gold nanoparticles incorporated zein biopolymer films [59]. The antibacterial assays testing at 100 μg/ml showed that the zones of inhibition were 13.9 mm and 14.2 mm against *Bacillus pumilus* and *Bacillus subtilis,* respectively, and 18.1 mm and 18.4 mm against *Shigella sonnei* and *Pseudomonas aeruginosa* [59].

3.6.2 Protein/Organic Filler Nanocomposites

The organic fillers are usually made from biodegradable biopolymers, including starch, chitin, cellulose, etc. Compared with inorganic fillers, these naturally occurring materials forming in nature during the life cycles of green plants, animals, bacteria, and fungi are attracting more and more attention for their eco-friendly nature.

Natural polysaccharide-based nanocrystals are attracting more and more interest from both academic institutions and industries, due to the low cost, low density, high specific strength and modulus, low energy consumption, and renewable character. The polysaccharide nanocrystals investigated now mainly include starch, chitin, and cellulose. Both cellulose and chitin nanocrystals exhibit rod-like morphology and are therefore often called whiskers. The starch nanocrystals usually exhibit a platelet-like morphology. Starch nanocrystals can be incorporated into a soy protein matrix to form bionanocomposites [60–62], a dramatic reinforcing effect that was observed due to the hydrogen bonding network through the matrix by the nanocrystals. The nanocrystal content as well as the aspect ratio are the key parameters influencing the mechanical properties.

To further increase interaction between the nanocrystal filler and protein matrix, a strong chemical bonding can be formed instead of intramolecular hydrogen bonds [63]. The chemical cross-linking network through the reactions between the nanocrystals and protein matrix highly improved the mechanical properties of the composite films.

Microfibrillated cellulose (MFC) with diameters around 50–60 nm, an average length of 485 ± 2 μm, and an aspect ratio (*L/d*) of 8800 [64] exhibited a much higher aspect ratio compared with cellulose whiskers, which is beneficial to the mechanical properties of the nanocomposites. When incorporated into soy protein resin [65], the fracture stress and Young's modulus increased significantly, while the fracture strain did not show a significant decrease, indicating the increased toughness.

In addition, electrospun cellulose nanofibrous mats (CNMs) [66], electrospun epoxidized natural rubber (NR) nanofibers [67], citric acid-modified starch nanoparticles [68], phthalic anhydride modified soy protein [69], etc., have been incorporated into protein matrix to exhibit reinforcing or toughening effect. These green protein nanocomposites could be used in agriculture and packaging fields, such as edible films, food packaging, and one-off packaging.

Ethylmethacrylate-based nanoparticles were incorporated in stearic acid modified casein-based polymers at 1% and 5% ratio to produce nanocomposites with improved thermal and mechanical properties, as compared to the neat polymers [70]. Seaweed extract can also be incorporated in gelatin and casein films as an active ingredient for the development of biofunctional films [71]. Transglutaminase (TGase) and different concentrations of nanocrystalline cellulose (NCC) (0–15% wt.) of whey protein concentrate (WPC) have been incorporated into the WPC matrix to prepare a WPC–NCC composite film with improved mechanical properties. The tensile strength of composite film increased from 1.3 MPa to 3.15 MPa as NCC increased to 15% wt. of WPC [72].

Antimicrobial zein films were prepared by incorporating cinnamon essential oil (CEO) and chitosan nanoparticles (CNPs) at 2% and 4% (w/w) amount. The films showed antimicrobial properties investigated against *E. coli* and *S.aureus* [73]. Nanofibril films were prepared by using konjacglucomannan (KGM) and zein via electrospinning technology. Further curcumin was incorporated into KGM/zein nanofibril films [74]. Antioxidant films were also prepared by incorporating curcumininto sodium caseinate (NaCas)-zein composite nanoparticles with an average particle size and ζ-potential around 131–184 nm and − 40 mV, respectively [75].

3.7 Protein-Based Blends as Packaging Materials

Blending with other polymers to improve the mechanical and water-resistant properties of protein materials is an effective method with high production efficiency. Both natural and synthetic polymers could be used as the blending components.

3.7.1 Protein/Natural Polymer Blends

Naturally occurring polymers, or biopolymers, cover a wide range of materials from plant- or animal-derived protein, polysaccharides like starch, cellulose, and chitosan, lignin, natural rubber, and so on. The compatibility between protein and natural polymers are often fine for the hydrogen bonding formed between the polar groups on them.

Cellulose is one of the most abundant biopolymers. The key to prepare cellulose/protein blend films is use of a suitable cosolvent. The reported cosolvents included NaOH/thiourea aqueous solution [76, 77], NaOH/urea solution [78], ionic liquid (IL) [79], etc. The blend films showed good oxygen and carbon dioxide barriers, which would be beneficial to the application in food packaging areas.

Some water-soluble biopolymers, such as agar [19], gelatine [80] etc., could be used to prepare protein blends by solution casting or thermomolding methods. The processing methods may significantly influence the structure of the blends and therefore pose an effect on the properties. Generally speaking, the casting films possessed more homogeneous interfaces compared with the molding films, leading to superior mechanical properties compared to molding films.

Some biopolymer derivatives, such as carboxymethyl cellulose (CMC) [81], carboxymethylkonjacglucomannan (CMKGM) [82], and hydroxyethyl cellulose (HEC) [83], could be used to prepare protein blends with high mechanical strength.

Most of the natural polymers blending with natural proteins are hydrophilic, and therefore the moisture resistance of the resulting blends could not be improved significantly. Natural rubber (NR) is a hydrophobic and elastic natural polymer. The blending of proteins with NR would highly improve the toughness and water resistance [22].

Zhuang et al. [84] reported the formation of casein/pectin blends and it has been found that pectin triggers the formation of casein microparticles during solution casting. Casein particles were formed surrounding the matrix of pectin. The incorporation of *Ascophyllum nodosum* extract on the physicochemical and antioxidant properties of gelatin and sodium caseinate films were proposed for the development of biofunctional films [71]. Blend films with three different pectin (P)/gelatine (G) ratios were prepared: P(1)/G(3.75), P(1)/G (1), and P(3.75)/G(1) (w/w), as well as a pectin film. All the blended films exhibited better UV/light barrier properties as compared to a synthetic polyethylene film [85].

Multilayer edible zein films can be fabricated with the stacking order of zein as the outer layer, hybrid zein/gelatine as the middle layer, and hydrophilic gelatine as the inner layer. Tea polyphenol (TP) with a concentration gradient was then loaded into the middle and inner layer of the optimized multilayer film to achieve a prolonged release property. The TP loaded multilayer film exhibited preservation effects in fresh fruit and could be practically useful in the fresh fruit and vegetable packaging industry [86]. A heat sealable corn zein (CZ) layer laminated on SPI film was fabricated that can be used to package olive oil condiments for use with instant noodles. The higher oxygen barrier property of the CZ/SPI bilayer films resulted in reduced oxidative rancidity of olive oil packaged in the CZ/SPI film when compared to olive oil packaged in NY/mLLDPE films [87].

3.7.2 Protein/Synthetic Biopolymeric Blends

Compatibility is the most concerned parameter influencing the mechanical properties, especially for the poor compatibility between hydrophobic synthetic biopolymers and hydrophilic proteins [89, 90, 93]. Certain compatibilization methods were adopted to improve the compatibility between the two phases.

The blending moieties reported are mainly biodegradable synthetic biopolymers, including polycaprolactone (PCL) [88], PLA [89, 90] and poly(butylenes succinate) (PBS). Isocyanate as well as maleated copolymers [91, 92] are the most used compatibilizers.

The composite fibrous films were developed from zein and PLA by incorporating carvacrol at three different concentrations (5%, 10%, and 20%) using electrospinning. Carvacrol-loaded electrospun zein and PLA fibers are able to preserve bread samples, indicating that they are good candidates for active food packaging applications to extend the shelf life of whole wheat bread [27].

3.8 Conclusions

Films based on wheat gluten, soy, zein, and caesin proteins have good film-forming and gas-permeability properties. Every protein polymer has its own material specific characteristics for packaging purposes because of differences in the inherent properties. It has been reported that the contact angle of water on zein film is 55.4°, which is much higher than that of a film of SPI film because zein is a hydrophobic polymer and SPI is a hydrophilic polymer. To increase specific material properties in protein polymers, nanoparticles and additives are incorporated. Cross-linking with citric acid and tannic acid leads to an increase in tensile strength of the protein films from 3.5 MPa to 13 MPa. The tensile strength of the zein/CS/TiO_2 films was reported as 28.33 ± 1.53 MPa at 0.15 wt % TiO_2 nanoparticles. Soy protein films incorporated with mandelic and benzilic acid show higher water resistance as compared to neat SPI film. WVP, the main criteria of packaging material, is influenced by the hydrophobic or hydrophilic nature of the proteinous material, by the presence of voids or cracks, and by steric hindrance and tortuosity in the structure. POSS, hydroxyapatite, eggshell nanopowder, and carbon nanotubes have been incorporated in protein film to improve the material properties of proteinous film that can be exploited for packaging applications. The bactericidal activity of protein films containing TiO_2 nanoparticles has been found to be significant against gram negative *E.coli* and gram positive *S.aureus* bacteria.

References

1. Mkandawire M, Aryee AN, Resurfacing and modernization of edible packaging material technology. *Curr. Opin. Food Sci.* 2018; **19**: 104–112.
2. Guilbert S, Gontard N, Gorris L. Prolongation of the shelf life of perishable products using biodegradable films and coatings. *LWT–Food Sci. Technol.* 1996; **29**: 10–17.
3. Mujica-Paz H, Gontard N. Oxygen and carbon dioxide permeability of wheat gluten film: effect of relative humidity and temperature. *J. Agric. Food. Chem.* 1997; **45**(10): 4101–4105.
4. Gontard N, Thibault R, Cuq B, Guilbert S. Influence of relative humidity and film composition on oxygen and carbon dioxide permeabilities of edible films. *J. Agric. Food. Chem.* 1996; **44**(4): 1064–1069.
5. Tian H, Fu X, Zheng M, Wang Y, Li Y, Xiang A, Zhong WH. Natural polypeptides treat pollution complex: moisture-resistant multi-functional protein nanofabrics for sustainable air filtration. *Nano Res.* 2018; **11**(8): 4265–4277.
6. Roy S, Weller CL, Gennadios A, Zeece MG, Testin RF. Physical and molecular properties of wheat gluten films cast from heated film-forming solutions. *J. Food Sci.* 1999; **64**(1): 57–60.
7. Bonnaillie L, Zhang H, Akkurt S, Yam K, Tomasula P. Casein films: the effects of formulation, environmental conditions and the addition of citric pectin on the structure and mechanical properties. *Polymers* 2014; **6**(7): 2018–2036.
8. Rani S, Kumar R. A review on material and antimicrobial properties of soy protein isolate film. *J. Polym. Environ.* 2019; **27**(8): 1–16.
9. Kumar R, Wu Q, Wang L, Zhang L. Proteinous nanomaterials in packaging applications. In: Mohanty AK, Misra M, editors. *Packaging Nanotechnology*. American Scientific Publisher; 2012.
10. Olabarrieta I, Cho SW, Gällstedt M, Sarasua JR, Johansson E, Hedenqvist MS. Aging properties of films of plasticized vital wheat gluten cast from acidic and basic solutions. *Biomacromolecules* 2006; **7**(5): 1657–1664.
11. Nagano T, Hirotsuka M, Mori H, Kohyama K, Nishinari K. Dynamic viscoelastic study on the gelation of 7S globulin from soybeans. *J. Agric. Food. Chem.* 1992; **40**(6): 941–944.

12. Utsumi S, Kinsella JE. Structure-function relationships in food proteins: subunit interactions in heat-induced gelation of 7S, 11S, and soy isolate proteins. *J. Agric. Food. Chem.* 1985; **33**(2): 297–303.
13. Badley RA, Atkinson D, Hauser H, Oldani D, Green JP, Stubbs JM. The structure, physical and chemical properties of the soy bean protein glycinin *Biochim. Biophys. Acta* 1975; **412**(2): 214–228.
14. Petruccelli S, Anon MC. Soy protein isolate components and their interactions. *J. Agric. Food. Chem.* 1995; **43**(7): 1762–1767.
15. Guilbert, S. 1986. Technology and application of edible protective films. In: Mathlouthi M, editor. *Food Packaging and Preservation: Theory and Practice.* New York: Elsevier Applied Science Publishers Ltd.
16. Subirade M, Kelly I, Guéguen J, Pézolet M. Molecular basis of film formation from a soybean protein: comparison between the conformation of glycinin in aqueous solution and in films. *Int. J. Biol. Macromol.* 1998; **23**(4): 241–249.
17. Kunte LA, Gennadios A, Cuppett SL, Hanna MA, Weller CL. Cast films from soy protein isolates and fractions. *Cereal Chem.* 1997; **74**(2): 115–118.
18. Krochta JM. Proteins as raw materials for films and coatings: definitions, current status, and opportunities. In: Gennadios A, editor, *Protein-Based Films and Coatings*, CRC Press; 2002.
19. Tian H, Xu G, Yang B, Guo G. Microstructure and mechanical properties of soy protein/agar blend films: effect of composition and processing methods. *J. Food Eng.* 2011; **107**(1): 21–26.
20. Tian H, Wang Y, Zhang L, Quan C, Zhang X. Improved flexibility and water resistance of soy protein thermoplastics containing waterborne polyurethane. *Ind. Crops Prod.* 2010; **32**(1): 13–20.
21. Qu, L, Chen G, Dong S, Huo Y, Yin Z, Li S, Chen Y. Improved mechanical and antimicrobial properties of zein/chitosan films by adding highly dispersed nano-TiO_2. *Ind. Crops Prod.* 2019; **130**: 450–458.
22. Tian H, Zhang L, Wu Q, Wang X, Chen Y. Creation of hydrophobic materials fabricated from soy protein and natural rubber: surface, interface, and properties. *Macromol. Mat. Eng.* 2010; **295**(5): 451–459.
23. Tian H. Processing and properties of soy protein/silica nanocomposites fabricated in situ synthesis. *J. Compos. Mat.* 2012; **46**(4): 427–435.
24. Xiang A, Guo G, Tian H. Fabrication and properties of acid treated carbon nanotubes reinforced soy protein nanocomposites. *J. Polym. Environ.* 2017; **25**(3): 519–525.
25. Tian H, Guo G, Fu X, Yao Y, Yuan L, Xiang A. Fabrication, properties and applications of soy-protein-based materials: a review. *Int. J. Biol. Macromol.* 2018; **120**: 475–490.
26. Chevalier E, Assezat G, Prochazka F, Oulahal N. Development and characterization of a novel edible extruded sheet based on different casein sources and influence of the glycerol concentration. *Food Hydrocolloids* 2018; **75**: 182–191.
27. Altan A, Aytac Z, Uyar T. Carvacrol loaded electrospun fibrous films from zein and poly (lactic acid) for active food packaging. *Food Hydrocolloids* 2018; **81**: 48–59.
28. Selling G, Sessa DJ. Multivalent carboxylic acids to modify the properties of zein. *Ind. Crops Prod.* 2007; **25**(1): 63–69.
29. Kumar R, Wang L, Zhang L. Structure and mechanical properties of soy protein materials plasticized by thiodiglycol. *J. Appl. Polym. Sci.* 2009; **111**(2): 970–977.
30. Picchio ML, Linck YG, Monti GA, Gugliotta LM, Minari RJ, Igarzabal CIA. Casein films crosslinked by tannic acid for food packaging applications. *Food Hydrocoll.* 2018; **84**: 424–434.
31. Pietsch VL, Werner R, Karbstein HP, Emin MA. High moisture extrusion of wheat gluten: relationship between process parameters, protein polymerization, and final product characteristics. *J. Food Eng.* 2019; **259**: 3–11.
32. Nataraj D, Sakkara S, Meenakshi HN, Reddy N. Properties and applications of citric acid crosslinked banana fibre-wheat gluten films. *Ind. Crops Prod.* 2018; **124**: 265–272.
33. Huo W, Wei D, Zhu W, Li Z, Jiang Y. High-elongation zein films for flexible packaging by synergistic plasticization: Preparation, structure and properties. *J. Cereal Sci.* 2018; **79**: 354–361.
34. Kumar R, Zhang. 2008. Water-induced hydrophobicity of soy protein materials containing 2, 2-diphenyl-2-hydroxyethanoic acid. *Biomacromolecules* 2008; **9**(9): 2430–2437.

35. Kumar R, Anandjiwala RD, Kumar A. Thermal and mechanical properties of mandelic acid-incorporated soy protein films. *J. Therm. Anal. Calorim.* 2016; **123**(2): 1273–1279.
36. Kumar R, Rani P, Kumar KD. Soy protein isolate film by incorporating mandelic acid as well as through fermentation mediated by *Bacillus Subtilis*. *J. Renew. Mat.* 2019; **7**(2): 103–115.
37. Echeverría I, Eisenberg P, Mauri AN. Nanocomposites films based on soy proteins and montmorillonite processed by casting. *J. Membrane Sci.* 2014; **449**: 15–26.
38. Kumar R, Yakubu MK, Anandjiwala RD. Biodegradation of flax fiber reinforced poly lactic acid. *Express Polym. Lett.* 2010; **4**(7): 423–430.
39. Guo G, Zhang C, Du Z, Zou W, Tian H, Xiang A, Li H. Structure and property of biodegradable soy protein isolate/PBAT blends. *Ind. Crops Prod.* 2015; **74**: 731–736.
40. Padrão J, Machado R, Casal M, Lanceros-Méndez S, Rodrigues LR, Dourado F, Sencadas V. Antibacterial performance of bovine lactoferrin-fish gelatine electrospun membranes. *Int. J. Biol. Macromol.* 2015; **81**: 608–614.
41. Kumar R, Anjum KN, Rani S, Sharma K, Tiwary KP, Kumar KD. Material properties of ZnS nanoparticles incorporated soy protein isolate biopolymeric film. *Plast. Rubber Compos.* 2019; **48**: 448–455.
42. Kumar R, Praveen R, Rani S, Sharma K, Tiwary KP, Kumar KD. ZnSe nanoparticles reinforced biopolymeric soy protein isolate film. *J. Renew. Mat.* 2019; **7**(8): 749–761.
43. Chen P, Tian H, Zhang L, Chen Y, Wang X, Du Y. Structure and properties of soy protein/alumina hydrate nanocomposites fabricated via in situ synthesis. *J. Biobased Mat. Bioenergy.* 2008; **2**(3): 248–257.
44. Ai F, Zheng H, Wei M, Huang J. Soy protein plastics reinforced and toughened by SiO_2 nanoparticles. *J. Appl. Polym. Sci.* 2007; **105**(3): 1597–1604.
45. Han Y, Wang L. Improved water barrier and mechanical properties of soy protein isolate films by incorporation of SiO_2 nanoparticles. *RSC Adv.* 2016; **6**(113): 112317–112324.
46. Xia C, Zhang S, Shi SQ, Cai L, Garcia AC, Rizvi HR, D'Souza NA. Property enhancement of soy protein isolate-based films by introducing POSS. *Int. J. Biol. Macromol.* 2016; **82**: 168–173.
47. Rahman MM, Netravali AN, Tiimob BJ, Apalangya V, Rangari VK. Bio-inspired "green" nanocomposite using hydroxyapatite synthesized from eggshell waste and soy protein. *J. Appl. Polym. Sci.* 2016; **133**(22): 43477.
48. Rahman MM, Netravali AN, Tiimob BJ, Rangari VK. Bioderived "green" composite from soy protein and eggshell nanopowder. *ACS Sustainable Chem. Eng.* 2014; **2**(10): 2329–2337.
49. Alexandre M, Dubois P. Polymer-layered silicate nanocomposites: preparation, properties and uses of a new class of materials. *Mat. Sci. Eng. R Rep.* 2000; **28**(1–2): 1–63.
50. Chen P, Zhang L. Interaction and properties of highly exfoliated soy protein/montmorillonite nanocomposites. *Biomacromolecules* 2006; **7**(6): 1700–1706.
51. Kumar P, Sandeep KP, Alavi S, Truong VD, Gorga RE. Preparation and characterization of bio-nanocomposite films based on soy protein isolate and montmorillonite using melt extrusion. *J. Food Eng.* 2010; **100**(3): 480–489.
52. Yu J, Cui G, Wei M, Huang J. Facile exfoliation of rectorite nanoplatelets in soy protein matrix and reinforced bionanocomposites thereof. *J. Appl. Polym. Sci.* 2007; **104**(5): 3367–3377.
53. Wang Z, Zhang N, Wang HY, Sui SY, Sun XX, Ma ZS. The effects of ultrasonic/microwave assisted treatment on the properties of soy protein isolate/titanium dioxide films. *LWT-Food Sci. Technol.* 2014; **57**: 548–555.
54. Zhao S, Yao J, Fei X, Shao Z, Chen X. An antimicrobial film by embedding in situ synthesized silver nanoparticles in soy protein isolate. *Mater. Lett.* 2013; **95**: 142–144.
55. Ma J, An W, Xu Q, Fan Q, Wang Y. Antibacterial casein-based ZnO nanocomposite coatings with improved water resistance crafted via double in situ route. *Prog. Org. Coat.* 2019; **134**: 40–47.
56. Alizadeh-Sani M, Khezerlou A, Ehsani A. Fabrication and characterization of the bionanocomposite film based on whey protein biopolymer loaded with TiO_2 nanoparticles, cellulose nanofibers and rosemary essential oil. *Ind. Crops Prod.* 2018; **124**: 300–315.

57. Spasojević L, Katona J, Bučko S, Savić SM, Petrović L, Budinčić JM, Sharipova A. Edible water barrier films prepared from aqueous dispersions of zein nanoparticles. *LWT-Food Sci. Technol.* 2019; **109**: 350–358.
58. Kadam DM, Thunga M, Srinivasan G, Wang S, Kessler MR, Grewell D, Lamsal B. Effect of TiO_2 nanoparticles on thermo-mechanical properties of cast zein protein films. *Food Packag. Shelf Life* 2017; **13**: 35–43.
59. Suganya P, Vaseeharan B, Vijayakumar S, Balan B, Govindarajan M, Alharbi NS, Benelli G. Biopolymer zein-coated gold nanoparticles: synthesis, antibacterial potential, toxicity and histopathological effects against the Zika virus vector Aedesaegypti. *J. Photoch. Photobio. B Bio.* 2017; **173**: 404–411.
60. Lu Y, Weng L, Zhang L. Morphology and properties of soy protein isolate thermoplastics reinforced with chitin whiskers. *Biomacromolecules* 2004; **5**(3), 1046–1051.
61. Wang Y, Cao X, Zhang L. Effects of cellulose whiskers on properties of soy protein thermoplastics. *Macromol. Biosci.* 2006; **6**(7): 524–531.
62. Zheng H, Ai F, Chang PR, Huang J, Dufresne A. Structure and properties of starch nanocrystal-reinforced soy protein plastics. *Polym. Compos.* 2009; **30**(4): 474–480.
63. Zhang S, Xia C, Dong Y, Yan Y, Li J, Shi SQ, Cai L. Soy protein isolate-based films reinforced by surface modified cellulose nanocrystal. *Ind. Crops Prod.* 2016; **80**: 207–213.
64. Ortiz CM, Salgado PR, Dufresne A, Mauri AN. Microfibrillated cellulose addition improved the physicochemical and bioactive properties of biodegradable films based on soy protein and clove essential oil. *Food Hydrocolloids* 2018; **79**: 416–427.
65. Huang X, Netravali A. Biodegradable green composites made using bamboo micro/nano-fibrils and chemically modified soy protein resin. *Compos. Sci. Technol.* 2009; **69**(7–8): 1009–1015.
66. Chen G, Liu H. Electrospun cellulose nanofiber reinforced soybean protein isolate composite film. *J. Appl. Polym. Sci.* 2008; **110**(2): 641–646.
67. Kim JR, Netravali AN. One-step toughening of soy protein based green resin using electrospunepoxidized natural rubber fibers. *ACS Sustainable Chem. Eng.* 2017; **5**(6): 4957–4968.
68. Tian H, Xu G. Processing and characterization of glycerol-plasticized soy protein plastics reinforced with citric acid-modified starch nanoparticles. *J. Polym. Environ.* 2011; **19**(3): 582–588.
69. Guo G, Zhang C, Du Z, Zou W, Xiang A, Li H. Processing and properties of phthalic anhydride modified soy protein/glycerol plasticized soy protein composite films. *J. Appl. Polym. Sci.* 2015; **132**(28): 42221.
70. Sophia A, Retna AM. Stearic acid modified casein based nanocomposites with improved mechanical and thermal properties. *Mater. Today Proc.* 2018; **5**(2): 6247–6257.
71. Kadam SU, Pankaj SK, Tiwari BK, Cullen PJ, O'Donnell CP. Development of biopolymer-based gelatin and casein films incorporating brown seaweed *Ascophyllumnodosum* extract. *Food Packag. Shelf life* 2015; **6**: 68–74.
72. Jiang S, Zhang T, Song Y, Qian F, Mu G. Mechanical properties of whey protein concentrate based film improved by the coexistence of nanocrystalline cellulose and transglutaminase. *Int. J. Biol. Macromol.* 2019; **126**, (1): 1266–1272
73. Vahedikia N, Garavand F, Tajeddin B, Cacciotti I, Jafari SM, Omidi T, Zahedi Z. Biodegradable zein film composites reinforced with chitosan nanoparticles and cinnamon essential oil: physical, mechanical, structural and antimicrobial attributes. *Colloids Surf. B* 2019; **177**: 25–32.
74. Wang L, Mu RJ, Li Y, Lin L, Lin Z, Pang J. Characterization and antibacterial activity evaluation of curcumin loaded konjacglucomannan and zeinnanofibril films. *LWT–Food Sci. Technol.* 2019; **113**: 108293.
75. Wang L, Xue J, Zhang Y. Preparation and characterization of curcumin loaded caseinate/zeinnanocomposite film using pH-driven method. *Ind. Crops Prod.* 2019; **130**: 71–80.
76. Chen Y, Zhang L. Blend membranes prepared from cellulose and soy protein isolate in NaOH/thiourea aqueous solution. *J. Appl. Polym. Sci.* 2004; **94**(2): 748–757.

77. Chen Y, Zhang L, Gu J, Liu J. Physical properties of microporous membranes prepared by hydrolyzing cellulose/soy protein blends. *J. Membrane Sci.* 2004; **241**(2): 393–402.
78. Yang Q, Lue A, Qi H, Sun Y, Zhang X, Zhang L. Properties and bioapplications of blended cellulose and corn protein films. *Macromol. Biosci.* 2009; **9**(9): 849–856.
79. Wu RL, Wang XL, Wang YZ, Bian XC, Li F. Cellulose/soy protein isolate blend films prepared via room-temperature ionic liquid. *Ind. Eng. Chem. Res.* 2009; **48**(15): 7132–7136.
80. Guerrero P, Stefani PM, Ruseckaite RA, De la Caba K. Functional properties of films based on soy protein isolate and gelatin processed by compression molding. *J. Food Eng.* 2011; **105**(1): 65–72.
81. Su JF, Huang Z, Yuan XY, Wang XY, Li M. Structure and properties of carboxymethyl cellulose/soy protein isolate blend edible films crosslinked by Maillard reactions. *Carbohydr. Polym.* 2010; **79**(1): 145–153.
82. Wang L, Xiao M, Dai S, Song J, Ni X, Fang Y, Jiang F. Interactions between carboxymethylkonjacglucomannan and soy protein isolate in blended films. *Carbohydr Polym.* 2014; **101**: 136–145.
83. Zhao Y, He M, Zhao L, Wang S, Li Y, Gan L, Chen Y. Epichlorohydrin-cross-linked hydroxyethyl cellulose/soy protein isolate composite films as biocompatible and biodegradable implants for tissue engineering. *ACS Appl. Mater. Interfaces* 2016; **8**(4): 2781–2795.
84. Zhuang Y, Sterr J, Kulozik U, Gebhardt R. Application of confocal Raman microscopy to investigate casein micro-particles in blend casein/pectin films. *Int. J. Biol. Macromol.* 2015; **74**: 44–48.
85. Sartori T, Feltre G, do Amaral Sobral PJ, da Cunha RL, Menegalli FC. "Properties of films produced from blends of pectin and gluten. *Food Packag. Shelf Life* 2018; **18**: 221–229.
86. Xia C, Wang W, Wang L, Liu H, Xiao J. Multilayer zein/gelatin films with tunable water barrier property and prolonged antioxidant activity. *Food Packag. Shelf Life* 2019; **19**: 76–85.
87. Cho SY, Lee SY, Rhee C. Edible oxygen barrier bilayer film pouches from corn zein and soy protein isolate for olive oil packaging. *LWT-Food Sci. Technol.* 2010; **43**(8): 1234–1239.
88. Zhong Z, Sun XS. Properties of soy protein isolate/polycaprolactone blends compatibilized by methylene diphenyldiisocyanate. *Polymer* 2001; **42**(16): 6961–6969.
89. Liu B, Jiang L, Liu H, Zhang J. Synergetic effect of dual compatibilizers on in situ formed poly (lactic acid)/soy protein composites. *Ind. Eng. Chem. Res.* 2010; **49**(14): 6399–6406.
90. Fang K, Wang B, Sheng K, Sun XS. Properties and morphology of poly (lactic acid)/soy protein isolate blends. *J. Appl. Polym. Sci.* 2009; **114**(2): 754–759.
91. Sasmal A, Sahoo D, Nanda R, Nayak P, Nayak PL, Mishra JK, Yoon JY. Biodegradable nanocomposites from maleated polycaprolactone/soy protein isolate blend with organoclay: preparation, characterization, and properties. *Polym. Compos.* 2009; **30**(6): 708–714.
92. Zhu R, Liu H, Zhang J. Compatibilizing effects of maleated poly (lactic acid) (PLA) on properties of PLA/Soy protein composites. *Ind. Eng. Chem. Res.* 2012; **51**(22): 7786–7792.
93. Ilyas RA, Sapuan SM, Atikah MSN, Asyraf MRM, Ayu Rafiqah S, Aisyah HA, Mohd Nurazzi N, Norrahim MNF. Effect of Hydrolysis Time on the Morphological, Physical, Chemical, and Thermal Behavior of Sugar Palm Nanocrystalline Cellulose (Arenga Pinnata (Wurmb.) Merr). *Textile Research Journal*. 2020. In print. 10.1177/0040517520932393

4

Chitin/Chitosan Based Films for Packaging Applications

J.M. Moura, B.S. Farias, T.R.S. Cadaval, and L.A.A. Pinto
School of Chemistry and Food, Federal University of Rio Grande–FURG, RS, Rio Grande, 96203–900, Brazil

4.1 Introduction

Great effort has been performed by various research groups in the formulation and development of biodegradable packaging [1–3]. The main reason for this attention is the fact that there are several concerns about the alarming decrease in natural resources regarding the production of petroleum-based synthetic polymers, in addition to the environmental problems caused by improper disposal of these non-biodegradable packaging [1, 4, 5]. Hence, natural polymers have been under extensive research to become suitable alternatives to synthetic products. The application of these biopolymers has increased significantly in various sectors [1, 5, 6].

Chitosan is a biopolymer, which has been researched for approximately three decades. This biopolymer has been considered an excellent film-forming agent and has gained growing interest, which enables a wide spectrum of applications in the food, biomedical, and chemistry industries [7]. However, as with other bio-based films, chitosan films lack important properties, such as mechanical and barrier properties, which are quality parameters required to meet industrial standards for packaging applications [5, 8]. Therefore, it is known that the physicochemical properties of chitosan films should be improved to enhance their quality and economic feasibility over synthetic materials [1, 9].

Chitosan has antimicrobial and antioxidant activities, which enhances its application in the food packaging industry [7, 10]. This polysaccharide also has the ability to interact with many compounds. Hence, a variety of other compounds has been used in conjunction with

Bio-Based Packaging: Material, Environmental and Economic Aspects, First Edition. Edited by S.M. Sapuan and R.A. Ilyas

chitosan to produce blend and composite films in order to overcome the issues mentioned previously and, in addition, to improve some biological properties, such as antimicrobial and antioxidant ones [11–15]. Chitosan can encapsulate different active agents and release them at a controlled rate, which can provide effective protection against bacteria and fungi that cause food spoilage [12, 16]. Another possibility is the introduction of hydrophobic substituents on the chitosan glycosidic groups, which may increase the overall hydrophobicity of the films [2]. In addition, this molecule type plays an important role in controlling the release of water-insoluble drugs or antimicrobial agents. Therefore, it is of great importance for many applications, not only in food packaging but also in the biomedical field [12, 16]. Furthermore, different types of oils and plant extracts can be used to enhance the antioxidant properties of chitosan films [1, 2]. The minimum concentration of the substances should be estimated to avoid significant modifications in film physicochemical properties, providing an effective protection for the packaged material [1].

The studies regarding chitosan-based films for packaging applications have been mainly conducted for food preservation. The production, processing, transport, and storage can make food products more susceptible to microbial spoilage and biochemical reactions, which can result in their deterioration. Therefore, the physicochemical and biological characteristics of chitosan-based films should be evaluated in order to preserve the nutritional and sensory properties of food [17, 18]. The mechanical properties, tensile strength (TS) and elongation, are important to protect the integrity of the product. Water vapor permeability (WVP) plays a key role in measuring the water transfer between the products and the surrounding environment. The WVP value can be directly related to the water activity of the food, controlling its shelf life [1, 12, 18]. The antimicrobial property plays a critical role in providing mechanisms against food-borne pathogens, which ensures food safety for a longer time [2, 19]. The antioxidant property has a profound effect on reducing the oxidation rate on food, such as enzymatic browning in fruits and vegetables and lipid peroxidation in meat [20, 21].

According to the web of science, about 7000 records (2014–2019) can be found considering chitosan-based packaging and chitosan-based films. Indeed, these manuscripts mainly address their elaboration, characterization, and possible applications. Hence, the current chapter aims to discuss and summarize the recent advances that were performed to improve the physicochemical and biological aspects of chitosan-based films in order to enhance their quality and overcome their drawbacks toward a feasible use in the packaging industry. Furthermore, this chapter also aims to explore future perspectives considering the possible packaging applications for chitosan-based films.

4.2 Chitin and Chitosan

Chitin is a polysaccharide with linear chains composed of 2-acetamide-2-deoxyβ-D-glucopyranose units, which are linked by glycosidic bonds $\beta(1\rightarrow4)$ [7]. Chitin is the second most abundant organic substance in the biosphere, surpassed only by cellulose, but chitin surpasses the latter in terms of replacement rate, which is twice as high as cellulose [22]. Chitin is found in the skeletal structure of invertebrates, such as arthropods, annelids, mollusks, and coelenterates, and in the cell walls of diatoms and some fungi [12]. Depending on the organism, chitin adopts different polymorphic structures called

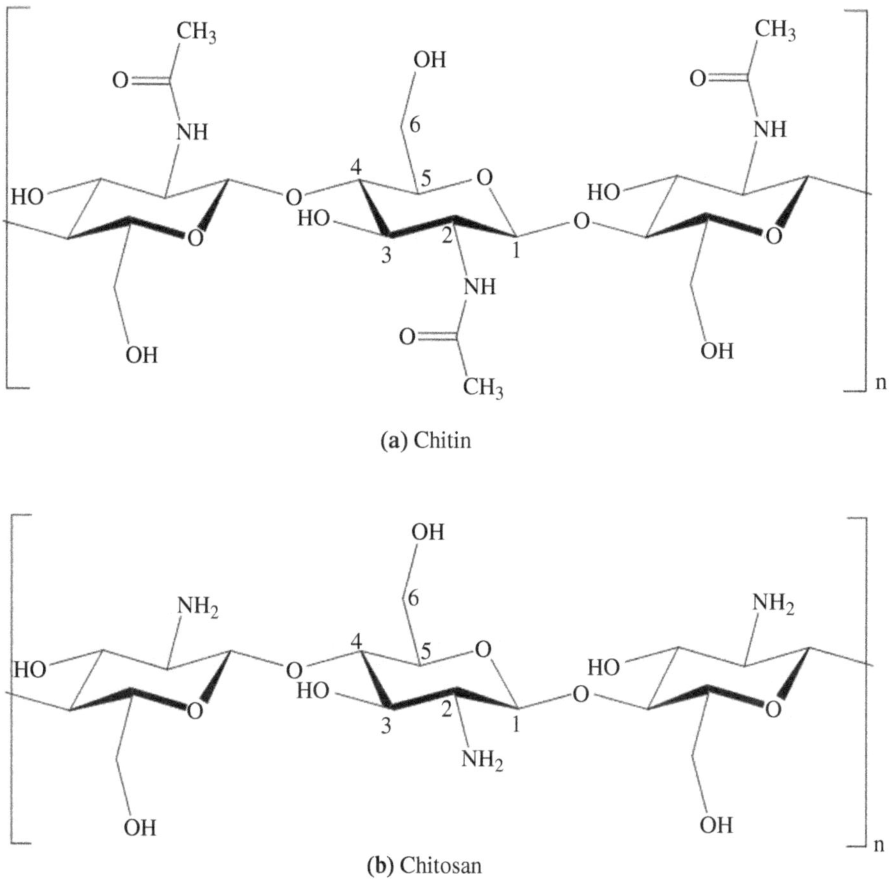

Figure 4.1 *Chemical structures: (a) chitin and (b) chitosan.*

α-, β-, and γ-chitin. The different polymorphic structures of chitin correspond to different arrangements in the solid state [23]. The α-chitin is found in rigid and resistant structures, such as the arthropod cuticle. This chitin is usually associated with proteins, inorganic materials, or both [22]. The β- and γ-chitin occur in flexible structures, but are also resistant. The α-chitin is the most abundant form and is considered the most stable, considering that the conversion of β- and γ-chitin in the first one is irreversible [22].

Chitosan is usually produced by alkaline deacetylation of α-chitin. Chitin and chitosan are important linear polysaccharides consisting, respectively, of 2-acetamide-2-deoxy-β-D-glucopyranose (GlcNAc) and 2-amino-2-deoxy-β-D-glucopyranose (GlcN) linked by β(1→4) [12, 24]. Figure 4.1 shows the molecular structures of chitin and chitosan.

Chitin samples contain a high content of GlcNAc, which is insoluble in water and common organic solvents. When the N-acetylation (defined as the average number of N–acetyl-D-glucosamine units per 100 monomers and expressed as a percentage) is lower than 50%, chitin becomes soluble in acidic aqueous solutions ($pH < 6.0$) and is called chitosan. Hence, the term chitosan represents a group of fully or partially deacetylated chitin. Chitosan solubilization can be attributed to protonation of the $-NH_2$ in the C-2 of D-glucosamine units in acid medium, which results in the conversion into a polyelectrolyte [25]. The polycationic character of chitosan in acid medium is due to its weak base characteristic; thus, its amino groups are easily protonated. Moreover, the hydroxyl groups of carbons 3 and 6 can also be deprotonated, which increases chitosan reactivity [12].

Chitin is an insoluble and chemically nonreactive material and thus it has much less application than chitosan [12, 25]. On the other hand, chitosan solubility allows a larger number of applications in almost all areas of human and health life (agriculture, medicine, industrial engineering, etc.). Biodegradability, biocompatibility, bioadhesivity, polycationic, antioxidant, antifungal, antibacterial, immunoadjuvant, antithrombogenic, anticholesteric, film-forming, moisturizing agent, adsorption, nontoxicity, and nonallergic are some of the basic properties of chitosan [3, 12, 16]. Deacetylation degree (DD), molecular weight (MW), and, consequently, the chitosan functional properties depend on the process used in the conversion of chitin into chitosan [24]. Therefore, chitosan characteristics and the film preparation parameters directly influence the physicochemical and biological properties, including thermal, barrier, antimicrobial, antioxidant, and mechanical properties [10, 16, 24]. Briefly, several characteristics must be considered depending on the final application of the film, such as the molecular weight and deacetylation degree of chitosan, acid type, acid concentration in chitosan solubilization, and the effect that other materials, which lead to the physicochemical and biological properties of the chitosan films.

4.3 Physicochemical and Biological Properties of Chitosan-Based Films

Chitosan-based films have a wide range of applications due to their good physicochemical and biological properties, which depend on several factors, as mentioned previously [3, 12]. For the composite films, several operating parameters play an important role in the miscibility and interaction among the compounds in the solution. These parameters comprise temperature, time, and stirring rate. Moreover, it is also important to evaluate whether there is a need to add other additives in order to enhance the solvation among molecules. Chitosan characteristics, as well as operating parameters, affect the physicochemical and biological properties, including TS, elongation, barrier properties, wettability, thermal stability, roughness, antimicrobial, and antioxidant activities [1, 2, 10]. These properties establish the application of chitosan-based films in a variety of fields, such as food packaging, adsorption studies, biomedical applications, and agriculture, as shown in Table 4.1.

4.3.1 Mechanical and Barrier Properties

Tensile strength (TS) and elongation (E) of chitosan-based films are directly related to chitosan molecular weight (MW) and deacetylation degree (DD). Moura et al. [24] studied 16 alkaline deacetylation reaction conditions, also produced films from these different chitosan types, and evaluated their characteristics regarding the mechanical and water vapor permeability properties (WVP). The authors reported that chitosan types with higher MW and lower DD showed suitable TS and E. This behavior was attributed to the increase of intermolecular interactions during film formation, considering that longer chains with small

Table 4.1 Some applications of chitosan-based films.

Solvent	Additive	Application	Reference
Acetic acid	Glycerol and proanthocyanidins	Antimicrobial and antioxidant packaging	Bi et al. [19]
Acetic acid	*Herba Lophatheri* extract	Antimicrobial and antioxidant packaging	Wang et al. [15]
Acetic acid	Xylan	Antimicrobial packaging	Kamdem et al. [14]
Acetic acid	Ion Vanadate	Dye adsorption	Rodrigues et al. [26]
Sulfuric acid	—	Cu^{2+} adsorption	Marques et al. [27]
Phosphoric acid	Glycerol	L- and D-Tyrosine adsorption	Takara et al. [28]
Acetic acid	Copaiba oil	Wound dressing	Debone et al. [29]
Acetic acid	Glycerol–honey	Wound dressing	Escárcega-Galaz et al. [30]
Acetic acid	Tannic acid	Wound dressing	Kaczmarek et al. [31]
Lactic acid Acetic acid Formic acid	—	Biopesticide	Hassan and Chang [32]
Acetic acid	Hydrogen peroxide and Tween 80	Antifungal agent for seed preservation	Silva-Castro et al. [13]
Hydrochloric acid	Sodium hydroxide	Antioxidant agent for strawberry production	Rahman et al. [33]

branches resulted in a higher packing efficiency. Moreover, the higher MW of chitosan showed the lower WVP, because as packing efficiency increased, the intermolecular spaces reduced.

Besides the MW and DD, other factors influence the mechanical and barrier properties of chitosan-based films. Most films are developed by chitosan dissolution into the acetic acid solution, although other acids could be used alone or in combination with acetic acid, aiming at increasing the degree of cross-linking among chitosan chains. Uddin et al. [11] produced chitosan films, with cellulose nanofibers and boric acid, like flame retardants. The authors discussed the fact that boric acid promoted an increase in TS and a decrease in E, which was due to the formation of bonds between boron and nitrogen through the chitosan structure. Moreover, the interactions between hydroxyl groups and boron resulted in several cross-links, which promoted less flexibility among chitosan chains.

Other researchers using different acids reported the same behavior. Nataraj et al. [12] produced chitosan films with acetic acid and different concentrations of citric acid as cross-linkers (1%, 3%, and 5%). The results pointed out that citric acid of 3% increased TS and decreased polymer flexibility and, consequently, reduced the E. This behavior could be related to the cross-linking network that was established among chitosan molecules. Regarding water sorption, cross-linking has not shown significant results. Thus, the authors treated the films with an alkaline solution in order to reduce solubility, which is required for application in food packaging. The alkali treatment was effective and significantly decreased the films hydrophilicity, but it also reduced TS and E compared to the control film. During treatment with NaOH, it is likely that some bonds between the amino and hydroxyl groups of chitosan break due to the chitosan deprotonation. Figure 4.2 shows the

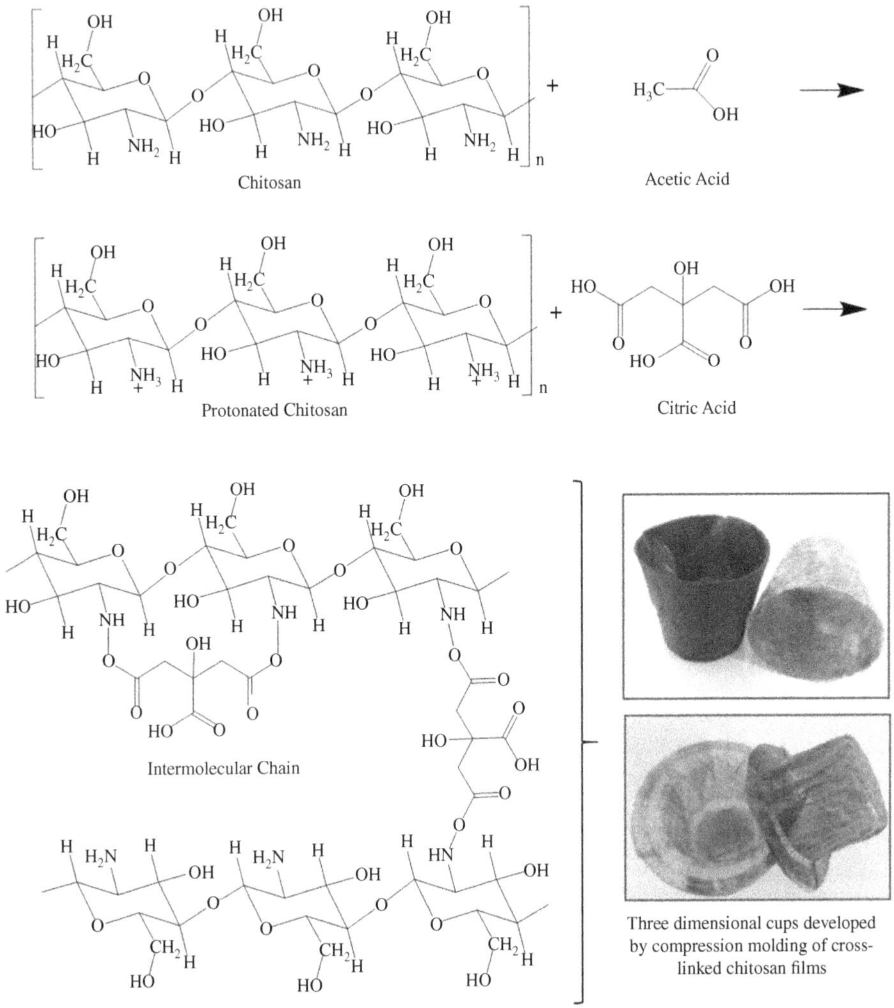

Figure 4.2 Scheme of the cross-linking reaction between chitosan and citric acid.

scheme of chitosan cross-linking with citric acid, as well as images of packaging produced after alkali treatment.

Different types of polyols can be used to produce chitosan-based films due to the low flexibility of chitosan films produced with acid cross-linking agents. However, the addition of polyols usually leads to a reduction in TS and an increase in WVP. The polyols interact with the hydroxyl groups of chitosan rather than the NH_3^+, as shown in Figure 4.3, which results in the weakening of hydrogen interactions among chitosan molecules. Ma et al. [7] investigated the influence of different polyols, such as glycerol, xylitol, and maltitol. The authors reported that the addition of polyols promoted a decrease in TS and an improvement in ductility.

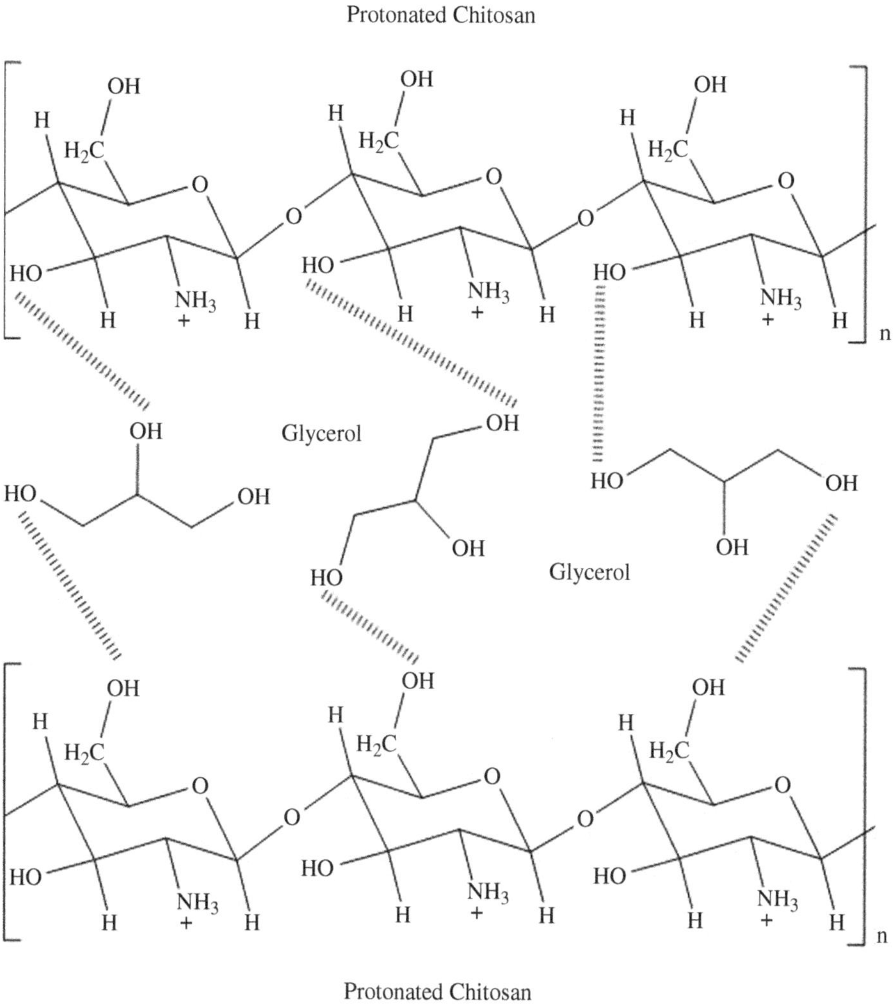

Figure 4.3 *Scheme of interactions between chitosan and glycerol.*

While polyols improve elongation, they lead to films that are more permeable, hygroscopic, and soluble due to the increased amount of hydroxyl groups in the film and the spacing of these structures among the polymer chains. Therefore, the introduction of animal and vegetable oils have attracted considerable attention in the elaboration of composite films, which can introduce hydrophobic agents that do not compromise the good mechanical properties promoted by the addition of polyols. Figure 4.4 shows the scanning electron microscopy (SEM) images of the chitosan film surface with olive oil and animal fat. This figure shows that the film surface with olive oil (with a higher amount of unsaturated fatty acids) is more homogeneous and the oil capsules have regular sizes and a spherical shape, whereas the film surface with animal fat (with a higher amount of saturated fatty acids) is more heterogeneous and the oil capsules are quite irregular.

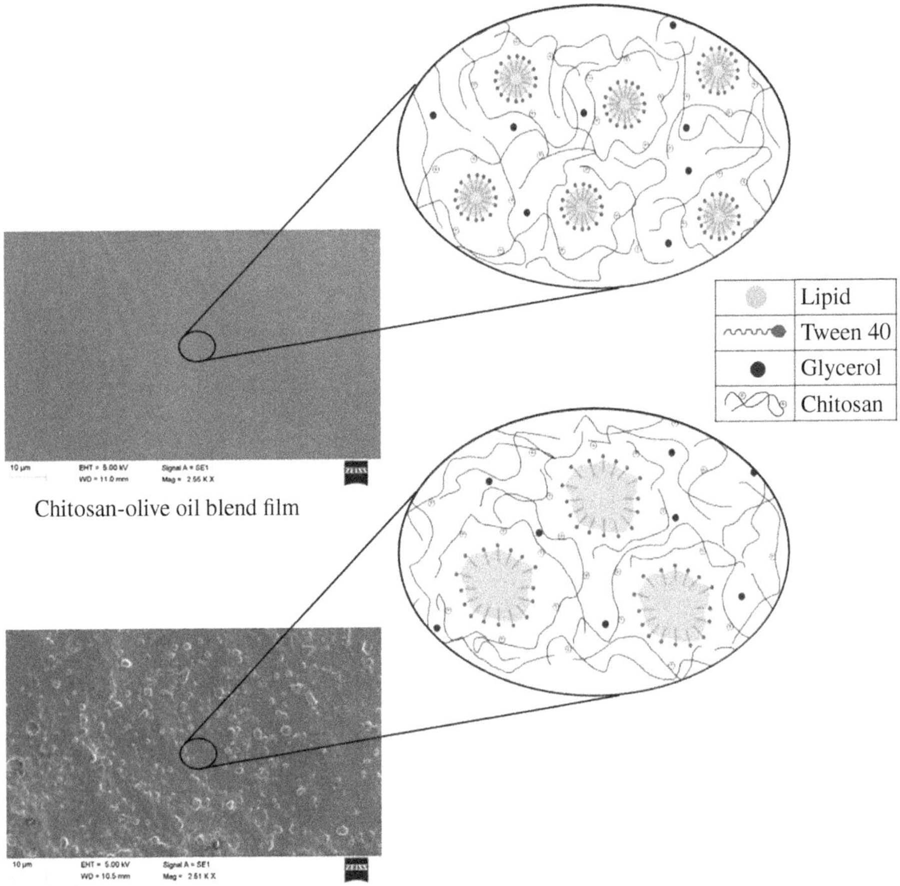

Figure 4.4 *Interactions between chitosan and oils with different degrees of unsaturation (olive oil and animal fat).*

Akyuz et al. [2] studied the effect of oil addition with different degrees of unsaturation (olive, corn, and sunflower oils, butter and animal fats) and verified their influence on mechanical and permeation properties in films composed of chitosan, glycerol, and Tween 40. The olive oil showed twice higher TS than the animal fat, but nevertheless the E values for both films were equal. This is due to the differential miscibility between the lipids used in the preparation of the solution. According to Pereira et al. [34], in the development of nanocapsules or microcapsules with bioactive compounds partial coalescence of the oil within the solution can occur. The coalescence was favored with a higher degree of lipid saturation. Regarding hydrophobicity, the film with olive oil resulted in a larger contact angle than the film with animal fat. Therefore, chitosan film with olive oil was shown to be less permeable to water. Thus, the more homogeneous the distribution of microcapsules or nanocapsules within the films, the better are their mechanical and permeability properties.

The antioxidants are other additives widely used in chitosan-based films. Souza et al. [1] tested the incorporation of natural antioxidants into the chitosan matrix. The control film was prepared by dissolving the chitosan powder in acetic acid solution, and the glycerol was added as plasticizer. The authors studied five essential oils (EOs) (ginger, rosemary, sage, tea tree, and thyme EO) and six different hydroalcoholic extracts (HAE) (ginger, rosemary, sage, black tea, green tea, and kenaf leaves HAE). The results indicated that the TS of the films with EO were slightly higher than the control film. This behavior could be related to the interactions between the hydroxyl groups of the oils and chitosan amides, which resulted in significant changes in mechanical properties, while the films with HAE had not shown differences in mechanical properties.

Valizadeh et al. [35] investigated the effect of cinnamon essential oil (CEO), oleic acid (OA), and glutaraldehyde (GL) on chitosan-carboxymethyl cellulose (chitosan-CMC) films. The cross-linking of chitosan-CMC films by GL increased approximately three times the TS compared to pure chitosan-CMC film. In fact, the covalent bonds developed after the cross-linking could have enhanced the cohesive forces (such as hydrogen bonds and Van der Waals forces) within the matrix. Nevertheless, the incorporation of CEO in the composite films decreased the TS, which could be an indication that some CEO compounds weakened the bond (between aldehyde groups from GL and functional groups from chitosan-CMC) in the cross-linked film. The addition of OA decreased, even more, the TS compared with the addition of CEO. This bond weakening occurred due to the incompatibility between OA and the hydrophilic chains from the polymers. In addition, the E increased significantly by incorporating CEO and OA, which was attributed to enhanced discontinuities within the polysaccharide structure. The WVP of chitosan-CMC films decreased after the incorporation of GL due to the formation of cross-linked networks. Despite the hydrophobic characteristics, the inclusion of CEO and AO significantly increased the WVP of the films due to the changes in their microstructures. The presence of micropores can increase the diffusion of water molecules throughout the film. The film porosity and the increase in WVP could be associated with covalent interactions (glycosidic bonds) between the essential oil and/or OA constituents with the biopolymer chains.

Sani et al. [36] developed biodegradable films for the release of antimicrobial agents. The antibacterial film was produced using chitosan with zinc oxide (ZnO) nanoparticles and Melissa essential oil. The results revealed increases in TS and in E due to the addition of ZnO nanoparticles and Melissa essential oil into the polymer structure. The reason for this was the uniform dispersion of ZnO nanoparticles through the chitosan polymer chains, which led to a strong interaction between chitosan and ZnO by ionic bonding. Furthermore, the addition of ZnO nanoparticles and Melissa essential oil decreased WVP. The justification for this was the strong intermolecular interaction among Melissa essential oil, chitosan, and ZnO nanoparticles, which resulted in the reduction of intermolecular distances and the formation of more compact films. In addition, the hydrophobic compounds present in Melissa essential oil also favored the reduction of WVP.

Both mechanical and permeation properties are very important in the chitosan-based film manufacturing. Therefore, it is reasonable to argue that the influence of each component should be carefully evaluated, in addition to the types of interactions between these compounds and chitosan.

4.3.2 Antimicrobial Properties

Chitosan-based films have antimicrobial properties, which allow a wide range of applications, especially in the food industry [2]. Chitosan-based coatings have been developed mainly in the preservation of fruits and vegetables and, to a lesser extent, in the pharmaceutical industry. Akyuz et al. [2] investigated the antimicrobial activity of chitosan films with the addition of different oils and animal fat (olive, corn and sunflower oils, butter and animal fats) against human pathogenic bacteria (*Escherichia coli, Salmonella typhmurium, Proteus microbilis, Proteus vulgaris, Peseudomonas aeruginosa, Enterobacter aerogenes, Staphylococcus aureus, Streptococcus mutans, and Bacillus thuringiensis*). Interestingly, the antimicrobial activities of all chitosan films tested were greater than the control (solvent only). This may have occurred due to the entrapment of oily compounds into chitosan film, which improved film stability and led to an antimicrobial activity efficiency.

Bi et al. [19] prepared pure chitosan films and chitosan films with the addition of proanthocyanidins (PA). The authors evaluated the antimicrobial activity against *E. coli, Salmonella, S. aureus, and Listeria monocytogenes*. The authors observed that both films showed antimicrobial activity. However, the control film (pure chitosan) showed a lower inhibition zone, because the antimicrobial activity was caused by an increase in the cellular permeability from the interactions between chitosan and bacterial cell membrane. The chitosan films with the addition of PA showed a higher inhibition zone. This fact was related to the characteristic O-heterocyclic structure in PA, which can bind to bacterial cell wall polymers and, subsequently, inhibit cell wall synthesis. Moreover, PA was capable of forming complexes with metals (e.g. iron) through their orthodiphenol groups, which reduced the bioavailability of mineral micronutrients that are essential for microbial growth. Furthermore, all films showed higher antimicrobial activity against Gram-positive bacteria compared to Gram-negative bacteria, which is mainly due to the different cell wall structures of these bacteria.

Indumathi et al. [37] used chitosan/cellulose acetate phthalate (CAP) films with ZnO nanoparticles for the preservation of black grapefruits. According to the authors, the films extended the shelf life of black grapefruits by up to nine days and showed biodegradability ranging from 30% to 50% in 28 days. Gómez–Estaca et al. [38] produced chitosan films with demineralized chitin extract (SME), which was used to package raw salmon as a ready-to-eat and calcium-enriched product. According to the authors, SME did not show toxicity and can be considered a functional food because it has a high content of calcium and fatty acids. Chitosan films with SME showed good performance as packaging material for raw salmon pieces. The packaged fish improved calcium and magnesium contents and reduced the saturated fatty acid intake. In addition, the presence of SME on chitosan film increased its antimicrobial activity and prolonged the shelf life of salmon.

Jaiswal et al. [39] produced carrageenan (Carr)/chitosan films with sulfur nanoparticles (SNPs) and grapefruit seed extract (GSE) for wound healing. The *in vivo* wound healing effect of films was tested using artificially injured male Sprague–Dawley rats. The composite film showed antibacterial activity strong enough to destroy *Staphylococcus epidermis* and *E. coli* in three hours of incubation and showed high biocompatibility against mouse fibroblasts (L929 cell lines). Furthermore, the film showed an excellent healing effect (1.3% wound area after two weeks) compared to the control group (31% wound area after two weeks) through the *in vivo* wound healing test.

The mentioned researches revealed the versatility of chitosan as a polymeric matrix in film production, especially for the food and pharmaceutical industries, and also the good compatibility of chitosan with antimicrobial agents. In addition to its antifungal and antibacterial characteristics, it may prevent food-borne pathogens and wound infections.

4.3.3 Antioxidant Characteristics

Chitosan films can also be used as antioxidant packaging with a controlled release property to prevent food oxidation. Benbettaïeb et al. [40] studied the release kinetics of some natural antioxidants (ferulic acid, caffeic acid, and tyrosol) from edible fish gelatin/chitosan films immersed in 96% ethanol (a food simulant). According to the authors, the release of tyrosol from films was much faster than the release of ferulic acid, caffeic acid, and a mixture of ferulic and caffeic acids. Tyrosol release was achieved within three hours, while the release of other antioxidants required up to 25 days (600 hours). The tyrosol diffusion coefficient was 170 times higher than that observed for other active compounds, although it showed the lowest antioxidant activity. Caffeic acid and the mixture of ferulic and caffeic acids showed a lower release percentage (20% and 40%, respectively) and antioxidant activity of 90% at equilibrium. On the other hand, the films with ferulic acid showed 82% antioxidant activity and an intermediate release percentage (51%). The high retentions of ferulic and caffeic acid can be explained by the greater ability of these antioxidants to establish charge–charge interactions with the polymer network (and fewer interactions with 96% ethanol). This is related to the negative groups (carboxylate groups) of the phenolic acids and the positive groups of fish gelatin and/or chitosan (ammonium groups). Chitosan may interact with the phenolic ring of ferulic or caffeic acid, mainly by weak interactions, such as dipole–dipole, dipole–charge, and hydrogen bonding. This study suggests that the release rate and antioxidant activity of the films depend on the antioxidant structure and the chemical interactions they establish with the polymer matrix.

Kurek et al. [41] produced chitosan/CMC films with blueberry and red grape extracts. The blueberry extract presented a higher total phenolic content, but the films showed a lower antioxidant activity in relation to the red grape skin extract. Crizel et al. [42] evaluated the antioxidant activity of chitosan-based films with the addition of olive pomace flour and olive pomace microparticles against nut oxidation. Both flour and microparticles increased the antioxidant activity of the films. They were also effective as protective packaging against nut oxidation for 31 days.

Lee et al. [43] developed chitosan-based films with clove essential oil (CEO) in addition to halloysite nanotubes (HNTs) to stabilize oil droplets. The antioxidant activity of the films was verified by the total phenolic content. In migration studies, the CEO release from the film was faster in 50% ethanol (simulator for oil-in-water emulsions and alcoholic foods) than in 10% and 95% ethanol, which are simulators for aqueous and fatty foods, respectively.

Qin et al. [44] developed active and intelligent packaging composed of chitosan, silver nanoparticles (AgNPs), and purple corn extract with a high amount of anthocyanins (PCE). The authors reported that chitosan films with AgNPs and PCE promoted remarkable antioxidant and antimicrobial properties due to the synergistic effect between PCE and AgNPs. Moreover, chitosan/AgNPs/PCE films could act as intelligent pH indicators by changing the film coloration at different pH buffers due to the presence of anthocyanins. This pH-sensitive character could be related to the structural transformation of the anthocyanins in PCE. The chitosan/PCE film showed a purple color at pH 3–6 and a blue color at pH 7–10, which corresponded to the carbinol pseudobase and quinoidal forms of anthocyanins, respectively. Furthermore, the color alteration of chitosan/AgNPs/PCE films was slightly different from chitosan/PCE films, which could be caused by the interactions between AgNPs and PCE, and, consequently, structural transformation of anthocyanins. The compatibility between filler and biopolymer is important, as it would affect the mechanical and barrier properties of films [44, 45].

4.4 Conclusion and Future Perspectives

This chapter considered the application and compatibility of chitosan with various materials in the preparation of active and intelligent packaging from biodegradable films. The major advantages were discussed concerning the production and use of chitosan-based films, in place of packaging from non-renewable sources, which causes environmental damage. The studies mentioned in this chapter pointed out that several factors influence the physicochemical and biological properties of chitosan-based films. The molecular weight and deacetylation degree of chitosan are very important characteristics because they directly affect the interaction between chitosan with other materials, such as biopolymers, oils and/or fats, herbal extracts, metals, metal oxides, carbon additives, antifungals, antibacterials, and antioxidants.

Researches have shown that cross-linking agents increase the TS of chitosan-based film but reduce its elongation. In addition, the polyols enhance the elongation of the chitosan-based film but reduce its TS and increase its WVP. Moreover, the researches have indicated that oils and/or fats with higher amounts of unsaturated fatty acids increase TS and reduce WVP of the film. However, regarding the use of essential oils, it is reasonable to discuss that there is an inaccuracy in the results considering the reports related to the physicochemical properties of the chitosan-based films. This lack of precision could be due to the range of molecules that can be attached to the chitosan structure in addition to the several chemical interactions that can occur between them. Thus, further research should be carried out to elucidate and overcome these difficulties. Furthermore, regarding the antimicrobial and antioxidant properties, the researches have demonstrated that the inclusion of different molecules could increase these properties by adding a new mechanism to the pre-existing ones from chitosan.

The number of research papers regarding the production of chitosan-based films is very extensive due to the versatility and outstanding properties that chitosan has in film and packaging design. Therefore, further studies should be directed toward new sources of chitin, new methods of chitin-to-chitosan conversion, and new film processing protocols involving economical and natural compounds. Moreover, the addition of new molecules should also be investigated to allow chitosan-based films to compete with commercially available packaging materials.

References

1. Souza, V.G.L., Fernando, A.L., Pires, J.R.A., et al. 2017. "Physical properties of chitosan films incorporated with natural antioxidants." *Ind. Crop. Prod.* **107**: 565–572. doi:10.1016/j.indcrop.2017.04.056
2. Akyuz, L., Kaya, M., Ilk, S., et al. 2018. "Effect of different animal fat and plant oil additives on physicochemical, mechanical, antimicrobial and antioxidant properties of chitosan films." *Int. J. Biol. Macromol.* **111**: 475–484. doi:10.1016/j.ijbiomac.2018.01.045
3. Dhall, R. K., Alam, M. S. 2019. "Biodegradable Packaging." In: *Encyclopedia of Renewable and Sustainable Materials*, 1–18. Cham: Elsevier. doi: 10.1016/b978–0–12–803581–8.11516–4
4. Lebreton, L., Andrady, A. 2019. "Future scenarios of global plastic waste generation and disposal." *Palgrave Commun.* **5**: 1–11. doi: 10.1057/s41599–018–0212–7
5. Rahman, R., Sood, M., Gupta, N., et al. 2019. "Bioplastics for food packaging: a review." *Int. J. Curr. Microbiol. App. Sci.* **8**: 2311–2321. doi: 10.20546/ijcmas.2019.803.274
6. John, G., Nagarajan, S., Vemula, P.K., et al. 2019. "Natural monomers: a mine for functional and sustainable materials – occurrence, chemical modification and polymerization." *Prog. Polym. Sci.* **92**: 158–209. doi:10.1016/j.progpolymsci.2019.02.008
7. Ma, X., Qiao, C., Wang, X., et al. 2019. "Structural characterization and properties of polyols plasticized chitosan films." *Int. J. Biol. Macromol.* **135**: 240–245. doi:10.1016/j.ijbiomac.2019.05.158
8. Thakhiew, W., Champahom, M, Devahastin, S., et al. 2015. "Improvement of mechanical properties of chitosan-based films via physical treatment of film-forming solution." *J. Food Eng.* **158**: 66–72. doi:10.1016/j.jfoodeng.2015.02.027
9. Mujtaba, M., Morsi, R.E., Kerch, G., et al. 2018. "Current advancements in chitosan–based film production for food technology; A review." *Int. J. Biol. Macromol.* **121**: 889–904. doi:10.1016/j.ijbiomac.2018.10.109
10. Huang, T., Qian, Y., Wei, J., et al., 2019. "Polymeric antimicrobial food packaging and its applications." *Polymers* **11**: 1–18. doi:10.3390/polym11030560
11. Uddin, K.M.A., Ago, M., Rojas, O.J. 2017. "Hybrid films of chitosan, cellulose nanofibrils and boric acid: flame retardancy, optical and thermo–mechanical properties." *Carbohydr. Polym.* **177**: 13–21. doi:10.1016/j.carbpol.2017.08.116
12. Nataraj, D., Sakkara, S., Meghwal, M., et al. 2018. "Crosslinked chitosan films with controllable properties for commercial applications." *Int. J. Biol. Macromol.* **120**: 1256–1264. doi:10.1016/j.ijbiomac.2018.08.187
13. Silva-Castro, I., Diez, J.J., Martín-Ramos, P., et al. 2018. "Application of bioactive coatings based on chitosan and propolis for pinus spp. protection against *Fusarium circinatum*." *Forests* **9**: 1–14. doi:10.3390/f9110685
14. Kamdem, D.P., Shen, Z., Nabinejad, O. 2019. "Development of biodegradable composite chitosan-based films incorporated with xylan and carvacrol for food packaging application." *Food Packag. Shelf Life* **21**: 100344. doi:10.1016/j.fpsl.2019.100344
15. Wang, L., Guo, H., Wang, J., et al. 2019. "Effect of Herba Lophatheri extract on the physicochemical properties and biological activities of chitosan film." *Int. J. Biol. Macromol.* **133**: 51–57. doi:10.1016/j.ijbiomac.2019.04.067
16. Gutiérrez, T.J. 2017. "Chitosan applications for the food industry." In: Ahmed, S., Ikram, S. (eds) *Chitosan: Derivatives, Composites and Applications*, 183–232. Publisher: Wiley–Scrivener Publisher. doi:10.1002/9781119364849.ch8

17. Han, J.W., Garcia, L.R., Qian, J.P., et al. 2018. "Food packaging: a comprehensive review and future trends." *Compr. Rev. Food Sci. Food Saf.* **17**: 860–877. doi:10.1111/1541-4337.12343
18. Wang, H., Qiang, J., Ding, F. 2018. "Emerging chitosan-based films for food packaging applications." *J. Agric. Food Chem.* **66**: 395–413. doi:10.1021/acs.jafc.7b04528
19. Bi, F., Zhang, X., Bai, R., et al. 2019. "Preparation and characterization of antioxidant and antimicrobial packaging films based on chitosan and proanthocyanidins." *Int. J. Biol. Macromol.* **134**: 11–19. doi:10.1016/j.ijbiomac.2019.05.042
20. Liu, J, Liu, S, Zhang, X, et al. 2019. "Effect of gallic acid grafted chitosan film packaging on the postharvest quality of white button mushroom (*Agaricus bisporus*)." *Postharvest Biol. Technol.* **147**: 39–47. doi: 10.1016/j.postharvbio.2018.09.004
21. Vieira, B.B., Mafra, J.F., Bispo, A.S.R., et al. 2019. "Combination of chitosan coating and clove essential oil reduces lipid oxidation and microbial growth in frozen stored tambaqui (*Colossoma macropomum*) fillets." *LWT Food Sci. Technol.* **116**: 108546. doi:10.1016/j.lwt.2019.108546
22. Deringer, V.L., Englert, U., Dronskowski, R. 2016. "Nature, strength, and cooperativity of the hydrogen–bonding network in α-chitin." *Biomacromolecules* **17**: 996–1003. doi:10.1021/acs.biomac.5b01653
23. Pighinelli, L., Broquá, J., Zanin, B.G., et al. 2019. "Methods of chitin production a short review." *Am. J. Biomed. Sci. Res.* **1**: 307–314. doi: 10.34297/AJBSR.2019.03.000682
24. Moura, J.M., Farias, B.S., Rodrigues, D.A.S., et al. "Preparation of chitosan with different characteristics and its application for biofilms production." *J. Polym. Environ.* **23**: 470–477. doi: 10.1007/s10924–015–0730–y
25. Roy, J. C., Salaün, F., Giraud, S., et al. 2017. "Solubility of chitin: solvents, solution behaviors and their related mechanisms." In: Xu, Z. (ed) *Solubility of Polysaccharides*, 109–127. IntechOpen. doi:10.5772/intechopen.71385
26. Rodrigues, D.A.S., Moura, J.M., Dotto, G.L., et al. 2018. "Preparation, characterization and dye adsorption/reuse of chitosan–vanadate films." *J. Polym. Environ.* **26**: 2917–2924. doi: 10.1007/s10924–017–1171–6
27. Marques, J.S., Pereira, M.R., Sotto, A., et al. 2018. "Removal of aqueous copper (II) by using crosslinked chitosan films." *React. Funct. Polym.* **134**: 31–39. doi:10.1016/j.reactfunctpolym.2018.10.009
28. Takara, E.A., Vega–Hissi, E.G., Garro–Martinez, J.C., et al. 2019. "About endothermic sorption of tyrosine on chitosan films." *Carbohydr. Polym.* **206**: 57–64. doi:10.1016/j.carbpol.2018.10.102
29. Debone, H.S., Lopes, P.S., Severino, P., et al. 2019. "Chitosan/copaiba oleoresin films for would dressing application." *Int. J. Pharm.* **555**: 146–152. doi:10.1016/j.ijpharm.2018.11.054
30. Escárcega-Galaz, A.A., Sánchez-Machado, D.I., López-Cervantes, J., et al. 2018. "Mechanical, structural and physical aspects of chitosan–based films as antimicrobial dressings." *Int. J. Biol. Macromol.* **116**: 472–481. doi:10.1016/j.ijbiomac.2018.04.149
31. Kaczmarek, B., Nadolna, K., Owczarek, A., et al. 2019. "The characterization of thin films based on chitosan and tannic acid mixture for potential applications as wound dressings." *Polym. Test.* **78**: 106007. doi:10.1016/j.polymertesting.2019.106007
32. Hassan, O., Chang, T. 2017. "Chitosan for eco–friendly control of plant disease." *Asian J. Plant Pathol.* **11**: 53–70. doi: 10.3923/ajppaj.2017.53.70
33. Rahman, M., Mukta, J.A., Sabir, A.A., et al. 2018. "Chitosan biopolymer promotes yield and stimulates accumulation of antioxidants in strawberry fruit." *PLoS One* **13**: 1–14. doi:10.1371/journal.pone.0203769
34. Pereira, L.J.B., Garcia–Rojas, E.E. 2015. "Multiple emulsions: formation and application in microencapsulation of bioactive components." *Ciência Rural* **45**, 155–162. doi: 10.1590/0103–8478cr20140315
35. Valizadeh, S., Naseri, M., Babaei, S., et al. 2019. "Development of bioactive composite films from chitosan and carboxymethyl cellulose using glutaraldehyde, cinnamon essential oil and oleic acid." *Int. J. Biol. Macromol.* **134**: 604–612. doi:10.1016/j.ijbiomac.2019.05.071

36. Sani, I. K., Pirsa, S., Tağı, Ş. 2019. "Preparation of chitosan/zinc oxide/Melissa officinalis essential oil nano–composite film and evaluation of physical, mechanical and antimicrobial properties by response surface method." *Polym. Test.* **79**: 106004. doi:10.1016/j.polymertesting.2019.106004
37. Indumathi, M.P., Sarojini, K.S., Rajarajeswari, G.R. 2019. "Antimicrobial and biodegradable chitosan/cellulose acetate phthalate/ZnO nano composite films with optimal oxygen permeability and hydrophobicity for extending the shelf life of black grape fruits." *Int. J. Biol. Macromol.* **132**: 1112–1120. doi:10.1016/j.ijbiomac.2019.03.171
38. Gómez-Estaca, J., Alemán, A., López-Caballero, M.E., et al. 2018. "Bioaccessibility and antimicrobial properties of a shrimp demineralization extract blended with chitosan as wrapping material in ready–to–eat raw salmon." *Food Chem.* **276**: 342–349. doi:10.1016/j.foodchem.2018.10.031
39. Jaiswal, L., Shankar, S., Rhim, J.W. 2019. "Carrageenan–based functional hydrogel film reinforced with sulfur nanoparticles and grapefruit seed extract for wound healing application." *Carbohydr. Polym.* **224**: 115191. doi:10.1016/j.carbpol.2019.115191
40. Benbettaïeb, N., Tanner, C., Cayot, P., et al. 2018. "Impact of functional properties and release kinetics on antioxidant activity of biopolymer active films and coatings." *Food Chem.* **242**: 369–377. doi:10.1016/j.foodchem.2017.09.065
41. Kurek, M., Hlupić, L., Elez Garofulić, I., et al. 2019. "Comparison of protective supports and antioxidative capacity of two bio–based films with revalorised fruit pomaces extracted from blueberry and red grape skin." *Food Packag. Shelf Life* **20**: 100315. doi:10.1016/j.fpsl.2019.100315
42. Crizel, T.M., Rios, O.A., Alves, V.D., et al. 2018. "Active food packaging prepared with chitosan and olive pomace." *Food Hydrocoll.* **74**: 139–150. doi:10.1016/j.foodhyd.2017.08.007
43. Lee, M.H., Kim, S.Y., Park, H.J. 2018. "Effect of halloysite nanoclay on the physical, mechanical, and antioxidant properties of chitosan films incorporated with clove essential oil." *Food Hydrocoll.* **84**: 58–67. doi:10.1016/j.foodhyd.2018.05.048
44. Qin, Y., Liu, Y., Yuan, L., et al. 2019. "Preparation and characterization of antioxidant, antimicrobial and pH–sensitive films based on chitosan, silver nanoparticles and purple corn extract." *Food Hydrocoll.* **96**: 102–111. doi:10.1016/j.foodhyd.2019.05.017
45. Ilyas RA, Sapuan SM, Atikah MSN, Asyraf MRM, Ayu Rafiqah S, Aisyah HA, Mohd Nurazzi N, Norrahim MNF. Effect of Hydrolysis Time on the Morphological, Physical, Chemical, and Thermal Behavior of Sugar Palm Nanocrystalline Cellulose (Arenga Pinnata (Wurmb.) Merr). *Textile Research Journal*. 2020. In print. doi: 10.1177/0040517520932393.

5

Perspectives for Chitin/Chitosan Based Films as Active Packaging Systems on a Food Product

Ewelina Jamróz[1], Piotr Kulawik[2], and Fatih Özogul[3]

[1]Department of Chemistry, Faculty of Food Technology, University of Agriculture, Balicka St. 122,PL-30-149 Cracow, Poland

[2]Department of Animal Products Technology, Faculty of Food Technology, University of Agriculture in Cracow, Cracow, 30-149, Poland

[3]Department of Seafood Processing Technology, Faculty of Fisheries, Cukurova University, Balcalı, Adana, 01330, Turkey

5.1 Introduction

More and more attention is being focused on the production of active packaging materials that could be an alternative to synthetic petroleum-based polymers. Although biopolymer materials are becoming popular in many industries, they are also often characterized by weak physicochemical properties that do not meet industry standards. One of the biggest challenges for scientists is to develop biodegradable packaging materials that will protect food products against oxidative degradation and UV radiation and inhibit microbiological growth while maintaining good physical and mechanical properties. Only this type of packaging is able to replace synthetic materials [1].

After cellulose, chitosan is the second most commonly used biopolymer. Chitosan with a degree of deacetylation of about 50% becomes soluble in an acidic environment due to the protonation of NH_2 groups in the polymer and carries positive surface charges, becoming a natural cationic biopolymer. Chitosan obtained from various sources differs in the degree

Bio-Based Packaging: Material, Environmental and Economic Aspects, First Edition. Edited by S.M. Sapuan and R.A. Ilyas

of deacetylation, molecular weight, and functional properties. In addition, it is known for its biocompatibility, biodegradation, and low toxicity [2]. The most common source of chitosan is biowaste from aquatic organisms (shrimps, crabs, lobsters, etc.). Alternative sources of chitosan are terrestrial organisms, such as insects and some microorganisms, e.g. fungi [3]. Derived from shrimp, crab, and other crustacean wastes, it has found an application in the production of films, nanofibers, pastes, gels, capsules, powders, granules, and hydrogels [2].

Active food protective agents can be mixed with a film-forming material that results in the controlled release of the active ingredient, providing protection against the bacteria that cause food spoilage. The film-forming chitosan is a suitable polymer for the incorporation of this type of additive [1]. There is much work that studies the effects of the addition of active agents (such as plant extracts, nanofillers, etc.) on the functional properties of chitosan films [4–7].

This chapter focuses on chitosan as an active packaging material and comprehensively discusses the effects of major factors on the properties of chitosan-based films. Also, current developments as well as potential applications in the food industry of active chitosan packaging material are mentioned in this chapter.

5.2 The Effect of the Incorporation of Chitosan on the Properties of Films

Food packaging is intended to provide a protective barrier against undesirable external and internal factors. Active packaging has become one of the most important new packaging technology trends, involving the inclusion of emitters or absorbers into the packaging material. Types of active packaging include oxygen, ethylene, moisture, or carbon dioxide emitters as well as antioxidant and antimicrobial packaging [8].

Chitosan films can act as active packaging materials due to their biological properties. The different types of active chitosan packaging systems with functional properties and their potential application has been shown in Figure 5.1. Chitosan, due to its antibacterial properties, is considered a promising packaging material. However, it is necessary to optimize polysaccharide concentration to ensure antimicrobial activity without imparting an undesirable taste and odor [9]. Its antimicrobial activity depends on many factors, such as the degree of deacetylation, molecular weight, degree of substitution, physical form and the structure of the cell membrane of microorganisms [10]. Chitosan has an intrinsic antimicrobial activity that is associated with the presence of positively charged amino groups that react with negatively charged cellular membranes of microorganisms. This type of interaction causes the outflow of intracellular components from microbial cells [11].

The increase in antimicrobial activity of chitosan may be affected by its chemical modification. Modified polymers of chitosan show greater antimicrobial activity than native chitosan. The increase in antimicrobial activity is related to the chemical structures of the substituents used (e.g. three synthesized 1,3-azlactone derivatives) during modification, which facilitates the penetration of the bacterial cell membrane [12].

Various active ingredients such as nanofillers, metals, drugs, extracts, etc., can be included to improve the antimicrobial properties of chitosan films [10]. The addition of rapeseed protein hydrolysate to chitosan films improves its antimicrobial properties against *Escherichia coli (E. coli)* (from 19.23 to 21.79 mm of the inhibition zone), *Staphylococcus*

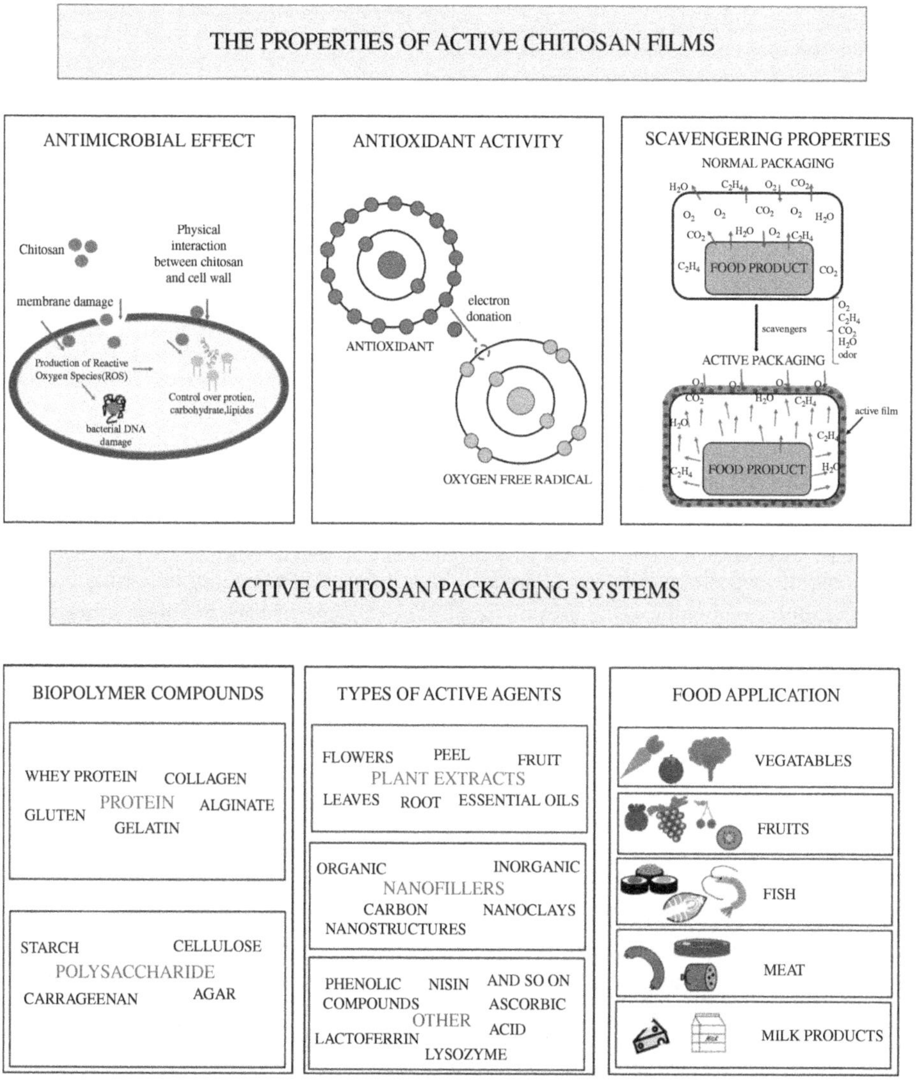

Figure 5.1 *The active properties of chitosan as one of the components of active packaging systems.*

aureus (S. aureus) (from 20.82 mm to 23.44 mm of the inhibition zone), and *Bacillus subtilis* (from 20.10 mm to 25.52 mm of the inhibition zone) [13]. In addition, it should be noted that the rapeseed protein hydrolysate does not have antimicrobiological activity against studied pathogens. This phenomenon may be related to the synergistic effect of both components. In recent years, metal nanoparticles (NPs) have been added to biopolymer films due to their antimicrobial properties. Jayaramudu and coworkers obtained chitosan films with the addition of Cu NPs and CuO NPs. The obtained results indicate that films with CuO NPs display a much better antibacterial activity than nanocomposites with Cu NPs. In addition, films with CuO NPs had a better inhibition efficiency against *E.coli* than

Bacillus. The inhibition capacity may be dependent on the structure of the bacterial cell wall and on the functionality of the NPs [14].

Chitosan films as an active packaging material can also prevent the oxidation of food, due to its antioxidant activity. However, its weaknesses include poor solubility, which hinders further applications. To overcome this problem and increase antioxidant activity, chemical modifications can be used. Urea-functionalized chitosan derivatives with urea groups bearing nitrogen-containing heterocycles showed higher antioxidant power measured by hydroxyl-radical scavenging and superoxide-radical scavenging tests than native chitosan [15]. Also, carboxymethylation of chitosan followed by γ-ray treatment improves their antioxidant activity. Moreover, coating peach fruits with such prepared carboxymethyl chitosan has a positive influence on prolonging shelf life [16]. Chitosan with a low molecular weight has stronger antioxidant properties than chitosan with a high molecular weight. This polymer and related nanofibers can reduce oxidative stress of the organism, which can result in the reduction of the occurrence of several stress oxidation related diseases such as metaboli syndrome or cardiovascular disease [2].

There are many examples of the impact of active chitosan packaging systems with scavenging properties on the quality of product storage. Carvalho and coworkers obtained chitosan coatings with trans-cinnamaldehyde, which improved the quality of freshly cut melons. The chitosan coating acted as a physical gas exchange barrier, resulting in a reduced respiration rate and consequently O_2-dependent processes, while the aldehyde acted as a free radical inhibitor and an enzyme [17]. Chitosan films with titanium dioxide NPs (TiO_2 NPs) exhibit photodegradation of ethylene. TiO_2 NPs exposed to ultraviolet light produce strong reactive oxygen species (ROS), including hydroxyl radicals and peroxide ions that react with ethylene, inducing the photodegradation [18].

The quality of food storage is influenced by the barrier properties of chitosan films. The absorption coefficient and the water vapor permeability increase after increasing the molecular weight of the chitosan [1]. The addition of cellulose and lignocellulose nanofibers and *Carum copticum* essential oil had an impact on the improvement of the film barrier properties (water vapor permeability [WVP] by up to ~42%, ~44%, and ~21%, respectively) [19]. The significant improvement of water vapor, oxygen, and carbon dioxide permeability of chitosan films with essential oil and nanoclay content has also been reported [20, 21]. Moreover, chitosan films with graphene oxide nanosheets with improved barrier, thermal, and mechanical properties make a good candidate for various packaging materials [22].

5.3 Blends of Chitosan and Other Biopolymers

Chitosan has poor flexibility, so it is important to apply plasticizers to give the film the ability to be lengthened. Ma et al. reported that the addition of plasticizers (glycerol, xylitol, and maltitol) to chitosan films results in the regulation of the microstructure, resulting in a reduction in tensile strength while improving their stretchability [23]. An alternative to this solution is to combine chitosan with other biopolymers to obtain films with improved properties. The physical and mechanical properties of the film have been improved due to the addition of other polymers, such as starch [24], cellulose [25], agar [26], carrageenan [27], gelatin [28], collagen [29] wheat gluten [30], alginate [31], and whey protein [32].

Mixing chitosan with other biopolymers often improves the physicochemical properties of the obtained films. The addition of chitosan to the starch film resulted in a reduction in water vapor transmission rates and an increase in tensile strength and elongation at the break [33]. Ren et al. determined the effect of adding chitosan to starch films. The addition of chitosan into starch films resulted in an increase in solubility, tensile strength, elongation at the break, and changes in color. In addition, a decrease in the Young modulus (YM) and WVP and a degree of crystallinity were observed [34]. The same results were observed in carrageenan-chitosan films [27]. Chen et al. concluded that the chitosan/wheat gluten material can be characterized by better moisture absorption and a lower diffusivity than chitosan and a higher extensibility and better antimicrobial activity than wheat gluten [30]. Blending chitosan with gelatin in appropriate ratios (1 : 3 and 1 : 1) results in the improved barrier and mechanical properties of chitosan/gelatin films [35]. Chitosan was used to improve the properties in fish myofibrillar protein films. Due to its hydrophobic properties, it not only influenced the gas barrier properties of the film but also positively influenced the mechanical properties, solubility, swelling, UV barrier properties, and thermal stability [36]. The addition of chitosan to the whey protein film reduced the oxygen and carbon dioxide permeability by 35% and 21%, respectively, and increased the water vapor permeability by threefold [37].

5.4 Characterization of Chitosan Films with Nanofillers

Recently, chitosan-based nanocomposites have become very popular among the scientific community. The addition of nanofillers to chitosan films changes their physical and chemical properties. In addition, the improved antimicrobial properties of chitosan films have an increased spectrum of packaging applications. The nanofillers, i.e. NPs of metallic or metallic oxides, carbon, clay, and organic nanofillers, are used to make nanocomposite packaging materials. The general definition of nanofillers says that they can have different shapes and sizes, but their individual size cannot exceed 100 nm [6]. The most recent advances regarding the properties of nanocomposite chitosan films are summarized in Table 5.1.

The addition of different types of nanofillers to chitosan films affects their properties. Liu and coworkers added nanosilica into the chitosan matrix, which was supposed to affect water and heat resistance of tested films. Nanosilica has been shown to act as a filler of empty spots in the chitosan matrix thanks to the presence of hydrogen bonds, which correlates with the improved mechanical and thermal properties of the film. In addition, nanosilica has improved the water resistance of chitosan films [51]. Similar behavior was observed in chitosan films with halloysite nanotubes. The electrostatic interactions and hydrogen bonds occurring between components resulted in improved mechanical properties [52]. The use of NPs as an additive to chitosan films arouses considerable interest among scientists. This includes the addition of sulfur, silver, selenium zinc oxide, and zinc oxide NPs, resulting in modification of antimicrobiological and physicochemical properties of chitosan films [38–41]. For example, the addition of AgNPs to the chitosan matrix improved the mechanical, thermal, antimicrobial, and barrier properties of the film. In addition, different degrees of an AgNPs release profile were observed from the matrix [53].

The addition of nanofillers to chitosan films has been tested in many areas of possible applications. Improvement of the properties of chitosan films is related to the interfacial

Table 5.1 *The effect of the addition of nanofillers on functional properties of chitosan films.*

Type of nanofillers	The effect on the functional film's properties	Reference
Metallic nanofillers		
Sulfur nanoparticles (SNPs)	• ↑ in TS (by up to ~22%), EM (by up to ~23%) and WCA (by up to ~6%) • ↓ in WVP (by up to ~14%) and EAB (by up to ~61%) • ↑ in antimicrobial properties against *E. coli* (from 7 to 0 log CFU/ml after 6 h) and *Listeria monocytogenes* (from 7 to 0 log CFU/ml after 12 h)	[38]
Selenium nanoparticles (SeNPs)	• ↓ in TS by ~24% • ↑ in EAB by ~2000% • ↑ in electrical conductivity (from 0.00017 to 0.00055 S cm^{-1})	[39]
Silver (Ag) nanowires	• ↑ in TS (by up to ~62%) • Very good conductivity parameters • ↑ in antimicrobial activity against *E. coli* (from 0 to 5.3 mm of the inhibition zone) and *B. subtilis* (from 0 to 9.6 mm of the inhibition zone)	[40]
Metal oxide nanofillers		
Zinc oxide nanoparticles (ZnONPs)	• ↑ in TS by up to ~225% • ↓ in EAB by up to ~56% and WVTR by up to~7% • ↑ of dielectric constant and conductivity • ↑ in antimicrobial efficiency against *E. coli* and *S. aureus*	[41]
Organic nanofillers		
Flax cellulose nanocrystals	• no changes in WVP values and in thermal stability • ↑ in TS by up to ~24% and in EAB by up to ~22% • ↑ in antimicrobial activity against *E. coli, Pseudomonas aeruginosa, S. aureus, L. monocytogenes* and *Enterococcus faecalis*	[42]
Cellulose nanocrystals from: • -corncob • -corn husk • -coconut shell • -wheat bran	• the highest values of water activity were presented for corn husk cellulose nanocrystals (0.658) and the lowest for wheat bran cellulose nanocrystals (0.601) • the highest values of WC were presented for wheat bran cellulose nanocrystals (20.86%) and the lowest for coconut shell cellulose nanocrystals (15.13%) • the highest values of TS were presented for corncob cellulose nanocrystals (11.43 MPa) and coconut shell cellulose nanocrystals (11.38 MPa) • ↑ in EAB by 137% (coconut shell cellulose nanocrystals)	[43]

Table 5.1 *(continued)*

Type of nanofillers	The effect on the functional film's properties	Reference
Clay nanofillers		
Halloysite nanotubes	• ↑ in thermal stability • ↑ in TS by ~34% and YM by ~21%	[44]
Laponite	• good affinity to its substrate (the apparent Michaelis–Menten constant for the sensor was found to be 0.16 mM) • high sensitivity (674 mA M^{-1} cm^{-2} for catechol) • remarkable long-term stability in storage (it retains 88% of the original activity after 60 days).	[45]
Carbon nanofillers		
Reduced graphene oxide	• high electrical activity (conductivity of 1.2 S m^{-1}) • strong mechanical properties (YM 6.3 GPa, TS 206 MPa and EAB 6.5%)	[46]
Graphene oxide nanosheets	• no changes in WC • ↑ in TS by ~119%, WVTR by ~56%, OP by ~65% • ↓ in EAB by ~21% • ↑ in thermal stability	[22]
Nanofillers hybrids		
Montmorillonite-CuO	• ↑ in TS (by up to ~59%), EAB (by up to ~52%) • ↓ in WVP (by up to ~55%) and OTR (by up to ~32%) • ↑ in antimicrobial activity against *E. coli, P. aeruginosa, S. aureus* and *Bacillus cereus*	[47]
Montmorillonite-AgNPs	• Silver nanoparticles (AgNPs) improved growth inhibition of *E. coli* (from 0 to 8 mm of the inhibition zone) and *B. subtilis* (from 0 to 9 mm of the inhibition zone) • montmorillonite alone had no influence on antimicrobial activity	[48]
Cellulose nanofibrils/ nanosilica	• ↑ in TS by up to ~65% • Increasing the content of nanosilica increased the thermal stability	[49]
Chitin nanofibers/AgNPs	• addition of 1% AgNPs had a negative effect on the physical properties of the film • addition of chitin nanofibers resulted in a reduced solubility (by ~2%), TS (by ~152%) and YM (by ~239%) • reduction in WVP by ~38%	[50]

Abbreviations: TS, tensile strength; EAB, elongation at break; YM, Young Modulus; WVP, water vapor permeability; WVTR, water vapor transition rate; OTR, oxygen transition rate; WS, water solubility; WC, water content; WCA, water contact angle.

interactions between functional groups derived from the biopolymer and nanofillers; therefore close attention should be paid to the dispersal of nanofillers within the chitosan matrix. Also, an environmentally friendly and nontoxic approach to obtain nanofillers should be considered. Currently, the biggest challenge for scientists is to develop and apply nanocomposite biopolymer films that have no negative impact on the environment and health [6].

5.5 Preparation of Chitosan Films with Active Compounds

Active packaging is intended to prevent adverse changes in the quality of food or even to improve the quality and extend the period of safe storage of the product through active interaction with the product or the atmosphere inside the package. Active packaging is defined as systems that actively cause changes in the packaged food and cause an extension of the shelf life. They also guarantee or significantly improve the microbiological safety of the products. Active packaging is divided mainly according to their function: antibacterial, antioxidant, oxygen-absorbing, carbon dioxide-absorbing, water-absorbing, ethylene-absorbing, ethanol-emitting, and odor-releasing/absorbing packaging [11].

There are a lot of research studies that determine the effect of the addition of the active substance on the properties of chitosan films. Kaya and coworkers [54] obtained chitosan films with the addition of stem, leaf, and seed methanol extracts of *Pistacia terebinthus*. The addition of extracts into chitosan films resulted in an improvement in the antioxidant and antimicrobial properties against *Salmonella typhimurium (S. typhimurium), E. coli, Bacillus thuringiensis, Pseudomonas aeruginosa,* and *Proteus vulgaris*. In addition, antiquorum sensing activity against the *Chromobacterium violaceum (C. violaceum)* was determined. All chitosan films with extracts inhibited violacein production and showed a greater activity than chitosan films and commercial antibiotics such as gentamicin [54]. The addition of turmeric into the chitosan films not only positively influenced the physical properties, such as tensile strength (by ~49%), fracture strength (by ~81%), and toughness (by ~73%), but also improved the antimicrobial activity of tested films against *S. aureus* (from 5.5 to 1 log during the three-hour exposure period) and *S. typhimurium* (from 5.5 to 4 log during the three-hour exposure period) [55]. In a different study, the apple peel polyphenols were incorporated into the chitosan matrix, significantly improving the physical, antioxidant, and antimicrobial properties of the films, but worsening the mechanical properties [56]. Akyuz et al. added different animal fats with different unsaturation degrees and plant oil additives (olive, corn, and sunflower oil) into the chitosan matrix. Chitosan films with olive oil showed the best mechanical and hydrophobic properties. Chitosan films with oils with a high content of unsaturated fatty acids had a higher antioxidant activity than chitosan films with oils with high levels of saturated fatty acids [57]. The addition of cinnamon bark oil improved the antioxidant activity of the film from 6 to 14.5 times compared to the control. Additionally, the obtained films can protect human erythrocytes from free radical attack by up to 80% [58].

Active packaging based on chitosan and biodegradable monomers has also been recently developed. This type of biodegradable binary grafted chitosan films have been used to store the apple and guava fruits. After 18 days of storage at 40–45 °C, the fruits wrapped in binary grafted chitosan films were still red, with no signs of microbiological spoilage. However, the fruits without films were brown and covered with mold [59]. Ascorbic acid was used as an anti-bronzing agent to chitosan films. The obtained coatings protected the color, taste and smell of pomegranate arils and extended the length of the shelf life of the fruit by up to 21 days at 5 °C [60]. Chitosan films with lysozyme addition have been used as coatings on the large yellow croaker (*Larimichthys crocea*) stored at 4 °C for 15 days. The obtained results indicate that the active coating effectively inhibits the growth of microorganisms and lipid oxidation. During the whole period of storage in the refrigerator, fish with active coatings did not exceed the critical value of 7 log cfu/g of the total viable count during the storage period, whereas fish without coatings exceeded this value on day 15 of storage [61].

5.6 Chitosan-Based Films as Packaging Material Systems

Chitosan can act as a packaging material for various types of food products. There are many studies that emphasize the huge contribution of this type of packaging material to the storage of products. Fresh fillets of hake (*Merluccius merluccius*) and sole (*Solea solea*) were wrapped in chitosan films by in-air and under-vacuum methods, and stored for 15 days at 4 °C. The presence of the film resulted in a reduction of the final bacteria population by approximately 1.6 and 3.8 log units in air and vacuum-packed samples, respectively [62].

One of the active packaging groups are chitosan films with the addition of nanofillers. Chitosan film with TiO_2 NPs shows a better photodegradation of ethylene than chitosan film, as it was studied during climacteric fruit storage, where qualitative changes are regulated by ethylene. The results obtained suggest that the nanocomposite film, in which the cherry tomatoes were wrapped, exhibited photodegradation of the ethylene after exposure to UV light, which consequently delayed the ripening process and the occurrence of changes in the quality of vegetables [63]. Active chitosan films with monolaurine caused a reduction in the population of *L. monocytogenes* in cheese by 2.4–2.3 logs after 14 days of storage [64].

The addition of plant byproducts to chitosan films is also a promising strategy in the food industry to extend the quality of stored products. The incorporation of a cotton lavender extract or *Santolina chamaecyparissus* L. solids byproduct into the chitosan film contributed to an improvement of the antifungal and antioxidant properties of the active coatings, and this type of material had a positive effect on the "*Manchego*" cheese storage quality [65].

Chitosan films were neutralized using sodium hydroxide and the neutralization process improved the barrier against water and oxygen. The conservative properties of chitosan films on chilled meat were used and the performed analyses indicated an improvement of the barrier properties against water and oxygen, as well as a reduction in the drip loss of meat, which may affect the inhibition of oxidation processes [66]. Table 5.2 shows recent studies of different agents incorporated into the chitosan matrix and their effect on various food model products.

The main challenges in the area of active packaging materials include the profitability of the production of this type of packaging for food products. Additives that are incorporated into the biopolymer matrix should be harmless to the environment and nontoxic. The complexity of different chitosan forms and the number of possible applications also provides a challenge for the legislative bodies, which regulate the possible use of new packaging materials. On one hand, the law regulations should be flexible enough to avoid the necessity for the acquisition of large amounts of permits, which would result in extended periods of time from the invention of the packaging to the moment when it would be possible to use it on the market. On the other hand, the legislative bodies need to ensure the safety of the packaging materials for the final consumer [5–7, 83].

Table 5.2 *Recent examples of the application of chitosan films as active packaging systems on a food product.*

Active compound	Properties of the film	Effect on food product	Reference
Plant extract			
Mango leaf extract	• -↑ in antioxidant activity by up to ~35% (DPPH method), ~594% (FRAP method) and ~ 336% (ABTS method)	*Cashew nuts*: • -56% higher oxidation resistance during 28 days of storage	[67]
Grape seed extract	• ↑ in WVP by ~11% and EAB by ~32% • ↓ in TS by ~58% • ↑ in antioxidant activity (DPPH method and total phenolic content) • ↑ in antimicrobial activity against *E. coli, L. monocytogenes, S. aureus* and *P. aureginosa*	*Vacuum packed chicken breast fillets*: • increased stability of pH during storage • effective reduction of the growth of mesophilic aerobic bacteria (from 3.4 to 5.5 log after 15 days) and coliforms (from 1.8 to 2.0 log after 15 days) • -reduction of the oxidation value during storage	[68]
Peanut skin extract Pink pepper residue extract	• -films with peanut skin extract had higher antioxidant activity than films with pink pepper residue extract (DPPH, ABTS and superoxide radical)	*Ground chicken meat*: • -no changes in pH and color values of chicken during storage • -films with pink pepper residues extract showed higher protection against lipid oxidation • improvement in inhibition of the growth of psychotropic microorganisms • the addition of pink pepper residues extract into films caused greater spectrum of properties (antimicrobial and antioxidant activity)	[69]
Kombucha tea	• -↓ in WVP by ~48% • -↑ in WS by ~144% and antioxidant activity by ~856% (DPPH method) • -↑ in antimicrobial activity against *S. aureus* (from 8.7 to 17.7 mm of inhibition zone) and *E. coli* (from 6.2 to 20.1 mm of inhibition zone)	*Minced beef meat*: • -extending the storage period to 6 d (decrease in the total number of bacteria from 4.9 log to 4.1 log on the 6th day) • -after 6 d of storage, the lowest TBA (0.31 mg of malondialdehyde kg^{-1}) values were noted in the film with the highest kombucha tea concentration • -sensory analysis indicates that films with the highest concentration of kombucha tea (3%) effectively delay the formation of unpleasant odors	[70]

Table 5.2 *(continued)*

Active compound	Properties of the film	Effect on food product	Reference
Essential oil			
Ginger essential oil	• ↑ in antimicrobial activity against *E. coli* and *S. aureus* • ↓ in TS by up to ~7% • No changes in EAB	*Steak of barracuda (Sphyraena jello)*: • inhibition of total volatile basic nitrogen formation • extending the shelf-life by 8 d • -sensory evaluation indicated the acceptability of steaks wrapped with films with ginger oil until 20 d of storage, while steaks wrapped with chitosan and ethylene vinyl alcohol films were discarded at day 12	[71]
Eucalyptus essential oil	• ↑ in antimicrobial activity against *S. aureus* (4.2 log reduction), *E. coli* (4.0 log reduction), *B. cereus* (4.5 log reduction) and *S. entertidis* (4.7 log reduction) • ↓ in WC by ~44%, WS by ~39%, TS by ~28% • ↑ in EAB by ~66% and WVP by ~159%	*Sliced sausage*: • the highest antimicrobial activity against *L. monocytogenes* observed for films with the highest concentration of essential oil (1.0 log reduction)	[72]
Nanofillers			
TiO_2 NPs	• ↑ in WVP by up to ~28% and TS by up to ~50%	*Tomato fruits*: • photodegradation of ethylene, leading to delayed maturation • the lowest CO_2 production rate was recorded in tomatoes stored in TiO_2 film during 3 d of storage • lengthening the time of fruit firmness • reduction of weight loss by ~20% during storage • the lowest rate of lycopene changes and total soluble solid content • the content of ascorbic acid kept stable during storage	[18]

Table 5.2 *(continued)*

Active compound	Properties of the film	Effect on food product	Reference
TiO_2 NPs	• ↑ in TS by ~90% and EAB by ~69% • ↓ of transmission in the visible light region • complete sterilization of *E. coli, S. aureus, C. albicans* and *Aspergillus niger*) took place within 12 h	*Red grape*: • extension of fruit shelf-life by 7 d	[73]
AgNPs with laponite clay	• ↑ in TS (by up to ~150%), solubility (by up to ~90%) and OTR (by up to ~100%)	*Litchi*: • prolonged shelf-life of litchi fruits by 2 d (visible traces of molds after 5 d of storing litchi fruits packed in polyethylene films	[74]
AgNPs and gelatin	• ↓ in TS by up to ~27% • ↑ in EAB by up to ~51%	*Red grapes*: • extension the shelf-life by 14 days	[75]
Chitosan thymol NPs and quinoa protein	• ↓ in WVP by up to ~20%	*Blueberries and cherry tomatoes*: • inhibition of weight loss (at 9 d, for blueberries- by up to ~8% lower than control films, cherry tomatoes - by up to 2.1% lower than control films)	[76]
Nanosilica		*Loquat (Eriobotrya japonica Lindl.) fruit*: • delaying internal browning and loss of weight • increase in glucose (by up to ~13%) and fructose (by up to ~34%) content • improvement of superoxide dismutase (SOD), catalase (CAT) and ascorbate peroxidase (APX) activities • lower level of malondialdehyde (MDA) (by up to ~91%) and membranes permeability (by up to ~79%)	[77]

Table 5.2 *(continued)*

Active compound	Properties of the film	Effect on food product	Reference
Other compounds Gallic acid		*White button mushrooms (Agaricus bisporus)*: • lower degree of browning (0.92) than polyethylene (1.06) and chitosan films (0.98) after 15 days of storage • lower respiration rate (80.2%) than polyethylene and chitosan films (89.7%) after 15 days of storage • very good water vapor and CO_2 transmission rates • the weight loss of mushrooms was larger than in polyethylene films, but was in the range of market values • lower malondialdehyde content, electrolyte leakage rate, superoxide anion production rate and hydrogen peroxide content than chitosan and polyethylene films	[78]
Procyanidins		*Fresh blueberry*: • no significant differences in weight loss, pH and dry matter • improvement in antiradical activity (DPPH and ABTS assays) • significantly higher yeast and mold growth inhibition	[79]
Shrimp extract	—	*Raw salmon*: • no significant changes in hardness • the levels of total viable counts, *Enterobacteriaceae*, *Pseudomonas* and specific spoilage bacteria counts decreased during 19 days storage (2–4 log reduction) • improvement in calcium and magnesium content and reduction of estimated saturated fatty acids intake	[80]

Table 5.2 (continued)

Active compound	Properties of the film	Effect on food product	Reference
Carvacol and caprylic acid	—	*Shrimps*: • reduction of total viable counts (3.5 log reduction) • inhibition total volatile basic nitrogen formation (19 and 36 mg/100 g in coated samples after 10 days and in uncoated samples after 6 d respectively) • delayed weight loss during storage • according to the sensory evaluation, uncoated shrimps were unacceptable after 6 d	[81]
Nisin, cellulosic paper and ZnO NPs	• ↑ in SI by 38% and WS by 286% • due to the presence of nisin, the films exhibited durable antibacterial activity at 6 and 22 °C during storage	*Ultra-filtered cheese*: • reduction in the number of *L.monocytogenes* (3 log reduction) • preparation of the film by a sol–gel method allowed to decelerate nisin diffusion into the cheese, which contributed to the improvement of the quality of the stored cheese	[82]

Abbreviations: TVB-N, Total volatile base nitrogen; TPC, Total viable count; SI, swelling index; TS, tensile strength; EAB, elongation at break; YM, Young Modulus; WVP, water vapor permeability; WVTR, water vapor transition rate; OTR, oxygen transition rate; WS, water solubility; WC, water content; DPPH (2,2-diphenyl-1-picryl-hydrazyl-hydrate) free radical method (antioxidant assay); ABTS, (2,2′-azino-bis(3-ethylbenzothiazoline-6-sulfonic acid) radical cation decolorization method (antioxidant assay); FRAP- Ferric Reducing Antioxidant Power method (antioxidant assay).

5.7 Conclusions

In recent years, a huge development in packaging materials based on proteins and polysaccharides has been observed, which is caused by environmental problems resulting from the presence of an increasing number of packaging based on petroleum. Biocompatibility, biodegradability, and antimicrobial properties of chitosan indicate that this polysaccharide can be used as packaging material. Enriched with various additives (e.g. natural extracts, nanofillers, etc.) it acquires completely new active properties. Despite the fact that there is a huge number of studies regarding the production and characteristics of various types of chitosan-based films, its industrial use is still limited. Currently, research should be focused on developing a scheme for obtaining cheap chitosan packaging material that would eliminate the use of the expensive and environmentally unfriendly chemicals that are used in the production of synthetic films.

References

1. Mujtaba M, Morsi RE, Kerch G, Elsabee MZ, Kaya M, Labidi J, et al. Current advancements in chitosan-based film production for food technology; a review. *Int. J. Biol. Macromol.* 2019;**121**:889–904.
2. Anraku M, Gebicki JM, Iohara D, Tomida H, Uekama K, Maruyama T, et al. Antioxidant activities of chitosans and its derivatives in *in vitro* and *in vivo* studies. *Carbohydr. Polym.* 2018;**199**:141–9.
3. Song C, Yu H, Zhang M, Yang Y, Zhang G. Physicochemical properties and antioxidant activity of chitosan from the blowfly *Chrysomya megacephala* larvae. *Int. J. Biol. Macromol.* 2013;**60**:347–54.
4. Mir SA, Dar BN, Wani AA, Shah MA. Effect of plant extracts on the techno-functional properties of biodegradable packaging films. *Trends Food Sci. Technol.* 2018;**80**:141–54.
5. H.P.S Abdul Khalil, Saurabh CK, A.S. Adnan, Nurul Fazita MR, Syakir MI, Davoudpour Y, et al. A review on chitosan-cellulose blends and nanocellulose reinforced chitosan biocomposites: properties and their applications. *Carbohydr. Polym.* 2016;**150**:216–26.
6. Sharma B, Malik P, Jain P. Biopolymer reinforced nanocomposites: a comprehensive review. *Mater. Today Commun.* 2018;**16**:353–63.
7. Aider M. Chitosan application for active bio-based films production and potential in the food industry: review. *LWT Food Sci. Technol.* 2010;**43**(6):837–42.
8. Dey A, Neogi S. Oxygen scavengers for food packaging applications: a review. *Trends Food Sci. Technol.* 2019;**90**:26–34.
9. Kim KW, Min BJ, Kim Y-T, Kimmel RM, Cooksey K, Park SI. Antimicrobial activity against food-borne pathogens of chitosan biopolymer films of different molecular weights. *LWT Food Sci. Technol.* 2011;**44**(2):565–9.
10. Perinelli DR, Fagioli L, Campana R, Lam JKW, Baffone W, Palmieri GF, et al. Chitosan-based nanosystems and their exploited antimicrobial activity. *Eur. J. Pharm. Sci.* 2018;**117**:8–20.
11. Mousavi Khaneghah A, Hashemi SMB, Limbo S. Antimicrobial agents and packaging systems in antimicrobial active food packaging: an overview of approaches and interactions. *Food Bioprod. Process.* 2018;**111**:1–19.
12. Azmy EAM, Hashem HE, Mohamed EA, Negm NA. Synthesis, characterization, swelling and antimicrobial efficacies of chemically modified chitosan biopolymer. *J. Mol. Liq.* 2019;**284**:748–54.
13. Zhang C, Wang Z, Li Y, Yang Y, Ju X, He R. The preparation and physiochemical characterization of rapeseed protein hydrolysate-chitosan composite films. *Food Chem.* 2019;**272**:694–701.
14. Jayaramudu T, Varaprasad K, Pyarasani RD, Reddy KK, Kumar KD, Akbari-Fakhrabadi A, et al. Chitosan capped copper oxide/copper nanoparticles encapsulated microbial resistant nanocomposite films. *Int. J. Biol. Macromol.* 2019;**128**:499–508.
15. Zhang J, Tan W, Wei L, Chen Y, Mi Y, Sun X, et al. Synthesis of urea-functionalized chitosan derivatives for potential antifungal and antioxidant applications. *Carbohydr. Polym.* 2019;**215**:108–18.
16. Elbarbary AM, Mostafa TB. Effect of γ-rays on carboxymethyl chitosan for use as antioxidant and preservative coating for peach fruit. *Carbohydr. Polym.* 2014;**104**:109–17.
17. Carvalho RL, Cabral MF, Germano TA, de Carvalho WM, Brasil IM, Gallão MI, et al. Chitosan coating with trans-cinnamaldehyde improves structural integrity and antioxidant metabolism of fresh-cut melon. *Postharvest Biol. Technol.* 2016;**113**:29–39.
18. Siripatrawan U, Kaewklin P. Fabrication and characterization of chitosan-titanium dioxide nanocomposite film as ethylene scavenging and antimicrobial active food packaging. *Food Hydrocoll.* 2018;**84**:125–34.

19. Jahed E, Khaledabad MA, Almasi H, Hasanzadeh R. Physicochemical properties of *Carum copticum* essential oil loaded chitosan films containing organic nanoreinforcements. *Carbohydr. Polym.* 2017;**164**:325–38.
20. Shen Z, Kamdem DP. Development and characterization of biodegradable chitosan films containing two essential oils. *Int. J. Biol. Macromol.* 2015;**74**:289–96.
21. Zhang L, Wang H, Jin C, Zhang R, Li L, Li X, et al. Sodium lactate loaded chitosan-polyvinyl alcohol/montmorillonite composite film towards active food packaging. *Innovative Food Sci. Emerg. Technol.* 2017;**42**:101–8.
22. Ahmed J, Mulla M, Arfat YA, Thai T LA. Mechanical, thermal, structural and barrier properties of crab shell chitosan/graphene oxide composite films. *Food Hydrocoll.* 2017;**71**:141–8.
23. Ma X, Qiao C, Wang X, Yao J, Xu J. Structural characterization and properties of polyols plasticized chitosan films. *Int. J. Biol. Macromol.* 2019;**135**:240–5.
24. Wu H, Lei Y, Lu J, Zhu R, Xiao D, Jiao C, et al. Effect of citric acid induced crosslinking on the structure and properties of potato starch/chitosan composite films. *Food Hydrocoll.* 2019:**105208**.
25. Lin S, Chen L, Huang L, Cao S, Luo X, Liu K. Novel antimicrobial chitosan–cellulose composite films bioconjugated with silver nanoparticles. *Ind. Crop. Prod.* 2015;**70**:395–403.
26. Elhefian EA, Nasef MM, Yahaya AH. Preparation and characterization of chitosan/agar blended films: part 2. Thermal, mechanical, and surface properties. *E-J. Chem.* 2012;**9**(2).
27. Shahbazi M, Rajabzadeh G, Ettelaie R, Rafe A. Kinetic study of κ-carrageenan degradation and its impact on mechanical and structural properties of chitosan/κ-carrageenan film. *Carbohydr. Polym.* 2016;**142**:167–76.
28. Uranga J, Puertas AI, Etxabide A, Dueñas MT, Guerrero P, de la Caba K. Citric acid-incorporated fish gelatin/chitosan composite films. *Food Hydrocoll.* 2019;**86**:95–103.
29. Bhuimbar MV, Bhagwat PK, Dandge PB. Extraction and characterization of acid soluble collagen from fish waste: development of collagen-chitosan blend as food packaging film. *J. Environ. Chem. Eng.* 2019;**7**(2):102983.
30. Chen F, Monnier X, Gällstedt M, Gedde UW, Hedenqvist MS. Wheat gluten/chitosan blends: a new biobased material. *Eur. Polym. J.* 2014;**60**:186–97.
31. Li K, Zhu J, Guan G, Wu H. Preparation of chitosan-sodium alginate films through layer-by-layer assembly and ferulic acid crosslinking: film properties, characterization, and formation mechanism. *Int. J. Biol. Macromol.* 2019;**122**:485–92.
32. Brink I, Šipailienė A, Leskauskaitė D. Antimicrobial properties of chitosan and whey protein films applied on fresh cut Turkey pieces. *Int. J. Biol. Macromol.* 2019;**130**:810–7.
33. Xu YX, Kim KM, Hanna MA, Nag D. Chitosan–starch composite film: preparation and characterization. *Ind. Crop. Prod.* 2005;**21**(2):185–92.
34. Ren L, Yan X, Zhou J, Tong J, Su X. Influence of chitosan concentration on mechanical and barrier properties of corn starch/chitosan films. *Int. J. Biol. Macromol.* 2017;**105**:1636–43.
35. Mohammadi R, Mohammadifar MA, Rouhi M, Kariminejad M, Mortazavian AM, Sadeghi E, et al. Physico-mechanical and structural properties of eggshell membrane gelatin- chitosan blend edible films. *Int. J. Biol. Macromol.* 2018;**107**:406–12.
36. Batista JTS, Araújo CS, Peixoto Joele MRS, Silva JOC, Lourenço LFH. Study of the effect of the chitosan use on the properties of biodegradable films of myofibrillar proteins of fish residues using response surface methodology. *Food Packag. Shelf Life* 2019;**20**:100306.
37. Di Pierro P, Sorrentino A, Mariniello L, Giosafatto CVL, Porta R. Chitosan/whey protein film as active coating to extend ricotta cheese shelf-life. *LWT Food Sci. Technol.* 2011;**44**(10):2324–7.
38. Shankar S, Rhim J-W. Preparation of sulfur nanoparticle-incorporated antimicrobial chitosan films. *Food Hydrocoll.* 2018;**82**:116–23.
39. Kalishwaralal K, Jeyabharathi S, Sundar K, Selvamani S, Prasanna M, Muthukumaran A. A novel biocompatible chitosan–selenium nanoparticles (SeNPs) film with electrical conductivity for cardiac tissue engineering application. *Mater. Sci. Eng. C* 2018;**92**:151–60.

40. Shahzadi K, Wu L, Ge X, Zhao F, Li H, Pang S, et al. Preparation and characterization of bio-based hybrid film containing chitosan and silver nanowires. *Carbohydr. Polym.* 2016;**137**:732–8.
41. Mujeeb Rahman P, Abdul Mujeeb VM, Muraleedharan K, Thomas SK. Chitosan/nano ZnO composite films: enhanced mechanical, antimicrobial and dielectric properties. *Arab. J. Chem.* 2018;**11**(1):120–7.
42. Mujtaba M, Salaberria AM, Andres MA, Kaya M, Gunyakti A, Labidi J. Utilization of flax (*Linum usitatissimum*) cellulose nanocrystals as reinforcing material for chitosan films. *Int. J. Biol. Macromol.* 2017;**104**:944–52.
43. de Andrade MR, Nery TBR, de Santana e Santana TI, Leal IL, Rodrigues LAP, de Oliveira Reis JH, et al. Effect of cellulose nanocrystals from different lignocellulosic residues to chitosan/glycerol films. *Polymers.* 2019;**11**(4):658.
44. De Silva RT, Pasbakhsh P, Goh KL, Chai S-P, Ismail H. Physico-chemical characterisation of chitosan/halloysite composite membranes. *Polym. Test.* 2013;**32**(2):265–71.
45. Fan Q, Shan D, Xue H, He Y, Cosnier S. Amperometric phenol biosensor based on laponite clay–chitosan nanocomposite matrix. *Biosens. Bioelectron.* 2007;**22**(6):816–21.
46. Wang X, Bai H, Yao Z, Liu A, Shi G. Electrically conductive and mechanically strong biomimetic chitosan/reduced graphene oxide composite films. *J. Mater. Chem.* 2010;**20**(41):9032–6.
47. Nouri A, Yaraki MT, Ghorbanpour M, Agarwal S, Gupta VK. Enhanced antibacterial effect of chitosan film using montmorillonite/CuO nanocomposite. *Int. J. Biol. Macromol.* 2018;**109**:1219–31.
48. Gabriel JS, Gonzaga VAM, Poli AL, Schmitt CC. Photochemical synthesis of silver nanoparticles on chitosans/montmorillonite nanocomposite films and antibacterial activity. *Carbohydr. Polym.* 2017;**171**:202–10.
49. Ismail MY, Patanen M, Sirviö JA, Visanko M, Ohigashi T, Kosugi N, et al. Hybrid films of cellulose nanofibrils, chitosan and nanosilica—structural, thermal, optical, and mechanical properties. *Carbohydr. Polym.* 2019;**218**:87–94.
50. Jafari H, Pirouzifard M, Khaledabad MA, Almasi H. Effect of chitin nanofiber on the morphological and physical properties of chitosan/silver nanoparticle bionanocomposite films. *Int. J. Biol. Macromol.* 2016;**92**:461–6.
51. Liu Y, Cai Z, Sheng L, Ma M, Xu Q. Influence of nanosilica on inner structure and performance of chitosan based films. *Carbohydr. Polym.* 2019;**212**:421–9.
52. Liu M, Zhang Y, Wu C, Xiong S, Zhou C. Chitosan/halloysite nanotubes bionanocomposites: structure, mechanical properties and biocompatibility. *Int. J. Biol. Macromol.* 2012;**51**(4):566–75.
53. Kadam D, Momin B, Palamthodi S, Lele SS. Physicochemical and functional properties of chitosan-based nano-composite films incorporated with biogenic silver nanoparticles. *Carbohydr. Polym.* 2019;**211**:124–32.
54. Kaya M, Khadem S, Cakmak YS, Mujtaba M, Ilk S, Akyuz L, et al. Antioxidative and antimicrobial edible chitosan films blended with stem, leaf and seed extracts of *Pistacia terebinthus* for active food packaging. *RSC Adv.* 2018;**8**(8):3941–50.
55. Kalaycıoğlu Z, Torlak E, Akın-Evingür G, Özen İ, Erim FB. Antimicrobial and physical properties of chitosan films incorporated with turmeric extract. *Int. J. Biol. Macromol.* 2017;**101**:882–8.
56. Riaz A, Lei S, Akhtar HMS, Wan P, Chen D, Jabbar S, et al. Preparation and characterization of chitosan-based antimicrobial active food packaging film incorporated with apple peel polyphenols. *Int. J. Biol. Macromol.* 2018;**114**:547–55.
57. Akyuz L, Kaya M, Ilk S, Cakmak YS, Salaberria AM, Labidi J, et al. Effect of different animal fat and plant oil additives on physicochemical, mechanical, antimicrobial and antioxidant properties of chitosan films. *Int. J. Biol. Macromol.* 2018;**111**:475–84.
58. Lopez-Mata MA, Ruiz-Cruz S, Silva-Beltrn NP, Ornelas-Paz JdJ, Ocano-Higuera VM, Rodrguez-Felix F, et al. Physicochemical and antioxidant properties of chitosan films incorporated with cinnamon oil. *Int. J. Polym. Sci.* 2015;**2015**:8.
59. Kumar D, Kumar P, Pandey J. Binary grafted chitosan film: synthesis, characterization, antibacterial activity and prospects for food packaging. *Int. J. Biol. Macromol.* 2018;**115**:341–8.

60. Özdemir KS, Gökmen V. Extending the shelf-life of pomegranate arils with chitosan-ascorbic acid coating. *LWT Food Sci. Technol.* 2017;**76**:172–80.
61. Wu T, Ge Y, Li Y, Xiang Y, Jiang Y, Hu Y. Quality enhancement of large yellow croaker treated with edible coatings based on chitosan and lysozyme. *Int. J. Biol. Macromol.* 2018;**120**:1072–9.
62. Fernández-Saiz P, Sánchez G, Soler C, Lagaron JM, Ocio MJ. Chitosan films for the microbiological preservation of refrigerated sole and hake fillets. *Food Control* 2013;**34**(1):61–8.
63. Kaewklin P, Siripatrawan U, Suwanagul A, Lee YS. Active packaging from chitosan-titanium dioxide nanocomposite film for prolonging storage life of tomato fruit. *Int. J. Biol. Macromol.* 2018;**112**:523–9.
64. Lotfi M, Tajik H, Moradi M, Forough M, Divsalar E, Kuswandi B. Nanostructured chitosan/ monolaurin film: preparation, characterization and antimicrobial activity against listeria monocytogenes on ultrafiltered white cheese. *LWT- Food Sci. Technol.* 2018;**92**:576–83.
65. Ortiz de Elguea-Culebras G, Bourbon AI, Costa MJ, Muñoz-Tebar N, Carmona M, Molina A, et al. Optimization of a chitosan solution as potential carrier for the incorporation of *Santolina chamaecyparissus* L. solid by-product in an edible vegetal coating on 'Manchego' cheese. *Food Hydrocoll.* 2019;**89**:272–82.
66. Chang W, Liu F, Sharif HR, Huang Z, Goff HD, Zhong F. Preparation of chitosan films by neutralization for improving their preservation effects on chilled meat. *Food Hydrocoll.* 2019;**90**:50–61.
67. Rambabu, Bharath, Fawzi, Loke P, Hernández H. Mango leaf extract incorporated chitosan antioxidant film for active food packaging. *Int. J. Biol. Macromol.* 2019;**126**:1234–43.
68. Sogut E, Seydim AC. The effects of chitosan and grape seed extract-based edible films on the quality of vacuum packaged chicken breast fillets. *Food Packag. Shelf Life* 2018;**18**:13–20.
69. Serrano-León JS, Bergamaschi KB, Yoshida CMP, Saldaña E, Selani MM, Rios-Mera JD, et al. Chitosan active films containing agro-industrial residue extracts for shelf life extension of chicken restructured product. *Food Res. Int.* 2018;**108**:93–100.
70. Ashrafi A, Jokar M, Mohammadi Nafchi A. Preparation and characterization of biocomposite film based on chitosan and kombucha tea as active food packaging. *Int. J. Biol. Macromol.* 2018;**108**:444–54.
71. Remya S, Mohan CO, Bindu J, Sivaraman GK, Venkateshwarlu G, Ravishankar CN. Effect of chitosan based active packaging film on the keeping quality of chilled stored barracuda fish. *J. Food Sci. Technol.* 2016;**53**(1):685–93.
72. Azadbakht E, Maghsoudlou Y, Khomiri M, Kashiri M. Development and structural characterization of chitosan films containing Eucalyptus globulus essential oil: potential as an antimicrobial carrier for packaging of sliced sausage. *Food Packag. Shelf Life* 2018;**17**:65–72.
73. Zhang X, Xiao G, Wang Y, Zhao Y, Su H, Tan T. Preparation of chitosan-TiO2 composite film with efficient antimicrobial activities under visible light for food packaging applications. *Carbohydr. Polym.* 2017;**169**:101–7.
74. Wu Z, Huang X, Li Y-C, Xiao H, Wang X. Novel chitosan films with laponite immobilized Ag nanoparticles for active food packaging. *Carbohydr. Polym.* 2018;**199**:210–8.
75. Kumar S, Shukla A, Baul PP, Mitra A, Halder D. Biodegradable hybrid nanocomposites of chitosan/gelatin and silver nanoparticles for active food packaging applications. *Food Packag. Shelf Life* 2018;**16**:178–84.
76. Medina E, Caro N, Abugoch L, Gamboa A, Díaz-Dosque M, Tapia C. Chitosan thymol nanoparticles improve the antimicrobial effect and the water vapour barrier of chitosan-quinoa protein films. *J. Food Eng.* 2019;**240**:191–8.
77. Song H, Yuan W, Jin P, Wang W, Wang X, Yang L, et al. Effects of chitosan/nano-silica on postharvest quality and antioxidant capacity of loquat fruit during cold storage. *Postharvest Biol. Technol.* 2016;**119**:41–8.
78. Liu J, Liu S, Zhang X, Kan J, Jin C. Effect of gallic acid grafted chitosan film packaging on the postharvest quality of white button mushroom (*Agaricus bisporus*). *Postharvest Biol. Technol.* 2019;**147**:39–47.

79. Mannozzi C, Tylewicz U, Chinnici F, Siroli L, Rocculi P, Dalla Rosa M, et al. Effects of chitosan based coatings enriched with procyanidin by-product on quality of fresh blueberries during storage. *Food Chem.* 2018;**251**:18–24.
80. Gómez-Estaca J, Alemán A, López-Caballero ME, Baccan GC, Montero P, Gómez-Guillén MC. Bioaccessibility and antimicrobial properties of a shrimp demineralization extract blended with chitosan as wrapping material in ready-to-eat raw salmon. *Food Chem.* 2019;**276**:342–9.
81. Wang Q, Lei J, Ma J, Yuan G, Sun H. Effect of chitosan-carvacrol coating on the quality of Pacific white shrimp during iced storage as affected by caprylic acid. *Int. J. Biol. Macromol.* 2018;**106**:123–9.
82. Divsalar E, Tajik H, Moradi M, Forough M, Lotfi M, Kuswandi B. Characterization of cellulosic paper coated with chitosan-zinc oxide nanocomposite containing nisin and its application in packaging of UF cheese. *Int. J. Biol. Macromol.* 2018;**109**:1311–8.
83. Ilyas RA, Sapuan SM, Atikah MSN, Asyraf MRM, Ayu Rafiqah S, Aisyah HA, Mohd Nurazzi N, Norrahim MNF. Effect of hydrolysis time on the morphological, physical, chemical, and thermal behavior of sugar palm nanocrystalline cellulose (Arenga Pinnata (Wurmb.) Merr). *Textile Research Journal.* 2020. In print. 10.1177/0040517520932393

6

Pectin-Based Bionanocomposite Coating for Food Packaging Applications

Dr. M. Vishnuvarthanan

Department of Printing Technology, College of Engineering Guindy, Anna University, Chennai, 600 025, Tamil Nadu, India

6.1 Introduction

In a recent scenario, packaging plays a very major and important role that protects and improves the shelf life of a product. The material handling would be very tough and ineffective without packaging and can be considered as a coordinated system for the product, designed for efficient delivery with high quality from the producer to the end user through goods managing, transportation, etc. In the modern world, the growth of the consumer market is effectively based on the innovations and efficiency of the packaging sectors. In current industrialized societies, packaging plays an important role in various fields, particularly in automotive, medical, food, electronics, etc.

Among them, the largest user of packaging materials is in the food packaging industry. The main aim is to protect the food in an efficient manner that satisfies both the consumer requirements and industry and also to maintain the safety of the food while minimizing the negative impact on the environment. There are several types of packaging materials that are used in the food packaging sector. The most common materials used are glass and metal. However, in the modern period, in food packaging applications plastics have become the material of choice because of their low cost and functional advantages over conventional materials.

Bio-Based Packaging: Material, Environmental and Economic Aspects, First Edition. Edited by S.M. Sapuan and R.A. Ilyas

6.2 Polymers in Food Packaging

Different types of polymers like polyvinylidene chloride, polypropylene, polyester, polyethylene, polystyrene, polyamide, and ethylene vinyl alcohol are being used in food packaging. Polypropylene (PP) is the most common plastic material used in food packaging applications. It has a clear glossy film with high strength and good barrier properties and changes in environmental conditions do not have any major influence on its properties. It is resistant to many chemical solvents and has been considered as the most important packaging material in the food sector and can easily be recycled. However, it has a poor wetting property due to its low surface energy. Hence, it is important to modify the surface characteristics of the polypropylene film, based on the property requirements in food packaging applications.

6.3 Surface Modification of Polymers

The plasma treatment on a polymer surface has been identified as an effective treatment to improve the polymer surface properties [1]. It is a process that improves the bonding characteristics by increasing the surface energy of the polymers. In general, the material wettability is greatly influenced by the presence of polar groups and the morphology of the surface [2]. By using different types of gases, the polar groups adhered to the polymer surface. In comparison with other gases, the oxygen was found to be the efficient gas to achieve suitable properties for a particular application [3]. Hence, the oxygen plasma surface modification technique is the most preferable method for coating the antimicrobial compounds on the plastic film [4].

6.4 Antimicrobial Packaging

Antimicrobial packaging is a form of active packaging that decreases or hinders the growth of microorganisms after packaging [5]. There are different approaches that can be taken to make the plastic packaging materials inhibit or retard the growth of microorganisms. The antimicrobial packaging materials can be developed either by coating antimicrobial materials on the surface of the packaging materials or by blending them into the polymeric materials. However, the blending method is limited due to the property limitations in thermal stability of the antimicrobial materials during the extrusion process or due to the incompatibility of the polymer with the antimicrobial material [6]. To prepare an antimicrobial packaging film, several coating materials have been used. They are prepared by adding different fillers with the coating materials in order to increase the properties. Mostly, naturally occurring materials such as carrageenan, pectin, and nisin are used as approved antimicrobial materials for food packaging applications.

6.5 Biopolymers

In recent times, the demand for biopolymers in food packaging has been greatly increased [7]. Several types of biopolymers have been used as packaging material since they are renewable, abundant, and environmentally friendly. The use of biopolymers in the field of food packaging is to improve the shelf life and to maintain the quality of the food and they should have good barrier and antimicrobial properties in addition to the mechanical property and film-forming ability. For food packaging applications, the carbohydrate-based biopolymers attracted much attention among the other biopolymers because of their excellent film-forming property, availability, and sustainability [8].

6.6 Pectin

Pectin is one of the most important carbohydrate-based biopolymers and attracted much attention in the field of food packaging, due to its processing behavior, is environmentally friendly, is a good barrier, and its mechanical properties. It is a complex anionic polysaccharide material consisting of linear chains of homogalacturonan with a complex structure and is low in cost. Hence, it can be used in the food and pharmaceutical industries due to its nontoxicity, bioadhesion, and biodegradability. The most challenging subject is to enhance the property of the biopolymers for food packaging applications with the required properties with respect to food safety and shelf life improvement [9]. The use of pectin films in food packaging applications is limited because of the inadequate barrier and its mechanical properties. In order to enhance the properties of pectin films, it should be reinforced with nanofillers such as nanoclay and nanometals [10].

6.7 Bionanocomposites

It has been recommended that the intrinsic inadequacy of biopolymer-coated packaging material may be overcome by nanocomposite technology. The nanocomposite shows an increase in mechanical and barrier properties compared to their neat polymers and other conventional composites. The bionanocomposites are expected to resist the stress of thermal conditions, storage, and transportation. In addition to this, because of their superior mechanical and barrier properties it may reduce the usage of source materials.

6.8 Nanoclay

Clay-based nanocomposites gave rise to an amazing interest from the food packaging industry. Nanoclays such as saponite, hectorite, and montmorillonite (MMT) are promising materials with high-aspect ratios and surface area. Because of their unique platelet-like structure, they have good barrier and mechanical properties. Among the nanoclays, the montmorillonite K10 clay (MMTK10) has been used widely because of its high surface area and ion exchange capacity. It also acts as a carrier for other nanoparticles [11]. The MMTK10 has a layered structure consisting of negatively charged aluminosilicates with a catalytic proton and is an acid treated clay. MMTK10 has a surface area of 210 - 280 $m^2\ g^{-1}$.

6.9 Silver Nanoparticles

In active food packaging, the protection of packaged foods against microbial contamination and to extend the shelf life, biopolymers with functional properties have been used [12].

Among the nanometals, silver nanoparticles (AgNPs) have attracted great attention in the food packaging sectors due to their exceptional antimicrobial activity against microorganisms [13]. The antimicrobial activity of the silver nanoparticles completely depends on the size, shape, stabilizing agents, and the methods of preparation. Compared to physical and chemical methods, the silver nanoparticles synthesis by biological methods is considered to be more effective, convenient, and environmentally friendly. For large-scale production, green synthesis of AgNPs from the plant extract is well recognized.

There is no report on nanoparticle synthesis from *Digitalis purpurea* plant extract. It is a plant species plentifully present in the UK, Germany, and in the southern parts of India, especially in the state of Tamil Nadu. In India, the flourishing digitalis plants are collected, cut, dried, and carefully packed for selling to the drug manufacturer. This plant is also known as foxglove. It can be of interest for green synthesis as it contains a huge amount of medicinal sources. There are numerous diversities of foxglove available, which differ in flower color and size. The medicinally important glycosides are found in the plant, a primary value of the foxglove. Digitalis is the most effective cardiovascular drug offered for heart failure that is caused by arteriosclerosis and hypertension. It can also be used medicinally as a heart regulator. The synthesis of silver nanoparticles by bio methods using aqueous extracts of *D. purpurea*, characterization of silver nanoparticles, and investigations on antibacterial activity have been reported [14].

6.10 Pectin-Based Bionanocomposite Coating

6.10.1 Preparation and Coating of Pectin-Based Bionanocomposite

The pectin and MMTK10 nanoclay were pre-dried in a hot air oven for about 6 hours. For the preparation of nanocomposites, the amount of pectin to be used was optimized. The MMTK10 with different wt% of 1, 3, 5, and 7 were dispersed each separately in 100 ml of double distilled water. The solutions were stirred by a magnetic stirrer for 12 hours and then homogenized for 15 minutes at 5000 rpm to get a completely swollen clay mineral. Then 4 g of pectin powder was mixed with the prepared MMTK10 clay solution with 0.5 ml of glycerol along with 20 μg of AgNPs solution. Then it was heated for about 30 minutes at 60°C with vigorous mixing using a magnetic stirrer. The solutions were then sonicated for 45 minutes to avoid agglomeration. The prepared pectin nanocomposites with different wt% of 0, 1, 3, 5, and 7 using MMTK10 nanoclay were designated as PK0, PK1, PK3, PK5, and PK7, respectively.

The coating of the prepared nanocomposites on the surface of modified PP films was carried out by a rod coater. The pectin-based nanocomposite solutions were taken in a syringe and then poured on to the oxygen plasma treated polypropylene of dimension 15 cm × 10 cm. The coated PP films were dried at room temperature for 24 hours. The average coating thickness obtained was about 24 μm based on the specification of the selected rod for coating.

For the measurement of pectin nanocomposite coating thickness, the coated samples were analyzed by a scanning electron microscope using the cross-sectional method. The film thickness was measured and is shown in Figure 6.1. The value of coating thickness was found to be 24.1 μm.

Figure 6.1 *Coating thickness measurement by cross-sectional method.*

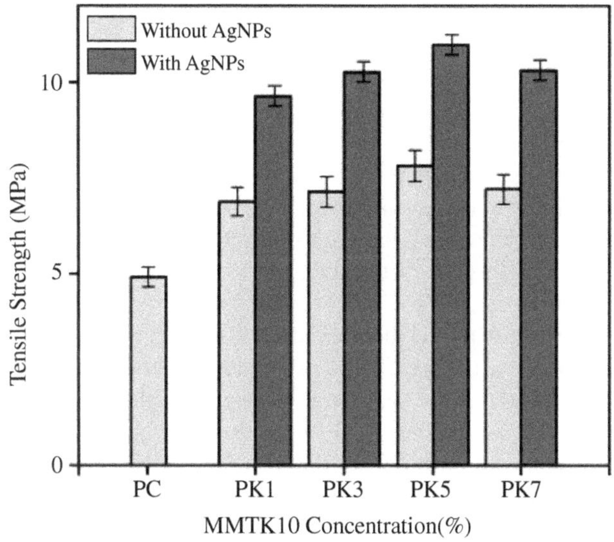

Figure 6.2 *Tensile strength of Pectin/MMTK10/AgNPs nanocomposite coated films.*

6.10.2 Tensile Strength

The tensile strength of the oxygen plasma surface modified PP coated with pectin-based nanocomposites are shown in Figure 6.2. The tensile strength of the coatings tend to rely strongly as a composite on the synthetic substrate [15]. The tensile strength of the uncoated PP film is 4.680 MPa. After the pectin coating, the value increased to 6.88 MPa. After the addition of 1 wt% of MMTK10, the tensile strength was significantly increased to 6.88 MPa. After the incorporation of AgNPs, the value was further increased to 9.64 MPa. For the 3 wt% of MMTK10 the value was 7.14 MPa and along with the AgNPs, it was 10.27 MPa. The maximum tensile strength value was observed along with the AgNPs for 5 wt% of MMTK10.

This also indicates that the value of tensile strength increases with the increase of concentration of nanoclays. The further increase of MMTK10 concentration decreased the tensile strength for the nanocomposite coating. This was caused due to the nanoclay agglomeration in the pectin matrix, which leads to the reduction of reinforcement action of nanofillers [16]. The increase in tensile strength of the nanocomposite coated films was mainly due to the physical attraction between the components. The adhesion strength results showed the good results for the attraction between the PP and pectin nanocomposite coating. The mechanical force formed in the pectin nanocomposites is mainly due to the physical forces between them.

6.10.3 Oxygen Transmission Rate

The oxygen transmission rate (OTR) of the uncoated oxygen plasma treated polypropylene film and the pectin nanocomposite coated films are shown in Figure 6.3.

The OTR of the uncoated oxygen plasma treated film was 1853.32 cc (m^2 day atm)$^{-1}$. After the pectin coating, the OTR decreased to 1658.7 cc (m^2 day atm)$^{-1}$. After the addition of 1 wt% of MMTK10 in pectin, the value of OTR decreased to 1419.8 cc (m^2 day atm)$^{-1}$. The incorporation of AgNPs in 1 wt% of MMTK10, it further reduced to 1390.36 cc (m^2 day atm)$^{-1}$. For the 3 wt% of MMTK10, the value further decreased to 995.27 cc (m^2 day atm)$^{-1}$. Along with the AgNPs, it was 972.99 cc (m^2 day atm)$^{-1}$. The incorporation of nanoclay or metallic nanoparticles in the polymeric matrix affects the gas permeability and these materials have the ability to act as barriers for oxygen and control the transport of oxygen to extend the shelf life of the food [17]. The value of OTR decreased continuously after the addition of MMTK10 and AgNPs with pectin.

The highest reduction of OTR was achieved for 5 wt% of MMTK10 and the value was 761.6 cc (m^2 day atm)$^{-1}$. After the incorporation of AgNPs, the value was 743.74 cc (m^2 day atm)$^{-1}$. This may be due to the existence of the more tortuous path created by MMTK10 and AgNPs that can go around the impenetrable platelets [18]. The dispersion of clay mineral in the polymer matrix with the AgNPs may also develop a further efficient tortuous path for oxygen transmission. Due to the agglomeration of MMTK10 at 7 wt%, the value increased to 1048.1 cc (m^2 day atm)$^{-1}$. The MMTK10 showed good dispersion in the pectin

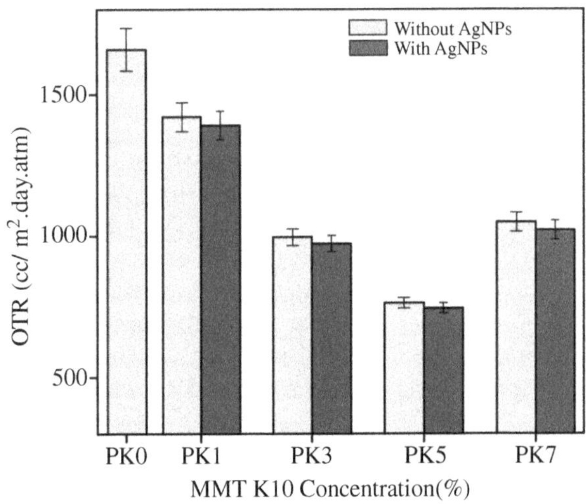

Figure 6.3 OTR of Pectin/MMTK10/AgNPs nanocomposite coated films.

at lower % levels and can increase the tortuousity of the diffusion path and have a significant improvement in barrier properties. The aspect ratio of the layered silicates is very high as compared to the other fillers used for the preparation of nanocomposites. The effective aspect ratio of the layered silicate is reduced because of agglomeration and a decrease in the length of the diffusion path.

6.10.4 Water Vapor Transmission Rate

The water vapor transmission rate (WVTR) of the pectin-coated oxygen plasma treated PP and the pectin nanocomposites coated PP films are shown in Figure 6.4.

The WVTR of the uncoated PP film was 9.62 g (m^2 day)$^{-1}$ and after the coating of pectin it decreased to 8.61 g (m^2 day)$^{-1}$. After the addition of 1 wt% of MMTK10, the WVTR value was 7.41 g (m^2 day)$^{-1}$. After the incorporation of AgNPs in 1 wt % of MMTK10, the value further decreased to 7.32 g (m^2 day)$^{-1}$. Reduction of the WVTR value was due to the effect of MMTK10, along with the AgNPs. For the 3 wt% of MMTK10, the value of WVTR further decreased to 5.2 g (m^2 day)$^{-1}$. With the addition of AgNPs, it showed the value of 5.09 and 5.78 g (m^2 day)$^{-1}$. The uniform dispersion of nanoparticles in the polymer matrix on the surface of the film decreases the permeability.

The MMTK10 and other layered silicates have the ability to improve the water vapor barrier for the synthetic and the biopolymer-based nanocomposites [19]. The efficient reduction of water vapor values were obtained for the pectin with 5 wt% of MMTK10 and the value was 4.04 g (m^2 day)$^{-1}$. Along with the AgNPs, it was further decreased to 3.93 g (m^2 day)$^{-1}$. It was found that there was a significant increase in WVTR with an increase in the clay content above 5 wt% of MMTK10. The value of WVTR was increased to 5.68 g (m^2 day)$^{-1}$ of 7 wt% of MMTK10. This was observed to be similar to the OTR results, and could be attributed to the aggregated structure of pectin nanocomposites

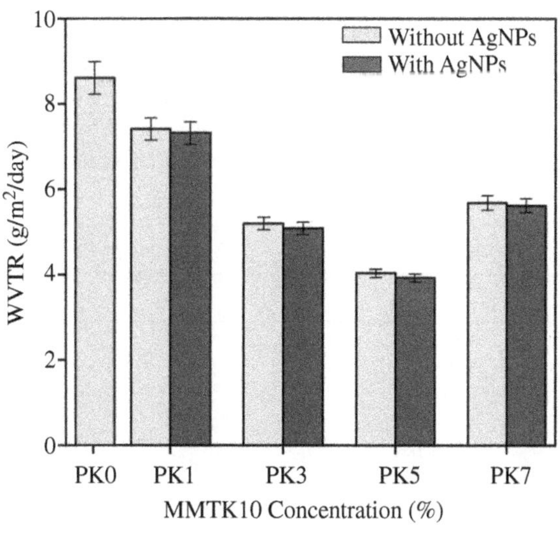

Figure 6.4 WVTR of Pectin/MMTK10/AgNPs nanocomposite coated films.

rather than the exfoliated material. Here, the materials acted as a discontinuous barrier for the diffusion of water vapor in the film matrix and thus the path length of the diffusion decreased. It resulted in an increase in the WVTR [20]. This result for the clays and nanoparticle agglomeration may present channels for water vapor to pass more readily [21].

6.10.5 Surface Color and Opacity

The surface color of the pectin nanocomposite-coated PP film is shown in Table 6.1. This result shows that the pectin, MMTK10, and AgNPs significantly changed the surface color of the film after the coating. The considerable change in the lightness value (L) was noted after the coating of pectin with MMTK10 nanoclay. For the MMTK10 nanoclay, when the clay content increases, the L value decreases. The L value ranged from 92.49 to 77.25 for MMTK10 where the uncoated PP film showed the value of 94.45.

The MMTK10 is more influenced by the value of L. There was a minor difference in a* value by an increase in the MMTK10 content. Similar results were obtained in b* value. From the results, it was observed that by increasing the concentration of MMTK10, the L* value decreased but the value of a* and b* increased. After the addition of AgNPs in the pectin coating, the value of L* was significantly reduced and the a* and b* values increased. This indicates that there is an increase of the redness and yellowness nature of the pectin nanocomposite coating [22]. This is mainly due to the brown color of the AgNPs caused because of the plasmonic effect. The UV penetration through the film is prevented by the brown color of the AgNPs, which can be properly used for UV protection.

The opacity values of oxygen plasma treated PP and the pectin nanocomposite coated PP films are shown in Figure 6.5. The opacity value for uncoated PP is 9.44%. After the pectin nanocomposite coating on PP, the opacity value increased due to the presence of light blockage particles. After the pectin coating, the opacity value increased to 12.2%. The value of opacity was higher after the incorporation of MMTK10. For the 1 wt% of MMTK10, the value was 24.12%. After the addition of AgNPs with 1 wt% of MMTK10, the value increased to 49.79%. It was noted that the opacity value increased by the addition of MMTK10. The maximum value of 25.28% was obtained for 7 wt% of MMTK10 and with the AgNPs it was 51.68%. The increase of nano clay content in the pectin coating showed the higher value of opacity, which was due to the strong scattering of light. The higher

Table 6.1 *Surface color values of Pectin/MMTK10/AgNPs nanocomposite coated PP films.*

	L*	a*	b*	ΔE
PCPP	92.49	−1.27	2.92	3.69
PK1	77.25	−0.51	7.97	19.06
PK3	76.06	−0.69	8.21	20.23
PK5	75.90	−0.73	8.67	20.57
PK7	75.41	−0.79	8.84	21.09
PK1 + AgNPs	49.92	3.12	10.61	46.03
PK3 + AgNPs	49.04	3.01	10.79	46.91
PK5 + AgNPs	48.56	2.54	10.86	47.35
PK7 + AgNPs	48.11	2.32	11.06	47.82

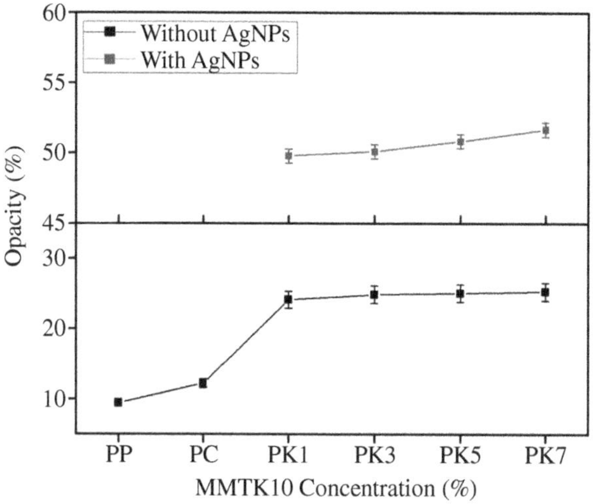

Figure 6.5 *Opacity of Pectin/MMTK10/AgNPs nanocomposite coated films.*

opacity value was obtained after the addition of AgNPs, which was due to the prevention of light transmission in the pectin by the dispersion and color of the nanoparticles [23]. Based on the results, the film is desirable to be used in food packaging due to the prevention of light penetration, which avoids light-driven reactions such as discoloration, lipid oxidation, off flavor formation, and nutrient loss [24].

6.10.6 Contact Angle Analysis

The wettability effect of MMTK10 and also the AgNPs on pectin nanocomposite coated polypropylene films were characterized by contact angle measurements. The contact angle values for the pectin and pectin nanocomposite coated PP films are shown in Figure 6.6.

The contact angle of the surface modified PP film is 60°. After the pectin coating this was reduced to 52.1° due to the hydrophilic nature of pectin. It may be mainly attributed to the functional carboxyl and hydroxyl groups of pectin. The measured contact angle for pectin coating is in good agreement with the previously reported literature [22]. The contact angle value reduced to 50.2° after the addition of 1 wt% of MMTK10. After the incorporation of AgNPs, the values increased to 51.4° for MMTK10. Further, the value reduced to 48.6° of 3 wt% of MMTK10. The minimal reduction of contact angle was obtained at 5 wt% for the nanoclay. The value increased after the addition of AgNPs to both 3 wt% and 5 wt%.

The contact angle was further reduced by the increase in MMTK10 content in the pectin. The decrease in contact angle value was due to the hydrophilic nature of MMTK10 as confirmed by the earlier report [25]. The value of contact angle increased after the addition of AgNPs with the nanoclay. This result shows the hydrophobic nature of AgNPs. These results infer that the pectin nanocomposite coating with AgNPs has a relatively higher contact angle compared to the coated film without the AgNPs. This is because of the hydrophobic character of the AgNPs and not of the nanoclay.

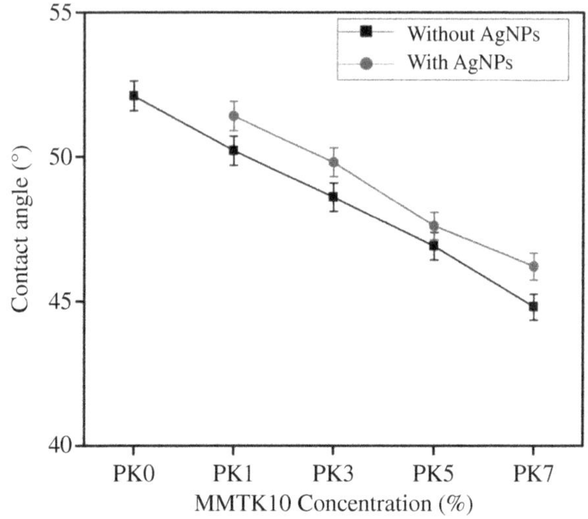

Figure 6.6 *Contact angle of Pectin/MMTK10/AgNPs nanocomposite coated PP films.*

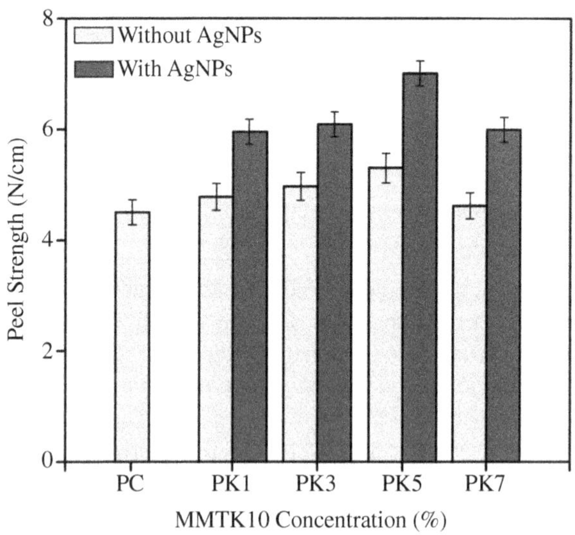

Figure 6.7 *Peel strength of Pectin/MMTK10/AgNPs nanocomposite coated films.*

6.10.7 Coating Adhesion Strength

The T-peel strength of the pectin coated and pectin nanocomposites coated PP was determined and the results are shown in the Figure 6.7. The peel strength of the uncoated oxygen plasma surface modified PP is 3.48 N cm^{-1}. After the pectin coating it increased to 4.5 N cm^{-1}. After the addition of 1 wt% of MMTK10, the peel strength was increased to 4.78 N cm^{-1}. The value was further increased to 5.96 N cm^{-1} after incorporation of AgNPs along with the 1 wt% of nanoclay. The peel strength value gradually increased with an increase in concentration of MMTK10. The maximum peel strength value was obtained as 7.01 N cm^{-1} for 5 wt% of MMTK10 along with the AgNPs. This was due to the increase of surface roughness and surface energy caused by oxygen plasma treatment. The key factor for the increase in adhesion strength is due to surface roughness. The oxygen plasma treatment made the PP surface polar in nature and thus created the adhesion between the PP surface and the coated pectin nanocomposites due to the van der Waals interaction. There was no significant increase of peel strength observed with an increase in the MMTK10 concentration in pectin along with the AgNPs. The high value was achieved because of the exfoliated nature of MMTK10. Furthermore, the interlocking of the polymer matrix, nanoclay, and AgNPs into the cavities, pores, and asperities of the solid surface are also the main factors in determining the adhesive strength. The mechanical interlocking can be attributed to the increase in interfacial area due to surface roughness by plasma treatment and created the pitted structure on the polymer.

6.10.8 Antimicrobial Properties

The antimicrobial activity of pectin-based nanocomposite coated PP was investigated by the agar well diffusion method against both gram-negative and gram-positive microorganisms using *E. coli* and *S. aureus*. The antimicrobial activities of the coated films were determined by the zone of inhibition and the results of the film samples are shown in Figure 6.8. The pectin-coated oxygen plasma surface modified PP film showed slight activity against both Gram-positive (DI = 3 mm) and Gram-negative (DI = 4 mm) bacteria. After the addition of MMTK10 with pectin, there was no change in the zone of inhibition. After the incorporation of AgNPs in the pectin along with MMT K10, it showed more prominent activity against both the *E. coli* (DI = 20 mm) and *S. aureus* (DI = 14 mm) due to the nanosize with a very high surface area. The difference in the antimicrobial activity between Gram-negative and Gram-positive bacteria might be due to the difference in the structures of their cell walls [22]. The Gram-positive bacteria are composed of three-dimensional thick peptidoglycan layers consisting of linear polysaccharide chains cross-linked with short peptides with a complex wall structure, which may prevent the clay and silver nanoparticles from penetrating. However, in Gram-negative bacteria, the AgNPs penetrate easily due to the thin

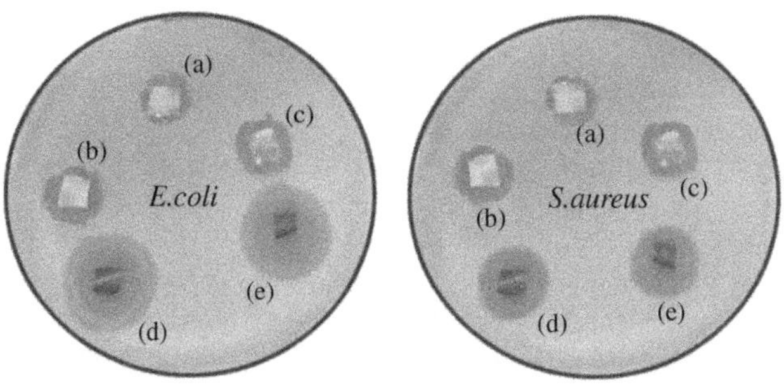

Figure 6.8 Antimicrobial properties of pectin nanocomposite coated films: (a) Pectin, (b) PK5, (c) PK7, (d) PK5-AgNPs, (e) PK7-AgNPs.

peptidoglycan layer on the outer membrane of the bacteria [26]. Many mechanistic actions for the antimicrobial activity of silver nanoparticles have been reported, though the exact mechanism has not been established until now. The interaction of silver nanoparticles disrupts the cell membrane preventing the replication of DNA and leads to cell death [27]. The cell wall and surface protein disruption is caused by the binding of the negatively charged bacterial cell membrane with the positively charged AgNPs, which leads to cell death. The enzyme inactivation is caused by AgNPs penetration into the bacteria and produces H_2O_2 and leads to cell death.

The pectin nanocomposite-coated PP films composed of MMTK10 have not exhibited any antimicrobial property. However, along with the AgNPs, the antimicrobial activity has greatly improved for both Gram-positive and Gram-negative organisms. The addition of MMTK10 has significantly improved the barrier properties, which have a high potential in the area of food packaging applications. In addition, it had been proven that the addition of nanofiller within the polymer composite could enhanced the mechanical and barrier properties of composites [25, 27].

6.11 Conclusions

The present work has focused on the preparation of pectin/MMTK10/AgNPs bionanocomposites and coating on the polypropylene, which was surface modified with oxygen plasma. The decreasing value of the contact angle represents the hydrophilic nature of the samples. After the addition of MMTK10 and AgNPs, the surface color and opacity of the coatings were changed considerably. Incorporation of MMTK10 and AgNPs in the pectin results in the increase of tensile strength and reduction of OTR and WVTR. The maximum reduction was achieved at 5 wt% of MMTK10 and was due to the exfoliated nature. Similarly, the peel strength increased after the addition of MMTK10 and AgNPs. The pectin nanocomposite coating showed good antimicrobial activity against *E. coli* and *S. aureus*.

References

1. Peyroux J, Dubois M, Tomasella E, Petit E, Flahaut D. Enhancement of surface properties on commercial polymer packaging films using various surface treatment processes (fluorination and plasma). *Applied Surface Science.* 2014; **315**:426–31.
2. Kostov KG, Nishime TM, Hein LR, Toth A. Study of polypropylene surface modification by air dielectric barrier discharge operated at two different frequencies. *Surface and Coatings Technology.* 2013; **234**:60–6.
3. Vesel A, Junkar I, Cvelbar U, Kovac J, Mozetic M. Surface modification of polyester by oxygen and nitrogen plasma treatment. *Surface and Interface Analysis: An International Journal devoted to the development and application of techniques for the analysis of surfaces, interfaces and thin films.* 2008; **40**(11):1444–53.
4. Vishnuvarthanan M, Rajeswari N. Effect of mechanical, barrier and adhesion properties on oxygen plasma surface modified PP. *Innovative food science & emerging technologies.* 2015;**30**:119–126.
5. Gan I, Chow WS. Antimicrobial poly (lactic acid)/cellulose bionanocomposite for food packaging application: a review. *Food packaging and shelf life.* 2018;**17**:150–61.
6. Theapsak S, Watthanaphanit A, Rujiravanit R. Preparation of chitosan-coated polyethylene packaging films by DBD plasma treatment. *ACS applied materials & interfaces.* 2012; **4**(5):2474–82.

7. Duncan TV. Applications of nanotechnology in food packaging and food safety: barrier materials, antimicrobials and sensors. *Journal of colloid and interface science.* 2011; **363**(1):1–24.
8. Rhim JW, Wang LF. Preparation and characterization of carrageenan-based nanocomposite films reinforced with clay mineral and silver nanoparticles. *Applied Clay Science.* 2014;**97**: 174–81.
9. Nešić A, Gordić M, Davidović S, Radovanović Ž, Nedeljković J, Smirnova I, Gurikov P. Pectin-based nanocomposite aerogels for potential insulated food packaging application. *Carbohydrate polymers.* 2018; **195**:128–35.
10. Shankar S, Rhim JW. Amino acid mediated synthesis of silver nanoparticles and preparation of antimicrobial agar/silver nanoparticles composite films. *Carbohydrate polymers.* 2015; **130**:353–63.
11. Shameli K, Ahmad MB, Jazayeri SD, Shabanzadeh P, Sangpour P, Jahangirian H, Gharayebi Y. Investigation of antibacterial properties silver nanoparticles prepared via green method. *Chemistry Central Journal.* 2012; **6**(1):73.
12. Alkan D, Yemenicioğlu A. Potential application of natural phenolic antimicrobials and edible film technology against bacterial plant pathogens. *Food Hydrocolloids.* 2016; **5**:1–10.
13. Solomon MM, Umoren SA. In-situ preparation, characterization and anticorrosion property of polypropylene glycol/silver nanoparticles composite for mild steel corrosion in acid solution. *Journal of colloid and interface science.* 2016; **462**:29–41.
14. Vishnuvarthanan M, Rajeswari N. Plant mediated greener approach for synthesis of silver nanoparticles from *Digitalis purpurea* plant and its antibacterial activity. *International Journal of Nanoparticles.* 2017;**9**(3):166–79.
15. Khwaldia K, Arab Tehrany E, Desobry S. Biopolymer coatings on paper packaging materials. *Comprehensive reviews in food science and food safety.* 2010; **9**(1):82–91.
16. Swain S, Sharma RA, Bhattacharya S, Chaudhary L. Effects of nano-silica/nano-alumina on mechanical and physical properties of polyurethane composites and coatings. *Trans. Electr. Electron. Mater.* 2013;**14**(1):1–8.
17. Abreu AS, Oliveira M, de Sá A, Rodrigues RM, Cerqueira MA, Vicente AA, Machado AV. Antimicrobial nanostructured starch based films for packaging. *Carbohydrate polymers.* 2015; **129**:127–34.
18. Kasirga Y, Oral A, Caner C. Preparation and characterization of chitosan/montmorillonite K10 nanocomposites films for food packaging applications. *Polymer composites.* 2012; **33**(11):1874–82.
19. Bae HJ, Park HJ, Hong SI, Byun YJ, Darby DO, Kimmel RM, Whiteside WS. Effect of clay content, homogenization RPM, pH, and ultrasonication on mechanical and barrier properties of fish gelatin/montmorillonite nanocomposite films. *LWT-Food Science and Technology.* 2009; **42**(6):1179–86.
20. Johansson C, Clegg F. Effect of clay type on dispersion and barrier properties of hydrophobically modified poly (vinyl alcohol) - bentonite nanocomposites. *Journal of Applied Polymer Science.* 2015; **132**(28).
21. Shankar S, Tanomrod N, Rawdkuen S, Rhim JW. Preparation of pectin/silver nanoparticles composite films with UV-light barrier and properties. *International journal of biological macromolecules.* 2016; **92**:842–9.
22. Kanmani P, Rhim JW. Physicochemical properties of gelatin/silver nanoparticle antimicrobial composite films. *Food chemistry.* 2014; **148**:162–9.
23. Ramos ÓL, Reinas I, Silva SI, Fernandes JC, Cerqueira MA, Pereira RN, Vicente AA, Poças MF, Pintado ME, Malcata FX. Effect of whey protein purity and glycerol content upon physical properties of edible films manufactured therefrom. *Food Hydrocolloids.* 2013;**30**(1):110–22.
24. Corrales T, Larraza I, Catalina F, Portolés T, Ramírez-Santillán C, Matesanz M, Abrusci C. In vitro biocompatibility and antimicrobial activity of poly (ε-caprolactone)/montmorillonite nanocomposites. *Biomacromolecules.* 2012;**13**(12):4247–56.
25. Priyadarshini S, Gopinath V, Priyadharsshini NM, MubarakAli D, Velusamy P. Synthesis of anisotropic silver nanoparticles using novel strain, Bacillus flexus and its biomedical application. *Colloids and Surfaces B: Biointerfaces.* 2013; **102**:232–7.

26. Vishnuvarthanan M, Rajeswari N. Food packaging: pectin–laponite–Ag nanoparticle bionanocomposite coated on polypropylene shows low O_2 transmission, low Ag migration and high antimicrobial activity. *Environmental Chemistry Letters.* 2019;**17**(1):439–45.
27. Raffi M, Hussain F, Bhatti TM, Akhter JI, Hameed A, Hasan MM. Antibacterial characterization of silver nanoparticles against E. coli ATCC-15224. *Journal of materials science and technology.* 2008; **24**(2):192–6.
28. Ilyas RA, Sapuan SM, Atikah MSN, Asyraf MRM, Ayu Rafiqah S, Aisyah HA, Mohd Nurazzi N, Norrahim MNF. Effect of Hydrolysis Time on the Morphological, Physical, Chemical, and Thermal Behavior of Sugar Palm Nanocrystalline Cellulose (Arenga Pinnata (Wurmb.) Merr). *Textile Research Journal.* 2020. In print. doi: 10.1177/0040517520932393.

7

Nanocomposite: Potential Nanofiller for Food Packaging Applications

Rafeeya Shams[1], Qurat ul eain Hyder Rizvi[2], Aamir Hussain Dar[3], Ishrat Majid[3], and Shafat Khan[3]

[1] *Sher-e-Kashmir University of Agricultural Sciences and Technology, Jammu, India*

[2] *Eternal University, Baru, HP, Sahib, India*

[3] *Department of Food Technology Islamic, University of Science and Technology, Jammu and Kashmir, Awantipora, India*

7.1 Introduction

The need and demand for minimum processed, ready-to-eat fresh, easily prepared food products, food trade globalization, and distribution from central processing possess the foremost challenges for food quality as well as food safety. Presently, various studies are being conducted to develop nonthermal processing methods (such as HPP, PEF, IR, etc.) as an alternative for thermal processing techniques to extend shelf life and to keep food products fresh for longer. Although some of these techniques may decontaminate food, these are expensive and energy-intensive, and hence have relatively restricted commercial applications. However, the appropriate packaging technique can be used to diminish food losses and make food products sound and safe. The demand for the production of renewable polymers from natural resources has been increased due to the need to minimize the quantity of plastic waste in the environment. Biopolymers, as raw material, can be used as a substitute for the development of biodegradable packaging material to plastic obtained from petroleum. To maintain the food quality, it is essential to select the correct materials

Bio-Based Packaging: Material, Environmental and Economic Aspects, First Edition. Edited by S.M. Sapuan and R.A. Ilyas

and appropriate technologies for the production of the packaging. Current trends include the production of packaging that interacts with food, called active biofilms or active packaging. Nanocomposites having antimicrobial properties can be largely used to minimize the contamination caused by microbial growth during the storage as well as processing of food and thus improves the food safety and extends the shelf-life. A nanocomposite is a multiphase material obtained by combining two or more components, such as a matrix (continuous phase) and a discontinuous nano-dimensional phase with a minimum of one nano-sized dimension (less than 100 nm). Two approaches are frequently used to form nanomaterials, "top-down" and "bottom-up." In the first approach, nanometric structures are produced by size reduction of bulk materials, such as grinding, laser, and chemical abrasion.

In contrast, the second allows nanostructures to be obtained from individual atoms or molecules having the capability of self-assembling, which consists of chemical reactions of reducing the metal ion of salt through a reducing agent [1]. The nano-dimensional phase is divided into three categories depending on the number of nanosized dimensions. Nanospheres or nanoparticles have three dimensions in the nanoscale. Both nanowhiskers (nanorods) and nanotubes have two nanometric dimensions, but nanotubes are hollow and nanowhiskers are solid in shape. Nanosheets or nanoplatelets have only single nano-sized dimension [2]. Due to the larger surface to volume ratios, nanoparticles exhibit different physical, chemical, mechanical, thermal, electrical, and optical properties compared to those of micro- and macro-scale particles. In food applications, nanotechnology is being used to enhance taste, improve the availability of nutrients, storage, preservation, processing, as well as packaging [3, 4]. Recent research emphasis has been on the use of nanotechnology in food packaging applications.

7.2 Nanofillers

At the nano-scale level, the size of the filler or nano phase is considerably decreased, causing an increase in the surface area of the fillers, which is preferred as bionanocomposites rely on the high surface area of the nanosized fillers, resulting in the large interfacial or boundary area between the matrix or biopolymer and nanofiller. Various types of nanosized fillers (less than 100 nm) are being used to improve the performance of biopolymers.

The nanofillers that are mostly used for food packaging can be categorized into nanoparticles, nanorods, nanofibrils, as well as nanotubes. The nanosized fillers may be either organic or inorganic, including clay (such as montmorillonite [MMT] and kaolinite), natural antimicrobial agents (e.g. nisin), natural biopolymers (e.g. chitosan), metal (e.g. silver), and metal oxides (e.g. TiO_2), carbon nanotubes (CNTs), and graphene nanosheets.

Among them, clays and silicates are significantly used due to their layered structures. They are plentiful, inexpensive, simple to process, and offer considerable improvements [5]. There are three essential polymer-clay morphologies, such as tactoid (or phase separated), intercalated, as well as exfoliated [6].

In the tactoid structure, mostly occurring in microcomposites, the polymer chains and the clay gallery are immiscible as they have poor affinity toward each other. Nanocomposite structures do not have this morphology [2]. In ideal polymer-clay nanocomposites, a higher affinity exists between the polymer and clay, resulting in exfoliated structures where polymer chains penetrate into the interlayer space of the clay, and thus make single sheets.

If the clay displays a moderate affinity for the polymer, it will result in intercalated structures [7]. In the literature, various other particle fillers are also being used, such as silver, titanium dioxide, zinc oxide, copper, and copper oxides, CNTs as well as graphene nanoplates (GNPs).

It has been reported that GNPs can form nanocomposites with enhanced heat resistance and barrier properties that are promising in food packaging applications [8]. GNPs, 20–60 nm in thickness and 0.5 to 25 mm in diameter, dispersed in poly(methyl methacrylate) (PMMA), were recently reported to increase the glass transition temperature for PMMA by 30°C at 1–5 wt% loading [8]. GNPs were dispersed by high-speed shearing methods.

CNTs, another carbon-based nanofiller, are also considerably used due to their intrinsic mechanical and electrical properties. Enhancements in modulus and strength of 30% and 15%, respectively, are reported for 1 wt % loading of functionalized single-walled CNTs in epoxy. However, its use in nanocomposites to date has been limited due to their high cost, difficulty in dispersion, as well as processing [7]. Copper, copper oxide, zinc oxide, titanium dioxide, and silver are mostly used for their antimicrobial properties [9].

7.2.1 Nanoclays

Nanoclays are the most studied nanofillers, because of their easy availability, less cost, better performance, and improved processability. The first publications regarding the applications of polymer-nanoclay composites to food packaging date from the 1990s [10]. The clays for nanocomposites are mostly bidimensional platelets with very small thicknesses (mostly around 1 nm) and several micrometers in length.

In contrast to the typical tactoid structure of microcomposites (conventional composites), where the polymer and the clay tactoids are immiscible [11, 12], the interaction between layered silicates and polymers produces two types of nanoscale composites, such as intercalated nanocomposites, resulting from penetration of polymer chains into the interlayer region of the clay, thus producing an ordered multilayer structure with alternating polymer/inorganic layers [13] and exfoliated nanocomposites, where extensive polymer penetration is involved, with the clay layers delaminated and randomly dispersed in the polymer matrix [12]. Exfoliated nanocomposites possess the best properties because of their optimal clay–polymer interactions [11, 14]. The commonly used clay is montmorillonite (MMT), having the general chemical formula Mx(Al4-xMgx) $Si_8O_{20}(OH)_4$. MMT is a representative of 2 : 1 layered phyllosilicates, with platelets having two layers of tetrahedral silica sheets filled with a central octahedral alumina sheet [13]. These clays have a moderate negative surface charge, which is essential to describe the interlayer spacing [2].

The surface negative charge imbalance is balanced by exchangeable cations (mostly Na^+ and Ca^{2+}). The parallel layers are bound with each other by weak electrostatic forces [15]. MMT is a tremendous reinforcing filler, due to its high surface area and large aspect ratio, ranging from 50 to 1000 [16]. The hydrophilic nature of the surface of most clays is responsible for its less dispersion in organic matrices [17]. Organoclays, obtained by interactions of clays and organic compounds, are widely used in polymer nanocomposite applications. A suitable organophilization is important for effective exfoliation of clays in most polymeric matrices, as organophilization lowers the energy of clays and enhances their compatibility with organic polymers [18]. Organomontmorillonite (oMMT) are obtained by exchanging inorganic cations of MMT with organic ammonium ions, enhancing compatibility of MMT with organic polymers [19], resulting in a more even organization of the layers, and reducing the water uptake by the resulting nanocomposite [20]. The enhanced barrier properties of polymer–clay nanocomposites are mostly due to improved tortuosity of the diffusive path for permeants, forcing them to travel a longer path to diffuse through the film [14]. The rise in path length is a function of the aspect ratio of the clay and the volume fraction of the filler in the composite. Nielsen's models are effectively used to expect the permeability of systems at clay loadings of less than 1%, but some experimental data has shown less permeabilities than expected at higher loadings [14]. Beall [21] proposed a new model to predict permeability of nanocomposites focused on the polymer–clay interface as an additional governing factor of the tortuous path, thus providing a correction factor to Nielsen's model. Clays are also used to enhance the mechanical strength of biopolymers [22], though they may reduce the polymer elongation [23].

7.2.2 Silica (SiO_2)

Silica nanoparticles ($nSiO_2$) have been reported to improve mechanical and/or barrier properties of several polymer matrices. Wu et al. [24] found that the addition of $nSiO_2$ into a polypropylene (PP) matrix improved tensile properties of the material – not only strength and modulus, but also elongation. Enhancements in tensile characteristics such as elongation has also been reported for a starch matrix resulting from $nSiO_2$ addition [25], and it was also observed that $nSiO_2$ addition reduced the water absorption by starch. Jia et al. [26] produced nanocomposites of PVOH with $nSiO_2$ by radical copolymerization of vinyl silica nanoparticles and vinyl acetate. The nanocomposites had enhanced mechanical as well as thermal properties as compared to that of pure PVOH, due to strong interactions between $nSiO_2$ and the polymer matrix through covalent bonding. Tang et al. [27] produced starch/PVOH/$nSiO_2$ biodegradable films. The tensile strength and water resistance of the films were enhanced due to an increase in the $nSiO_2$ content. The miscibility and compatibility between film components were improved due to a rise in the intermolecular hydrogen bonds, as well as formation of C-O-Si groups, between $nSiO_2$ and starch or $nSiO_2$ and PVOH. Zhang and Rong [28] found that the tailorability of the composites can be enhanced due to the presence of grafting polymers on the surface of $nSiO_2$, which means that different species of grafting monomers result in different interfacial interactions as well as tensile properties.

7.2.3 Silver

Huang et al. [1] found that nanosized particles of silver (Ag) find a wide range of applications, such as food packaging. Stable Ag nanoparticles can be *ex situ* produced from a regular borohydride decrease of Ag+ ions and then dispersed into a polymerizable formulation, or we can say that nanoparticles can be *in situ* made in a polymerizable medium from precursors with improved dispersion ability. Ag has been known to have strong antimicrobial activity against a number of bacteria, such as *Enterococcus faecalis*, *Vibrio* cholera, and *Staphylococcus aureus*, and among fungi, such as *Aspergillus niger* and *Candida albicans* [29]. Additionally, Ag nanoparticles can also catalyze the absorption and decomposition of ethylene released from fruit metabolism, which can be postulated as a blocker for ethylene. Fruit and vegetable rotting and ripening caused by ethylene gas can, therefore, be retarded with an increase in shelf life of a product [1]. Mastromatteo et al. [30] observed the synergetic effect of Ag with modified atmosphere packaging with 50% of CO_2 and 50% N_2 to increase the shelf life of Fiordilatte packaged with and without the traditional covering liquid. The results showed that the synergistic effect of Ag nanoparticles in the coating may represent a valid preservation strategy to improve the diffusion of dairy products beyond the local market, due to the robust antimicrobial effect against *Pseudomonas* spp., *Escherichia coli*, and Enterobacteriaceae.

7.2.4 Gold

Gold nanoparticles have been shown to be a potential agent in food packaging. The proposal for the production of gold nanoparticles is based on the decrease of gold salts in the solution having the ionic silsesquioxane as a stabilizer. This stabilizer is important for the nanoparticle synthesis as it contains quaternary ammonium groups that stimulate water solubility and are recognized for their antimicrobial and inhibitory effect, which makes it an essential component for the preparation of bioactive films [31]. Pagno et al. [31], synthesized active biofilms of quinoa (*Chenopodium quinoa*) starch in combination with gold nanoparticles. The results suggest that the presence of gold nanoparticles enhances the optical, mechanical, and morphological characteristics of biofilm. Also, the active biofilms have strong antibacterial activity against food-borne pathogens, with inhibition percentages of 98% against *S. aureus* and 99% against *E. coli*.

7.2.5 Metal Oxide

Metal oxide nanoparticles also find a wide range of applications. Their effective application depends on controlled synthesis and solution-phase techniques that allow a higher degree of control over products. Metal oxide nanoparticles are mostly produced by the sole gel method in which the reaction is interrupted before gelation. The nanoparticle characteristics are determined by the nucleation, growth, and aging processes. The commonly used metal oxides in food packaging include metal oxide materials, namely titanium dioxide (TiO_2), magnesium oxide (MgO), zinc oxide (ZnO), and copper oxide (CuO) [32]. Sarojini et al. [33] have studied the application of ZnO in mahua oil-based polyurethane (PU) and chitosan, which shows enhanced hydrophobicity, antibacterial (against *S. aureus* and *E. coli*), and barrier characteristics of the film.

7.2.6 Zinc Oxide

Zinc oxide (ZnO) nanoparticles can be produced by various techniques such as precipitation, sonochemical, thermal decomposition, and microemulsion methods [34]. Once ZnO nanoparticles are formed, they tend to bind with each other to form agglomerated particles, which decreases their functional properties. It is essential to stabilize these nanoparticles using capping or stabilizing agents. As a stabilizing agent, melanin is a multifunctional biologically active compound [38]. Nanoparticles of ZnO possess antibacterial activity against *S. aureus* and *Listeria monocytogenes*, through the exploration of their inhibition and deactivation of cell development. [37]. Besides antibacterial activity, ZnO could enhance barrier properties and hydrophobicity of the film as Sarojini et al. [33] revealed in carrots.

7.2.7 Titanium Dioxide

Nano-sized titanium dioxide (TiO_2) particles are synthesized by various methods, in which solegel processing is the most common. TiO_2 is a commonly studied oxide nanoparticle because of its UV blocking characteristics as it is an effective short-wavelength light absorber with great photostability. Different from other nanoparticles, the antimicrobial activity of TiO_2 occurs via photocatalysts and, thus, the antimicrobial activity is only available in the presence of UV light. TiO_2 nanoparticles are active against food-borne pathogens, namely *L. monocytogenes*, *Salmonella choleraesuis*, and *Vibrio parahaemolyticus*, in the presence of UV radiation [29]. Xie and Hung [60] observed the addition of TiO_2 in three biodegradable polymers (cellulose acetate – CA, polycaprolactone – PCL and polylactic acid – PLA) and analyzed their photocatalytic bactericidal property. Results suggest that TiO_2 embedded in PCL and PLA composite films did not show a high bactericidal property against *E. coli*. Also, TiO_2 embedded CA film is potentially used as antimicrobial food packaging. Metak [35] showed that the combination of TiO_2 with Ag finds a wide range of applications in fresh foods and has effective antibacterial activity against *S. aureus*, coliforms, *E. coli*, and *L. monocytogenes*.

7.2.8 Copper Oxide

Copper oxide (CuO) properties rise with a decrease in size, when it is expected that finely dispersed bioactive Cu nanoparticles provide a better disinfectant effect. Due to its reduced size, it possesses a higher mobility of released Cu ions, which allows them to interact closely with bacterial membranes [39]. CuO oxidation by the reduction of borohydride indicates antimicrobial activity in ammonia-contained media [1]. Lomate et al. [40] prepared a packaging with CuO nanoparticles to increase the shelf-life of Peda (an Indian sweet dairy product). It was observed that Cu nanoparticles uniformly dispersed into the low-density polyethylene (LDPE) matrix and showed enhanced mechanical properties. The nanocomposite film also shows a higher antimicrobial activity against Gram-positive and Gram-negative food deteriorating microbes.

7.2.9 Chitosan Nanostructures

Chitosan is a linear polysaccharide and copolymer and is an effective antibacterial agent against a wide variety of microorganisms. Chitosan nanoparticles are usually prepared by interaction of oppositely charged macromolecules. The formation mechanism depends on electrostatic interaction between amine groups of chitosan and the negatively charged groups of a polyanion, such as tripolyphosphate (TPP) [65]. Chitosan nanostructures disperse well in biopolymers and can be potentially used as antimicrobial agents in edible or biodegradable food packaging systems. Various studies focus on antimicrobial effects of nanostructures in combination with chitosan and other antimicrobial agents [41]. Wang et al. [57] developed a polymer based on chitosan–ZnO supported by polyvinyl acetate (PVA) and verified it against *E. coli* and *C. albicans* in varying composite concentrations, where it was found to be an effective antimicrobial agent.

7.2.10 Carbon Nanotubes

CNTs are cylindrical in shape with nanoscale diameters consisting of one-atom thick single-wall nanotubes, or several concentric tubes known as multiwall nanotubes. They can be effectively used in food packaging to improve the mechanical properties of polymeric matrices and also to employ influential antimicrobial effects, as the long and thin CNTs puncture the microbial cells, resulting in their irreversible damage [1]. Yu et al. [63] used poly(3-hydroxybutyrate-co-3hydroxyvalerate) (PVHB) in combination with CNTs and found that the presence of CNTs can improve tensile strength and enhance the thermal stability of the polymer. With such enhanced properties, there is a potential to produce packaging for food and beverage applications.

7.3 Nanocomposites in Active Packaging

Nanofillers, like silver, zinc oxide, and magnesium oxide, have antimicrobial or antioxidant properties. The addition of such nanofillers in polymer or biopolymer matrices is responsible for inhibiting or retarding the effect on microbial growth and thus retard spoilage of food [1]. The major purpose for active packaging systems is to increase the shelf life of food products. They are also synthesized to enhance the safety of food and to reduce food waste [42]. Antimicrobial nanocomposite films are used due to their antimicrobial characteristics because of natural antimicrobial agents and their suitable structural integrity, which results from the barrier properties due to the nanocomposite matrix [46]. Nanoscale materials have a higher surface-to-volume ratio than microscale counterparts and are thus able to attach to a vast number of biological molecules, which is responsible for its improved efficiency [43]. Earlier findings have recognized potential applications of nanocomposites as growth inhibitors [44], bacteriocides, and antibiotic carriers [45]. Apart from its packaging purpose, nanocomposites are also used as delivery systems by helping the migration of functional additives, like minerals, probiotics, and vitamins, into the food [47]. These controlled release packagings are examples of a nanomaterial application in active packaging. Many antimicrobial nanocomposites used for food packaging have been developed from silver and have intense toxicity to a large number of microorganisms [48]. Authors have recommended various mechanisms for the antimicrobial property of silver nanoparticles, like rising cell permeability through attachment to the cell surface and making pits in the membranes [51], Ag^+ release [36] and damaging DNA followed by nanoparticle penetration inside the bacterial cell [49]. Active nanocomposites are commonly used as an alternative to conventional active plastic technologies or active sachets to increase food safety and quality. The term "active nanocomposite" refers to a plastic composite (polymer blend) consisting of an active, nanostructured material that confers an activity on the plastic matrix [62]. At least one of the dimensions of the active nanostructured material must be less than 100 nm in size. Nanoclays can also be used as active agent carriers. The efficiency of the active agent is improved due to its higher dispersion in the polymeric matrix and, hence, it is exposed more effectively to the substance on which it is required to act.

7.4 Nanocomposites in Intelligent Packaging

Intelligent packaging effectively communicates with the consumer and provides information about the condition of the food product through the food chain. Such packagings monitor, trace, as well as record inner and outer variations that are occurring in the product or its environment [1]. Application of reactive components in the form of nanoparticles and creating so-called nanosensors, the packaging will be able to respond to environmental changes, like temperature, the presence of oxygen, spoiled products, and microbial contamination [53]. Through the addition of nanosensors into food packaging, detection of various chemical components, pathogens, and toxins in food is possible. This accomplishes the necessity for providing exact expiration dates, which, in various cases, are not suitable for the products due to incorrect estimations of the product's condition during storage. Hence, they give a real-time status to the food's freshness/condition [54]. These sensors offer various advantages over time-consuming and costly conventional detection techniques, like high-performance liquid chromatography (HPLC). This includes high-throughput detection, cost effectiveness, speed, ease of operation, and less power requirements [56].

7.5 Nanomaterial Migration into the Food Matrix

Migration is one of the serious issues in food packaging. Migration is the unintended transfer of packaging materials into the food. This problem may affect the safety of food and, consequently, the health of the consumer [55]. It may also be responsible for objectionable organoleptic changes in food that is in contact with the packaging material, such as migration of TiO_2 into the lipid matrix, causing rancidity [32]. However, in some active packaging, nanomaterials are released deliberately. To evaluate the possibility of the nanomaterial's incorporation into food packaging and also assessing the associated safety problems, various migration tests are carried out under controlled conditions. The analysis condition depends on the type of food matrix or food simulant being tested [1]. Different factors affect the migration of nanomaterials into foodstuffs. These factors can be attributed to the nanoparticle's characteristics (such as solubility, particle size, concentration, molecular weight, and diffusivity in the polymer), environmental conditions (temperature, mechanical stress), condition of the food (pH, composition), packaging characteristics (viscosity and polymer structure), and the time of contact [1]. For example, elevated temperatures and low pH values raise the solubility of metal nanomaterials in aqueous solution, resulting in an increase in their migration [52]. Lesser molecular weight polymers having more free volume hasten the migration rate and diffusivity of the nanoparticles [50]. There is also a reverse relationship between the migration rate of a system and the nanoparticles size [42]. If the nature of the food is compatible with the type of packaging, the food itself may be absorbed into the polymer matrix, enlarging the gaps between the polymer chains and thus raising the rate of migration [1]. For instance, fats having more affinity for polyethylene (PE) and polypropylene (PP) may be absorbed by the packaging and lead to a rise in plastic mobility and higher migration rates [1]. The method of preservation of foods is also essential and can affect the migration of nanoparticles. It is also observed that microwave heating causes structural changes to the packaging, thus increasing the migration of silver ions [58]. High surface-to-volume ratios in nanomaterials prompts an active surface chemistry followed by probably undesirable chemical reactions. This may be difficult if their presence in the packaging increases the formation of unpredicted reaction byproducts during processing [59]. Nanoparticles may also affect the migration of other packaging components, either increasing or decreasing their migration rate. To carry out a migration analysis, the best method is to use actual food matrices, though it may be very tedious and difficult to directly evaluate the amount of migration into real food matrices due to the complex composition of foodstuffs. Using natural food stimulants is an alternative method used to measure the specific and overall migration of different packaging substances into foods [61]. The rate and amount of mass transport in the selected food stimulants must be the same as those that occur in the food matrix [64].

7.6 Commercial Aspects of Food Packaging

Nanocomposites have attracted significant research interest due to their potential application in various industrial sectors including transportation, defense systems, cosmetics, pharmaceutical delivery system, energy storage, food packaging, etc. The market for nanocomposites is growing rapidly and is expected to reach above $5100 million by 2020 and $7475 million by 2022. The clay-based nanocomposite segment was reported to be a global player in the nanocomposite market in 2017 [66]. Biodegradable and bio-based plastic materials have the potential to move the packaging industry from conventional packing to the next level of intelligent packaging and biodegradation packaging [67]. Recently, the European Commission released its communication "A European Strategy for Plastics in a Circular Economy" to ensure all plastic packaging material in the European Union market should be reusable or recycled by 2030 and making the European plastics system more resource efficient. In 2007, the global market share of starch (bioplastics) was 50% followed by polylactic acid (PLA). The global opportunity analysis and industry forecast reported that the starch-based bioplastic market is expected to reach US$ 561 million by 2023 with a compound annual growth rate of 3.7% from 2017 to 2023. Currently, bioplastics represent about 1% of the about 320 million tonnes of plastic produced annually. However, the demand and market for bioplastic materials are rising and global bioplastic production is predicted to reach 2.44 million tonnes in 2022.

7.7 Conclusion and Future Trends

The increasing scientific and industrial interest on the applications of nanomaterials in the packaging sector led to the development of numerous reinforcement materials. To protect food products from external factors and to maintain the quality and safety of the food packed inside, it is important to satisfy the barrier and mechanical properties of nanocomposite films. The incorporation of nanofillers like MMT, silver, and ZnO to the biopolymers including starch and PLA has potential to create new and innovative bionanocomposite materials with enhanced characteristics and performance. Nanofillers are able to enhance the thermal, mechanical, and barrier properties and to exhibit other desired functions and applications in food packaging, like antimicrobial agents, oxygen scavengers, and biosensors. Bionanocomposite materials are important alternatives to be used for food packaging applications rather than that of conventional plastics, especially those synthesized from petroleum. It provides a chance to create eco-friendly and sustainable food packaging materials to substitute for non-biodegradable materials. Though the nanocomposite is the outcome of recent research with validated results, there is still a question on safety issues with the possibility of diffusion/migration of nanomaterials from the packaging material to the food. There is limited scientific data concerning the migration of nanoparticles from the packaging material. Future research needs to concentrate on the potential toxicity and enhancement of water resistance characteristics of the nanocomposite materials and environmental safety.

References

1. Huang JY, Li X, Zhou W. Safety assessment of nanocomposite for food packaging application. *Trends Food Sci. Technol.*, 2015; **45**: 187–199.
2. Alexandre M, Dubois P. Polymer-layered silicate nanocomposites: preparation, properties and uses of a new class of materials. *Mater. Sci. Eng. R*, 2000; **28**: 1–63.
3. Bhushani J, Anu, Kurrey NK, Anandharamakrishnan C. Nanoencapsulation of green tea catechins by electrospraying technique and its effect on controlled release and in-vitro permeability. *J. Food Eng.*, 2017; **199**:82–92.
4. Kathiravan T, Moses J A, Anandharamakrishnan C. Nano food fortification. *Everyman's Sci.*, 2017; **4**:171–175.
5. Azeredo H, Mattoso LHC, Wood D, Williams TG, Avena-Bustillos RJ, McHugh TH. Nanocomposite edible films from mango puree reinforced with cellulose nanofibers. *J. Food Sci.*, 2009; **74**(5): 31–5.
6. McGlashan S A, Halley PJ. Preparation and characterisation of biodegradable starch-based nanocomposite materials. *Polym. Int.*, 2003; **52**(11): 1767–1773.
7. Arora A, Padua, GW. Review: nanocomposites in food packaging. *J. Food Sci.*, 2010; **75**(1): 43–9.
8. Ramanathan T, Abdala A, Stankovich S, Dikin D, Herrera-Alonso M, Piner R. Functionalized graphene sheets for polymer nanocomposites. *Nat. Nanotechnol.*, 2008; **3**(6): 327–331.
9. Duncan TV. Applications of nanotechnology in food packaging and food safety: barrier materials, antimicrobials and sensors. *J. Colloid Interface Sci.*, 2011; **363**(1): 1–24.
10. Ray S, Easteal A, Quek S Y, Chen X D. The potential use of polymer-clay nanocomposites in food packaging. *Int. J. Food Eng.*, 2006; **2** (4): artigo 5.
11. Ludueña L N, Alvarez V A, Vasquez A. Processing and microstructure of PCL/clay nanocomposites. *Mater. Sci. Eng., A*, 2007; (**460–461**): 121–129.
12. Alexandre B, Langevin D, Médéric P, Aubry T, Couderc H, Nguyen Q T, Saiter A, Marais S. Water barrier properties of polyamide 12/montmorillonite nanocomposite membranes: structure and volume fraction effects. *J. Membr. Sci.*, 2009; **328**(1–2): 186–204.
13. Weiss J, Takhistov P, McClements D J. Functional materials in food nanotechnology. *J. Food Sci.*, 2006; **71** (9): 107–116.
14. Adame D, and Beall GW. Direct measurement of the constrained polymer region in polyamide/clay nanocomposites and the implications for gas diffusion. *Appl. Clay Sci.*, 2009; **42**: 545–552.
15. Tan W, Zhang Y, Szeto Y S, Liao L. A novel method to prepare chitosan/montmorillonite nanocomposites in the presence of hydroxyl-aluminum olygomeric cations. *Compos. Sci. Technol.*, 2008; **68**(14): 2917–2921.
16. Uyama H, Kuwabara M, Tsujimoto T, Nakano M, Usuki A, Kobayashi S. Green nanocomposite from renewable resources: plant oil-clay hybrid materials. *Chem. Mater.*, 2003; **15**: 2492–2494.

17. Kim K Y, Lim H J, Park S M, Lee S J. Synthesis and characterization of high impact polystyrene/organically modified layered silicate nanocomposites. *Polymer (Korea)*, 2003; **27**: 377–384.
18. Paiva L B, Morales A R, Díaz F R V. Argilas organofílicas: características, metodologias de preparação, compostos de intercalação e técnicas de caracterização. *Cerâmica*, 2008; **54**: 213–226.
19. Paul M A, Alexandre M, Degée P, Henrist C, Rulmont A, Dubois P. New nanocomposite materials based on plasticized poly(L-lactide) and organo-modified montmorillonites: thermal and morphological study. *Polymer*, 2003; **44**(2): 443–450.
20. Picard E, Gauthier H, Gérard J F, Espuche E. Influence of the intercalated cations on the surface energy of montmorillonites: consequences for the morphology and gas barrier properties of polyethylene/montmorillonites nanocomposites. *J. Colloid Interface Sci.*, 2007; **307**: 364–376.
21. Beall GW. New conceptual model for interpreting nanocomposite behavior, In: *Polymer-Clay Nanocomposites*, Pinnavaia, T. J. and Beall, G. W. (Ed.), Wiley, New York: 2000; 267–279.
22. Cyras V P, Manfredi LB, Ton-that M T, Vázquez A. Physical and mechanical properties of thermoplastic starch/montmorillonite nanocomposite films. *Carbohydr. Polym.*, 2008; **73**: 55–63.
23. Petersson L, Oksman K. Biopolymer based nanocomposites: comparing layered silicates and microcrystalline cellulose as nanoreinforcement. *Compos. Sci. Technol.*, 2006; **66**: 2187–2196.
24. Wu C L, Zhang M Q, Rong M Z, Friedrick K. Tensile performance improvement of low nanoparticles filled-polypropylene composites. *Compos. Sci. Technol.*, 2002; **62**: 1327–1340.
25. Xiong H G, Tang S W, Tang H L, Zou P. The structure and properties of a starch-based biodegradable film. *Carbohydr. Polym.*, 2008; **71**: 263–268.
26. Jia X, Li Y, Cheng Q, Zhang S, Zhang B. Preparation and properties of poly(vinyl alcohol)/silica nanocomposites derived from copolymerization of vinyl silica nanoparticles and vinyl acetate. *Eur. Polym. J.*, 2007; **43**: 1123–1131.
27. Tang S, Zou P, Xiong H, Tang H. Effect ofnano-SiO2 on the performance of starch/polyvinyl alcohol blend films. *Carbohydr. Polym.*, 2008; **72**: 521–526.
28. Zhang MQ, Rong M Z. Performance improvement of polymers by the addition of grafted nano-inorganic particles. *Chin. J. Polym. Sci.*, 2003; **21**(6): 587–602.
29. Almeida ACS, Franco EAN, Peixoto FM, Pessanha KLF, Melo NR. Aplicação de nanotecnologia em embalagens de alimentos, *Polímeros*, 2015; **25**, 89–97.
30. Mastromatteo M, Conte A, Lucera A, Saccotelli MA, Buonocore GG, Zambrini AV, Nobile MAD. Packaging solutions to prolong the shelf life of Fiordilatte cheese: bio-based nanocomposite coating and modified atmosphere packaging, *LWT Food Sci. Technol.*, 2015; **60**: 230–237.
31. Pagno CH, Costa TMH, Menezes EW, Benvenutti EV, Hertz PF, Matte CR, Tosati JV, Monteiro AR, Rios AO, Flôres SH. Development of active biofilms of quinoa (Chenopodium quinoa W.) starch containing gold nanoparticles and evaluation of antimicrobial activity. *Food Chem.*, 2015; **173**: 755–762.
32. Azeredo HMC. Antimicrobial nanostructures in food packaging, *Trends Food Sci. Technol.*, 2013; **30**: 56–69.
33. Sarojini KS, Indumathi MP, Rajarajeswari GR. Mahua oil-based polyurethane/chitosan/nano ZnO composite films for biodegradable food packaging applications, *Int. J. Biol. Macromol.*, 2019; **124**: 163–174.
34. Roy S, Rhim JW. Carrageenan-based antimicrobial bionanocomposite films incorporated with ZnO nanoparticles stabilized by melanin, *Food Hydrocolloids*, 2019; **90**: 500–507.
35. Metak AM. Effects of nanocomposites based nano-silver and nano-titanium dioxide on food packaging materials, *Int. J. Appl. Sci. Technol.*, 2015; **5**: 26–40.
36. Morones JR, Elechiguerra JL, Camacho A, Holt K, Kouri JB, Ramirez JT. The bactericidal effect of silver nanoparticles. *Nanotechnology*, 2005; **16**(10): 2346–53.
37. Morsy MK, Khalaf HH, Sharoba AM, El-Tanahi HH, Cutter CN. Incorporation of essential oils and nanoparticles in pullulan films to control foodborne pathogens on meat and poultry products. *J. Food Sci.*, 2014; **79**: 675–682.

38. Solano, F. Melanin and melanin related polymers as materials with biomedical and biotechnological applications – cuttlefish ink and mussel foot proteins as inspired biomolecules. *International Journal of Molecular Sciences*, 2017; **18** (7): 1561.
39. Conte A, Longano D, Costa C, Ditaranto N, Ancona A, Cioffi N, Scrocco C, Sabbatini L, Contò F, Nobile M A D. A novel preservation technique applied to fiordilatte cheese. *Innovative Food Sci. Emerg. Technol.*, 2013; **19**: 158–165.
40. Lomate GB, Dandi B, Mishra S. Development of antimicrobial LDPE/Cu nanocomposite food packaging film for extended shelf life of peda, *Food Packag. Shelf Life*, 2018; **16**: 211–219.
41. Medina E, Caro N, Abugoch L, Gamboa A, Díaz-Dosque M, Tapia C. Chitosan thymol nanoparticles improve the antimicrobial effect and the water vapour barrier of chitosan-quinoa protein films, *J. Food Eng.*, 2019; **240**: 191–198.
42. Cushen M, Kerry J, Morris M, Cruz-Romero M, Cummins E. Nanotechnologies in the food industry–recent developments, risks and regulation. *Trends Food Sci. Technol.*, 2012; **24**(1): 30–46.
43. Laskin AI, Bennett JW, Gadd GM. *Advances in Applied Microbiology*: Academic Press, 2003.
44. Cioffi N, Torsi L, Ditaranto N, Tantillo G, Ghibelli L, Sabbatini L. Copper nanoparticle/polymer composites with antifungal and bacteriostatic properties. *Chem. Mater.*, 2005; **17**(21): 5255–5262.
45. Gu H, Ho P, Tong E, Wang L, Xu B. Presenting vancomycin on nanoparticles to enhance antimicrobial activities. *Nano Lett.*, 2003; **3**(9): 1261–3.
46. Rhim JW, Ng PK. Natural biopolymer-based nanocomposite films for packaging applications. *Crit. Rev. Food Sci. Nutr.*, 2007; **47**(4): 411–433.
47. Rodriguez F, Sepulveda HM, Bruna J, Guarda A, Galotto MJ. Development of cellulose eco nanocomposites with antimicrobial properties oriented for food packaging. *Packag. Technol. Sci.*, 2013; **26**(3): 149–60.
48. Liau S, Read D, Pugh W, Furr J, Russell A. Interaction of silver nitrate with readily identifiable groups: relationship to the antibacterialaction of silver ions. *Lett. Appl. Microbiol.*, 1997; **25**(4): 279–283.
49. Li Q, Mahendra S, Lyon DY, Brunet L, Liga MV, Li D. Antimicrobial nanomaterials for water disinfection and microbial control: potential applications and implications. *Water Res.*, 2008; **42**(18): 4591–4602.
50. Schmidt B, Katiyar V, Plackett D, Larsen EH, Gerds N, Koch CB. Migration of nanosized layered double hydroxide platelets from polylactide nanocomposite films. *Food Addit. Contam. Part A Chem. Anal. Control Expo Risk Assess*, 2011; **28**(7): 956–966.
51. Sondi I, Salopek-Sondi B. Silver nanoparticles as antimicrobial agent: a case study on *E. coli* as a model for Gram-negative bacteria. *J. Colloid Interface Sci.*, 2004; **275**(1): 177–182.
52. Song H, Li B, Lin QB, Wu HJ, Chen Y. Migration of silver from nanosilver–polyethylene composite packaging into food simulants. *Food Addit. Contam., A*, 2011; **28**(12): 1758–1762.
53. Bouwmeester H, Dekkers S, Noordam MY, Hagens WI, Bulder AS, De Heer C. Review of health safety aspects of nanotechnologies in food production. *Regul. Toxicol. Pharm.*, 2009; **53**(1): 52–62.
54. Liao F, Chen C, Subramanian V. Organic TFTs as gas sensors for electronic nose applications. *Sens. Actuators, B*, 2005; **107**(2): 849–55.
55. Torres A, Guarda A, Moraga N, Romero J, Galotto MJ. Experimental and theoretical study of thermodynamics and transport properties of multilayer polymeric food packaging. *Eur. Food Res. Technol.*, 2012; **234**(4): 713–722.
56. Azeredo HM. Nanocomposites for food packaging applications. *Food Res. Int.*, 2009; **42**(9): 1240–1253.
57. Wang Y, Zhang Q, Zhang C, Li P. Characterisation and cooperative antimicrobial properties of chitosan/nano-ZnO composite nanofibrous membranes. *Food Chem.*, 2012; **132**: 419–427.
58. Echegoyen Y, Nerin C. Nanoparticle release from nano-silver antimicrobial food containers. *Food Chem. Toxicol.* 2013; **62**: 16–22.
59. Bradley EL, Castle L, Chaudhry Q. Applications of nanomaterials in food packaging with a consideration of opportunities for developing countries. *Trends Food Sci. Technol.*, 2011; **22**(11): 604–610.

60. Xie J, Hung YC. UV-A activated TiO2 embedded biodegradable polymer film for antimicrobial food packaging application. *LWT Food Sci. Technol.*, 2018; **96**: 307–314.
61. Busolo MA, Lagaron JM. Oxygen scavenging polyolefin nanocomposite films containing an iron modified kaolinite of interest in active food packaging applications. *Innovative Food Sci. Emerg. Technol.*, 2012; **16**: 211–7.
62. Yam K L, Lee DS. *Emerging Food Packaging Technologies, Principles and Practice*. Elsevier; 2012.
63. Yu HY, Qin ZY, Sun B, Yang XG, Yao JM. Reinforcement of transparent poly(3hydroxybutyrate-co-3-hydroxyvalerate) by incorporation of functionalized carbon nanotubes as a novel bionanocomposite for food packaging, *Compos. Sci. Technol.*, 2014; **94**: 96–104.
64. Hernandez M P, Catala R, Gavara R. Simple method for the selection of the appropriate food simulant for the evaluation of a specific food/packaging interaction. *Food Addit. Contam.*, 2002; **19**(1): 192–200.
65. Zhao LM, Shi LE, Zhang ZL, Chen JM, Shi DD, Yang J, Tang ZX. Preparation and application of chitosan nanoparticles and nanofibers, *Braz. J. Chem. Eng.*, 2011; **28**: 353–362.
66. Plastemart. Packaging sector – the highest consumer of polymer nanocomposites drives global demand. 2016; Retrieved June 19, 2018.
67. Ilyas RA, Sapuan SM, Atikah MSN, Asyraf MRM, Ayu Rafiqah S, Aisyah HA, Mohd Nurazzi N, Norrahim MNF. Effect of Hydrolysis Time on the Morphological, Physical, Chemical, and Thermal Behavior of Sugar Palm Nanocrystalline Cellulose (Arenga Pinnata (Wurmb.) Merr). *Textile Research Journal*. 2020. In print. doi: 10.1177/0040517520932393.

8

Nanocellulose Reinforced Polypropylene and Polyethylene Composite for Packaging Application

Mohd Nor Faiz Norrahim[1], Tengku Arisyah Tengku Yasim-Anuar[2], S.M. Sapuan[3], R.A. Ilyas[4,5], Mohd Idham Hakimi[2], Syed Umar Faruq Syed Najmuddin[6], and Mohd Azwan Jenol[2]

[1]*Research Centre for Chemical Defence (CHEMDEF), Universiti Pertahanan Nasional Malaysia, 57000 Kuala Lumpur, Malaysia*

[2]*Department of Bioprocess Technology, Faculty of Biotechnology and Biomolecular Sciences, Universiti Putra Malaysia, 43400 Serdang Selangor, Malaysia*

[3]*Advanced Engineering Materials and Composites Research Centre (AEMC), Department of Mechanical and Manufacturing Engineering, Universiti Putra Malaysia, 43400, Serdang, Selangor, Malaysia*

[4]*School of Chemical and Energy, Faculty of Engineering, Universiti Teknologi Malaysia, 81310 UTM Johor Bahru, Johor, Malaysia*

[5]*Centre for Advanced Composite Materials, Universiti Teknologi Malaysia, 81310 UTM Johor Bahru, Johor, Malaysia*

[6]*Laboratory of Vaccines and Immunotherapeutics (LIVES), Institute of Bioscience, Universiti Putra Malaysia, 43400, Serdang, Selangor, Malaysia*

8.1 Introduction

Rapid human population encourages the use of plastics and indirectly raise their production to meet the current market demand. Plastics have excellent properties and have been

Bio-Based Packaging: Material, Environmental and Economic Aspects, First Edition. Edited by S.M. Sapuan and R.A. Ilyas

beneficial for a wide range of applications over the years. For example, plastics are widely used for food packaging, pharmaceuticals, chemicals, detergents, cosmetics, compost bags, grocery bags, shipping bags, cutlery, plates, and toys [1]. Versatility, low production cost, and good mechanical and thermal properties are among the key features of plastics, thus making them favorable materials to be used in daily life [2].

Polypropylene (PP) and polyethylene (PE) are among the most important and widely used polyolefins for packaging materials, mainly due to their low production cost, design flexibility, and recyclability [3, 4]. Recently, numerous studies have been conducted to improve the properties of these polymers with the aim to produce new polymeric materials for specific applications [5].

Polymeric composite materials consist of a matrix phase and reinforcement materials such as commercial and traditional fibers [6]. A reinforcing material is usually incorporated to improve chemical and physical properties of polymer composites. Eco-friendly types of fibers such as oil palm biomass (OPB), kenaf, hemp, flax, jute, and sisal have had a pioneering role in composite technology and have been used for a large number of applications in automotive, construction, marine, electronic, packaging, and aerospace industries [7]. Among these fibers, cellulose and especially nanocellulose are attracting much attention. Currently, the study of nanocellulose has attracted a tremendous level of attention, as revealed by the increasing number of scientific contributions and industrial investments [8].

Nanocellulose is an interesting renewable material for many industrial applications, such as for composites, biosensors, membranes, biomedicines, electronics, and optical device development [9]. Reinforcement of nanocellulose as a nanoscale filler in composites has been extensively studied, and such applications are expected to increase the opportunities of the conventional petroleum-based composite with new or improved properties [10–12]. The development of nanocellulose-based composites has attracted great interest owing to their unique and excellent features, including abundancy, high renewability, high strength, eco-friendliness, and low weight [13]. Owing to its high stiffness, nanocellulose has been used as a reinforcement material to increase the mechanical strength of other polymers like thermoplastics (PP, PE, polyvinyl chloride [PVC]), thermosets (polyurethane [PU]), and biopolymers (polylactic acid [PLA]) [14–19].

Nevertheless, there are several limitations on the development of PP- and PE-nanocellulose composites, with the main challenge coming from the hydrophilicity of nanocellulose, which is incompatible with hydrophobic polymers including PP and PE. The differences in water properties will greatly affect the dispersion of nanocellulose in the polymer matrix [20]. Besides that, the preparation method of composites also plays an important role as it may influence the final properties of the material [17, 20, 21]. All of these factors are important and need to be considered for the future direction of nanocellulose-based composites. To the best of our knowledge, there is a lack of reports critically discussing the approach needed to overcome these limitations for PP- and PE-nanocellulose composites.

Therefore, this review addresses several important findings to reduce limitations of nanocellulose toward PP and PE composites. It begins with an introduction to plastic packaging and nanocellulose, followed by a comprehensive discussion on the processing methods and factors influencing the properties of PP- and PE-nanocellulose composites. Several research findings have been combined to provide readers with a comprehensive overview.

8.2 Plastic Packaging

Packaging is a means of ensuring the safe delivery of a product to the consumer at the right time and at the optimum cost [22]. Plastic packaging needs to consider the physical, ambient, and human environment. Failure to consider all these three environments during package development will result in poorly designed packages, indirectly increase operational cost, receive consumer complaints, and even avoidance or rejection of the product by the consumer. Plastic packaging is used to pack various materials such as foods, electronic devices, construction material as well as for shipping and storing of materials. Plastics are widely used for shipping and storage as they may save energy, space, and money. A plastic container generally requires less energy to be manufactured than other kinds of packaging such as metal. In fact, the use of a plastic container may avoid high consumption of fuel, as a low weight container requires less fuel for transportation compared to a heavier container. Moreover, plastic also helps in reducing the cost as shipping damage can be avoided and the need for additional packaging materials can be eliminated.

There are several requirements that need to be considered for plastic packaging, such as their flexibility, mechanical strength, barrier, light weight, lack of odor, hygiene, recyclability, resistance to hot and cold temperatures, resistance to oil and fats, good barrier properties, sealing capability, and low cost [1, 2]. However, different applications of plastics may require different properties.

The main concern of plastic packaging is to protect and preserve products from several factors, such as oxidizing agents and microorganisms. Currently, polyolefins such as PP, PE, and polyesters have excellent characteristic as they have good tensile strength and are able to block oxidizing agents [2]. It has been reported that food packaging is the largest part most accounted for in the industry and contributes to a large number of total packaging waste [23].

Several roles in food packaging have been reported by Marsh and Bugusu [23] such as for protection, containment, marketing, traceability, convenience, and temper indication. Protection enables the preservation of the food products from chemical, biological, and physical deterioration factors. During the distribution process, the quality of the food products could deteriorate if a suitable package is not being used. Therefore, plastic packaging contributes by extending food shelf life as it can help in maintaining the quality and safety of food products. Moreover, plastic packaging is also used as a containment to reduce food waste. Food waste is mainly caused by inadequate preservation, storage, and transportation. In addition, packaging is also used for marketing and to deliver information of the packed products. The package can be designed to enhance product presentation and to differentiate one product from another in market competition.

A package could also be used for product tracking. Traceability is a way to track the product throughout the production, processing, and distribution process. Traceability promises three objectives, including (1) improving supply management, (2) facilitating food safety and quality, and (3) differentiating and commercializing foods with subtle or undetectable quality attributes. Moreover, the package should be considered for its convenient features, such as handling, ease of access, disposability, and others. Convenience packaging will add value and competitive advantage to products in the market. The packages must be able to reduce or eliminate the risk of tampering and adulteration by banding, special membranes, breakaway closures, and special printing on bottle liners or composite cans.

8.3 Nanocellulose

An increasing amount of research has been done in recent years to evaluate the feasibility of plant biomass as a feedstock for producing value-added products such as nanocellulose. Nanocellulose including nanofiber can be described as a fiber with a diameter of 100 nm or less, an extremely high specific area, and high porosity with excellent pore interconnectivity [24]. The total global accessable market for nanocellulose enabled products was estimated at US$65 million in 2015, projected to grow to around $530 million by 2021 [25]. In the case of Malaysia, there is a great potential to produce nanocellulose mainly due to the infinite number of OPB that could be derived. Various OPB such as oil palm mesocarp fiber, oil palm empty fruit bunch, and oil palm fronds are suitable candidates for the development of bio-based products such as biocomposites, biofuels, biosugars, and particularly nanocellulose [26–28]. Moreover, nanotechnology was identified as one of the new growth engines for the Malaysia New Economic Model. The Malaysian government has listed four key sectors of nanotechnology development, which are:

(a) Food and agriculture
(b) Energy and environment
(c) Wellness, medical, and healthcare
(d) Electronics, devices, and systems

Nanotechnology was also listed in the "Eleventh Malaysia Plan" as one of the strategic researches, and this involves research in nanocellulose. Nanocellulose consists of CNF, microfibrillated cellulose (MFC), CNC, and BNC [15, 29, 30]. Different approaches were used to extract these three types of nanocellulose, resulting in different properties of size, crystallinity, surface chemistry, and mechanical properties [30]. CNF is normally produced by several approaches, such as mechanical, physical, physiochemical, or by combination, while CNC could only be produced by acid treatment, which requires strong acid hydrolysis [31]. CNF or MFC fiber have a similar meaning, which refers to a nanostructured, long threadlike bundles of cellulose stabilized laterally by hydrogen bonds between the hydroxyl groups and the oxygen atoms of adjacent molecules [32, 33]. Normally, MFC has a larger diameter size as compared to CNF, while CNC has a rod-like, highly crystalline, short structure, with a rectangular cross-section. Their dimensions depend on the native cellulose source material, hydrolysis time, and temperature [34].

On the other hand, BNC is produced by microorganisms [30]. *Gluconacetobacter xylinum* being the most efficient amongst cellulose-producing microorganisms. Different from CNF and CNC, which may require pretreatment to remove lignin and hemicellulose before hydrolysis, BNC is synthesized as pure cellulose and characterized by average diameters of 20–100 nm and micrometer lengths, entangled to form stable network structures.

The differences between the morphological structures of CNC, CNF, and BNC are shown in Figure 8.1. CNF has a fibrillated structure and is linked between fibers (Figure 8.1b), with long and flexible linked fibers around 20 nm wide and several micrometers in length. They consist of alternating crystalline and amorphous domains. However, CNC (Figure 8.1a) are rigid rod-like crystals with diameters in the range of 10–20 nm and lengths of a few hundred nanometers [35].

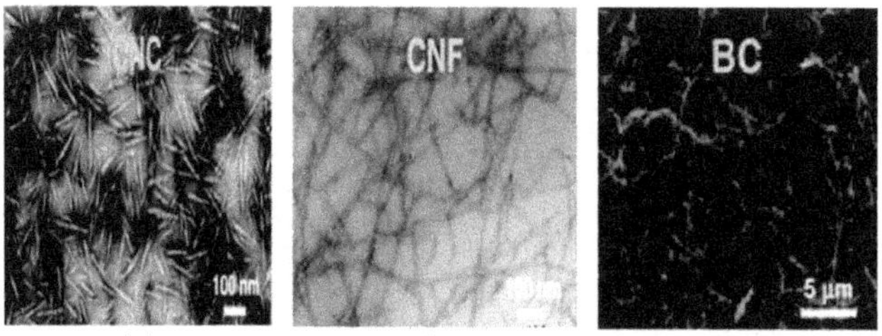

(a) (b) (c)

Figure 8.1 *Micrograph of (a) cellulose nanocrystals, (b) cellulose nanofibers, and (c) bacterial nanocellulose. Source: Abitbol et al. [30].*

8.4 Polypropylene and Polyethylene Nanocellulose Composites

Composites with improved mechanical properties from nanocellulose have been successfully produced [36] but the use of nanocellulose as a reinforcing material in composites is still relatively low [31]. Owing to the high stiffness of nanocellulose, plant-based fillers can be used to increase the mechanical strength of general-purpose thermoplastics. Several studies were carried out to evaluate the influence of nanocellulose on the properties of packaging plastics to meet current market demands. Furthermore, there have been significant advances in the development of plant-plastic composites [37]. PP and PE composites containing nanocellulose received much attention. Most findings show a great improvement of the composite properties as compared to the neat polymer.

Table 8.1 shows an example of research done on composites consisting of PP and PE blended with nanocellulose. There are numerous sources of nanocellulose such as OPB, wood, kenaf, and tunicin. Most of them derived from the biological waste and this can be considered as a good approach to utilize waste for a new value-added material such as nanocellulose. Usually, only a small amount of nanocellulose is incorporated in a polymer matrix to produce a composite. Despite the low content as a reinforcement material, nanocellulose can give an observable effect on the composite compared to other micro-sized materials [46].

8.5 Compatibility Between Nanocellulose with Polyethylene and Polypropylene Matrices

Despite the numerous advantages offered by nanocellulose when it is being incorporated in PP and PE composites, as discussed earlier, there are also some drawbacks. The chemical compatibility between nanocellulose and polymer plays a major role in effecting the dispersion of particles in the matrix as well as adhesion between these phases [18]. The major constraint for nanocellulose-based composites is their wettability properties. Nanocellulose is hydrophilic, whereas most polymers including PP and PE are hydrophobic and this creates adhesion problems at the interface of the composites, resulting in poor dispersion of nanocellulose, which indirectly affects the mechanical properties of composites afterwards. Nevertheless, this issue can be solved by introducing a compatibilizer during compounding processing.

Table 8.1 PP- and PE-nanocellulose composites.

Materials	Nanocellulose source	Nanocellulose content (%)	References
PE/CNF	Commercial (Daicel Chemical Industries, Ltd., Japan)	1–10	[38]
PE/thermoplastic starch/CNF	Wheat straws	6–14	[39]
PP/CNF	Commercial (J. Rettenmaier and Söhne GMBH Co., Germany)	2–10	[40]
PP/CNC	Commercial (Avicel type GP1030, from FMC Biopolymer)	0–15	[19]
PE/CNC	Commercial (FMC Biopolymers [Avicel Lattice NT-100])	5–15	[17]
CNF/PP	Kenaf bast fibers	5	[41]
PE/CNC	Curauá fibers	3–9	[42]
PE/CNF	Oil palm biomass	1–5	[43]
PP/CNF	Wood	1–10	[37]
PP/CNC	Tunicin	6	[44]
PE/CNF	Eucalyptus pulp	0–100	[45]
PP/CNF	Commercial (Process Development Center at The University of Maine)	2	[6]

A compatibilizer is a polymeric interfacial agent that is able to modify interfacial properties of a blend of immiscible materials by making them bind to each other more tightly during composite processing. A compatibilizer is able to offer several advantages over the immiscible polymer blends. Using a compatibilizer in PP- and PE-nanocellulose composites can eliminate the interface problem as it can help the blend materials to bond or interact together tightly. Its primary function is to reduce the interfacial tension between incompatible polymeric components of a blend, thus allowing finer dispersion of a mutually incompatible pair, as well as leading to stabilization of morphology against its destructive modification during subsequent high stress or strain processing such as during the melt-processing, e.g. injection molding [47]. Another function of a compatibilizer is to enhance the adhesion between a liquid and solid phase, and facilitating the stress transfer, which then improves the mechanical properties of the composite. According to Prachayawarakorn et al. [48], a compatibilizer also should be able to form entanglements or segmental crystallization within the polymer matrices to produce polymer blends with desirable properties. Usually, the compatibilizer, a block or a graft, is a copolymer consisting of plural immiscible polymeric segments. The interaction occurs between the compatibilizer and polymers at either covalent, ionic, or hydrogen bonds [49, 50].

Graft copolymers are some of the compatibilizers that can be added in the PP- and PE-nanocellulose blends. Graft copolymers consist of polymeric side chains where the repeating units are grafted to a strand of repeating unit. The addition of grafted compatibilizers enables enhanced adhesion of polymer blends, thus enhancing the mechanical properties as well. Most PP- and PE-nanocellulose composites recorded a substantial improvement in elongation at break and tensile toughness compared to the corresponding binary blends.

Grafted maleic anhydride is also one of the of reactive grafted compatibilizers. It is a common compatibilizer used to blend PP and PE with any natural fibers including nanocellulose

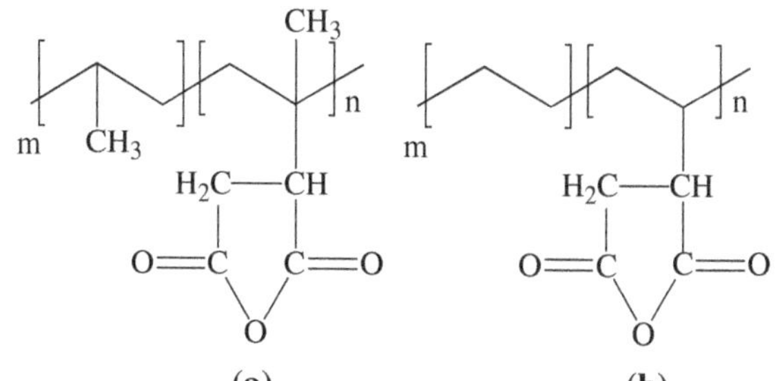

Figure 8.2 *Chemical structure of (a) PP-grafted-maleic anhydride and (b) PE-grafted-maleic anhydride.*

[19, 21, 39, 43, 51, 52]. Figure 8.2 shows the chemical structure of PP grafted maleic anhydride (PP-g-MA) and polyethylene grafted maleic anhydride (PE-g-MA).

Li et al. [53] reported that the addition of 3 wt% maleic anhydride grafted polyethylene (PE-*g*-MA) in a blend of PE and cotton-CNF was found to improve the dispersion of CNF in the PE matrix, as no agglomeration of CNF was recorded. This supports the function of PE-*g*-MA as a compatibilizer in creating the interaction between two different wettability materials: hydrophobic PE matrix and hydrophilic cotton-CNF. It was also concluded that the grafted maleic anhydride had influenced the flexibility of the blends, thus proving its function as a compatibilizer at the matrix interface [54].

The use of a coupling agent in PP- and PE-nanocellulose composites has also been reported. Studies by Yakkan et al. [6] revealed that thermal stability and mechanical properties of composites were improved by adding 0.1 wt% FUSE® (commercial coupling agent). The tensile strength improved from 14 MPa to around 20 MPa. In comparison to the neat PP, the Young's (elastic) modulus of the 0.1 wt% FUSE containing PP-nanocellulose composites was 87% higher and the tensile strength was 38% higher.

Another possible approach to obtain better nanocellulose dispersion in a thermoplastic is by using a water-soluble polymer as a carrier for nanocellulose. Water-soluble polymers such as starch and poly-vinyl alcohol (PVA) are expected to encapsulate single nanocellulose, providing better dispersion in the polymer matrix and formation of strong hydrogen bonds as water is evaporated. Kiziltas et al. [55] studied the possibility of using thermoplastic starch as a carrier system for nanocellulose in the PP matrix. It was found that the flexural modulus and strength of the composites increased compared to the neat PP. Later on, Kiziltas et al. [38] demonstrated that the melt processing of composites with better dispersion and good mechanical properties can be successfully developed using a novel carrier system of nanocellulose, which is PVA.

To summarize, compatibilizers exhibit interfacial activity in heterogeneous polymer blends and their compatibilizing effects are attributed to an enhancement in interfacial adhesion, which then lead to an enhancement of mechanical properties of the composite.

8.6 Processing Method of PP- and PE-Nanocellulose Composites

The processing method also has an important impact on the final properties of the PP- and PE-nanocellulose composites. It was shown that the adhesion and dispersion of

Table 8.2 Production methods of PP- and PE-nanocellulose composites.

Materials	Production method	References
PE/CNF	Melt blending and hot compression	[38]
PP/CNF	One-pot and hot compression	[52]
PE/thermoplastic starch/CNF	Melt blending and injection compression	[39]
PP/CNF	Melt blending	[40]
PP/CNC	Solvent casting	[19]
PE/CNC	Melt blending and hot compression	[17]
CNF/PP	Melt blending and hot compression	[41]
PE/CNC	Melt blending and hot compression	[42]
PE/CNF	One-pot and hot compression	[43]
PP/CNF	Melt blending and injection compression	[37]
PP/CNC	Solvent casting	[44]
PE/CNF	Solvent casting and hot compression	[45]
PP/CNF	Melt blending and hot compression	[6]
PE/CNC	Melt blending and hot compression	[56]

nanocellulose and the polymer matrix are important in order to obtain good mechanical properties of composites [38, 43]. According to Kiziltas et al. [38], the degree of dispersion of nanocellulose in a thermoplastic matrix strongly depends on the conditions of the melt compounding and processing techniques to prepare the composites. The solvent casting method provides a uniform dispersion of nanocellulose in a polymer matrix. Nevertheless, the process is not industrially applicable and is not suitable for PP- and PE-nanocellulose composites. Melt compounding is more practical for scaling up to an industrial level. Some reported examples of processing methods of PP- and PE-nanocellulose composites are summarized in Table 8.2.

8.6.1 Solvent Casting

Nanocellulose composites can be prepared by evaporation or the solvent casting technique. This method requires a mixing and drying process. Even though this technique is the most successful technique for producing a homogenous mixture of nanocellulose composites, it is less suitable to be used in large-scale production [57]. The selection of a suitable solvent to dissolve PP or PE plays an important part in the process. Bahar et al. [19] successfully produced PP-nanocellulose composite using this approach by dissolving the PP in toluene.

8.6.2 Melt Compounding

Composite production by the melt-blending method has gained interest, especially in large-scale manufacturing since the final product can be easily shaped by compression molding [58]. However, there is a lack of studies done on this technique, especially on the production of nanocellulose composites. Melt blending could be done by internal melt blending or extrusion. There are some challenges to maintain the nano-size of cellulose in extrusion processes [52, 59]. The hardest challenges that need to overcome are the difficulty of feeding the material into the extruder and the tendency of nanocellulose to aggregate [60]. As a practice, nanocellulose needs to freeze-dry prior to compounding

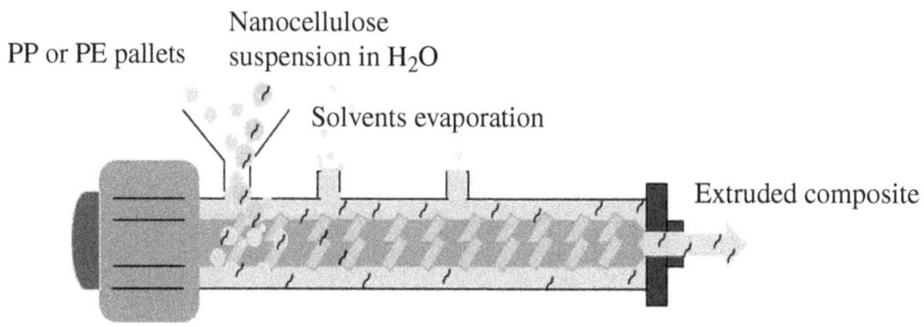

Figure 8.3 *Schematic image of extrusion process with liquid feeding. Source: Herrera et al. [61].*

with the polymer. This led to severe aggregation of nanocellulose. The nanocellulose has a high surface area and has a tendency to aggregate when dried. One suggested way to overcome these problems is by practicing liquid feeding during blending. Figure 8.3 shows the schematic image of the preparation of PP- or PE-nanocellulose composites by the extrusion process using the liquid feeding technique. Liquid feeding involves dispersing the nanocellulose in a liquid medium such as water, plasticizer, and solvents, and then pumping the prepared suspension into the extruder during blending. Water and solvents were removed during the process through evaporation. This approach has been done by Norrahim [20] and Yasim-Anuar [62] for the preparation of PP- and PE-nanocellulose composite, respectively. Interestingly, the mechanical properties obtained were similar to the composites prepared with the freeze-dried nanocellulose prepared composites.

8.6.3 Injection and Compression Molding

Injection molding is a method used to produce composites with complex shapes and excellent surface smoothness. This method is suitable to be used for polymer granules or mixtures of polymer granules with reinforcements, which will be heated within a metallic barrel. For composites, injection molding was applied after the reinforcements were incorporated in the polymer matrix by other compounding methods, typically extrusion and internal melt blending [13, 63]. The main role of injection molding is to mold the final shape of the composites [64].

Similar to injection molding, a compression molding was also applied after polymers and reinforcements were compounded by other methods. This method involves the use of stationary and movable molds to produce composites with a desirable shape [65]. Composites are placed between the molds and then selected heat and pressure are applied to obtain the desired shaped composites [66].

8.6.4 One-Pot

Conventional nanocellulose composites production requires two separate unit operations for the processing: (i) a cellulose nanofibrillation unit and (ii) a polymer nanocomposites compounding unit [43, 52]. This two-unit process is less efficient in terms of energy usage and material handling compared to a one-unit process. Moreover, the presence of downtime in between the two steps may affect the overall productivity of the product. The possibility

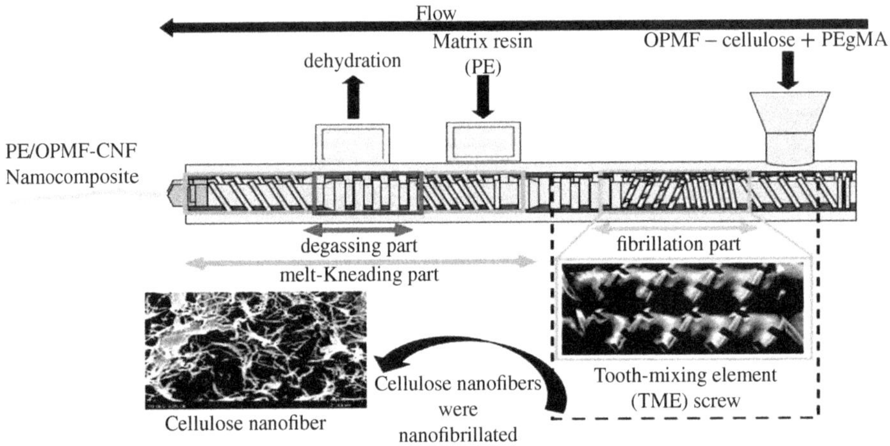

Figure 8.4 Schematic design of one-pot nanofibrillation and melt compounding processes by twin-screw extruder. Source: Yasim-Anuar et al. [43].

of combining nanofibrillation and the blending process in extrusion was rarely studied. Hietala et al. [67] also discovered the efficiency of the twin-screw extrusion process to combine the fibrillation of cellulose into nanocellulose in the same step as to blend with starch. However, no fibrillation was observed even though good dispersion of fibers in the composites matrix was obtained. It was identified that the low viscosity of the starch–fiber mixtures due to a high water content is the probable reason why the nanofibrillation was not successful.

The one-pot process has been successfully used to produce the PP- and PE-nanocellulose composites and has been proven to be feasible [9, 43, 52, 67, 68]. The one-pot process for composite production has several advantages for commercialization, such as being able to process a high solid content mixture, is a continuous process, and has a high production volume. These advantages have not been achieved yet in any other method described in the literature so far. Figure 8.4 shows the schematic diagram of the one-pot processing method used to produce nanocellulose-based composites.

8.7 Factors Influencing the Performance of the PP- and PE-Nanocellulose Composites

Factors such as chemical composition and dispersion of nanocellulose in the PP and PE matrix will also influence the properties of composites. The interface between nanocellulose and polymer matrix as a stress transfer medium will determine composite performances. Well-dispersed nanocellulose in a polymer matrix resulted in an iso-strain condition between the fiber and the matrix, both materials remaining in the elastic range. For instance, failures in composites are more likely due to the low stress transfer and accumulation of interfacial cracks which allow little or no stress transfer from the matrix to the nanocellulose and vice versa [69].

8.7.1 Drying Effect of Nanocellulose

The dispersion of the hydrophilic nanocellulose in PP and PE composites is difficult, since nanocellulose has a large specific surface area and is highly possible to aggregate in the composites. By increasing the degree of dispersion of nanocellulose in the polymer matrix, the mechanical performance of the composites will be improved as well. Since nanocellulose production normally was prepared by wet-state fibrillation treatments, a drying process is necessary to make it usable for composites processing. Although freeze-drying is a useful method to reduce the aggregations of nanocellulose, it requires vacuum-like conditions and long processing times [70]. There is also an issue on the high cost with regard to the use of freeze-drying on an industrial scale. According to Norrahim [20], slurry of CNF was mixed with PP by melt blending and resulted in a well-distributed CNF in the polymer matrix. The ideal drying method involves evaporating the water in the nanocellulose suspension by heating. Since water is intensely evaporated in melted PP and PE, nanocellulose would be dried during the process and better dispersion of CNF can be obtained. The formation of nanocellulose aggregations can be reduced by slowly adding the nanocellulose to the melted PP or PE [20, 62].

8.7.2 Chemical Composition of Nanocellulose

There were studies reported on the PP-composites reinforced with lignocellulose nanofibers [20, 37]. The results showed improved properties of the composites obtained as compared to the neat PP. Interestingly, the presence of the remaining hemicellulose was known to improve the nanocellulose production as compared to cellulose fiber as a raw material. However, the presence of remaining hemicellulose causes a slight reduction to the mechanical properties of composites as compared to composites with pure nanocellulose [20].

8.8 Characteristics of the PP- and PE- Nanocellulose Composites

Generally, material for packaging is counted for properties such as flexibility, strength, lightness, and impermeability. However, different applications of plastics may require different properties. It is important to evaluate the characteristics of PP- and PE-nanocellulose composites. Several studies have reported the reinforcement effect of nanocellulose in PP and PE matrices on their mechanical, thermal, and crystallinity properties, as shown on Table 8.3.

Mechanical performance such as tensile strength and moduli of composites are important parameters in determining the quality of nanocellulose-based composites. According to Norrahim et al. [1], mechanical properties of plastic packaging varied from low to high values; hence, selection of plastic for any application will be determined by its properties. For example, a less-strength plastic with a low tensile strength was usually used for light applications such as bread wrapping. Since tensile strength indicates the maximum load that the material can sustain, a higher tensile strength value is suitable for packaging heavy products as the package material can resist a direct pull. This is especially important to avoid transportation damage. By referring to Table 8.3, several researchers have reported that the addition of nanocellulose on polymer matrices led to an increment in tensile and flexural

Table 8.3 Properties of PP- and PE- nanocellulose composites.

Materials	Properties				References
	Mechanical	Crystallinity	Thermal	Physical	
PE/CNF	Tensile and modulus properties increased 25% and 40%, respectively. Flexural properties also increased	–	Thermal stability slightly decreased	–	[38]
PP/CNF	Tensile and flexural properties increased	Increased	Not affected	Transparent	[52]
PE/thermoplastic starch/CNF	Tensile and elongation at break declined; whereas, the Young's modulus was increased	–	–	–	[39]
PP/CNF	Tensile properties increased	–	–	–	[40]
PP/CNC	Tensile increased by 70–80% as compared to neat PP	Crystallinity increased 50%	Higher thermal degradation and thermal conductivity	Transparent	[19]
PE/CNC	Tensile and modulus properties increased	–	–	Homogeneous appearance	[17]
CNF/PP	Tensile and modulus properties were greatly increased	–	–	–	[41]
PE/CNC	Tensile properties decreased	Increased	Not affected	Translucent	[42]
PE/CNF	Tensile and flexural properties increased	Increased	Slightly decreased		[43]
PP/CNF	Tensile, flexural and impact properties increased	Increased	Improved crystallization temperature	–	[37]
PE/CNF	Modulus properties was greatly increased	Increased	Not affected	Transparent	[45]
PP/CNF	Tensile and modulus properties increased by 87%	–	–	–	[6]
PE/CNC	Modulus properties increased	Crystallinity increased	Not affected	Similar to neat film appearance	[56]

properties, especially on the modulus of materials [19, 43, 52, 56]. Norrrahim et al. [52] and Yasim-Anuar et al. [43] have reported that both PP and PE containing 3 wt% CNF exhibited a very high increment on the tensile and flexural modulus than that of the neat polymer. This pronounced increase in the effect on the modulus is assigned to the geometry and stiffness of the CNF. Moreover, a highly crystalline CNF also contributed to the increment on tensile and flexural properties [52]. The mechanical properties are greatly influenced by the dispersion and fiber alignment of nanocellulose in the polymer matrix. Aggregation of nanocellulose could lead a poor stress transfer in the polymer matrix that resulted in the reduction of the mechanical properties as nanocellulose concentration increased. The improvement of the mechanical strength of PP- and PE-nanocellulose composite could help in reducing cost by preventing shipping damage and eliminates the need for additional packaging materials.

The crystallinity characteristics of the composites also plays an important role in determining the final properties of the material. From Table 8.3, the incorporation of nanocellulose into PP or PE not only improved their mechanical performance but also indirectly increased the crystallinity value as compared to the neat polymers. Nanocellulose acts as a nucleation agent for crystallization and hence helps in improving the crystallinity index of composites afterwards [37, 71]. The crystallinity properties were also correlated with the thermal properties of composites.

Several studies have revealed that the addition of nanocellulose is able to improve the performance of the materials with respect to thermal properties. Thermal gravimetry analysis (TGA) and differential scanning calorimetry (DSC) are the common analyses done to investigate the thermal properties of any materials, including nanocellulose-based composites. Referring to Table 8.3, it can be seen that nanocellulose-based composites are thermally stable compared to the neat polymer [72]. This is mainly due to a higher ratio of crystalline regions to amorphous regions in the nanocellulose-based composites compared to the neat polymer [32]. Higher crystallinity results in a higher thermal stability, which results in a higher decomposition temperature. In addition to this, the thermal properties of nanocellulose composites such as the crystallization temperature T_c and ΔH_m are also affected by the presence of nanocellulose. The T_c and ΔH_m values of nanocellulose composites improved as the crystallinity improved [37].

Transparency of the packaging material is also one of the important criteria that needs to be considered. By using clear packaging for food products, consumers can assess food freshness to determine whether the food is good quality or vice versa. Visual appearance examination is usually conducted to observe the transparency of the composites [73].

Permeability is one of the important criteria that needs to be considered in packaging. The use of packaging materials can be justified for preventing the deterioration of products due to physical, biochemical, and microbiological factors. Additionally, packaging should also provide a sufficient barrier against oxygen, water vapor, grease, and microorganisms [74]. According to Ferrer et al. [75], nanocellulose is known to have a very low oxygen permeability (0.666 cc μm (m^2 day kPa)$^{-1}$) than PE (50–200 cc μm (m^2 day kPa)$^{-1}$). Moreover, nanocellulose also has a higher water vapor transmission rate (234 g (m^2 day)$^{-1}$) as compared to PE (16.8 g (m^2 day)$^{-1}$). Thus, it is expected that the presence of nanocellulose could influence the properties of composites obtained [76]. However, to the best of our knowledge, there is a lack of study reported on the permeability of PP- and PE-nanocellulose composites. Nanocellulose materials could be a great potential to sustainably enhance oxygen and water vapor barrier properties when used as coatings or fillers in composites or as self-standing thin films.

8.9 Conclusion and Future Recommendations

All in all, several important factors on developing the PP- and PE-nanocellulose composites for packaging applications were discussed. Nanocellulose has been proven as a great green material to improve the performance of PP and PP matrices, especially with regard to mechanical performance. However, to take advantage of the great potential of these materials, several aspects as discussed above should not be disregarded. It can be concluded that the major challenges of incorporating nanocellulose in polymer matrices is coming from the wettability differences between these materials, which will give the effects to the dispersion and mechanical properties of composites.

In addition, all important criteria that is important for the future direction of PP- and PE-nanocellulose composites, especially for packaging applications, need to be pointed out. Firstly, permeability of the PP- and PE-nanocellulose composites are rarely being studied. This property is important, especially for packaging applications. Moreover, functionalizing the nanocellulose for active packaging applications must also be explored to meet the current demand. Lastly, the production of PP- and PE-nanocellulose composites by using a one-pot process has been shown to be the most promising method for packaging production. Therefore, focus should be given to scale-up experimentation on this method for the commercialization aspect of PP- and PE-nanocellulose composites.

References

1. Norrahim MNF, Ariffin H, Hassan MA, Ibrahim NA, Nishida H. Performance evaluation and chemical recyclability of a polyethylene/poly(3- hydroxybutyrate-co-3-hydroxyvalerate) blend for sustainable packaging. *RSC Adv.* 2013;**3**(46):24378–88.
2. Camann A, Dragsbeak K, Krol S, Sandgren J, Song D. Properties, recycling and alternatives to PE bags. *J. Elastomers Plast.* 20101–132.
3. Polymer Solutions News Team. Top types of polyolefins: The most common kind of plastics. 2018 [cited 2019 Dec. 15]. Available at: https://www.polymersolutions.com/blog/top-types-of-polyolefins-the-most-common-kind-of-plastics.
4. Marcilla A, Garcia-Quesada JC, Lopez M, Gil E. Study of the behavior of blends of a poly (hydroxybutyrate-valerate) copolymer, polypropylene, and SEBS. *J. Appl. Polym. Sci.* 2009;**113**(5):3187–95.
5. Moritomi S, Watanabe T, Kanzaki S. Polypropylene compounds for automotive applications. *Sumitomo Kagaku.* 2010;**1**(1):1–16.
6. Yakkan E, Uysalman T, Atagür M, Sever K, Seydibeyoğlu MÖ. Nanocellulose-polypropylene nanocomposites enhanced with coupling agent. *Bartın Orman Fakültesi Derg.* 2018;**20**(3):491–502.
7. Sen T, Reddy HNJ. Various industrial applications of hemp, kinaf, flax and ramie natural fibres. *Int. J. Innov. Manag. Technol.* 2011;**2**(3):192.
8. Trache D. Nanocellulose as a promising sustainable material for biomedical applications. *AIMS Mater. Sci.* 2018;**5**(2):201–5.
9. Ho TTT, Abe K, Zimmermann T, Yano H. Nanofibrillation of pulp fibers by twin-screw extrusion. *Cellulose* 2015;**22**(1):421–33.

10. Cherian BM, Leao AL, de Souza SF, Thomas S, Pothan LA, Kottaisamy M. Cellulose nanocomposites for high-performance applications. In: *Cellulose Fibers: Bio-and Nano-Polymer Composites*. Springer; 2011. p. 539–87.
11. Ben Azouz K, Ramires EC, Van den Fonteyne W, El Kissi N, Dufresne A. Simple method for the melt extrusion of a cellulose nanocrystal reinforced hydrophobic polymer. *ACS Macro Lett.* 2011;**1**(1):236–40.
12. Lee S-H, Teramoto Y, Endo T. Cellulose nanofiber-reinforced polycaprolactone/polypropylene hybrid nanocomposite. *Compos. A Appl. Sci. Manuf.* 2011;**42**(2):151–6.
13. Kargarzadeh H, Mariano M, Huang J, Lin N, Ahmad I, Dufresne A, et al. Recent developments on nanocellulose reinforced polymer nanocomposites: A review. *Polymer (Guildf)*. 2017;**132**:368–93.
14. Li X, Gong Y. Design of polymeric nanofiber gauze mask to prevent inhaling PM2. 5 particles from haze pollution. *J. Chem.* 2015.
15. Kalia S, Dufresne A, Cherian BM, Kaith BS, Avérous L, Njuguna J, et al. Cellulose-based bio- and nanocomposites: a review. *Int. J. Polym. Sci.* 2011; **2011**:1–35.
16. Azeredo HMC, Rosa MF, Mattoso LHC. Nanocellulose in bio-based food packaging applications. *Ind. Crops Prod.* 2017;**97**:664–71. doi: 10.1016/j.indcrop.2016.03.013
17. Sapkota J, Natterodt JC, Shirole A, Foster EJ, Weder C. Fabrication and properties of polyethylene/cellulose nanocrystal composites. *Macromol. Mater. Eng.* 2017;**302**(1):1600300.
18. dos Santos FA, Iulianelli GC V, Tavares MIB. The use of cellulose nanofillers in obtaining polymer nanocomposites: properties, processing, and applications. *Mater. Sci. Appl.* 2016;**7**(05):257.
19. Bahar E, Ucar N, Onen A, Wang Y, Oksüz M, Ayaz O, et al. Thermal and mechanical properties of polypropylene nanocomposite materials reinforced with cellulose nano whiskers. *J. Appl. Polym. Sci.* 2012;**125**(4):2882–9.
20. Norrahim MNF. *Superheated Steam Pretreatment of Oil Palm Biomass for Improving Nanofibrillation of Cellulose and Performance of Polypropylene/Cellulose Nanofiber Composites*. Universiti Putra Malaysia; 2018.
21. Pandey JK, Lee S, Kim H, Takagi H, Lee CS, Ahn SH. Preparation and properties of cellulose-based nano composites of clay and polypropylene. *J. Appl. Polym. Sci.* 2012;**125**(S1):E651–60.
22. Coles R, McDowell D, Kirwan MJ. *Food Packaging Technology*. Vol. **5**. CRC Press; 2003.
23. Marsh K, Bugusu B. Food packaging—roles, materials, and environmental issues. *J. Food Sci.* 2007;**72**(3):R39–55.
24. Chen W, Yu H, Liu Y, Chen P, Zhang M, Hai Y. Individualization of cellulose nanofibers from wood using high-intensity ultrasonication combined with chemical pretreatments. *Carbohydr. Polym.* 2011;**83**(4):1804–11. doi: 10.1016/j.carbpol.2010.10.040
25. Zon Research Analysis. Global nanocellulose (nano-crystalline cellulose, nano-fibrillated cellulose and bacterial nanocellulose) market set for rapid growth, to Reach around USD 530.0 million by 2021 [Internet]. 2018 [cited 2019 Dec. 15]. Available at: https://www.marketresearchstore.com/news/global-nanocellulose-market-223
26. Wu Q, Qiang TC, Zeng G, Zhang H, Huang Y, Wang Y. Sustainable and renewable energy from biomass wastes in palm oil industry: a case study in Malaysia. *Int. J. Hydrogen Energy* 2017;**42**(37):23871–7. Available at: 10.1016/j.ijhydene.2017.03.147
27. Nordin NIAA, Ariffin H, Andou Y, Hassan MA, Shirai Y, Nishida H, et al. Modification of oil palm mesocarp fiber characteristics using superheated steam treatment. *Molecules*. 2013**18**(8):9132–46.
28. Fahma F, Iwamoto S, Hori N, Iwata T, Takemura A. Effect of pre-acid-hydrolysis treatment on morphology and properties of cellulose nanowhiskers from coconut husk. *Cellulose* 2011;**18**(2):443–50.
29. Kang Y, Ahn Y, Lee SH, Hong JH, Ku MK, Kim H. Lignocellulosic nanofiber prepared by alkali treatment and electrospinning using ionic liquid. *Fibers Polym.* 2013;**14**(4):530–6.
30. Abitbol T, Rivkin A, Cao Y, Nevo Y, Abraham E, Ben-Shalom T, et al. Nanocellulose, a tiny fiber with huge applications. Vol. **39**, *Current Opinion in Biotechnology*. Elsevier Ltd; 2016. p. 76–88. doi: 10.1016/j.copbio.2016.01.002

31. Rueda L. Cellulose nanocrystals/polyurethane nanocomposites. Study from the viewpoint of microphase separated structure. 2013(August 2015).
32. Lee JJS, Endo T, Kim N. Dimension change in microfibrillated cellulose from different cellulose sources by wet disk milling and its effect on the properties of PVA nanocomposite. 2015;495–506.
33. Azizi Samir MAS, Alloin F, Dufresne A. Review of recent research into cellulosic whiskers, their properties and their application in nanocomposite field. *Biomacromolecules* 2005;**6**(2):612–26.
34. Habibi Y, Lucia LA, Rojas OJ. *Cellulose Nanocrystals : Chemistry, Self-Assembly , and Applications.* 2009;
35. Peng X, Ren J, Zhong L, Sun R. Nanocomposite films based on xylan-rich hemicelluloses and cellulose nanofibres with enhanced mechanical properties. *Biomacromolecules.* 2011;3321–9. Available at: http://www.ncbi.nlm.nih.gov/pubmed/21817784
36. Mohamad Haafiz MK, Eichhorn SJ, Hassan A, Jawaid M. Isolation and characterization of microcrystalline cellulose from oil palm biomass residue. *Carbohydr. Polym.* 2013;**93**(2):628–34
37. Iwamoto S, Yamamoto S, Lee S-H, Endo T. Solid-state shear pulverization as effective treatment for dispersing lignocellulose nanofibers in polypropylene composites. *Cellulose.* 2014;**21**(3):1573–80.
38. Kiziltas A, Nazari B, Kiziltas EE, Gardner DJS, Han Y, Rushing TS. Cellulose nanofiber-polyethylene nanocomposites modified by polyvinyl alcohol. *J. Appl. Polym. Sci.* 2016;**133**(6).
39. Alidadi-Shamsabadi M, Behzad T, Bagheri R, Nari-Nasrabadi B. Preparation and characterization of low-density polyethylene/thermoplastic starch composites reinforced by cellulose nanofibers. *Polym. Compos.* 2015;**36**(12):2309–16.
40. Yang H-S, Gardner DJ. Morphological characteristics of cellulose nanofibril-filled polypropylene composites. *Wood Fiber Sci.* 2011;**43**(2):215–24.
41. Ghaemi F, Tahir PM, Abdullah LC, Kargarzadeh H. Comparative study of cellulose nanofiber and carbon nanofiber effects as reinforcement fillers on mechanical properties of polypropylene composites. In: *AIP Conference Proceedings.* AIP Publishing; 2017. p. 20019.
42. Oliveira de Castro D, Frollini E, Ruvolo-Filho A, Dufresne A. "Green polyethylene" and curauá cellulose nanocrystal based nanocomposites: effect of vegetable oils as coupling agent and processing technique. *J. Polym. Sci. Part B Polym. Phys.* 2015;**53**(14):1010–9.
43. Yasim-Anuar TAT, Ariffin H, Norrahim MNF, Hassan MA, Tsukegi T, Nishida H. Sustainable one-pot process for the production of cellulose nanofiber and polyethylene/cellulose nanofiber composites. *J. Cleaner Prod.* 2019;**207**:590–9.
44. Ljungberg N, Bonini C, Bortolussi F, Boisson C, Heux L, Cavaillé J-Y. New nanocomposite materials reinforced with cellulose whiskers in atactic polypropylene: effect of surface and dispersion characteristics. *Biomacromolecules* 2005;**6**(5):2732–9.
45. Maia THS, Larocca NM, Beatrice CAG, de Menezes AJ, de Freitas Siqueira G, Pessan LA, et al. Polyethylene cellulose nanofibrils nanocomposites. *Carbohydr. Polym.* 2017;**173**:50–6.
46. Okpala CC. The benefits and applications of nanocomposites. *Int. J. Adv. Eng. Technol.* 2014;**12**:18.
47. Farzadfar A, Khorasani SN, Khalili S. Blends of recycled polycarbonate and acrylonitrile–butadiene–styrene: comparing the effect of reactive compatibilizers on mechanical and morphological properties. *Polym. Int.* 2014;**63**(1):145–50.
48. Prachayawarakorn J, Khamsri J, Chaochanchaikul K, Sombatsompop N. Effects of compatibilizer type and rubber-wood sawdust content on the mechanical, morphological, and thermal properties of PVC/LDPE blend. *J. Appl. Polym. Sci.* 2006;**102**(1):598–606.
49. Benhamou K, Kaddami H, Magnin A, Dufresne A, Ahmad A. Bio-based polyurethane reinforced with cellulose nanofibers: a comprehensive investigation on the effect of interface. *Carbohydr. Polym.* 2015;**122**:202–11.
50. Hassan ML, Mathew AP, Hassan EA, Fadel SM, Oksman K. Improving cellulose/polypropylene nanocomposites properties with chemical modified bagasse nanofibers and maleated polypropylene. *J. Reinf. Plast. Compos.* 2014;**33**(1):26–36.

51. Ayaz O, Ucar N, Bahar E, Oksuz M, Ucar M, Onen A, et al. Production and analysis of composite nanofiber and heat applied nanofiber. *J. Text Eng.* 2012;**19**(85).
52. Norrahim MNF, Ariffin H, Yasim-Anuar TAT, Ghaemi F, Hassan MA, Ibrahim NA, et al. Superheated steam pretreatment of cellulose affects its electrospinnability for microfibrillated cellulose production. *Cellulose*. 2018;**25**(7):3853–9. doi: 10.1007/s10570-018-1859-3
53. Li J, Song Z, Li D, Shang S, Guo Y. Cotton cellulose nanofiber-reinforced high density polyethylene composites prepared with two different pretreatment methods. *Ind. Crops Prod.* 2014;**59**:318–28.
54. Clasen SH, Müller CMO, Pires ATN. Maleic anhydride as a compatibilizer and plasticizer in TPS/PLA blends. *J. Braz. Chem. Soc.* 2015;**26**(8):1583–90.
55. Kiziltas A, Han Y, Gardner DJ, Neivandt D, Rushing TS, Nader JW. Functionalized carrier systems for cellulose nanofibrils designed for polymer composites. In: *TAPPI International Conference on Nanotechnology for Renewable Materials*, Arlington, VA. 2011.
56. De Menezes AJ, Siqueira G, Curvelo AAS, Dufresne A. Extrusion and characterization of functionalized cellulose whiskers reinforced polyethylene nanocomposites. *Polymer (Guildf)* 2009;**50**(19):4552–63.
57. Chang F, Lee S, Toba K, Nagatani A, Endo T. Bamboo nanofiber preparation by HCW and grinding treatment and its application for nanocomposite. *Wood Sci. Technol.* **46**(1–3) 2012;393–403.
58. Herrera N, Mathew AP, Oksman K. Plasticized polylactic acid/cellulose nanocomposites prepared using melt-extrusion and liquid feeding : mechanical , thermal and optical properties. *Compos. Sci. Technol.* 2015;**106**:149–55. doi: 10.1016/j.compscitech.2014.11.012
59. Oksman K, Aitomäki Y, Mathew AP, Siqueira G, Zhou Q, Butylina S, et al. Review of the recent developments in cellulose nanocomposite processing. *Compos. A Appl. Sci. Manuf.* 2016;**83**:2–18.
60. Oksman K, Skrifvars M, Selin J-F. Natural fibres as reinforcement in polylactic acid (PLA) composites. *Compos. Sci. Technol.* 2003;**63**(9):1317–24.
61. Herrera Franco PJ, Valadez-González A. Fiber-matrix adhesion in natural fiber composites. *Natural Fibers, Biopolymers, and Biocomposites*. 2005;177–230.
62. Yasim-Anuar TAT. Cellulose nanofiber from oil palm mesocarp fiber and their utilization as reinforcement material in low density polyethylene composites. 2018.
63. Mahmood H, Moniruzzaman M, Yusup S, Muhammad N, Iqbal T, Akil HM. Ionic liquids pretreatment for fabrication of agro-residue/thermoplastic starch based composites: a comparative study with other pretreatment technologies. *J. Cleaner Prod.* 2017;**161**:257–66.
64. Wang L, Ando M, Kubota M, Ishihara S, Hikima Y, Ohshima M, et al. Effects of hydrophobic-modified cellulose nanofibers (CNFs) on cell morphology and mechanical properties of high void fraction polypropylene nanocomposite foams. *Compos. A Appl. Sci. Manuf.* 2017;**98**:166–73.
65. Chakrabarty G, Vashishtha M, Leeder D. Polyethylene in knee arthroplasty: a review. *J. Clin. Orthop. Trauma* 2015;**6**(2):108–12.
66. Khanam PN, Al Maadeed MAA. Processing and characterization of polyethylene-based composites. *Adv. Manuf. Polym. Compos. Sci.* 2015;**1**(2):63–79.
67. Hietala M, Rollo P, Kekäläinen K, Oksman K. Extrusion processing of green biocomposites: compounding, fibrillation efficiency, and fiber dispersion. *J. Appl. Polym. Sci.* 2014;**131**(6):1–9.
68. Nishida H, Yamashiro K, Tsukegi T. Biomass composites from bamboo-based micro/Nanofibers. In: *Handbook of Composites from Renewable Materials* VK Thakur, MK Thakur, MR Kessler, eds, Scrivener Publ LLC. 2017;339–61.
69. Budiman BA, Triawan F, Adziman F, Nurprasetio IP. Modeling of stress transfer behavior in fiber-matrix composite under axial and transverse loadings. *Compos. Interfaces* 2017;**24**(7):677–90.
70. Peng BL, Dhar N, Liu HL, Tam KC. Chemistry and Applications of Nanocrystalline Cellulose and its Derivatives: A Nanotechnology Perspective. 2011;9999:1–16.
71. Quillin DT, Caulfield DF, Koutsky JA. Crystallinity in the polypropylene/cellulose system. I. Nucleation and crystalline morphology. *J. Appl. Polym. Sci.* 1993;**50**(7):1187–94.

72. Rahman R, Lai J, Hui C, Hamdan S. Physical, mechanical, thermal and morphology properties of biodegradable polymer nanocomposites and its comparison. 2017;03005:1–5.
73. Iwamoto S, Nakagaito AN, Yano H, Nogi M. Optically transparent composites reinforced with plant fiber-based nanofibers. *Appl. Phys. A* 2005;**81**(6):1109–12.
74. Nair SS, Zhu JY, Deng Y, Ragauskas AJ. High performance green barriers based on nanocellulose. *Sustainable Chem. Processes* 2014;**2**(1):23.
75. Ferrer A, Pal L, Hubbe M. Nanocellulose in packaging : advances in barrier layer technologies. *Ind. Crops Prod.* 2017;**95**:574–82. doi: 10.1016/j.indcrop.2016.11.012
76. Ilyas RA, Sapuan SM, Atikah MSN, Asyraf MRM, Ayu Rafiqah S, Aisyah HA, Mohd Nurazzi N, Norrahim MNF. Effect of Hydrolysis Time on the Morphological, Physical, Chemical, and Thermal Behavior of Sugar Palm Nanocrystalline Cellulose (Arenga Pinnata (Wurmb.) Merr). *Textile Research Journal*. 2020. In print. doi: 10.1177/0040517520932393.

9

Green Food Packaging from Nanocellulose-Based Composite Materials

Abdel Rehim M.H.
Packaging Materials Department, Division of Chemical Industries Research, National Research Centre, 33 El Behooth St., Dokki, Giza, Egypt

9.1 Introduction

The rising interest in the global environment forced researchers to focus their efforts on finding and establishing new eco-friendly methods for the synthesis of materials, especially those in daily use that represent a clear danger for the environment. Food packaging materials based on the petrochemical industry exemplify a serious environmental problem. Being nonrenewable and nonbiodegradable, millions of tons of food packaging wastes are thrown into the oceans every year, drastically affecting marine animals. Because of these problems, there is an imperative need for novel biodegradable food packaging materials based on renewable and sustainable resources. It should also be taken into consideration that the market for packaging materials is growing as the global consumer already needs annually US$400 billion–US$500 billion [1].

Cellulose is a naturally abundant polymer since it is the main constituent of plants and some animal species such as tunicate [2]. Its annual production is 75 billion tons [3] of which only 2 billion tons are annually used in different industrial applications [4]. Cellulose is a linear biopolymer composed of β-D-glucopyranose repeating units [5]. The cellulosic fibers are bonded by lignin and contain amorphous and crystalline parts. Cellulose is mostly used as a packaging material in the form of paper, wrapping material, containers, and

Bio-Based Packaging: Material, Environmental and Economic Aspects, First Edition. Edited by S.M. Sapuan and R.A. Ilyas

flexible and rigid packaging [6]. This is due to the benefits of using cellulose in packaging, such as its low cost, light weight, and sustainability [7]. Lately, nanostructured cellulose has been applied as a reinforcing material that adds interesting properties to the hosting polymer matrix [8–10]. Moreover, it can be applied as a coating and self-standing packaging film [11]. These features draw a lot of attention to prepare biocompatible and biodegradable packaging material with less dependency on petrochemical sources [12].

9.2 Synthesis of Cellulose Nanostructures

Recently, isolation of microfibrillated cellulose (MFC), crystalline nanocellulose (CNC), and nanoscale fibrils from cellulose have paved the way to using them in more advanced applications due to their versatile and attractive features. Among these are their large surface area and good ability to bond to hydrogen as well as being from a renewable resource [13]. The cellulose chains are synthesized to form nanofibers in different ways, such as chemical, mechanical, and enzymatic methods. The formed cellulosic nanofibers (CNFs) are elongated bundles of molecules stabilized through hydrogen bonding. Nanofibers have nanosized diameters (2–20 nm) and lengths ranging from a few hundred nanometers up to a few micrometers [14–17]. Cellulose nanofibers are recognized as being very effective in reinforcing polymers due to interactions between the nanosized elements that form a percolated network connected by hydrogen bonds, provided there is a good dispersion of the nanofibers in the matrix [18]. Moreover, the presence of free reactive groups on the CNF chains offers the opportunity for their functionalization of CNF to achieve materials with new properties. The special hierarchical structure and the linear and symmetric packing of the chains that are stabilized by the strong intra- and interchain hydrogen bond network promotes the high axial stiffness and poor solubility of cellulose, as well as leads to the laborious delamination of the fibers to its nanoscale fibrils in CNF processes. The preparation of CNF is typically carried out by a mechanical method using homogenizer to increase the degree of fibrillation. To decrease the energy consumption and obtain the individual fibers, chemical [19, 20] or enzymatic [21, 22] pretreatments of the cellulosic raw materials are required. Saito et al. [23] used oxidation pretreatment during the preparation of CNF. In this procedure, cellulose fiber suspension is oxidized by 2,2,6,6-tetramethylpiperidine-1-oxyl radical (TEMPO) with NaClO as a primary oxidant and NaBr as an additional catalyst, at pH of 10–11 [19, 24]. As a result of this treatment, a C6 hydroxyl group is selectively converted into a carboxylate group and, consequently, the nanofibrils separate from the fibers due to the repulsion between the negatively charged carboxylate groups. Isolation of CNF using a green method is also described as it is an energy saver technique. Enzymatic pretreatment of cellulosic raw material, Endocluganases, which can attack only in the amorphous, disordered regions, have been frequently applied [21, 25, 26]. This procedure is typically followed by homogenization, which liberates the cellulose nanofibrils from the fiber wall fragments. It was reported that a reduction in energy consumption on using enzymatic pretreatment reaches over 91% without pretreatment [26]. Lately, isolation of CNF from bagasse pulp using immobilized cellulase has been reported [27]. Immobilization of the enzyme on gel desks improved its heat resistance and preserved most of its activity for six cycles. Cellulose nanocrystals (CNC) or whiskers are commonly prepared using the chemical method of acid hydrolysis of cellulosic materials [28] However, agricultural residues such as bagasse and rice straw

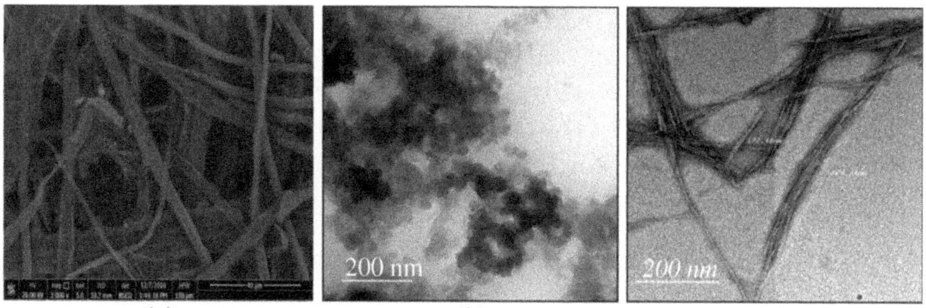

Figure 9.1 *(a) SEM image of cellulose fibers form rice straw, (b) TEM micrographs of spherical-shaped CNC, and (c) TEM micrograph of CNF. Both are prepared from the same source by a chemical method.*

are also used for CNC and CNF isolation [29, 30]. This CNC is characterized by its high crystallinity as the crystallinity index can reach 90%. Indeed, the crystallinity index can differ depending on the source of the nanocellulose (either nanofibrillated or nanocrystals) [31–33]. Spherical-shaped CNC or cellulose nanofibers can be obtained by acid hydrolysis of the cellulosic fibers from the same source, as shown in Figure 9.1. Nevertheless, other morphological aspects of CNC, such as length and width, are also varied depending on the source of CNC [34]. Other methods for CNC synthesis have also been reported, among them mechanical grinding, a high-pressure homogenizer, and cryo-crushing. It should be pointed out that the synthesis protocols should be adjusted to fit the used cellulosic raw material and to obtain the required morphology.

9.3 Modification of Nanocellulose

Surface modification of nanocellulose is essential since it is characterized by its hydrophilicity and poor moister resistance. Moreover, the chemical modification of CNC increases its dispersibility in the organic solvent or the polymer matrix. Surface modification can be either noncovalent through electrostatic adsorption of surfactant on the CNC surface or covalent [35, 36]. In the first technique, ionic or nonionic surfactant, such as cetyltetramethylammoniumbromide (CTAB), sorbitan monostearate, Beycostat A B09, and acid phosphate ester of ethoxylated nonylphenol, are successfully used [35, 37, 38]. Introduction of positive charges using ammonium salts as a substitute for the negative charges on the CNC are also described [39]. On the other and, covalent surface modification of CNC could be achieved through several methods, among them (i) TEMPO-mediated oxidation by selective oxidation of a hydroxyl group on C6 to the carboxyl group and (ii) taking advantage of the presence of many hydroxyl groups to introduce different functionalities, such as etherification, silylation, urethane group formation, or esterification [34, 40].

Mesquita et al. [41] described the functionalization of CNCs with methyl adipoyl chloride, where the reactive methyl ester end groups on the CNC surface reacted with chitosan amino groups. The obtained bionanocomposite showed enhancement in tensile properties and modulus by addition of 60% modified CNC to chitosan. Chemical surface grafting of CNC is used to enhance the dispersion of the nanoparticle in the reaction medium, taking into consideration the preservation of the integrity of the nanoparticles after the grafting process. A "grafting on to" strategy is reported in which coupling agents are used. A "grafting from" approach using radical polymerization techniques, such as ring-opening, atom transfer radical polymerization (ATRP), and single electron transfer radical polymerization are also described [42–44].

9.4 Properties of Nanocellulose-Based Nanocomposites

Nanocellulose is commonly used to enhance the mechanical properties of the packaging material. Generally, tensile strength and stiffening are affected by the presence of enough CNC to produces a continuous phase within the matrix [45, 46]. After the CNC optimum concentration, the mechanical properties are reduced significantly along with a decrease in transparency due to aggregation of the nanocellulose [47]. The interaction between the OH groups on the cellulose chains causes aggregation problems at higher concentrations. To overcome the aggregation problem, a surfactant is used to enhance the dispersion of CNC in the polymer matrix by covering the nanocellulose in order to hinder agglomeration. Moreover, the surfactant can increase the compatibility of nanocellulose and the hydrophobic polymer. Utilization of acid phosphate ester of ethoxylated nonylphenol as a surfactant to increase the dispersion of nanocellulose in polylactic acid is demonstrated [48, 49]. However, degradation in the PLA matrix was reported as a negative effect of the surfactant [38]. The most suitable processing medium for CNC is water due to its good disposability in it. Therefore, water-soluble polymers such as polyvinyl alcohol and polymer dispersions can be mixed with CNC and cast to obtain free-standing films after solvent evaporation. This simple procedure preserves the individuality state of the nanoparticles in the colloidal dispersion [48]. Bionanocomposites based on wheat gluten, CNC, and TiO_2 prepared by casting and then evaporation were reported [50]. The formed films showed enhanced mechanical properties and improved water sensitivity. Increasing concentration of CNC revealed an increase in the film thickness and elongation at break for corn plasticized films [51]. A combination of CNC and starch in a matrix formulation yielded films of high puncture resistance, which is a favorable property in packaging material. Nassima et al. [52] reported the synthesis of starch/CNC/carboxymethyl cellulose film of increasing elastic properties with a variation of CNC content. Barrier properties of food packaging materials against water vapor or gases are essential in extending the shelf life of packed food items [50]. CNC acts as a physical barrier that increases the diffusion bath of water vapor across the packaging film and hence improves water vapor permeability (WVP) [53]. Aulin et al. [54] reported that a large fraction of CNC enhances oxygen impermeability through the creation of longer diffusion paths. Development of bionanocomposites based on sugar palm nanocrystalline cellulose reinforced with starch has also been reported. A bionanocomposite of enhanced water barrier properties is obtained that can find application in recyclable packaging [55–58]. The addition of CNC and Ag nanoparticles to PLA yielded films of high barrier properties and antibacterial activity against *Staphylococcus aureus* and *Escherichia coli* [59]. Moreover, barrier properties and migration behavior of PLA loaded with CNC and CNC modified with a surfactant were studied. The results revealed that the presence of surfactant led to better dispersion of CNC in the polymer matrix and addition of 1 wt% s-CNC reduced water permeability by 34% [60].

In addition, low migration levels were noticed for the modified films. The thermal stability of bionanocomposites PLA/PHB/CNC is increased significantly with increasing ratios of the added filler in the polymer matrix, as it was reported that blends of PLA and polyhydroxy butyrate were degraded in one step [61]. This was attributed to the good interaction between the components of the formed nanocomposite. Sheets of polymethylmethacrylate loaded with different ratios of surface-modified CNC were prepared using the casting technique and their thermal and WVP properties were investigated [62]. It was found that by increasing the amount of CNC in the polymer matrix, both thermal stability and WVP were substantially improved. The layer-by-layer technique is another way to fabricate a multicomposite material by controlled adsorption from the solution or dispersion [63]. Properties essential for packaging material, such as barrier, antibacterial, or superhydrophobic, are largely improved or enhanced. In this technique, hydrogen bonding or electrostatic interaction between a polyanion and a polycation is established to build a multilayer film [64]. The simplicity of this approach enabled the researchers to apply it in the synthesis of nanocomposites based on nanocellulose. Multilayer composites of CNC with poly(diallyldimethylammonium chloride) were prepared and their morphology was studied employing AFM and SEM [65]. The micrographs showed uniform coverage and a packed structure. Fabrication of a tunable oxygen barrier packaging film has been described by Lee et al. [66], who used corona treatment to prepare a negatively charged surface polyethylene terphethalate sheet. Then a layer of chitosan and CNC is applied by dipping the sheet in dispersions of chitosan and CNC at different pH values. It was found that by adjusting the pH of the electrolytes, oxygen barrier properties of the fabricated sheet can be varied.

Coating is another tool used to fabricate a packaging material based on CNC with enhanced barrier properties. Coatings are usually used in fixable packaging through applying a thin layer on the surface of the substrate or sandwiching the filler between two substrates [67, 68]. Coating of paper with carboxymethylated MFC and investigation of barrier properties of coated paper at different relative humidity (RH) values were reported [69]. The results showed that the oxygen barrier properties are increased at low RH with values very close to those obtained for synthetic polymers. At high RH, the oxygen permeability increased due to the plasticizing effect of MFC and swelling of the fibers. Generally speaking, coating of paper and paper board with CNC was found to improve WVP by several researchers [70–72]. Paper sheets prepared from bagasse pulp have been modified through loading with different ratios of nanocellulose or coating with a layer of the nanoparticles. It was found that the thickness of the coating layer was increased by increasing the concentration of the coating formulation. Imaging of the coated paper sheets by SEM revealed that complete surface coverage with the closing of the pores could be attained due to good compatibility and strong adhesion between the sheet and coating materials [73]. However, investigation of air permeability and burst index for the coated paper sheets showed remarkable improvement in these properties by increasing the coating thickness.

Incorporation of another material such as starch or clay, with CNC in the polymer matrix, is also reported. Azeredo et al. [74] studied the effect of CNC and montmorillonite nanoclay (MMT) as a reinforcing agent in alginate–acerola puree films and coatings. The coated films showed WVP 30% lower than the uncoated films. Moreover, evaluation of the coating films after storage for seven days revealed that coatings containing CNC lowered fruit weight loss by 53%, ascorbic acid loss by 70%, and decay incidence by 46%. Incorporation of MMT with CNC in the coating formulation decreased weight and ascorbic acid loss. The addition of limonene as a plasticizer to polylactic acid/CNC (PLA) nanocomposite led to an increase in elastic modulus and elongation of the 3% CNC film from 38% to 272% without affecting film transparency [61].

9.5 Active Packaging Material

Current consumer demand toward safer, healthier, and long shelf-life food motivated the researchers to design more sustainable packaging materials such as active packaging. According to European Regulation (EC) No. 450/2009, active packaging has been developed to "deliberately incorporate components that would release or absorb substances into or from the packaged food or the environment surrounding the food" [75]. Therefore, active packaging is intended to extend shelf life and maintain the conditions of packed food [76]. Developing packaging materials to fulfill the improved protective function has been the core interest of many researchers [77–84]. Active packaging technologies involved antioxidants, antibacterial, ethylene, or water vapor absorbent systems and carbon dioxide releasing agents. Active packaging films containing silver nanoparticles (Ag Nps) are being widely studied due to the antibacterial properties of silver, although there are possible toxic effects from the nanoparticles [85, 86]. Production of PLA/CNC films loaded with Ag nanoparticles by melt extrusion have been described [48]. The modified films demonstrated antibacterial activity against Gram-positive and Gram-negative bacteria. The migration tests revealed that the levels are within the permissible limits while the release of the silver from the nanocomposite to the ethanol solution was easier in the case of using CNC treated with surfactant. Synthesis of Ag Nps using a green method and its incorporation in polyvinyl alcohol/CNC nanocomposite has been reported by Sarwar et al. [87] Starch-capped Ag Nps and CNC isolated from microcrystalline cellulose are loaded in different concentrations to a PVA matrix and nanocomposite films were prepared by the casting/atmospheric evaporation technique. It was found that the tensile strength of the obtained films is increased from 5.52 MPa to 12.32 MPa when filled with 8 wt% of CNC. Moisture retention capability evaluations after six hours revealed that the films have a good moisture retention capability irrespective of the concentration of CNC in the films. This was attributed to the hydrophilic nature of both PVA and nanocellulose owing to the presence of hydroxyl groups that can hold the water molecules inside the nanocomposite films. The highest antibacterial activity was observed for Gram-negative bacteria *Escherichia coli* (DH5-alpha). The cytotoxicity investigations showed that the nanocomposite films have no cytotoxicity effect on HepG2 and cell viability was more than 90%. George et al. [88] investigated the antibacterial properties of PLA/CNC/Ag nanocomposites against four strains of organisms commonly encountered in the food industry over different periods. The results showed high antibacterial activity toward the tested strains although low concentrations of active agents were used. Technological application of Ag Nps incorporated into hydroxymethyl cellulose (HPMC) has been investigated [89]. The average particle size was between 40 and 100 nm and the antibacterial properties of HPMC/Ag NPs thin films were evaluated based on the diameter of the inhibition zone in a disk diffusion test against *Escherichia coli* (*E. coli*) and *Staphylococcus aureus* (*S. aureus*). The antibacterial effect according to the diffusion test was more pronounced for Ag Nps

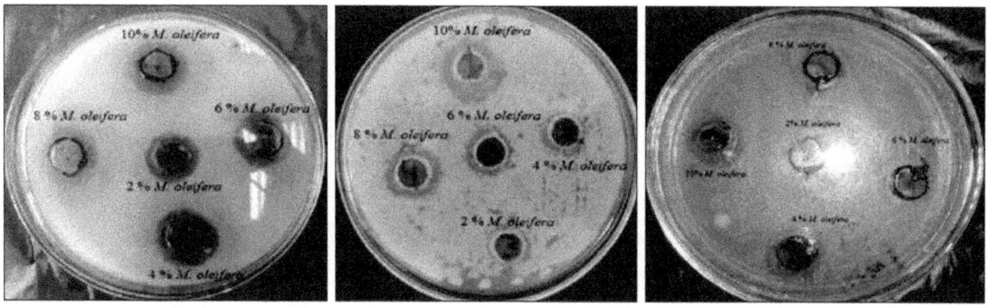

(a) (b) (c)

Figure 9.2 *Inhibition zones of the prepared films against different bacterial strains, for (a) Yersinia enterocolitica, (b) Staphylococcus aureus, and (c) Aspargillus niger.*

of size 40 nm. Other antibacterial agents were investigated, such as the work of Zhu et al., who prepared sausage casings from bacterial cellulose BC and ε-polylysine (ε-PL). The casing composites showed antibacterial activity against several bacteria and good tensile and barrier properties [90]. Abdel Rehim et al. [63] described the synthesis of polymethyl methacrylate loaded with CNC coated with polyaniline. The coated nanocomposites demonstrated high WVP and acquired both antioxidant and antibacterial properties due to the presence of a thin layer of the conductive polymer. Bionanocomposite films were prepared (in our laboratory) through the incorporation of chitosan (CS) and extracted oil in PVA/CNC films from *Moringa oleifera* leaves. The biofilms were tested against different pathogenic strains using the diffusion method (Figure 9.2). The results showed that the diameter of the inhibition zone was increased by increasing the concentration of the added extracted oil.

Food packaging films based on starch–chitosan and gelatin–chitosan were modified through the addition of different ratios of CNC and their mechanical, thermal, and barrier properties were examined and compared [91]. Both film formulations exhibited enhanced mechanical, WVP, and antibacterial properties by increasing concentrations of CNC and chitosan. Nevertheless, gelatin nanocomposites showed better transparency and lower air permeability. Moreover, the two-layer sample showed better food preservation and contact angle compared to the one-layer film, although the strength and transparency were slightly reduced. Sustainable active packaging made from cellulose nanofibrils and tannin was prepared. The strong films acquired antioxidant properties from the tannin as well as high air barrier properties by sixfold due to the hydrophobicity of tannin. Moreover, a slow release of the antioxidant was verified upon soaking with good solvent resistance [92]. Li et al. [93] described the synthesis of biohybrid films from cellulose nanofibrils and tannin extract. The hybrid films showed antioxidant properties and UV-absorbing ability in addition to a 15% increase in thermal stability and tensile stability of up to about 170 MPa. Moreover, tannin extract enhanced the optical clarity of the obtained bio-hybrid films.

9.6 Nanocellulose in Smart Packaging

Intelligent (smart) packaging systems are a new type of food packaging material with special labels incorporated into or attached to the packaging material in order to enhance monitoring product quality and to provide quick information about the product [94, 95]. Indirect visual indicators are commonly used as time/temperature indicators and leak detectors, while direct detectors are preferable since they provide precise information about the quality of the packed product [96]. Nanocellulose has been tailored and applied in the (bio)sensing technology whose aim is to provide analytical information related to different fields such as environmental, clinical, and food safety [97]. In this context, Esmaeili et al. [98] described the synthesis of a biosensor for glucose detection through integrating cellulose nanocrystals/polypyrrole composite with glucose oxidase. The newly prepared composite membrane showed a high sensing ability and dynamic response ranging from 1.0 to 20 mM of glucose but excluding interfering spices such as uric acid, ascorbic acid, and cholesterol. A humidity-detecting sensor was prepared through a combination of bacterial nanocellulose and poly(sulfobetaine methacrylate) in the form of free-standing film [99]. The obtained film revealed a higher transparency than pristine poly(sulfobetaine methacrylate), good mechanical properties, and a UV-blocking ability. Recently, nanocellulose and activated carbon nanocomposites were prepared by film casting as a biosensor for smart packaging applications [100]. The electrochemical conductivity of the nanocomposites was investigated by a linear sweep voltammogram. The results showed decreasing electrical properties by increasing nanocellulose content in the assembled electrode. Investigations of thermal stability and mechanical properties confirmed increased thermal stability and tensile strength from 0.3 to 4.78 MPa. The study concluded that 15% nanocellulose content in the nanocomposite is the optimum concentration for preparing a biosensor.

9.7 Future Trends and Conclusions

Nanocellulose has many advantages, such as its abundance and biorenewability; it is also characterized by its large surface area, hydrophilicity, and the ability to possess active functional groups. These unique advantages moved nanocellulose from a food additive for gelling or a thickening agent to versatile applications in many fields, among them biomedical, biosensing, and food packaging materials. Cellulose nanocrystals, bacterial nanocellulose, and cellulose nanofibers have been exploited to develop sustainable eco-friendly packaging. These efforts were motivated by global environmental concerns about the excessive use of plastics as packaging materials and the desire to find a dependable replacement for the petrochemical polymers. The nanocellulose polymer hybrids can acquire new properties, such as barrier and antioxidant properties, besides improving the thermal and mechanical properties of the host polymer. However, although CNC gains great attention from the researchers, more understanding and development of new technology for fast CNC production are still required. Nevertheless, a variety of promising strategies for implementation of nanocellulose in packaging materials have been demonstrated in this chapter. These results have shown that the developed bionanocomposites have acquired new properties such as gas or water vapor barrier ability, resistance to microorganisms, as well as gaining enhanced mechanical properties. Preparation of sensitive detectors based on surface-modified CNC is an effective approach to obtain sensing detectors with a low detection limit. Moreover, innovation in food packaging science has brought novel methods in preservation, transportation, and monitoring of any changes in foodstuffs and will lead to better management of food products. It is expected that the next generation of food packaging materials will offer new products of edible and biodegradable materials that can be commercialized. However, bacterial and powdered cellulose have been approved by the Food and Drug Administration (FDA) as nontoxic materials. Nanocellulose is still classified as having unidentified properties and thus can expose humans and the environment to harm. Nevertheless, a few studies have pointed to the assessment of environmental and safety risks related to the impending use of nanocellulose in food packaging materials and have confirmed low toxicity. More studies are required for new technologies for mass production of nanocellulose, an evaluation of its nutritional properties, and a safety test.

References

1. Nair SS, Zhu JY, Deng Y, Ragauskas AJ. High performance green barriers based on nanocellulose. *Sustainable Chem. Processes* 2014; **2**(23):1–7. doi:10.1186/s40508-014-0023-0.
2. Zhang D, Zhangn Q, Gao X, Piao G. A nanocellulose polypyrrole composite based on tunicate cellulose. *Int. J. Polym. Sci.* 2013; **2013**, Article ID 175609, 6 pages. doi:10.1155/2013/175609.
3. French AD, Bertoniere AR, Brown RM, Chanzy H, Gray D, Hattori K, Glasser K. In: Seidel, A. (Ed.), *Kirk-Othmer Encyclopedia of Chemical Technology*, 5. Wiley., New York 2004. Vol. **5**.
4. Keijsers ERP, Yilmaz G, van Dam JEG. The cellulose resource matrix. *Carbohydr. Polym.* 2013; **93** (1): 9–21. doi:10.1016/j.carbpol.2012.08.110
5. Dufresne A. Cellulose-based composites and nanocomposites. In: Belgacem MN, Gandini A, editors. *Monomers, Polymers and Composites from Renewable Resources*. Amsterdam, Netherlands: Elsevier; 2008.
6. Lee DS, Yam KL, Piergiovanni L. *Food Packaging Science and Technology*. CRC Press, Boca Raton, New York, 2008; pp. 243–274.
7. Nair SS, Wang SQ, Hurley DC. Nanoscale characterization of naturalfibers and their composites using contact-resonance force microscopy. *Composite, Part A* 2010; **41**: 624–631. doi:10.1016/j.compositesa.2010.01.009
8. Chen Y, Zhang Y, Xu C, Cao X. Cellulose nanocrystals reinforced foamed nitrile rubber nanocomposites. *Carbohydr. Polym.* 2015; **130**(5): 149–154. doi:10.1016/j.carbpol.2015.05.017
9. Santamaria-Echart A, Ugarte L, García-Astrain C, Arbelaiz A, Angeles Corcuera M, Ecciza A. Cellulose nanocrystals reinforced environmentally-friendly waterborne polyurethane nanocomposites. *Carbohydr. Polym.* 2016; **151** (20): 1203–1209. doi:10.1016/j.carbpol.2016.06.069.
10. Abral H, Ariksa J, Mahardika M, Handayani D, et al. Highly transparent and antimicrobial PVA based bionanocomposites reinforced by ginger nanofiber. *Polym. Test.*, 2019; **106186**. doi: 10.1016/j.polymertesting.2019.106186
11. Li F, Mascheroni E, Piergiovanni L. The potential of nanocellulose in the packaging field: a review. *Packag. Technol. Sci.* 2015; **28**: 475–508. doi: 10.1002/pts.2121
12. Ferrer A, Pal L, Hubb M. Nanocellulose in packaging: advances in barrier layer technologies. *Ind. Crops Prod.* 2017; **95**: 574–582. doi:10.1016/j.indcrop.2016.11.012
13. Klemm D, Kramer F, Moritz S, Lindstrm MA, Gray D. et al. Nanocelluloses: a new family of nature based materials. *Angew. Chem. Int. Ed. Engl.* 2011; **50**(24): 5438–5466. doi:10.1002/anie.201001273.
14. Azizi SA, Alloin F, Dufresne A. Review of recent research into cellulosic whiskers, their properties and their application in nanocomposite field. *Biomacromolecules* 2005; **6**: 612–626. doi:10.1021/bm0493685
15. Ilyas AR, Sapuan SM, Ibrahim R, Abral H, et al. Sugar palm (*Arenga pinnata* (Wurmb.) Merr) cellulosic fiber hierarchy: a comprehensive approach from macro to nano scale. *J. Mater. Res. Technol.* 2019; **8**(3): 2753–2766. doi:10.1016/j.jmrt.2019.04.011

16. Ilyas AR, Sapuan SM, Ishak MS, et al., Isolation and characterization of nanocrystalline cellulose from sugar palm fibers (*Arenga pinnata*), *Carbohydr. Polym.* 2017; **181**:1038–1051. doi:10.1016/j.carbpol.2017.11.045
17. Ilyas AR, Sapuan SM, Ishak MR, Zainudinb ES. Sugar palm nanofibrillated cellulose (*Arenga pinnata* (Wurmb.) Merr): effect of cycles on their yield, physic-chemical, morphological and thermal behavior. *Int. J. Biol. Macromol.* 2019; **123**(15):379–388. doi: 10.1016/j.ijbiomac.2018.11.124
18. Anglès MN, Dufresne A. Plasticized starch/tunicin whiskers nanocomposite materials. 2. Mechanical behavior. *Macromolecules*. **2001**; **34**: 2921–2931. doi: 10.1021/ma001555h
19. Saito T, Kimura S, Nishiyama Y, Isogai A. Cellulose nanofibers prepared by TEMPO-mediated oxidation of native cellulose. *Biomacromolecules* 2007; **8**: 2485–2491. doi:10.1021/bm0703970.
20. Isogai A, Saito T, Fukuzumi H. TEMPO-oxidized cellulose nanofibers. *Nanoscale* 2011; **3**: 71–85. doi: 10.1039/C0NR00583E.
21. Janardhnan S, Sain M. Isolation of cellulose microfibriles-an enzymatic approach. *BioResources* 2006; **1**: 176–188.
22. Henriksson, M. Cellulose nanofibril networks and composites. Preparation, structure and properties. Doctoral thesis, KTH, Sweden 2008.
23. Saito T, Isogai A. Wet strength improvement of TEMPO-oxidized cellulose sheets prepared with cationic polymers. *Ind. Eng. Chem. Res.* 2007; **46**(3): 773–780. doi: 10.1021/ie0611608.
24. Kimura R, Saito T, Nishiyama Y, Isogai A. TEMPO-oxidized cellulose nanofibrils prepared from various plant holocelluloses. *Funct. React. Polym.* 2014; **85**: 126–133. doi: 10.1016/j.reactfunctpolym.2014.06.011.
25. Henriksson M, Henriksson G, Berglund LA, Lindström T. An environmentally friendly method for enzyme-assisted preparation of microfibrillated cellulose (MFC) nanofibers. *Eur. Polym. J.* 2007; **43**: 3434–3441. doi:10.1016/j.eurpolymj.2007.05.038.
26. Ankerfors M, Microfibrillated cellulose: Energy-efficient preparation techniques and key properties, Licentiate thesis, KTH, Stockholm 2012.
27. Yassin, MA, Gad AM, Ghanem AF, Abdel Rehim MH. Green synthesis of cellulose nanofibers using immobilized cellulose. *Carbohydr. Polym.* 2019; **205**: 255–260. doi: 10.1016/j.carbpol.2018.10.040.
28. Favier V, Chanzy H, Cavaille JY. Polymer Nanocomposites reinforced by cellulose whiskers. *Macromolecules* 1995; **28**:6365–6367. doi: 10.1021/ma00122a053
29. Abdel Rehim MH, El-Samahy MA, Badawy AA, Mohram ME. Photocatalytic activity and antimicrobial properties of paper sheets modified with TiO_2/sodium alginate nanocomposites. *Carbohydr. Polym.* 2016; **148**: 194–199. doi: 10.1016/j.carbpol.2016.04.061
30. El-Samahy MA, Mohamed SAA, Abdel Rehim MH, Mohram ME. Synthesis of hybrid paper sheets with enhanced air barrier and antimicrobial properties for food packaging. *Carbohydr. Polym.* 2017; **168**: 212–219. doi: 10.1016/j.carbpol.2017.03.041
31. Alemdar A, Sain M. Isolation and characterization of nanofibers from agricultural residues –wheat straw and soy hulls. *Bioresour. Technol.* 2008; **99**(6): 1664–1671. doi: 10.1016/j.biortech.2007.04.029
32. Siqueira G, Tapin-Lingua S, Bras J, Da Silva Perez D, Dufresne A. Morphological investigation of nanoparticles obtained from combined mechanical shearing, and enzymatic and acid hydrolysis of sisal fibers. *Cellulose* 2010; **17**(6): 1147–1158. doi: 10.1007/s10570-010-9449-z
33. Heux L, Dinand E, Vignon MR. Structural aspects in ultrathin cellulose microfibrils followed by 13C CP-MAS NMR. *Carbohydr. Polym.* 1999; **40**(2): 115–124. doi: 10.1016/S0144-8617 (99)00051-X
34. Sinha A, Martin EM, Lim Ki-T, Carrier DJ, Han H, Zharov VP, Kim J-W. Cellulose nanocrystals as advanced "green" materials for biological and biomedical engineering. *J. Biosys. Eng.* 2015; **40**(4): 373–393. doi: 10.5307/JBE.2015.40.4.373
35. Zhou Q, Brumer H, Teeri TT. Self-organization of cellulose nanocrystals adsorbed with xyloglucan oligosaccharide-poly(ethylene glycol)-polystyrene triblock copolymer. *Macromolecules* 2009; **42**(15): 5430–5432. doi: 10.1021/ma901175j

36. Habibi Y, Lucia LA, Orlando J. Rojas. Cellulose nanocrystals: chemistry, self-assembly and applications. *Chem. Rev.* 2010; **110**(6): 3479–3500. doi: 10.1021/cr900339w.
37. Kim J, Montero G, Habibi Y, Hinestroza JP, Genzer J, Argyropoulos DS; Rojas OJ. Dispersion of cellulose crystallites by nonionic surfactants in a hydrophobic polymer matrix. *Polym. Eng. Sci.* 2009; **49**(10): 2054–2061. doi:10.1002/pen.21417.
38. Bondeson D, Oksman K. Dispersion and characteristics of surfactant modified cellulose whiskers nanocomposites. *Compos. Interfaces* 2007; **14**(7–9): 617–630. doi: 10.1163/156855407782106519.
39. Hasani M, Cranston ED, Westmana G, Gray DG. Cationic surface functionalization of cellulose nanocrystals. *Soft Matter* 2008; **4**: 2238–2244. doi.10.1039/B806789A.
40. Moon RJ, Martini A, Nairn J, Simonsen J, Youngblood J. Cellulose nanomaterials review: structure, properties and nanocomposites. *Chem. Soc. Rev.* 2011; **40**(7): 3941–3994. doi:10.1039/C0CS00108B.
41. Mesquita, JP, Donnici CL, Teixeira IF, Pereira FV. Bio-based nanocomposites obtained through covalent linkage between chitosan and cellulose nanocrystals. *Carbohydr. Polym.* 2012; **90**: 210–217. doi: 10.1016/j.carbpol.2012.05.025.
42. Goffin AL, Raquez JM, Duquesne E, Siqueira G, Habibi Y, Dufresne A, Dubois P. From interfacial ring-opening polymerization to melt processing of cellulose nanowhisker-filled polylactide based nanocomposites. *Biomacromolecules*. 2011; **12**: 2456–2465. doi: 10.1021/bm200581h.
43. Matyjaszewski K, Xia J. Atom transfer radical polymerization. *Chem. Rev.* 2001; **101**(9): 2921–2990. doi: 10.1021/cr940534g.
44. Zhang Z, Wang W, Xia H, Zhu J, Zhang W, Zhu X. Single electron transfer living radical polymerization (SET-LRP) of methyl methacrylate (MMA) with a typical RAFT agent as an initiator. *Macromolecules* 2009; **42**: 7360–7366.
45. Cho M-J, Park B-D. Tensile and thermal properties of nanocellulose-reinforced poly(vinyl alcohol) nanocomposites. *J. Ind. Eng. Chem.* 2011; **17**: 36–40. doi: 10.1016/j.jiec.2010.10.006.
46. Ljungberg N, Bonini C, Bortolussi F, Boisson C, Heux L, Cavaillé JY. New nanocomposite materials reinforced with cellulose whiskers in atactic popypropylene: effect of surface and dispersion characteristics. *Biomacromolecules*. 2005; **6**: 2732–2739. doi: 10.1021/bm050222v.
47. Abdollahi M, Alboofetileh M, Behrooz R, Rezaei M, Miraki R. Reducing water sensitivity of alginate bio-nanocomposite film using cellulose nanoparticles. *Int. J. Biol. Macromol.* 2013; **54**: 166–173. doi: 10.1016/j.ijbiomac.2012.12.016.
48. Fortunati E, Armentano I, Zhou Q, Iannoni A, Saino E, Visai L, Berglund LA, Kenny JM. Muntifunctional bionanocomposite films of poly(lactic acid): celulose nanocrystals and silver nanoparticles. *Carbohydr. Polym.*2012; **87**: 1596–1605. doi:10.1016/j.carbpol.2011.09.066.
49. Fortunati E, Peltzer M, Armentano I, Torre L, Jiménez A, Kenny JM. Effects of modified cellulose nanocrystals on the barrier and migration properties of PLA nano-biocomposites. *Carbohydr. Polym.* 2012; **90**: 948–956. doi: 10.1016/j.carbpol.2012.06.025.
50. Nahla AE-W, Hassan E, Abou-Zeid R, Dufresne A. Development of wheat gluten/nanocellulose/titanium dioxide nanocomposites for active food packaging. *Carbohydr. Polym.* 2015; **124**: 337–346. doi: 10.1016/j.carbpol.2015.01.076.
51. Alves JS, Reis KC, Menezes EGT, Pereira FV, Pereira J. Effect of cellulose nanocrystals and gelatin in corn starch plasticized films. *Carbohydr. Polym.* 2015; **115**: 215–222. doi: 10.1016/j.carbpol.2014.08.057.
52. Nassima ELM, Karima A, Abdellatif B, Mohamed Z, Aziz F, Abderrahim S, Mounir ELA. Bio-nanocomposite films reinforced with cellulose nanocrystals: rheology of film-forming solutions, transparency, water vapor barrier and tensile properties of films. *Carbohydr. Polym.* 2015; **129**:156–167. doi: 10.1016/j.carbpol.2015.04.051.
53. Yam KL, Lee DS. In: *Emerging Food Packaging Technologies: Principles and Practice*. Elsevier, Amsterdam, Netherlands 2012.
54. Rhim JW, Ng PKW. Natural biopolymer-based nanocomposite films for packaging applications. *Crit. Rev. Food Sci. Nutr.* 2007; **47**: 411–433. doi:10.1080/10408390600846366.

55. Aulin C, Gallsted M, Lindstrom T. Oxygen and oil barrier properties of microfibrillated cellulose films and coating. *Cellulose* 2010; **17**:559–574. doi: 10.1007/s10570-009-9393-y.
56. Ilyas RA, Sapuan SM, Ishak MR, Zainudin ES. Development and characterization of sugar palm nanocrystalline cellulose reinforced sugar palm starch bionanocomposites. *Carbohydr. Polym.* 2018; **202**: 186–202. doi:10.1016/j.carbpol.2018.09.002.
57. Ilyas RA, Sapuan SM, Ishak MR, et al., Sugar palm nanocrystalline cellulose reinforced sugar palm starch composite: degradation and water-barrier properties, *IOP Conf. Ser. Mater. Sci. Eng.*, 2018; **368**: 012006. doi:10.1088/1757-899X/368/1/012006.
58. Ilyas RA, Sapuan SM, Ibahim R, Abral H, et al. Effect of sugar palm nanofibrillated cellulose concentrations on morphological, mechanical and physical properties of biodegradable films based on agro-waste sugar palm (*Arenga pinnata (Wurmb.) Merr*) starch. *J. Mater. Res. Technol.* 2019; **8**(5): 4819–4830. doi: 10.1016/j.jmrt.2019.08.028.
59. Halimatul MJ, Sapuan SM, Jawaid M, et al. Effect of sago starch and plasticizer content on the properties of thermoplastic films: mechanical testing and cyclic soaking drying, *Polimery*, 2019;**64**: 32–41. doi: 10.14314/polimery.2019.6.5.
60. Fortunati E, Armentano I, Zhou Q, Iannoni A, Saino E, Visa L, Berglund LA, Kenny JM. Multifunctional bionanocomposite films of poly (lactic acid), cellulose nanocrystals and silver nanoparticles. *Carbohydr. Polym.* 2012; **87**:1596–1605. doi:10.1016/j.carbpol.2011.09.066.
61. Fortunati E, Luzi F, Puglia D, Dominici F, Santulli C, Kenny JM, Torre L. Investigation of thermo-mechanical, chemical and degradative properties of PLA-limonene films reinforced with cellulose nanocrystals extracted from Phormium tenax leave. *Eur. Polym. J.* 2014; **56**: 77–91. doi: 10.1016/j.eurpolymj.2014.03.030.
62. Arrieta MP, Fortunati E, Dominici F, Lopez J, Kenny JM. Bionanocomposite films based on plasticized PLA-PHB/cellulose nanocrystal blends. *Carbohydr. Polym.* 2015; **121**:265–275. doi: 10.1016/j.carbpol.2014.12.056.
63. Abdel Rehim MH, Yassin MA, Zahran H, Kamel S, Moharam ME, Turky G. Rational design of active packaging films based on polyaniline-coated polymethyl methacrylate/nanocellulose composites, *Polym. Bull.* 2019. doi:10.1007/s00289-019-02866-0.
64. Decher G. Fuzzy nano-assemblies: toward layered polymeric multi-composites. *Science* 1997; **277**(5330): 1232–1237. doi: 10.1126/science.277.5330.1232.
65. Li F, Mascheroni E, Piergiovanni L. The potential of nano cellulose in the packaging field: a review. *Packag. Technol. Sci.* 2015; **28**: 475–508. doi: 10.1002/pts.2121.
66. Podsiadlo P, YChoi SY, Shi B, Lee J, Cuddihy M, Kotov NA. Molecularly engineered nanocomposites: layer-by-layer assembly of cellulose nanocrystals. *Biomacromolecules* 2005; **6**(6): 2914–2918. doi: 10.1021/Bm050333u.
67. Li F, Biagioni P, Finazzi M, Tavazzi S, Piergiovanni L. Tunable green oxygen barrier through layer-by-layer self-assembly of chitosan and cellulose nanocrystals. *Carbohydr. Polym.* 2013; **92**(2): 2128–2134. doi: 10.1016/j. carbpol.2012.11.091.
68. Farris S, Piergiovanni L. Emerging coating technologies for food and beverage packaging materials. In: *Emerging Food Packaging Technologies: Principles and Practice*, KL Yam, DS Lee (eds.), Woodhead Publishing Ltd: Oxford, UK; 2012; 274–302.
69. Bastarrachea LJ, Wong DE, Roman MJ, Lin Z, Goddard JM. Active packaging coatings, *Coatings* 2015; **5**(4):771–791. doi: 10.3390/coatings5040771.
70. Aulin C, Gallstedt M, Lindstrom T. Oxygen and oil barrier properties of microfibrillated cellulose films and coatings. *Cellulose* 2010; **17**(3): 559–574. doi: 10.1007/s10570-009-9393-y.
71. Hult EL, Iotti M, Lenes M. Efficient approach to high barrier packaging using microfibrillar cellulose and shellac. *Cellulose* 2010; **17**(3): 575–586. doi: 10.1007/s10570-010-9408-8.
72. Sanchez-Garcia MD, Hilliou L, Lagaron JM. Morphology and water barrier properties of nanobiocomposites of k/i-hybrid carrageenan and cellulose nanowhiskers. *J. Agric. Food Chem.* 2010; **58**(24): 12847–12857. doi: 10.1021/Jf102764e.

73. da Silva JBA, Pereira FV, Druzian JI. Cassava starch-based films plasticized with sucrose and inverted sugar and reinforced with cellulose nanocrystals. *J. Food Sci.* 2012; **77**(6): N14–N19. doi: 10.1111/j.1750-3841.2012.02710.x.
74. Azeredo, HMC, Miranda KWE, Ribeiro HL, Rosa MF, Nascimento DM. Nanoreinforced alginate–acerola puree coatings on acerola fruits. *J. Food Eng.* 2012; **113**: 505–510. doi: 10.1016/j.jfoodeng.2012.08.006.
75. European Commission, 2009. EU Guidance to the Commission Regulation (EC) No. 450/2009 of 29 May 2009 on active and intelligent materials and articles intended to come into contact with food (version 1.0). Available at: https://ec.europa.eu/food/safety/chemical_safety/food_contact_materials_en. (accessed 24 October 2017).
76. Yildirim S, Röcker B, Pettersen MK, Nilsen-Nygaard J, Ayhan Z, Rutkaite R, Radusin T, Suminska P, Marcos B, Coma V. Active packaging applications for food. *Compr. Rev. Food Sci. F.* 2018; **17** (1): 165–199. doi: 10.1111/1541-4337.12322.
77. Singh P, Abas Wani A, Saengerlaub S. Active packaging of food products: recent trends. *J. Nutr. Food Sci.* 2011; **41**(4): 249–60. doi: 10.1108/00346651111151384.
78. Yildirim S. Active packaging for food biopreservation. In: C Lacroix, editor. *Protective Cultures, Antimicrobial Metabolites and Bacteriophages for Food and Beverage Biopreservation*. Cambridge, England: Woodhead Publishing Ltd. 2011; 460–89.
79. Arvanitoyannis IS, Oikonomou G. Active and intelligent packaging. In: IS Arvanitoyannis, editor. *Modified Atmosphere and Active Packaging Technologies*. Boca Raton, FL: CRC Press. , 2012; 628–654.
80. Pereira de Abreu DA, Cruz JM, Losada PP. Active and intelligent packaging for the food industry. *Food Rev. Int.* 2012; **28** (2): 146–187. doi: 10.1080/87559129.2011.595022.
81. Dobrucka R, Cierpiszewski R. Active and intelligent packaging food – research and development – a review. *Pol. J. Food Nutr. Sci.* 2014; **64** (1): 7–15. doi: 10.2478/v10222-012-0091-3.
82. Realini CE, Marcos B. Active and intelligent packaging systems for a modern society. *Meat Sci.* 2014; **98** (3): 404–19. doi: 10.1016/j.meatsci.2014.06.031.
83. Brockgreitens J, Abbas A. Responsive food packaging: recent progress and technological prospects. *Compr. Rev. Food Sci. F.* 2016; **15**(1): 3–15. doi: 10.1111/1541-4337.12174.
84. Kuorwel KK, Cran MJ, Orbell JD, Buddhadasa S, Bigger SW. Review of mechanical properties, migration, and potential applications in active food packaging systems containing nanoclays and nanosilver. *Compr. Rev. Food Sci. F.* 2015; **14**(4): 411–430. 10.1111/1541-4337.12139.
85. Kumar R, Münstedt H. Silver ion release from antimicrobial polyamide/silver composites. *Biomaterials* 2005; **26**: 2081–2088. doi: 10.1016/j.biomaterials.2004.05.030.
86. Kim YS, Song MY, Park JD, Song KS, Ryu HR, Chung YH, Chang HK, Lee JH, Oh KH, Kelman BJ, Hwang IK, Yu IJ. Subchronic oral toxicity of silver nanoparticles. *Part Fiber Toxicol.* 2010; **7**: 20. doi:10.1186/1743-8977-7-20
87. Sarwar MS, Khan Niazi MB, Jahan Z, Ahmad T, Hussain A. Preparation and characterization of PVA/nanocellulose/Ag nanocomposite films for antimicrobial food packaging. *Carbohydr. Polym.* 2018; **184**, 453–464. doi: 10.1016/j.carbpol.2017.12.068.
88. George M, Shen W-Z, Qi Z, Bhatnagar A, Montemagno C. Characterization of cellulose nanocrystals and PLA based thin films with either silver or antimicrobial peptide. *IOSR J. Polym. Text. Eng.* 2017; **4** (3): 08–24. doi: 10.9790/019X-04030824.
89. de Mouraa MR, Mattosob LHC, Zucolotto V. Development of cellulose-based bactericidal nanocomposites containing silver nanoparticles and their use as active food packaging. *J. Food Eng.* 2012; **109** (3) 520–524. doi: 10.1016/j.jfoodeng.2011.10.030.
90. Zhum H, Jia S, Yang H, Tang W, Jia Y, Tan Z. Characterization of bacteriostatic sausage casings: a composite of bacterial cellulose embedded with ε–polylysine. *Food Sci. Biotechnol.* 2010; **19**: 1479–1484. doi: 10.1007/s10068-010-0211-y.

91. Noorbakhsh-Soltani SM, Zerafat MM, Sabbaghi S. A comparative study of gelatin and starch-based nano-composite films modified by nano-cellulose and chitosan for food packaging applications. *Carbohydr. Polym.* 2018; **189**: 48–55. doi: 10.1016/j.carbpol.2018.02.012.
92. Missio AL, Mattos BD, Ferreira DF, Magalhães WLE, Bertuol DA, Gatto DA, Petutschnigg A, Tondi G. Nanocellulose-tannin films: from trees to sustainable active packaging. *J. Cleaner Prod.* 2018; **184**: 143–151. doi: 10.1016/j.jclepro.2018.02.205.
93. Li P, Sirviö JA, Haapala A, Khakaloc A, Liimatainen He. Anti-oxidative and UV-absorbing biohybrid film of cellulose nanofibrils and tannin extract. *Food Hydrocolloids*. 2019; **92**: 208–217. doi: 10.1016/j.foodhyd.2019.02.002.
94. Han J H, Ho CHL, Rodrigues ET. Intelligent packaging. In J. H. Han (Ed.), *Innovations in Food Packaging*. Elsevier Academic Press. 2005.
95. Ilyas RA, Sapuan SM, Atikah MSN, Asyraf MRM, Ayu Rafiqah S, Aisyah HA, Mohd Nurazzi N, Norrahim MNF. Effect of Hydrolysis Time on the Morphological, Physical, Chemical, and Thermal Behavior of Sugar Palm Nanocrystalline Cellulose (Arenga Pinnata (Wurmb.) Merr). *Textile Research Journal*. 2020. In print. doi: 10.1177/0040517520932393.
96. Gontard N. Active packaging for food processing and preservation. In *International Congress on Engineering and Food: ICEF*, 2004; 9, 7–11 March, Montpellier, France.
97. Golmohammadi H, Morales-Narváez E, Naghdi Ti, Merkoçi A. Nanocellulose in sensing and biosensing. *Chem. Mater.* 2017; **29** (13):5426–5446. doi: 10.1021/acs.chemmater.7b01170.
98. Esmaeili C, Abdi MM, Mathew AP, Jonoobi M, Oksman K, Rezayi M. Synergy effect of Nanocrystalline cellulose for the biosensing detection of glucose. *Sensors* 2015; **15**: 24681–24697. doi:10.3390/s151024681.
99. Vilela C, Moreirinha C, Domingues E, Figueiredo FML. Antimicrobial and conductive nanocellulose-based films for active and intelligent food packaging. *Nanomaterials*, 2019; **9** (7):. doi: 10.3390/nano9070980.
100. Sobhan A, Muthukumarappan K, Cen Z, Wei L. Characterization of nanocellulose and activated carbon nanocomposite films' biosensing properties for smart packaging. *Carbohydr. Polym.* 2019; **225**: 115189. doi:10.1016/j.carbpol.2019.115189.

10

Nanocellulose Polylactide-Based Composite Films for Packaging Applications

Dogan Arslan[1], Emre Vatansever[2], and Mohammadreza Nofar[1,2]

[1] *Metallurgical & Materials Engineering Department, Faculty of Chemical and Metallurgical Engineering, Istanbul Technical University, Maslak, Istanbul, 34469, Turkey*

[2] *Polymer Science and Technology Program, Istanbul Technical University, Maslak, Istanbul, 34469, Turkey*

10.1 Introduction

The demand for polymers in packaging applications such as food, beverage, and pharmaceutical industries has been dramatically increased due to providing high protection to their contents [1]. Most polymers that are used in packaging applications are currently petroleum-based, such as polyolefins, polyesters, and polyamides. This is due to their good processability, low cost, and excellent barrier properties [1, 2]. Since petroleum-based polymers cause severe environmental land fill, usage of biodegradable polymers are becoming more desirable. As a family of biodegradable polymers, poly(lactic acid) or polylactide (PLA) has been considered as an appropriate candidate for packaging applications as it is transparent, nontoxic, and possesses reasonable barrier properties. Moreover, it has high physical and mechanical properties comparable with the noted petroleum-based polymers [1, 3].

PLA is obtained from renewable resources with reduced carbon dioxide release and provides significant energy saving in production. PLA could be broken down through metabolism of microorganisms, leading to complete degradation under appropriate conditions of oxygen availability, temperature, and moisture [1, 2]. However, PLA suffers

Bio-Based Packaging: Material, Environmental and Economic Aspects, First Edition. Edited by S.M. Sapuan and R.A. Ilyas

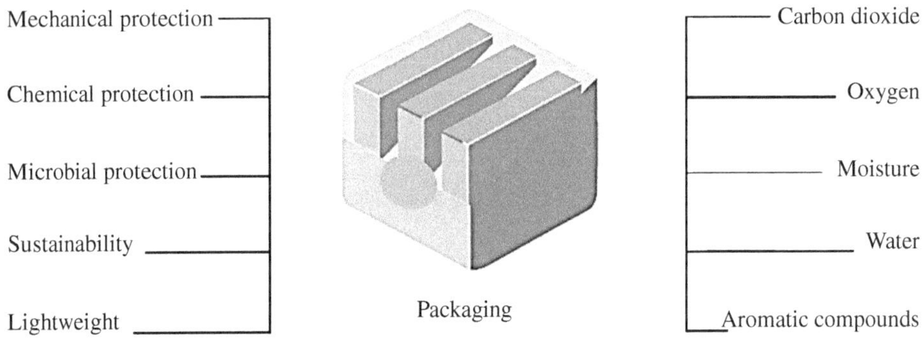

Figure 10.1 *Desirable properties of packaging materials in contact with various elements [1–3].*

from a series of drawbacks, such as slow crystallization [4] and low melt strength [5] which result in poor processability, formability [6], and foamability [7, 8]. These substantially limit its production and usage. Other shortcomings of PLA should also be resolved to make them applicable in various areas. The low toughness, brittleness, low heat deflection temperature, slow crystallization, and low melt strength of PLA could be improved through copolymerization and chemical modifications [9, 10], blending with other polymers [9, 11, 12], nanocomposite development with the addition of nanofillers [13], and plasticizing [14, 15]. The use of plasticizers could resolve the noted mechanical drawbacks but it could also negatively affect the PLA's barrier properties through improving molecular mobility. Moreover, blending PLA with other polymers or reinforcing with macroscopic level materials usually brings additional problems due to the adhesion at the interface and in terms of economic aspects while preventing development of fully bio-based and biodegradable systems for packaging. However, developing nanocomposite systems becomes prominent among these strategies, since it presents an innovative cost-effective solution to resolve the noted drawbacks of PLA while improving its barrier properties remarkably [16, 17].

Moreover, PLA must fulfill some conditions, by serving as selective barriers to oxygen, moisture uptake, losses of volatile aromas, maintaining food safety, and quality in packaging applications (Figure 10.1) [18]. In this context, the use of bio-based cellulose nanoparticles such as CNCs, CNFs, and BC have been proposed to develop fully green PLA-based biocomposites while they could resolve some of the noted drawbacks of PLA [19]. The low permeability of cellulose could be enhanced by the highly crystalline nature of cellulose nanoparticles, which are very promising as they could produce cheap, lightweight, and very strong nanocomposites [20]. Cellulose nanoparticles are considered effective to reinforce polymers due to the interactions between the nanosized particles that could form a percolated network connected by hydrogen bonds. This interconnection provides superior performance and extensive applications, thereby reducing the chain segmental mobility and thus diffusion of the penetrants [20, 21].

Many studies have been conducted that have incorporated nanocellulose into the PLA matrix, which has resulted in enhancing the strength and modulus of PLA as the crystal region in nanocellulose and exhibit superior mechanical properties, which allows the transfer of an applied load from the matrix through the interface. However, many articles have also been focused on barrier properties of PLA nanocellulose-based systems for packaging applications. This was through investigation of the PLA/nanocellulose nanocomposites of water vapor, liquid water, and oxygen barrier properties that are important in various applications of nanocellulose-based films in packaging [22]. It is reported that by developing PLA-based cellulose nanocomposites, it is possible to enhance the barrier properties of the PLA matrix. This is because the incorporation of nanocellulose creates a tortuous path against a penetrant through the matrix, thus preventing diffusion or penetration of gases or liquid substances [16, 17]. Moreover, nanocellulose in PLA could enhance the crystallizability of the matrix by acting as a crystal nucleating agent. This could also further enhance the barrier properties [16, 17].

The barrier enhancement efficiency of nanocellulose incorporated PLA nanocomposites is directly linked with the dispersion level of nanocellulose in the PLA matrix. However, the achievement of good dispersion of cellulose nanoparticles in polymers is a serious challenge to developing PLA nanocellulose nanocomposites due to the hydrophilic structure of cellulose particles [19]. Moreover, water uptake, water sensitivity, and gas permeability could also be a problem when immersed in liquid or a moist atmosphere in terms of barrier properties. However, the nanostructure of cellulose nanoparticles could allow these drawbacks to be altered via chemical treatment/surface modification of cellulose nanoparticles [21]. In general, it has been claimed that the presence of nanocellulose leads to improvements in elastic modulus, gas permeability, dimensional stability, and optical properties with lower loading levels with respect to the pure PLA [13, 23–25].

This chapter provides recent developments on the role of nanocellulose nanoparticles on the processing, mechanical, crystallization, and especially barrier properties of PLA and discusses potential applications of PLA nanocellulose nanocomposite systems in packaging areas.

10.2 Polylactide

Polylactide (PLA) is an aliphatic polyester biopolymer that could be produced from annually renewable resources such as corn, potato, molasses, tapioca, cane sugar, and rice [6, 26, 27]. Lactic acid from the fermentation product of these renewable sources is used as the monomer to synthesize PLA via ring-opening polymerization, polycondensation, or enzymatic polymerization [28–30]. Commercial PLAs are mostly developed based on ring-opening polymerization because in the direct condensation technique the presence of trace amounts of water limit the production of the desired molecular weight of PLA. Figure 10.2 shows the synthesis of PLA through various polymerizations methods [28, 29, 31].

There are two types of lactic acid stereoisomers (Figure 10.2); L-lactic acid is the most common and natural form of lactic acid, but D-lactic acid could also exist. The presence of two chiral carbon centers allows different stereoisomers of PLA to be produced, such as poly(L-lactide) (PLLA), poly(D-lactide), and poly(DL-lactide) (PDLLA) [26, 27, 30, 31]. The component isomers, processing route, molecular weight, and sources are the main factors affecting the properties of PLA [26]. The rigid thermoplastic structure of PLA could be semi-crystalline or totally amorphous depending on the stereochemistry of the polymer backbone. It is reported that D-lactic acid units optimize the crystallization kinetics of

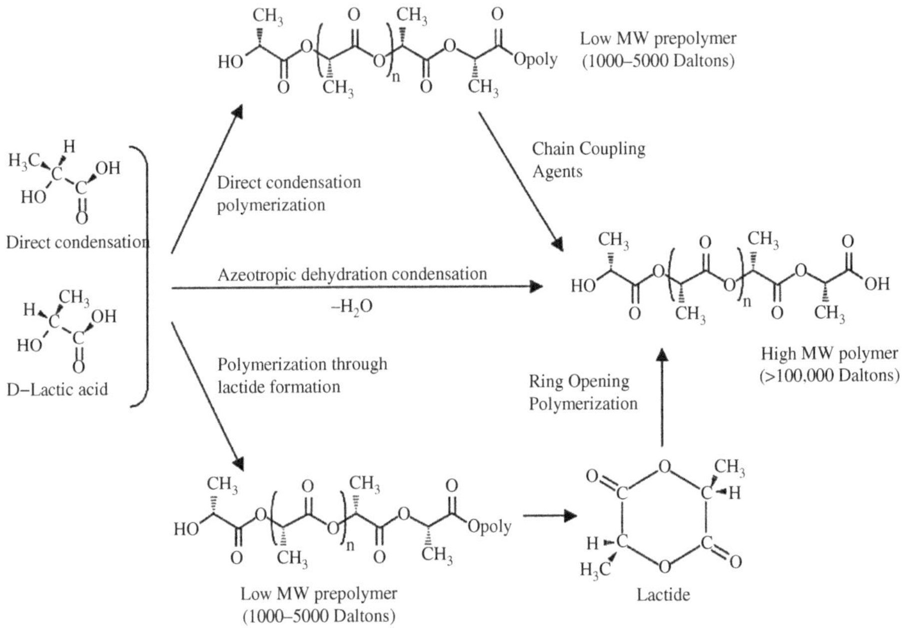

Figure 10.2 *Synthesis of PLA from L- and D-lactic acids [6].*

PLA [4, 32]. PLAs with a D-content greater than ~10% tend to be amorphous while those with lower D-contents are crystalline with lower optical purity [6]. Depending on the composition of the optically active enantiomers, PLA could crystallize in many forms that are effective on the melting temperature (T_m), the glass transition temperature (T_g), and barrier properties [4, 6].

Barrier properties of PLA have been studied with ultraviolet (UV) and visible light, which are important parameters in designing packaging products. It is reported that PLA shows no UV transmission at low ranges (190–220 nm), but above 225 nm 85% of the UV light is transmitted. Therefore, PLA films might require additives to block UV light transmission for dairy packaging applications [3]. PLA shows an oxygen transmission rate (OTR) value between 11000 and 36000 ml μm $(m^2\ day)^{-1}$ at 23°C, different relative humidity (RH), and a water vapor permeability (WVP) of 5250 g μm $(m^2\ day)^{-1}$ at 23°C and 90% RH. To become an alternative to other industrial polymers, approximately 85% and 92% reduction in OTR and WVP values are required, respectively [1, 33]. As also presented in Figure 10.3 when compared with other commodity polymers used in packaging applications, the barrier properties in PLA against oxygen, nitrogen, carbon dioxide, and methane gases are overall higher than those of polyethylene (PE) and comparable to polystyrene (PS) but lower than polyethylene terephthalate (PET) [34].

10.3 Nanocellulose Classification

Cellulose is one of the most abundant materials on the earth and is a structural component of the cell wall in plants and many life forms such as algae [2, 35, 36]. Cellulose

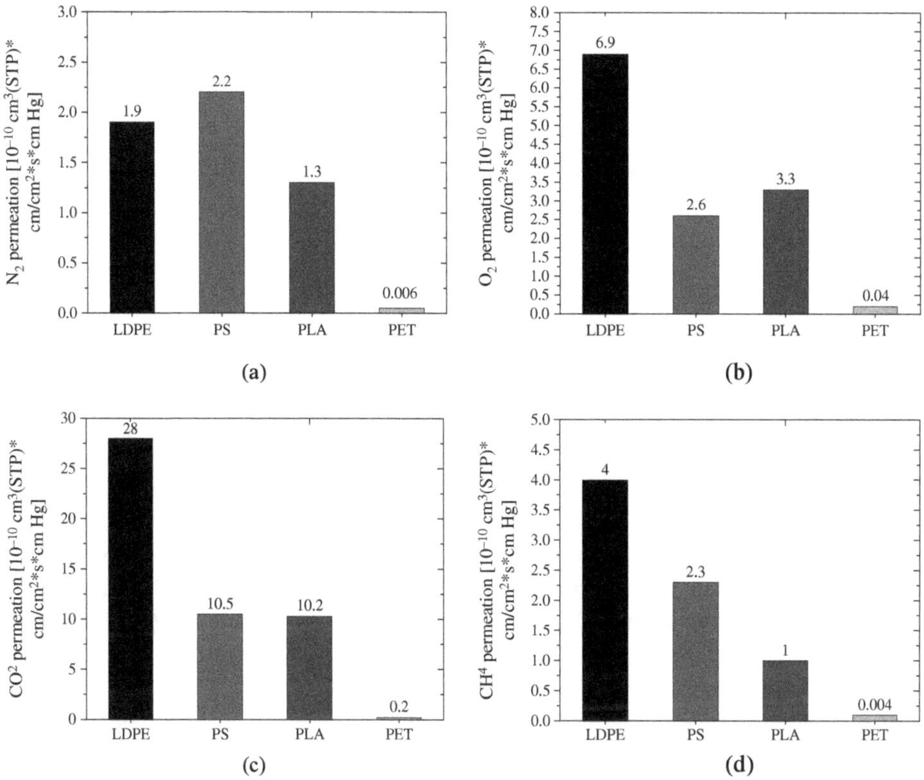

Figure 10.3 *Barrier properties of PLA in comparison to other common polymers at 30°C: (a) N_2, (b) O_2, (c) CO_2, and (d) CH_4 [34].*

can be obtained from many natural sources such as lingo-cellulosic biomass like wood, agricultural residues, cotton, flax, hemp, and sisal [37]. The molecular formula of cellulose is $C_6H_{10}O_5$, which is independent from its sources, but botanical origin, growth conditions, soil, climate circumstances, and age affect the chain length and chemical composition [36, 38]. The hierarchical structure of cellulose shown in Figure 10.4 contains both crystalline and amorphous phases due to the fact that it does not contain evenly distributed cellulose macromolecules [35, 36]. Cotton is the purest source of cellulose, which has a higher amount of crystallinity, while cellulose could exist in combination with lignin and hemicellulose, which affects purity and crystallinity in other sources [36, 37, 39]. Production of nanocellulose has recently attracted more attention as a green solution to develop fully green nanocomposites [19] due to its high abundancy, active and large surface area, and light weight [37]. Furthermore, it is reported that nanocellulose shows better tensile strength and stiffness than that of aramid and glass fiber, which are used to reinforce polymers [37]. However, nanocellulose has a strong tendency to aggregate due to the interaction between surface hydroxyl groups on their surface, providing the possibility of extensive chemical modification such as esterification, silylation, cationization, carboxylation, and polymer grafting, which aim to enhance the disperse ability [40]. Cellulose nanoparticles can be

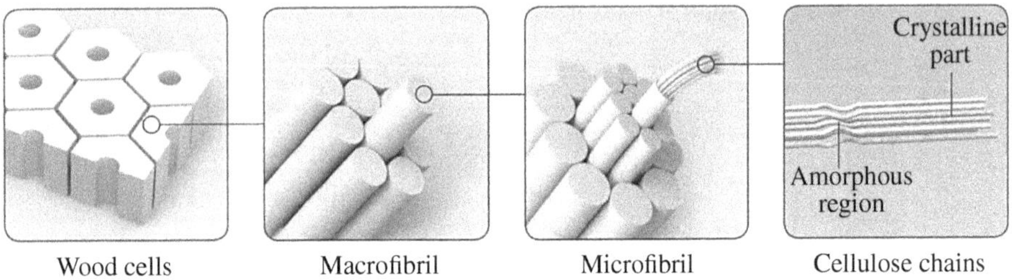

Figure 10.4 *Hierarchical structure of cellulose from plant [19].*

categorized based on their manufacturing processes and dimensions into three categories: cellulose nanocrystals (CNC is also known as CNW or NCC), CNFs, and BC [36, 37].

CNC has attracted significant attention as potentially reinforcing nanoparticles in different polymers during the last decade. Bleached wood pulp, cotton, and ramie fibers could be used as common resources for producing CNC [19]. It is worth mentioning that the origin and crystallinity of cellulose sources influence the size distribution of CNC [41]. A controlled chemically induced de-structuring strategy is used to produce CNC. Strong acid hydrolysis is the classical method to prepare CNC, but another technique has also recently been reported that includes an enzymatic hydrolysis treatment, TEMPO oxidation, and treatment with ionic liquids [2, 13, 21, 36]. CNC is extracted from cellulose sources by removing the amorphous region and preserving the highly crystalline structure [40, 42]. Since crystalline regions display a higher resistance to acid attack, hydronium ions create hydrolytic cleavage at the amorphous region, leading to individual crystallites during strong acid hydrolysis [21]. These rod-like nanocrystals present a diameter of 5–30 nm, a high aspect ratio, and a high modulus of around 140–220 GPa [42–44]. The nature of the acid is an important parameter for the chemical character of CNC. CNC derived from hydrolysis of different types of acids have different properties. For example, it is reported that CNC prepared by the action of hydrochloric acid has a lower de-aggregation ability than sulfuric acid. Because anionic sulfate ester groups are grafted on to the surface of hydroxyl groups via an esterification process, the presence of negatively charged groups promotes the dispersion of CNC [21, 40]. The high aspect ratio and good mechanical properties of CNC have made it a great candidate for the preparation of low-cost, light weight, and highly mechanical nanocomposites [45].

CNF could be isolated via top-down techniques, which are mechanical treatment procedures such as high-pressure homogenization and grinding [19, 21, 36, 37, 46]. These techniques are mainly energy intensive and without any enzymatic and/or chemical treatments [36, 37, 40] and are therefore treatment techniques that could be applied to reduce energy consumption [46]. CNF is obtained as a suspension with 2 wt.% and behaves in a gel-like fashion after mechanical disintegration of natural cellulose fibers along their long axis [13, 21]. The isolation condition, processing intensity, and source of cellulose are the most important parameters affecting the CNF properties and performance [20, 36]. CNF has a diameter in the range of 2–50 nm and an aspect ratio of around 50–100 depending on their origin [19, 36, 46]. Moreover, CNF has an interconnected network structure and its crystallinity ranges between 60% and 70% with 100–150 GPa elastic modulus [19, 21, 35, 36]. CNF also shows many intrinsic properties such as low density, high chemical reactivity, high strength, and easy production in large scales that make it attractive as a green nanocomposite, especially for packaging applications [37, 46].

BC is produced via bottom-up techniques and utilizes some species of bacteria, such as *Gluconacetobacter xylinus*, *Acetobacter hansenii*, and *Acetobacter pasterianus*, contrary to plant-based nanocelluloses [20, 36, 38, 41, 47, 48]. *Acetobacter* species are used to produce cellulose by converting sugar to cellulose via fermentation in large quantities [36, 38]. BC is used as an alternative to plant-based cellulose and requires fewer purification steps since it is free from wax, lignin, hemicellulose, and pectin [38, 47]. Moreover, BC shows a fine and pure network structure with 5–50 nm cross-section, up to several micrometer lengths, high-order crystallinity up to 90 wt.%, and superior mechanical properties with about a 114 GPa elastic modulus [47]. The structure of BC could be influenced by the type of bacterial species, the growing condition, and post-processing drying techniques [41]. Although BC has limitations in production, it is a promising green nanoparticle due to its pure structure, crystallinity, and large aspect ratio [19].

10.4 PLA/Nanocellulose Nanocomposites

10.4.1 Processing

The enhancements in mechanical and thermal properties of polymers using cellulose nanoparticles depends on the quality of nanocellulose dispersion in the polymers and the strength of adhesion at the polymer–filler interface [49]. Although cellulose nanoparticles have a big potential to produce fully biodegradable nanocomposite systems, there are also some drawbacks originating from its strong hydrophilic and polar characteristic [50, 51]. As well as the high surface energy of these nanoparticles, the presence of strong hydrogen bonding among cellulose nanoparticles leads to flocculation of nanoparticles in nonpolar PLA matrices, frustrating the dispersion of cellulose nanoparticles [50].

Melt mixing is a very effective method in producing industrial scale of polymer nanocomposites and is highly preferred in the context of sustainable development since it avoids the use of organic solvents [52, 53]. However, agglomeration of cellulose nanoparticles and incompatibility between PLA and nanocellulose are the main challenges in direct melt processing of PLA/nanocellulose nanocomposites due to the polar structure of nanocellulose [54, 55]. Moreover, the thermal degradation is also more possible during melt mixing under high temperature and shear rates [50]. A study conducted by Kamal and Khoskava [53] clearly indicated that direct melt mixing is not enough to surpass the hydrogen bonding between CNC to have a good CNC dispersion. Figure 10.5 compares the direct melt mixing with solution casting and illustrates poor de-agglomeration of CNC nanoparticles in the PLA matrix observed in direct melt mixing. However, it is also revealed that the change in the CNC preparation method through spray freeze drying could significantly influence the CNC dispersion in PLA with superior properties using direct melt mixing [53]. This is one of the rare studies showing significant enhancements in CNC dispersion and improvements in PLA properties.

Challenges during melt mixing have made solution casting or a combination of solution casting and melt mixing more often used in the literature. In solution casting, nanocellulose and PLA are dissolved in solvents and eventually after the casting the solvent would be evaporated to obtain PLA cellulose nanocomposites [4]. Thermal degradation could be

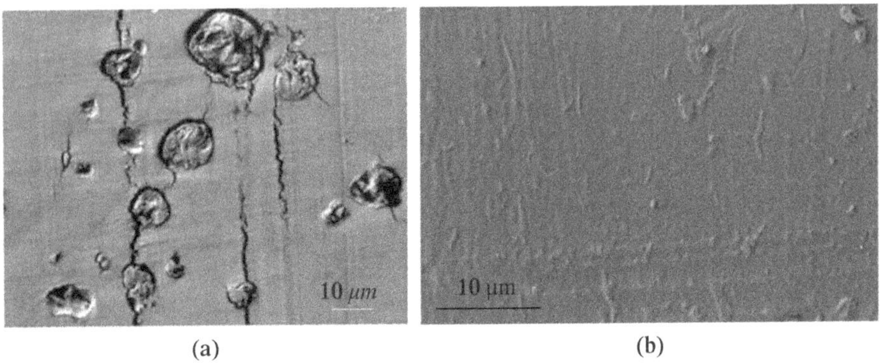

Figure 10.5 Scanning electron microscopy (SEM) images of (a) direct melt mixed PLA/CNC nanocomposites at 7 wt% [53] and (b) solution casted PLA/CNC nanocomposites at 4 wt% [56].

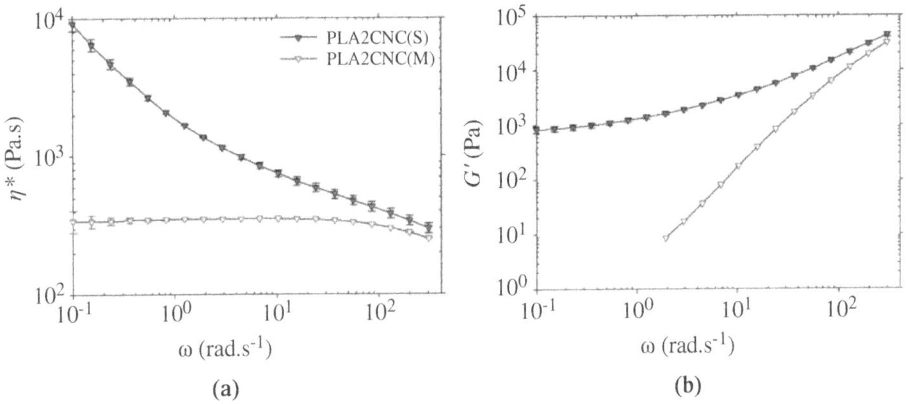

Figure 10.6 Comparison of (a) complex viscosity(η^*) and (b) storage modulus (G') of PLA/CNC nanocomposites at 2 wt% prepared via solution casting (S) and melt mixing (M) methods [57, 59].

eliminated during solution casting, but at the end there could still be some trapped solvent in the composites. As illustrated in Figure 10.5b, Bagheriasl et al. [56] prepared PLA/CNC nanocomposites through solution casting using DMF as the solvent and proved that good dispersion of nanoparticles could be accomplished contrary to melt mixing. Moreover, efficiency of solution casting over melt mixing could be observed from a rheological analysis, which is shown in Figure 10.6. Well dispersed nanoparticles increase the viscosity and storage modulus at low frequencies due to the formation of a network structure among cellulose nanoparticles [57, 58]. Vatansever et al. [59] also revealed that when using solution casting, the CNC dispersion in PLA could be improved in PLAs with lower molecular weight as the shorter PLA molecules could penetrate better among the dissolved CNC particles. They also revealed that PLAs with a low D-content that possess higher crystallizability could result in better CNC dispersion as the crystals could segregate the CNCs while preventing their re-agglomeration.

Different modification techniques have been studied to reduce the surface tension and increase the interfacial interaction among cellulose nanoparticles by using a compatibilizer,

(a) (b) (c)

Figure 10.7 Neat PLA film (a) and nanocomposite PLA films reinforced with (b) 10 wt% unmodified CNF and (c) acetylated CNF with (17% Acetylation) [62].

or surface chemical modification [56, 60, 61]. Ambrosio-Martín et al. [48] investigated the effect of surface modification of BC on its dispersion in PLA. BC is modified by grafting lactic acid oligomers (OLLA) on to a BC surface that enhanced the compatibilization between hydrophilic BC and hydrophobic PLA, which resulted in better BC dispersion. To enhance surface interaction between CNF and the PLA matrix, Tingaut et al. [62] modified the CNF surface via acetylation. Figure 10.7 compares the neat PLA film to modified and unmodified PLA/CNF nanocomposite films. It has been claimed that acetylation prevented CNF agglomeration and increased affinity between CNF and PLA. Martínez-Sanz et al. [25] used polyethylene glycol (PEG) as a compatibilizer, which acts as a bridge between BC and PLA, thus increasing surface adhesion. This resulted in a more homogeneous dispersion of BC in PLA. In a recent study, Gupta et al. [63] used lignin as a bio-based surfactant to coat the CNC surface (L-CNC) in order to avoid agglomeration within PLA. They showed that the use of a surfactant not only improves compatibility between phases and CNC dispersion within PLA but the resulting composites maintain the melt viscosity, melt strength contrary to neat PLA without sacrificing the optical transparency that was tested through the film blowing process, as shown in Figure 10.8.

Sanchez-Garcia et al. [64] studied the effect of two different CNC preparation methods, freeze drying and solvent exchange, on the final properties of solvent casted PLA/CNC nanocomposites. They claimed that CNC aggregates were found in PLA when using solvent exchanged CNC. The reason is that during the solvent exchange process agglomeration can occur and could not be hindered during the casting process. It is claimed that strong mixing after the solvent exchange process, such as high-pressure homogenization or sonication, could prevent the formation of CNC agglomerates. Therefore, as shown in Figure 10.9, PLA nanocomposite films prepared using freeze-dried CNC showed better transparency against the background due to better dispersion of CNC nanoparticles.

10.4.2 Mechanical Properties

To produce PLA/nanocellulose nanocomposites for packaging applications the mechanical properties are some of the most important assessments. For example, if these products are to be used in film packaging, high elongation at break and excellent tear resistance is required. On the other hand, rigidity of the product is an important criterion for food and beverage packaging containers.

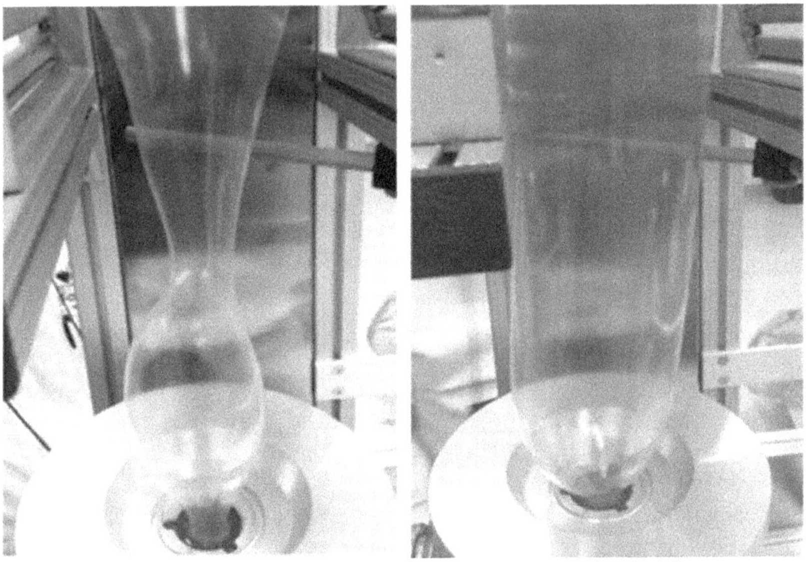

(a) (b)

Figure 10.8 *(a) Neat PLA showing unstable balloon and (b) PLA/L-CNC-0.3% composite with stable balloon [63].*

(a) (b)

Figure 10.9 *PLA films containing (a) 1 wt% of CNC-freeze dried and (b) 1 wt% of CNC-solvent exchanged [64].*

The effects of cellulose nanoparticles on tensile properties of PLA derived from various studies are presented in Table 10.1. In this table, the best achievements in tensile strength and modulus based on the nanocellulose type and content are collected. The majority of the studies reported that unmodified-CNCs reinforced PLA nanocomposites show limited enhancement in tensile modulus and strength [64, 66, 69, 70]. At low CNC contents (<5 wt%), unmodified CNCs improved the mechanical properties to some extent, owing to the high modulus of CNCs [56, 65]. However, some showed that when CNC contents are higher (>5 wt%) unmodified-CNCs could even suppress the mechanical properties of PLA [64, 68] due to the increased chance of CNC aggregates. On the other hand, unmodified-CNFs [82–84] indicated better enhancement within PLA compared to those

Table 10.1 *Effect of CNC on the mechanical properties of PLA.*

Researchers	Processing method	Type of cellulose nano-particles	Nano-particle content (wt%)	Modification	Other components (surfactant, compatibili-zerplasticizer, etc.)	Tensile modulus (GPa)		Tensile strength (MPa)		Strain at break (%)	
						PLA	PLA/NC	PLA	PLA/NC	PLA	PLA/NC
Oksman et al. [44]	Liquid feeding with TSE	CNC	5	—	PLA-MA	2.9	3.9	40.9	77.9	1.9	2.7
Oksman et al. [44]	Liquid feeding with TSE	CNC	5	—	PLA-MA PEG	2.9	2.6	40.9	48.4	1.9	17.8
Bondeson et al. [65]	TSE	CNC	5	—	—	2.65	2.7	62.8	55.5	19.5	9.7
Pei et al. [66]	Solution casting	CNC	2	Silylation	—	1.1	1.4	48.3	58.6	31.1	8.3
Fortunati et al. [67]	TSE	CNC	5	—	Beycostat A B09	2.4	4.4	43	46.1	90	18
Espino-Perez et al. [68]	Solution casting	CNC	2.5	CNC-ICN	—	—	—	40	52	5.6	5.6
Pracella et al. [69]	Hybrid	CNC	3	—	PVAc	2.18	2.61	48	55	4	2.3
Robles et al. [70]	TSE	CNC	1	Esterifiction		2.7	2.8	45	42	11	10
Arias et al. [52]	IM	CNC	3	—	PEO	3.4	3.5	61	60	2.7	2.3
Herrera et al. [71]	Liquid feeding	CNC	1	—	Triethyl citrate (TEC)	0.6	0.9	15.8	19.9	19.6	21.4
Xu et al. [72]	Solution casting	CNC	3	Acetylation	—	1.8	2.1	57	62	3.3	4.2
Dhar et al. [73]	Reactive TSE	CNC	1	PLA-g-CNC	DCP	0.4	2.1	35	51	8.4	1.9
Bagheriasl et al. [56]	Solution	CNC	3	—	—	2.9	3.3	65	67	2.9	2.3
Gwon et al. [74]	Solution casting	CNC	5	Urethanization	—	3.1	3.5	65	74	—	—
Mariano et al. [75]	SC + TSE	CNC	20	—	PEG-b-PLLA	2	2.2	40	33	3.5	1.9
N.Pal et al. [76]	Solution casting	CNC	1	—	—	2.6	3.5	28.6	34.2	2.2	1.3
Qian et al. [77]	Solution casting	CNC	4	Silylation		0.42	0.11	19.8	14	13	250
Yin et al. [78]	Hot press	CNC	1	CNC–TCT–HA	—	—	—	25	48	1.9	3.1
Yin et al. [79]	Hot press	CNC	1	CNC–TCT–DA	—	—	—	25	43.6	1.9	2.9
Shojaeiarani et al. [80]	SC + TSE	CNC	1	Esterification	—	3.3	4.7	57.4	59.9	3.5	3.6
Shojaeiarani et al. [81]	SC + TSE	CNC	3	—	—	3.3	3.6	59	63.8	3.5	4.7
Shojaeiarani et al. [81]	Spin coating+TSE	CNC	3	—	—	3.3	3.8	59	62.3	3.5	6
Iwatake et al. [82]	Three-roll mill	CNF	10	—	—	—	—	50	70	—	—
Okubo et al. [83]	SC+twin rotary roller mixer	CNF	2	—	—	3.5	4.75	45	53	—	—

(continued)

Table 10.1 *(continued)*

Researchers	Processing method	Type of cellulose nano-particles	Nano-particle content (wt%)	Modification	Other components (surfactant, compatibili-zerplasticizer, etc.)	Tensile modulus (GPa)		Tensile strength (MPa)		Strain at break (%)	
						PLA	PLA/NC	PLA	PLA/NC	PLA	PLA/NC
Jonoobi et al. [84]	SC + TSE	CNF	5	—		2.9	3.6	59	72	3.4	2.7
Kowalczyk et al. [61]	SC + IM	CNF	2	—		—	—	48	51	21	12
Eyholzer et al. [55]	TSE	CNF	5	Carboxy-methylation esterification	—	3.6	3.75	66	63	—	—
Kiziltas et al. [85]	IM	CNF	10	—	PHB (carrier system)	2.98	3.6	60	57	—	—
Feng et al. [86]	SC + IM	CNF	5	*In situ* chemical grafting	—	2.5	2.55	58	58.5	8	7
Lee et. al. [87] (2009)	SC	BC	5	Acetic, hexanoic, and dodecanoic acid	—	1.34	1.98	60.7	68.5	3.6	2.7
Quero et al. [88] (2012)	Hot press	BC	15	Glyoxalization of BC	MA-*g*-PLAand	2	3.2	27	39	23	1.4
Lee et. al [89] (2012)	SC + injection molding	BC	5	—	Bio-derived PLLA copolymer	4.08	4.71	63	67	1.7	1.9
Martínez-Sanz et al. [90] (2013)	SC	BC	3	Polyglycidyl methacrylate-Grafting	—	2.4	2.8	45	55	4.1	4
Blaker et al. [91] (2014)	Thermally induced phase separation+TSE+melt spiing	BC	2	Hexanoic and dodecanoic acid	—	6	5.7	127	141	41	17
Ambrosio-Martín et al. [48] (2015)	In situ melt poly-condensation	BC	5	Lactic acid oligomers grafting	—	1.36	1.66	45	46	5.9	2.7

SC = solution casting, TSE = twin screw extruder, IM = internal mixer, PLA-MA = maleated PLA, PVAc = poly (vinyl acetate), PEO = polyethylene oxide, PEG = polyethylene glycol, CNC-ICN = CNC modified with n-octadecyl-isocyanate, CNC-GMA = CNC grafted with glycidyl methacrylate, TCT = 2,4,6-trichloro-1,3,5-triazine, HA = hexylamine, DA = dodecylamine, PEG-b-PLLA = diblock polymer of PEG and PLLA, Im = imidazolium, DCP = dicumyl peroxide.

of CNCs and BCs, probably due to their higher aspect ratios and their more engaged fiber entanglement [92]. However, it could also be observed that unmodified CNFs affect the mechanical properties of PLA much less than their modified counterparts. It is known that the reinforcing effect of nanofillers is governed by the stress transfer mechanism. Therefore, in order to have superior mechanical properties, besides a good level of nanoparticle dispersion, interfacial interactions between PLA and cellulose nanoparticles need to be considerably higher. Poor interfacial interaction between PLA and unmodified cellulose could cause a nonuniform stress distribution in the nanocomposites, which could lead to a suppression in the improvement of mechanical properties. In this context, the surface of cellulose nanoparticles should be modified to improve the matrix–filler interfacial bonding, resulting in an increase in tensile properties of the PLA nanocellulose nanocomposites. Among those, the surface grafting of CNC with isocynates or urethanization [68, 74], silylation [66], acetylation [72, 93], esterification [70, 80], and surface grafting of cellulose nanoparticles with PLA [73] and PLLA [94] are most common and resulted in an enhancement in tensile strength and modulus. For instance, upon surface modification of BC with dodecanoic acid, 50% and 15%, improvements, respectively, were observed in the PLA tensile modulus and strength [87]. On the other hand, in some studies, ductility of PLA improved when the cellulose nanoparticle surface was modified. Bulota et al. [95] showed that upon 1 wt.% acetylated-CNF incorporation, the strain at break increased 30-fold compared to neat PLA. This was explained by the plastic deformation of pre-existing voids that behaved similar to caveated rubber particles within rubber toughened polymer systems.

Surfactants/compatibilizers, such as acid phosphate ester of ethoxylated nonylphenol [67] PEO (or PEG) [52, 96, 97], lauric arginate [98], lignin [27], PLA-*g*-MAH [96, 99], and PLA-*g*-GMA [69, 100] can also lead to a better compatibility between PLA and cellulose nanoparticles. Bondeson et al. [65] used an anionic surfactant consisting of an ester of phosphoric acid bearing alkylphenol tails to produce PLA/CNC nanocomposites through melt mixing and showed that a good level of CNC dispersion within PLA could be provided using this strategy. As a result, the tensile modulus of surfactant-coated nanocomposites improved by 18% compared to its uncoated counterparts. However, tensile strength and strain at the break of reinforced nanocomposites decreased compared to that of neat PLA. Meng et al. [101] improved the strain at break and toughness of PLA without sacrificing the tensile modulus and strength when they used epoxidized soybean oil (ESO) as plasticizer, together with CNF. It was reported that use of CNF (10 wt%) and ESO (5 wt%) enhanced the modulus more than 20% over that of neat PLA while the strain at break was increased threefold. In some studies, however, even though good dispersion could have been achieved, a negative effect of surfactants on mechanical properties of PLA was also observed [44, 52, 102, 103]. Moreover, the possible presence of residual solvent in PLA/CNC nanocomposites prepared by solution casting could also reduce the strength and modulus although they could act as a plasticizer [104].

Spinella et al. [105] showed that both mechanical and thermomechanical properties of PLA significantly improved when CNC was modified with lactic acid moieties (CNC-LA). As seen in Figure 10.10, the storage modulus of PLA was enhanced upon CNC-LA addition, especially at 70 °C, where this increase was much more evident. This increase of PLA/nanocellulose nanocomposites at higher temperatures has also been observed in other studies [63, 68].

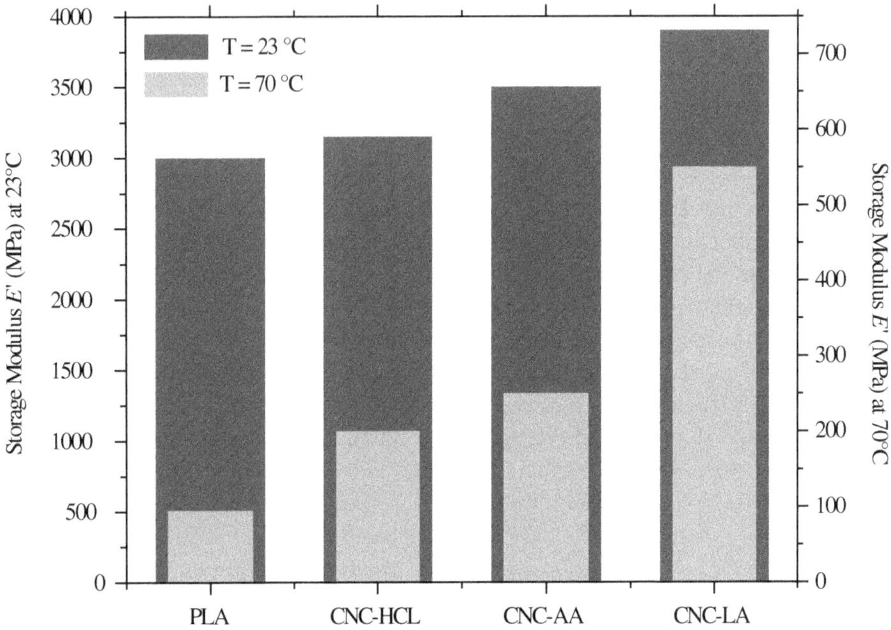

Figure 10.10 *Comparison of the effect of CNC surface chemistry on the storage modulus of PLA at 23°C and 70°C with 5 wt% CNC. Data adapted from [105] (CNC-AA: acetate modified CNC; CNC-LA: lactic acid modified CNC; and CNC-HCl: hydro chloric acid modified CNC)*

It is also known that there is a correlation between the crystallinity of the matrix and the mechanical behavior of the composites. As a general tendency, an increase in the crystalline amount of polymer matrix leads to an improvement in the tensile modulus and strength of the nanocomposites. This behavior was also observed for PLA/nanocellulose systems [63, 71, 106, 107]. Pei et al. [66] demonstrated that tensile modulus and strength of PLA was increased by 27% and 21%, respectively, upon the incorporation of 1 wt.% content of silylated CNC. The main reason for these enhancements was attributed to an increase in the crystalline amount of PLA.

The CNC percolation volume fraction and elastic modulus could be predicted from the modified Takayanagi model, which is named the Ouali model [108]. In some studies, it was shown (Figure 10.11) that the classical mechanical models such as those of Halpin-Kardos [109] and Halpin-Tsai [110] could fail to describe the experimental data as no interaction between nanofillers was assumed in these models [56, 63, 111, 112]. The Ouali model, however, is based on percolation theory, which describes the interactions among nanofillers at certain conditions that lead to the formation of a network structure that is the main reinforcing mechanism for nanocelluose-based nanocomposites.

The elastic modulus of the nanocomposite in the presence of such a percolated network could also be calculated as:

$$E_c = \frac{(1-2\phi\psi)E_m E_f + (1-\phi)\psi E_f^2}{(1-\phi)E_f + (\psi-\phi)E_m} \tag{10.1}$$

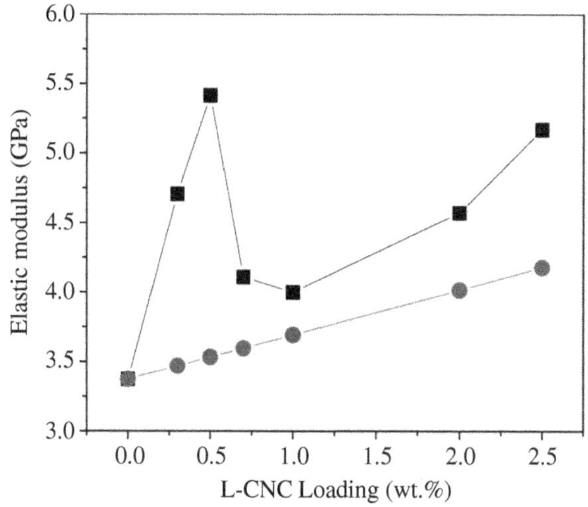

Figure 10.11 *Experimental and calculated values of storage modulus (using Halpin-Tsai model) of PLA/CNC nanocomposites as a function of CNC wt% content [63].*

$$\psi = 0 \; \phi \leq \phi_c \tag{10.2}$$

$$\psi = \phi\left(\frac{\phi - \phi_c}{1 - \phi_c}\right)^b \; \phi > \phi_c \tag{10.3}$$

where φ is the filler volume fraction, E_m, E_f, and E_c are the moduli of the polymer, fiber, and the composite, respectively, ψ is the volume fraction of the percolating filler network, and b is the critical percolation exponent and equal to a value of 0.4 for a three-dimensional network [43]. To calculate the critical volume fraction (ϕ_c) the following relation is used:

$$\phi_c = \frac{0.7}{\left(\frac{L}{w}\right)} \tag{10.4}$$

$P = L/w$ is the aspect ratio of the CNCs. A good correlation between the predicted model values and the experimental data of the PLA/CNC system was reported in various studies [56, 112]. However, the prediction and modeling of nanocomposite mechanical properties materials is difficult for CNF due to its inconsistent morphology, tangling, and flexible nature [113].

10.4.3 Crystallization Behavior

Crystallinity of thermoplastic polymers is one of the key factors to control various final properties such as degradation rate, thermal stability, optical, barrier, and mechanical properties of PLA that are crucial for packaging applications [114]. However, compared to many other thermoplastics, PLA has a slow crystallization rate, which limits production of PLA products through conventional processing methods such as injection molding, that generally includes fast cooling. In this context, CNC, CNF, and BC have been used as nucleating agents to speed up PLA crystallization. Ding et al. [115] showed that, as a result of the

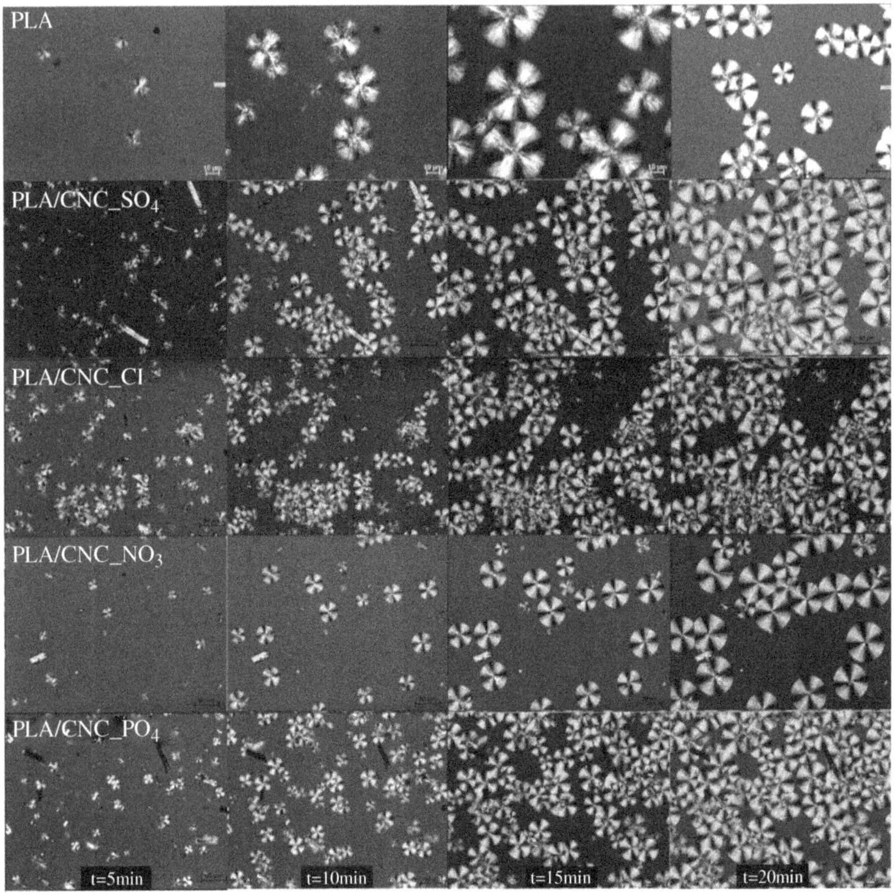

Figure 10.12 Polarized optical micrographs of neat PLA and PLA/CNC nanocomposites with CNCs prepared via different acid hydrolysis isothermally crystallized at 120°C, taken at different time intervals (5, 10, 15, and 20 minutes) [111].

decreased activation energy required for PLA crystallization, the crystallization rate of PLA significantly increased by addition of CNF with respect to neat PLA. Likewise, Kamal and Khoshkova [53] reported that during the isothermal melt crystallization at 100°C, the use of 3 wt% spray-freeze-dried CNC nanoparticles reduced the PLA's crystallization half time significantly from around 16.5 minutes to 2.2 minutes as a result of well-dispersed CNC. Dhar et al. [111] used H_3PO_4, HNO_3, H_2SO_4, and HCl for hydrolysis of CNC and showed that functionalization of CNC with various acids could influence the CNC aspect ratio and hence their nucleation efficiency. Figure 10.12 shows the PLA spherulites density variations during isothermal crystallization at 120°C for 5, 10, 15, and 20 minutes when using differently hydrolyzed CNCs in PLA. Compared to its counterparts, a greater number of spherulites were observed in PLA/CNC-Cl samples due to improved CNC dispersion as well as their lower aspect ratio. Moreover, even CNC-SO_4 and CNC-PO_4 have a similar aspect ratio, although CNC-PO_4 improved the nucleation better than CNC-SO_4, probably due to its more uniform dispersion in PLA.

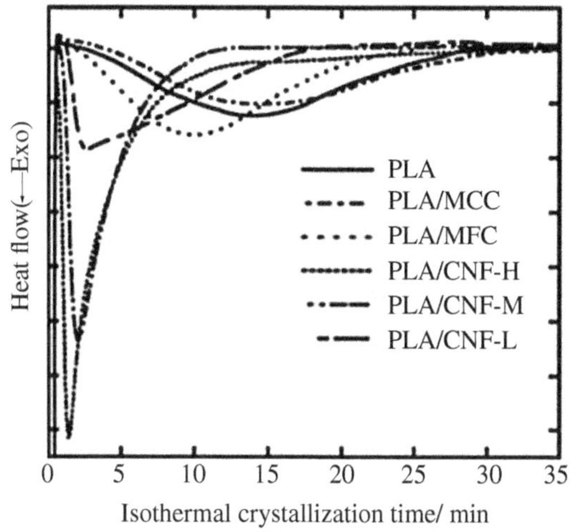

Figure 10.13 *Isothermal melt crystallization at 130°C of PLA with various types of cellulose [24].*

Kose et al. [24] studied the effect of cellulose particles with various geometries, including microcrystalline-cellulose (MCC), microfibrillated-cellulose (MFC), and CNF with three different diameters. These CNFs with low, medium, and high diameters are referred to as CNF-L, CNF-M, and CNF-H, respectively. The isothermal melt crystallization behaviors of these composites at 130°C are shown in Figure 10.13. CNF is shown to be more effective than MCC and MFC, but the reduction in the CNF diameter negatively influenced the PLA crystallization as the agglomeration tendency with the decrease in CNF diameter could increase.

10.4.4 Barrier Properties

Using cellulose nanoparticles could also improve the barrier properties of PLA by creating a tortuous pathway, which makes this biocomposite favorable for packaging applications. Figure 10.14 represents the schematic illustration of the tortuous path to permeation of the water and gas molecules due to the addition of cellulose nanoparticles. Besides its solid structure with a low permeable feature, cellulose nanoparticles could provide a strong polymer–particle interaction and create a solid network through the matrix that could eventually hinder the molecular mobility and hence the diffusion of penetrant [37]. It is important to note that the morphology, orientation, and good dispersion of the nanoparticles within the polymer matrix have a crucial effect on the barrier properties of the PLA [116]. The properties that highlight the cellulose nanoparticles for packaging applications are the degree of crystallinity, aspect ratio, surface polarity, and internal cohesion. These properties are beneficial in terms of barrier properties; for example, crystallinity improves the water resistance and decreases the molecular penetration although, due to the presence of hydroxyl groups in cellulose, they also possess low water resistance and moisture uptake [39]. It should be noted that barrier properties of polymers are mainly evaluated by OTR, WVP, and overall migration testing because these parameters have a noticeable

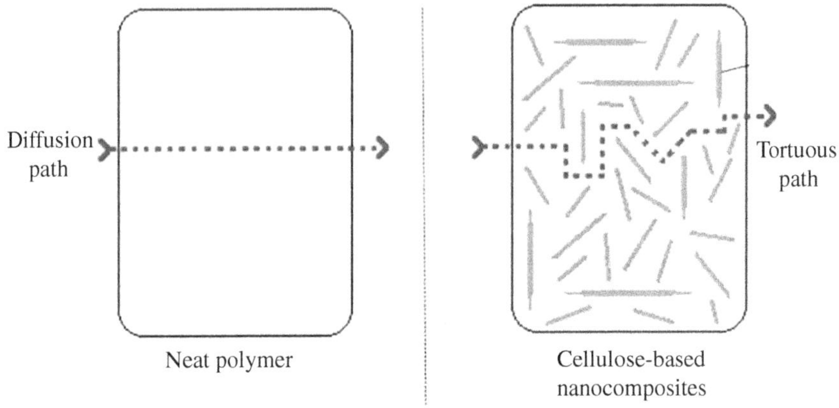

Figure 10.14 Schematic representation of the more tortuous path for water and gas molecules to diffuse due to addition of celluloses nanoparticles [116].

impact in deteriorative reactions [117]. In this context, Table 10.2 provides useful information about barrier properties of PLA as a result of incorporation of nanocellulose nanoparticles in the PLA matrix.

Many studies have investigated the barrier properties of PLA/CNC nanocomposites systems. Fortunati et al. [118] prepared PLA-CNC nanocomposites with surfactant modified (s-CNC) and unmodified CNC via solution casting at 1 and 5 wt%. They reported that the reduction rate at the OTR value was enhanced from 9% (PLA/CNC1) to 43% (PLA/CNC5) for the unmodified case and from 26% (PLA/s-CNC1) to 43% (PLA/s-CNC5) for the modified case [102, 118]. s-CNC incorporated PLA revealed better gas barrier behavior due to better dispersion of CNC nanoparticles, which create a more efficient tortuous pathway. They also reported that enhancement in WVP was only obtained in PLA with modified s-CNC. To determine the amount of nonvolatile substances that might migrate from PLA, an overall migration test was performed in isooctane and ethanol. It was reported that the migration level for both cases was much lower than the migration limits for food contact materials. Sanchez-Garcia et al. [64] reported that incorporation of CNC reduced the water and oxygen permeability of PLA up to 82–90%, respectively. On the other hand, Espino-Pérez et al. [68] investigated the barrier properties of PLA-CNC nanocomposites where *n*-octadecyl-isocyanate grafted CNC (CNC-ICN) and unmodified CNC were compared. There were no significant differences in oxygen permeability between neat PLA and PLA nanocomposites by incorporating modified and unmodified CNCs. The WVP of PLA was increased by increasing the unmodified CNC content, but no significant differences were found in cases of modified CNC. It was claimed that the hydrophilic structure of unmodified CNC in the PLA matrix could increase the water solubility coefficient, but hydrophobic grafting could counteract the expected effect of enhanced WVP in the case of modified CNC.

CNF is another type of cellulose nanoparticle that has been widely studied as a reinforcing material in the PLA matrix to improve barrier properties for packaging applications. Guzman et al. [33, 121] prepared the PLA-CNF nanocomposites at 1, 3, and 5 wt% and compared them with C30B (nanoclay). Although the incorporation of CNF slightly decreased

Table 10.2 Barrier properties of PLA nanocellulose nanocomposites.

References	Type of nanocelluloses	Nanoparticles content wt.%	Modifications	Barrier properties		Key findings
				Reduction in OTR (%)	Reduction in WVP (%)	
Fortunati et al. [118]	CNC	1, 5	Surface modification/surfactant	—	34	O_2 barrier enhancement and migration resistance
Martínez-Sanz et al. [60]	BC	1, 2, 3	Compatibizer/EVOH	—	43	Reduction at water permeability, but less reduction at oxygen permeability.
Espino-Pérez et al. [68]	CNC	2.5, 7.5, 15	Surface modification/grafting	—	26	Enhancement at water vapor permeability
Aulin et al. [119]	CNF	—	Surface coating/PEI	65–90	50	Outperformance at oxygen barrier properties
Arrieta et al. [103]	CNC	5	Surface modification/surfactant	—	—	Enhancement at water vapor and oxygen barrier properties
Arrieta et al. [102]	CNC	5	Plasticizer	—	—	Enhancement at oxygen barrier properties.
Ambrosio Martin et al. [48]	BC	0.5, 1, 3, 5	Surface modification/grafting	—	—	High resistance to both water vapor and oxygen were achieved.
Ambrosio-Martin et al. [120]	BC	0.5, 1, 3, 5	Ball-milling of freeze-dried BC	—	—	High oxygen barrier properties even at high humidity level.
Trifol et al. [121]	CNF	1, 3, 5	Surface modification/acetylation	64	46	Enhancement at water vapor and oxygen barrier properties

the transparency of PLA, CNF reduced the OTR to much lower values. Moreover, the incorporation of CNF also significantly reduced the WVP value compared to neat PLA, but this reduction was less than that in OTR, indicating that the hydrophilic structure of CNF makes it less efficient in reducing the WVP. In another study, Song et al. [122] tried to improve the compatibility of PLA and CNF via hydrophobic modification of CNF, hence improving the barrier properties of the nanocomposites. The hydrophobic modification of CNF was conducted via a free radical polymerization. It was claimed that the WVP values of the nanocomposites containing hydrophobic-modified CNF were lower than those of the control samples, indicating that hydrophobic modified CNF has a better compatibility and critical role in reducing WVP. Few studies claimed that incorporation of CNF enhanced the crystallinity of PLA [24]; hence the increase in crystallinity improved the barrier properties of PLA [123].

A few studies [25, 38, 48, 60, 90] investigated the barrier properties of PLA with respect to incorporation of BC. Martínez-Sanz et al. [25, 60, 90] studied the effect of modified BC on the barrier properties of PLA. Although incorporation of unmodified BC enhanced the oxygen barrier properties, modified BC has a higher barrier effect due to its better dispersion as a result of enhanced interfacial interaction with PLA. Further incorporation of BC and increasing levels of relative humidity could affect the barrier properties because it is highly dependent on the humidity of the environment. It is important to note that use of a compatibilizer could also increase the water sensitivity of the composites, which eventually could affect the barrier properties of nanocomposite films [19].

10.4.5 Applications

The incorporation of nanocellulose in PLA is linked to their properties, which include light weight, nontoxicity, high strength and modulus, dimensional stability, high optical transparency, low oxygen permeability, and biodegradable characteristics [124]. Their nanoscale dimension and ability to form a strong entangled network led to the emergence of new high-value applications [37]. Most of the studies developed a PLA nanocellulose nanocomposites system for packaging, presenting results on their transparency, barrier, and mechanical properties. Economic studies have also showed that PLA is an economically feasible material to use as a packaging polymer. However, in order to extend the use of PLA in packaging industries, the price level should be comparable to other commodity polymers [3]. Through the development of PLA nanocellulose nanocomposites, PLA films and packages could show better mechanical properties than PS and have properties comparable to those of PET [3, 39]. The transparency, stress crystallization upon drawing, barrier properties, and the folding ability of PLA allow it to be designed for many different packaging applications, such as short shelf life products for fruit, vegetables, and package applications including containers, drinking cups, salad cups, and lamination films [3, 125, 126].

10.5 Conclusion and Future Perspectives

Biopolymers have a great attraction, especially in the packaging industry, because of global environmental concerns and the difficulty in managing solid wastes from petroleum-based polymers. The incorporation of nanocellulose into PLA has created innovative and fully green nanocomposites with enhanced performance. Regardless of PLA's advantages, some drawbacks of PLA, like low toughness, slow crystallization kinetics, and low melt strength, could be eliminated by the addition of cellulose nanoparticles. However, the highly hydrophilic surface of nanocellulose and their dispersion in PLA is still a big challenge, especially when using direct melt mixing. Many studies have evaluated the performance of PLA nanocellulose nanocomposites by using different production techniques, such as solution casting and hybrid processes and/or the use of surfactants, compatibilizers, and surface chemical modifications, since the dispersion quality of nanocelluloses directly affects the performance of PLA nanocellulose nanocomposite systems. Through these studies, the incorporation of well-dispersed nanocellulose could increase the crystallinity, crystallization rate, and mechanical properties of PLA. The incorporation of cellulose nanoparticles together with the enhanced crystallinity of PLA also result in further enhancements in barrier properties against oxygen, water, and other volatile substances. With the successful development of PLA-based cellulose nanocomposites, they could be used in a wide range of applications and replace petroleum-based polymers, especially in packaging applications.

References

1. Malucelli G. "High barrier composite materials based on renewable sources for food packaging applications". 2017; 45–78.
2. Fortunati E., Luzi F., Yang W., Kenny J.M., Torre L., Puglia D. "Bio-based nanocomposites in food packaging". 2018; 71–110.
3. Auras R., Harte B., Selke S. "An overview of polylactides as packaging materials". *Macromol. Biosci.*, 2004; **4**: 835–864.
4. Saeidlou S., Huneault M.A., Li H., Park C.B. "Poly(lactic acid) crystallization". *Prog. Polym. Sci.*, 2012; **37**: 1657–1677.
5. Nofar M., Salehiyan R., Ray S. "Rheology of poly (lactic acid)-based systems". *Polym. Rev.*, 2019; **59**.3; 465–509.
6. Lim L.-T., Auras R., Rubino M. "Processing technologies for poly (lactic acid)". *Prog. Polym. Sci.*, 2008; **33**: 820–852.
7. Nofar M., Park C.B. "Poly (lactic acid) foaming". *Prog. Polym. Sci.*, 2014; **39**: 1721–1741.
8. Nofar M., Park C.B. *Polylactide Foams: Fundamentals, Manufacturing and Applications*. 1 : Elsevier; 2017.
9. Rasal R.M., Janorkar A.V., Hirt D.E. "Poly(lactic acid) modifications". *Prog. Polym. Sci.*, 2010; **35**: 338–356.
10. Cheng Y., Deng S., Chen P., Ruan R. "Polylactic acid (PLA) synthesis and modifications: a review". *Front. Chem. China*, 2009; **4**: 259–264.
11. Saini P., Arora M., Kumar M.N.V.R. "Poly(lactic acid) blends in biomedical applications". *Adv. Drug Delivery Rev.*, 2016; **107**: 47–59.
12. Nofar M., Sacligil D., Carreau P.J., Kamal M.R., Heuzey M.-C. "Poly (lactic acid) blends: processing, properties and applications". *Int. J. Biol. Macromol.*, 2019; **125**: 307–360.
13. Raquez J.-M., Habibi Y., Murariu M., Dubois P. "Polylactide (PLA)-based nanocomposites". *Prog. Polym. Sci.*, 2013; **38**: 1504–1542.
14. Martin O., Avérous L. "Poly(lactic acid): plasticization and properties of biodegradable multiphase systems". *Polymer*, 2001; **42**: 6209–6219.
15. Baiardo M., Frisoni G., Scandola M., Rimelen M., Lips D., Ruffieux K., et al. "Thermal and mechanical properties of plasticized poly (L-lactic acid)". *J. Appl. Polym. Sci.*, 2003; **90**: 1731–1738.

16. Arrieta M.P., Peponi L., López D., López J., Kenny J.M. "An overview of nanoparticles role in the improvement of barrier properties of bioplastics for food packaging applications". *Food Packaging*: Elsevier; 2017; 391–424.
17. Ramos M., Valdés A., Garrigós M. "Multifunctional applications of nanocellulose-based nanocomposites. *Multifunctional Polymeric Nanocomposites Based on Cellulosic Reinforcements*": Elsevier; 2016; 177–204.
18. Fortunati E., Peltzer M., Armentano I., Jiménez A., Kenny J.M. "Combined effects of cellulose nanocrystals and silver nanoparticles on the barrier and migration properties of PLA nano-biocomposites". *J. Food Eng.*, 2013; **118**:117–124.
19. Vatansever E., Arslan D., Nofar M. "Polylactide cellulose-based nanocomposites". *Int. J. Biol. Macromol.*, 2019; **137**: 912–938.
20. Khan A., Huq T., Khan R.A., Riedl B., Lacroix M. "Nanocellulose-based composites and bioactive agents for food packaging". *Crit. Rev. Food Sci. Nutr.*, 2014; **54**: 163–174.
21. Dufresne A. "Nanocellulose: a new ageless bionanomaterial". *Mater. Today*, 2013; **16**: 220–227.
22. Hubbe M.A., Ferrer A., Tyagi P., Yin Y., Salas C., Pal L., et al. "Nanocellulose in thin films, coatings, and plies for packaging applications: a review". *BioResources*, 2017; **12**, 2143–2233.
23. Camarero-Espinosa S., Boday D.J., Weder C., Foster E.J. "Cellulose nanocrystal driven crystallization of poly(D,L-lactide) and improvement of the thermomechanical properties". *J. Appl. Polym. Sci.*, 2015; **132**: 41607.
24. Kose R., Kondo T. "Size effects of cellulose nanofibers for enhancing the crystallization of poly(lactic acid)". *J. Appl. Polym. Sci.*, 2013;**128**:1200–1205.
25. Martínez-Sanz M., Lopez-Rubio A., Lagaron J.M. "High-barrier coated bacterial cellulose nanowhiskers films with reduced moisture sensitivity". *Carbohydr. Polym.*, 2013; **98**:1072–1082.
26. Mokhena T., Sefadi J., Sadiku E., John M., Mochane M., Mtibe A. "Thermoplastic processing of PLA/cellulose nanomaterials composites". *Polymers*, 2018; **10**: 1363.
27. Henton D.E., Gruber P., Lunt J., Randall J. "Polylactic acid technology". *Natural Fibers, Biopolymers, and Biocomposites*, 2005; **16**: 527–577.
28. Ren J. *Biodegradable Poly (Lactic Acid): Synthesis, Modification, Processing and Applications*: Springer Science & Business Media; 2011.
29. Auras R.A., Lim L.-T., Selke S.E., Tsuji H. *Poly (Lactic Acid): Synthesis, Structures, Properties, Processing, and Applications*: Wiley; 2011.
30. Garlotta D. "A literature review of poly (lactic acid)". *J. Polym. Environ.*, 2001; **9**: 63–84.
31. Di Lorenzo M.L., Androsch R. *Synthesis, Structure and Properties of Poly(Lactic Acid)*: Springer; 2018.
32. Nofar M., Ameli A., Park C.B. "The thermal behavior of polylactide with different D-lactide content in the presence of dissolved CO_2". *Macromol. Mater. Eng*, 2014; **299**: 1232–1239.
33. Guzman J.T., Szabo P., Daugaard A.E., Hassager O. "Hybrid nanocellulose/nanoclay composites for food packaging applications". 2016.
34. Hamad K., Kaseem M., Yang H., Deri F., Ko Y. "Properties and medical applications of polylactic acid: a review". *Express Polym. Lett.*, 2015; **9:435-455**
35. Klemm D., Heublein B., Fink H.P., Bohn A. "Cellulose: fascinating biopolymer and sustainable raw material". *Angew. Chem. Int. Ed.*, 2005; **44**: 3358–3393.
36. Mosley F. "Biodegradable polylactic acid cellulose nanocomposites". 2018
37. Bharimalla A.K., Deshmukh S.P., Vigneshwaran N., Patil P.G., Prasad V. "Nanocellulose-polymer composites for applications in food packaging: current status, future prospects and challenges". *Polym. Plast. Technol. Eng.*, 2016; **56**: 805–823.
38. Gitari B., Chang B.P., Misra M., Navabi A., Mohanty A.K.J.B. "A comparative study on the mechanical, thermal, and water barrier properties of PLA nanocomposite films prepared with bacterial nanocellulose and cellulose nanofibrils". *BioResources* 2019; **14**: 1867-1889.

39. Helanto K.E., Matikainen L., Talja R., Rojas O.J. "Bio-based polymers for sustainable packaging and biobarriers: a critical review". *BioResources*, 2019; **14**: 4902-4951.
40. Vilarinho F., Sanches Silva A., Vaz M.F., Farinha J.P. "Nanocellulose in green food packaging". *Crit. Rev. Food Sci. Nutr.*, 2018; **58**:1526–1537.
41. Reiniati I. "Bacterial cellulose nanocrystals: production and application". 2017;
42. Lin N., Dufresne A. "Nanocellulose in biomedicine: current status and future prospect". *Eur. Polym. J.*, 2014; **59**: 302–325.
43. Xu X., Liu F., Jiang L., Zhu J.Y., Haagenson D., Wiesenborn D.P. "Cellulose nanocrystals vs. cellulose nanofibrils: a comparative study on their microstructures and effects as polymer reinforcing agents". *ACS Appl. Mater. Interfaces*, 2013; **5**: 2999–3009.
44. Oksman K., Mathew A.P., Bondeson D., Kvien I. "Manufacturing process of cellulose whiskers/polylactic acid nanocomposites". *Compos. Sci. Technol.*, 2006; **66**: 2776–2784.
45. Sánchez-García M.D., Hilliou L., Lagarón J.M. "Morphology and water barrier properties of nanobiocomposites of κ/ι-hybrid carrageenan and cellulose nanowhiskers". *J. Agric. Food. Chem.*, 2010; **58**:12847–12857.
46. Nair S.S., Zhu J., Deng Y., Ragauskas A.J. "High performance green barriers based on nanocellulose". *Sustainable Chem. Processes*, 2014; **2**: 23.
47. Zhang X., Li W., Ye B., Lin Z., Rong J. "Studies on confined crystallization behavior of nanobiocomposites consisting of acetylated bacterial cellulose and poly (lactic acid)". *J. Thermoplast. Compos. Mater.*, 2013; **26**: 346–361.
48. Ambrosio-Martín J., Fabra M.J., Lopez-Rubio A., Lagaron J.M. "Melt polycondensation to improve the dispersion of bacterial cellulose into polylactide via melt compounding: enhanced barrier and mechanical properties". *Cellulose*, 2015; **22**: 1201–1226.
49. Mariano M., El Kissi N., Dufresne A. "Cellulose nanocrystals and related Nanocomposites: review of some properties and challenges". *J. Polym. Sci., Part B: Polym. Phys.*, 2014; **52**: 791–806.
50. Miao C.W., Hamad W.Y. "Cellulose reinforced polymer composites and nanocomposites: a critical review". *Cellulose*, 2013; **20**: 2221–2262.
51. Petersson L., Kvien I., Oksman K. "Structure and thermal properties of poly(lactic acid)/cellulose whiskers nanocomposite materials". *Compos. Sci. Technol.*, 2007; **67**: 2535–2544.
52. Arias A., Heuzey M.C., Huneault M.A., Ausias G., Bendahou A. "Enhanced dispersion of cellulose nanocrystals in melt-processed polylactide-based nanocomposites". *Cellulose*, 2015; **22**: 483–498.
53. Kamal M.R., Khoshkava V. "Effect of cellulose nanocrystals (CNC) on rheological and mechanical properties and crystallization behavior of PLA/CNC nanocomposites". *Carbohydr. Polym.*, 2015; **123**: 105–114.
54. Frone A.N., Berlioz S., Chailan J.F., Panaitescu D.M. "Morphology and thermal properties of PLA-cellulose nanofibers composites". *Carbohydr. Polym.*, 2013; **91**: 377–384.
55. Eyholzer C., Tingaut P., Zimmermann T., Oksman K. "Dispersion and reinforcing potential of carboxymethylated nanofibrillated cellulose powders modified with 1-hexanol in extruded poly(lactic acid) (PLA) composites". *J. Polym. Environ.*, 2012; **20**: 1052–1062.
56. Bagheriasl D., Carreau P.J., Riedl B., Dubois C. "Enhanced properties of polylactide by incorporating cellulose nanocrystals". *Polym. Compos.*, 2018; **39**: 2685–2694.
57. Bagheriasl D., Carreau P.J., Riedl B., Dubois C., Hamad W.Y. "Shear rheology of polylactide (PLA)-cellulose nanocrystal (CNC) nanocomposites". *Cellulose*, 2016; **23**: 1885–1897.
58. Arslan D ., Vatansever E, Sarul D.S., Kahraman Y., Gunes G., Durmus A., Nofar, M. "Effect of preparation method on the properties of polylactide/cellulose nanocrystal nanocomposites". *Polymer Composites*, 2020; 1-11; https://doi.org/10.1002/pc.25701
59. Vatansever E., Arslan D., Sarul D.S., Kahraman Y., Nofar, M. "Effect of the molecular weight and crystallization of polymer matrix on the CNC percolation network formation". *Int. J. Biol. Macromol.*, 2020; 154: 276-290

60. Martinez-Sanz M., Lopez-Rubio A., Lagaron J.M. "Optimization of the dispersion of unmodified bacterial cellulose nanowhiskers into polylactide via melt compounding to significantly enhance barrier and mechanical properties". *Biomacromolecules*, 2012; **13**: 3887–3899.
61. Kowalczyk M., Piorkowska E., Kulpinski P., Pracella M. "Mechanical and thermal properties of PLA composites with cellulose nanofibers and standard size fibers". *Composites, Part A*, 2011; **42**: 1509–1514.
62. Tingaut P., Zimmermann T., Lopez-Suevos F.J.B. "Synthesis and characterization of bionanocomposites with tunable properties from poly (lactic acid) and acetylated microfibrillated cellulose"*Biomacromolecules*. 2009; **11**: 454–464.
63. Gupta A., Simmons W., Schueneman G.T., Hylton D., Mintz E.A. "Rheological and thermo-mechanical properties of poly(lactic acid)/lignin-coated cellulose nanocrystal composites". *ACS Sustainable Chem. Eng.*, 2017;**5**:1711–1720.
64. Sanchez-Garcia M., Lagaron J. "On the use of plant cellulose nanowhiskers to enhance the barrier properties of polylactic acid". *Cellulose*, 2010;**17**:987–1004.
65. Bondeson D., Oksman K. "Dispersion and characteristics of surfactant modified cellulose whiskers nanocomposites". *Compos. Interfaces*, 2007;**14**:617–630.
66. Pei A.H., Zhou Q., Berglund L.A. "Functionalized cellulose nanocrystals as biobased nucleation agents in poly(L-lactide) (PLLA) – crystallization and mechanical property effects". *Compos. Sci. Technol.*, 2010;**70**:815–821.
67. Fortunati E., Armentano I., Zhou Q., Iannoni A., Saino E., Visai L., et al. "Multifunctional bionanocomposite films of poly(lactic acid), cellulose nanocrystals and silver nanoparticles". *Carbohydr. Polym.*, 2012;**87**:1596–1605.
68. Espino-Perez E., Bras J., Ducruet V., Guinault A., Dufresne A., Domenek S. "Influence of chemical surface modification of cellulose nanowhiskers on thermal, mechanical, and barrier properties of poly(lactide) based bionanocomposites". *Eur. Polym. J.*, 2013;**49**:3144–3154.
69. Pracella M., Hague M.M.U., Puglia D. "Morphology and properties tuning of PLA/cellulose nanocrystals bio-nanocomposites by means of reactive functionalization and blending with PVAc". *Polymer*, 2014;**55**:3720–3728.
70. Robles E., Urruzola I., Labidi J., Serrano L. "Surface-modified nano-cellulose as reinforcement in poly(lactic acid) to conform new composites". *Ind. Crops Prod.*, 2015;**71**:44–53.
71. Herrera N., Salaberria A.M., Mathew A.P., Oksman K. "Plasticized polylactic acid nanocomposite films with cellulose and chitin nanocrystals prepared using extrusion and compression molding with two cooling rates: effects on mechanical, thermal and optical properties". *Composites, Part A*, 2016;**83**:89–97.
72. Xu C.J., Chen J.X., Wu D.F., Chen Y., Lv Q.L., Wang M.Q. "Polylactide/acetylated nanocrystalline cellulose composites prepared by a continuous route: a phase interface-property relation study". *Carbohydr. Polym.*, 2016;**146**:58–66.
73. Dhar P., Tarafder D., Kumar A., Katiyar V. "Thermally recyclable polylactic acid/cellulose nanocrystal films through reactive extrusion process". *Polymer*, 2016;**87**:268–282.
74. Gwon J.G., Cho H.J., Chun S.J., Lee S., Wu Q., Li M.C., et al. "Mechanical and thermal properties of toluene diisocyanate-modified cellulose nanocrystal nanocomposites using semi-crystalline poly(lactic acid) as a base matrix". *RSC Adv.*, 2016;**6**:73879–73886.
75. Mariano M., Pilate F., de Oliveira F.B., Khelifa F., Dubois P., Raquez J.M., et al. "Preparation of cellulose nanocrystal-reinforced poly(lactic acid) nanocomposites through noncovalent modification with PLLA-based surfactants". *ACS Omega*, 2017;**2**:2678–2688.
76. Pal N., Dubey P., Gopinath P., Pal K. "Combined effect of cellulose nanocrystal and reduced graphene oxide into poly-lactic acid matrix nanocomposite as a scaffold and its anti-bacterial activity". *Int. J. Biol. Macromol.*, 2017;**95**:94–105.
77. Qian S.P., Sheng K.C. "PLA toughened by bamboo cellulose nanowhiskers: Role of silane compatibilization on the PLA bionanocomposite properties". *Compos. Sci. Technol.*, 2017;**148**:59–69.

78. Yin Y., Zhao L., Jiang X., Wang H., Gao W. "Poly(lactic acid)-based biocomposites reinforced with modified cellulose nanocrystals". *Cellulose*, 2017;**24**:4773–4784.
79. Yin Y., Zhao L., Jiang X., Wang H., Gao W. "Cellulose nanocrystals modified with a triazine derivative and their reinforcement of poly(lactic acid)-based bionanocomposites". *Cellulose*, 2018;**25**:2965–2976.
80. Shojaeiarani J., Bajwa D.S., Stark N.M. "Green esterification: A new approach to improve thermal and mechanical properties of poly(lactic acid) composites reinforced by cellulose nanocrystals". *J. Appl. Polym. Sci.*, 2018;**135**:46468.
81. Shojaeiarani J., Bajwa D.S., Stark N.M. "Spin-coating: A new approach for improving dispersion of cellulose nanocrystals and mechanical properties of poly (lactic acid) composites". *Carbohydr. Polym.*, 2018;**190**:139–147.
82. Iwatake A., Nogi M., Yano H. "Cellulose nanofiber-reinforced polylactic acid". *Compos. Sci. Technol.*, 2008;**68**:2103–2106.
83. Okubo K., Fujii T., Thostenson E.T. "Multi-scale hybrid biocomposite: Processing and mechanical characterization of bamboo fiber reinforced PLA with microfibrillated cellulose". *Composites, Part A*, 2009;**40**:469–475.
84. Jonoobi M., Harun J., Mathew A.P., Oksman K. "Mechanical properties of cellulose nanofiber (CNF) reinforced polylactic acid (PLA) prepared by twin screw extrusion". *Compos. Sci. Technol.*, 2010;**70**:1742–1747.
85. Kiziltas A., Nazari B., Erbas Kiziltas E., Gardner D.J., Han Y., Rushing T.S. "Method to reinforce polylactic acid with cellulose nanofibers via a polyhydroxybutyrate carrier system". *Carbohydr. Polym.*, 2016;**140**:393–399.
86. Feng J., Sun Y., Song P., Lei W., Wu Q., Liu L., et al. "Fire-resistant, strong, and green polymer nanocomposites based on poly(lactic acid) and core–shell nanofibrous flame retardants". *ACS Sustainable Chem. Eng.*, 2017;**5**:7894–7904.
87. Lee K.-Y., Blaker J.J., Bismarck A. "Surface functionalisation of bacterial cellulose as the route to produce green polylactide nanocomposites with improved properties". *Compos. Sci. Technol.*, 2009;**69**:2724–2733.
88. Quero F., Eichhorn S.J., Nogi M., Yano H., Lee K.-Y., Bismarck A. "Interfaces in cross-linked and grafted bacterial cellulose/poly(lactic acid) resin composites". *J. Polym. Environ.*, 2012;**20**:916–925.
89. Lee K.-Y., Tang M., Williams C.K., Bismarck A. "Carbohydrate derived copoly(lactide) as the compatibilizer for bacterial cellulose reinforced polylactide nanocomposites". *Compos. Sci. Technol.*, 2012;**72**:1646–1650.
90. Martínez-Sanz M., Abdelwahab M.A., Lopez-Rubio A., Lagaron J.M., Chiellini E., Williams T.G., et al. "Incorporation of poly(glycidylmethacrylate) grafted bacterial cellulose nanowhiskers in poly(lactic acid) nanocomposites: improved barrier and mechanical properties". *Eur. Polym. J.*, 2013;**49**:2062–2072.
91. Blaker J.J., Lee K.-Y., Walters M., Drouet M., Bismarck A. "Aligned unidirectional PLA/bacterial cellulose nanocomposite fibre reinforced PDLLA composites". *React. Funct. Polym.*, 2014;**85**:185–192.
92. Alemdar A., Sain M. "Biocomposites from wheat straw nanofibers: morphology, thermal and mechanical properties". *Compos. Sci. Technol.*, 2008;**68**:557–565.
93. Lin N., Huang J., Chang P.R., Feng J.W., Yu J.H. "Surface acetylation of cellulose nanocrystal and its reinforcing function in poly(lactic acid)". *Carbohydr. Polym.*, 2011;**83**:1834–1842.
94. Lizundia E., Fortunati E., Dominici F., Vilas J.L., León L.M., Armentano I., et al. "PLLA-grafted cellulose nanocrystals: role of the CNC content and grafting on the PLA bionanocomposite film properties". *Carbohydr. Polym.*, 2016;**142**:105–113.
95. Bulota M., Kreitsmann K., Hughes M., Paltakari J. "Acetylated microfibrillated cellulose as a toughening agent in poly(lactic acid)". *J. Appl. Polym. Sci.*, 2012;**126**:E449-E458.

96. Chen J., He H., Yu P., Jia Y., Meng S., Wang J. "Preparation and properties of poly(lactic acid)/cellulose nanocrystals nanocomposites compatibilized with maleated poly(lactic acid)". *Polym. Compos.*, 2018;**39**:3092–3101.
97. Zhang C.M., Salick M.R., Cordie T.M., Effingham T., Dan Y., Turng L.S. "Incorporation of poly(ethylene glycol) grafted cellulose nanocrystals in poly(lactic acid) electrospun nanocomposite fibers as potential scaffolds for bone tissue engineering". *Mater. Sci. Eng., C*, 2015;**49**:463–471.
98. Chi K., Catchmark J.M. "Enhanced dispersion and interface compatibilization of crystalline nanocellulose in polylactide by surfactant adsorption". *Cellulose*, 2017;**24**:4845–4860.
99. Hong J., Kim D.S. "Preparation and physical properties of polylactide/cellulose nanowhisker/nanoclay composites". *Polym. Compos.*, 2013;**34**:293–298.
100. Yang W., Dominici F., Fortunati E., Kenny J.M., Puglia D. "Melt free radical grafting of glycidyl methacrylate (GMA) onto fully biodegradable poly(lactic) acid films: effect of cellulose nanocrystals and a masterbatch process". *RSC Adv.*, 2015;**5**:32350–32357.
101. Meng X., Bocharova V., Tekinalp H., Cheng S., Kisliuk A., Sokolov A.P., et al. "Toughening of nanocelluose/PLA composites via bio-epoxy interaction: mechanistic study". *Mater. Des.*, 2018;**139**:188–197.
102. Arrieta M.P., Fortunati E., Dominici F., Rayón E., López J., Kenny J.M. "PLA-PHB/cellulose based films: Mechanical, barrier and disintegration properties". *Polym. Degrad. Stab.*, 2014;**107**:139–149.
103. Arrieta M.P., Fortunati E., Dominici F., Lopez J., Kenny J.M. "Bionanocomposite films based on plasticized PLA-PHB/cellulose nanocrystal blends". *Carbohydr. Polym.*, 2015;**121**:265–275.
104. Fortunati E., Luzi F., Puglia D., Petrucci R., Kenny J.M., Torre L. "Processing of PLA nanocomposites with cellulose nanocrystals extracted from Posidonia oceanica waste: innovative reuse of coastal plant". *Ind. Crops Prod.*, 2015;**67**:439–447.
105. Spinella S., Lo Re G., Liu B., Dorgan J., Habibi Y., Leclere P., et al. "Polylactide/cellulose nanocrystal nanocomposites: Efficient routes for nanofiber modification and effects of nanofiber chemistry on PLA reinforcement". *Polymer*, 2015;**65**:9–17.
106. Fortunati E., Armentano I., Zhou Q., Puglia D., Terenzi A., Berglund L.A., et al. "Microstructure and nonisothermal cold crystallization of PLA composites based on silver nanoparticles and nanocrystalline cellulose". *Polym. Degrad. Stab.*, 2012;**97**:2027–2036.
107. Qian S., Zhang H., Yao W., Sheng K. "Effects of bamboo cellulose nanowhisker content on the morphology, crystallization, mechanical, and thermal properties of PLA matrix biocomposites". *Composites, Part B*, 2018;**133**:203–209.
108. N. Ouali, J.Y. Cavaille, J. Perez. "Elastic, viscoelastic and plastic behavior of multiphase polymer blends, ". *Plast. Rubber Compos.* **16** (1991) 55,
109. Halpin J.C., Kardos J.L. "Moduli of crystalline polymers employing composite theory". *J. Appl. Phys.*, 1972;**43**:2235–2241.
110. Affdl J.C.H., Kardos J.L. "The Halpin-Tsai equations: a review". *Polym. Eng. Sci.*, 1976;**16**:344–352.
111. Dhar P., Bhasney S.M., Kumar A., Katiyar V. "Acid functionalized cellulose nanocrystals and its effect on mechanical, thermal, crystallization and surfaces properties of poly (lactic acid) bionanocomposites films: A comprehensive study". *Polymer*, 2016;**101**:75–92.
112. Dhar P., Tarafder D., Kumar A., Katiyar V. "Effect of cellulose nanocrystal polymorphs on mechanical, barrier and thermal properties of poly(lactic acid) based bionanocomposites". *RSC Adv.*, 2015;**5**:60426–60440.
113. Dufresne A. *Nanocellulose: From Nature to High Performance Tailored Materials*: Walter de Gruyter GmbH & Co KG; 2017.
114. Murariu M., Dubois P. "PLA composites: From production to properties". *Adv. Drug Delivery Rev.*, 2016;**107**:17–46.
115. Ding W., Kuboki T., Wong A., Park C.B., Sain M. "Rheology, thermal properties, and foaming behavior of high d-content polylactic acid/cellulose nanofiber composites". *RSC Adv.*, 2015;**5**:91544–91557.

116. Ferreira F.V., Dufresne A., Pinheiro I.F., Souza D.H.S., Gouveia R.F., Mei L.H.I., et al. "How do cellulose nanocrystals affect the overall properties of biodegradable polymer nanocomposites: A comprehensive review". *Eur. Polym. J.*, 2018;**108**:274–285.
117. Salmieri S., Islam F., Khan R.A., Hossain F.M., Ibrahim H.M.M., Miao C.W., et al. "Antimicrobial nanocomposite films made of poly(lactic acid)-cellulose nanocrystals (PLA-CNC) in food applications-part B: Effect of oregano essential oil release on the inactivation of *Listeria monocytogenes* in mixed vegetables". *Cellulose*, 2014;**21**:4271–4285.
118. Fortunati E., Peltzer M., Armentano I., Torre L., Jiménez A., Kenny J. "Effects of modified cellulose nanocrystals on the barrier and migration properties of PLA nano-biocomposites". *Carbohydr. Polym.*, 2012;**90**:948–956.
119. Aulin C., Karabulut E., Tran A., Wågberg L., Lindström T. "Transparent nanocellulosic multilayer thin films on polylactic acid with tunable gas barrier properties". *ACS Appl. Mater. Interfaces*, 2013;**5**:7352–7359.
120. Ambrosio-Martín J., Lopez-Rubio A., Fabra M.J., Gorrasi G., Pantani R., Lagaron J.M. "Assessment of ball milling methodology to develop polylactide-bacterial cellulose nanocrystals nanocomposites". *J. Appl. Polym. Sci.*, 2015;**132**: 41605.
121. Trifol J., Plackett D., Sillard C., Hassager O., Daugaard A.E., Bras J., et al. "A comparison of partially acetylated nanocellulose, nanocrystalline cellulose, and nanoclay as fillers for high-performance polylactide nanocomposites". *J. Appl. Polym. Sci.*, 2016;**133: 43257**.
122. Song Z., Xiao H., Zhao Y. "Hydrophobic-modified nano-cellulose fiber/PLA biodegradable composites for lowering water vapor transmission rate (WVTR) of paper". *Carbohydr. Polym.*, 2014;**111**:442–448.
123. Meriçer Ç., Minelli M., Angelis M.G.D., Giacinti Baschetti M., Stancampiano A., Laurita R., et al. "Atmospheric plasma assisted PLA/microfibrillated cellulose (MFC) multilayer biocomposite for sustainable barrier application". *Ind. Crops Prod.*, 2016;**93**:235–243.
124. Blanco A., Monte M.C., Campano C., Balea A., Merayo N., Negro C. Nanocellulose for industrial use. *Handbook of Nanomaterials for Industrial Applications*;2018. p. 74–126.
125. Farah S., Anderson D.G., Langer R. "Physical and mechanical properties of PLA, and their functions in widespread applications – A comprehensive review". *Adv. Drug Delivery Rev.*, 2016;**107**:367–392.
126. Ilyas RA, Sapuan SM, Atikah MSN, Asyraf MRM, Ayu Rafiqah S, Aisyah HA, Mohd Nurazzi N, Norrahim MNF. Effect of hydrolysis time on the morphological, physical, chemical, and thermal behavior of sugar palm nanocrystalline cellulose (Arenga Pinnata (Wurmb.) Merr). *Textile Research Journal*. 2020. In print. 10.1177/0040517520932393

11

Nanocellulose Composite Films for Packaging Applications

Latifah Jasmani[1], Sharmiza Adnan[1], Z.M.A. Ainun[2], S.M. Sapuan[3,4], and R.A. Ilyas[5,6]

[1] *Pulp and Paper Laboratory, Biomass Technology Programme, Forest Products Division, Forest Research Institute Malaysia, Selangor, Kepong, 52109, Malaysia*

[2] *Program of Pulp and Paper & Pollution Control, Laboratory of Biopolymer and Derivatives, Institute of Tropical Forestry and Forest Product (INTROP), Universiti Putra Malaysia, Selangor, Serdang, 43400, Malaysia*

[3] *Advanced Engineering Materials and Composites Research Centre (AEMC), Department of Mechanical and Manufacturing Engineering, Faculty of Engineering, Universiti Putra Malaysia (UPM), Serdang, Selangor, 43400, Malaysia*

[4] *Laboratory of Biocomposite Technology, Institute of Tropical Forestry and Forest Product (INTROP), Universiti Putra Malaysia, Serdang, Selangor, 43400, Malaysia*

[5] *School of Chemical and Energy, Faculty of Engineering, Universiti Teknologi Malaysia, 81310 UTM Johor Bahru, Johor, Malaysia*

[6] *Centre for Advanced Composite Materials, Universiti Teknologi Malaysia, 81310 UTM Johor Bahru, Johor, Malaysia*

11.1 Introduction

Packaging has become one of the most important and fast-growing industries around the globe, worth about US$ 400–500 billion [1]. The main purpose of packaging is to protect goods such as food, beverage, and medicine from deterioration. In addition, it also readily functions as a medium for product transportation from one place to another. In general, there are many types of packaging material available in the market, such as plastics, glass, aluminum, and paper. However, the paper-based packaging industry has received enormous attention recently in many countries due to environmental pressure by local authorities or

Bio-Based Packaging: Material, Environmental and Economic Aspects, First Edition. Edited by S.M. Sapuan and R.A. Ilyas

driven by national policy. The shift to more environmentally friendly packaging is also because of sustainability, renewability, and disposal issues. Petroleum-based packaging, i.e. plastic, continuously faces big issues of disposal as they can only be post-managed via incineration or land filling. Both methods have problems such that incineration of such nondegradable materials releases toxic gas whereas land filling requires big land and space. Therefore, alternative or sustainable packaging that can avoid such problems is very much desired immediately. Packaging material produced from renewable material such as cellulose is the most suitable green alternative to plastics as it is easier to dispose of and can degrade naturally over a short period of time. The fact that it (cellulose) is also abundant is another plus point worth exploring in many other basic and high end uses. Paper packaging is an extensive industry in which the product or process therefrom can be as basic as container or wrapping material or can be on a large scale, such as in the production of liner and corrugated material for boxes.

Despite its green attributes, cellulose-based packaging has limited application in areas where protection against water, oxygen, or oil is a prime requirement. In most cases to tackle this common need, cellulose-based packaging is coated with plastics or wax to improve the barrier properties. The use of such coating materials is not environmentally friendly and thus is clearly unjustified in the use of renewable material, i.e. cellulose, in the very beginning. As such, nanocellulose has the potential to become the next generation material to address this issue. This is because nanocellulose can be directly converted into a thin film or incorporated in a polymer composite to form a stronger film. Nanocellulose is cellulose at the molecular level and can be extracted from various lignocellulosic materials, marine animals, bacteria, and algae. Given the abundance and availability of cellulose worldwide, estimated at 75 billion generated annually [2], nanocellulose extracted from plants seems to give a better option. Nanocellulose has unique physicochemical properties such as: high mechanical strength, ease of surface modification, interesting optical properties (birefringence), a low thermal coefficient, and a high surface area. Due to its high surface area feature, nanocellulose forms a very dense network structure that provides tortuosity to diffusing molecules to pass through. This particular behavior is highly favorable in packaging, i.e. for a barrier property.

Nanocellulose has diversified uses in packaging in which it can be used as a strength additive in nanocomposite material, as a coating material on paper-based packaging, or can be formed into a thin film. An innovative approach using nanocellulose for paper-based packaging not only enhances the desired packaging properties but can also help to reduce the impact on the environment due to its high recyclability and reusability potential. The objective of this chapter is to give an overall look at the use of nanocellulose-based film for packaging purposes. Several review articles related to the use of nanocellulose in packaging field have been published, such as that by Turbak et al. [3].

11.2 Preparation of Nanocellulose

There are three types of nanocellulose that can be used for packaging applications, namely nanocrystalline cellulose, nanofibrillated cellulose, and bacterial cellulose. These types of nanocellulose are summarized in Table 11.1. Details of their preparation will be further discussed below.

Table 11.1 Summary of different types of nanocellulose.

Types of nanocellulose	Source	Preparation	Morphology
Nanocrystalline cellulose	Wood, crops, agricultural residues, algae, tunicin, bacteria	Acid hydrolysis	Shape: whisker, rod-like Diameter: 5–70 nm Length: 100–300 nm
Nanofibrillated cellulose	Wood, crops, agricultural residues, industrial residues	Disintegration of	Shape: entangled fibril Diameter: 5–60 m Length: several micrometers
Bacterial cellulose	Sugars and alcohols		Bacterial synthesis Diameter: 20–100 nm

Source: Adapted and reproduced from [4].

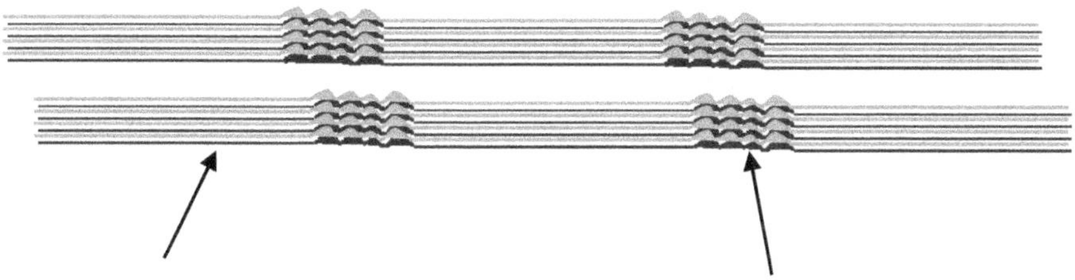

Figure 11.1 Crystalline and amorphous region in cellulose microfibril.

11.2.1 Nanocrystalline Cellulose

The cellulose microfibril structure consists of crystalline and amorphous or disordered regions (Figure 11.1). The amorphous region has a lower density and thus exhibits more free volume than the crystalline region [5]. Therefore, nanocrystalline cellulose is commonly prepared by subjecting lignin free cellulosic materials to acid hydrolysis in order to remove less dense amorphous domains, thus leaving the crystalline parts in a form of nanocrystalline cellulose suspension. The nanocrystalline cellulose appears as a whisker-like or rod-like shape when viewed under a high-resolution microscope, with an aspect ratio between 10 and 70. Due to its rigidity, the crystallinity of nanocrystalline cellulose is usually between 80% and 85%. The nanocrystalline extracted from different sources of materials and using different hydrolysis conditions would yield distinct morphologies and properties. Nanocrystalline cellulose isolated by sulfuric acid bears negatively charged sulfate half ester groups. The presence of this charged sulfate group nevertheless results in lower thermal stability of the nanocrystalline cellulose [6] compared to uncharged nanocrystals, which in turn might have an impact on the potential packaging application. In addition, the high crystallinity of nanocrystalline cellulose has the advantage in packaging such that it can reduce air and oxygen permeability due to its structure compactness.

On the other hand, the use of acid such as hydrochloric acid does not yield a charge to the surface. The presence of anionic groups on nanocrystalline cellulose results in a more stable suspension compared to uncharged cellulose. Anionic groups can also be imparted on the cellulose by the use of ammonium persulfate, phosphorylation, and difunctional carboxylic acids [7, 8]. The presence of charge contributes to suspension stability, which results in the ease of nanocrystalline dispersibility in a water medium. The nanocrystalline can be used as coating material functioning as a barrier material in packaging.

11.2.2 Nanofibrillated Cellulose

Nanofibrillated cellulose can be extracted from the same fiber used for nanocrystalline cellulose. The only difference is that of the extraction strategy, in which nanofibrillated cellulose can utilize different types of mechanical disintegration devices such as a high-pressure homogenizer, microfluidizer, ultrafine grinder, sonicator, ball milling apparatus, and refiner. Pre-treatment either by the chemical or enzymatic method is usually preferred prior to disintegration to reduce energy consumption. The nanofibrillated cellulose can be visualized as long, entangled, and flexible fiber or simply like spaghetti strands. Due to its length, nanofibrillated cellulose has a high aspect ratio and thus is very suitable for use in applications like composite, pulp, paper, and cement as the reinforcing agent. The presence of amorphous and crystalline parts contributes to flexibility and nanofibrillated cellulose can be used independently as a self-standing film.

11.2.3 Bacterial Cellulose

Bacterial cellulose is produced from the primary metabolic process of microbes. There are many bacteria that can produce cellulose but only species of *Acetobacter* from the genus Gluconacetobacter can produce a sufficient amount of cellulose for upscale quantities. This type of microorganism can convert glucose or sugar to pure cellulose. The synthesis of cellulose is an extracellular process that occurs between the outer and cytoplasma membrane by terminal complexes. This occurs in linear assembly of the pores at the surface of the bacterium. A well-defined cellulose mesh can be produced having 3–4 nm thickness and about 200–100 nm width.

11.3 Nanocellulose Barrier Property

Nanocellulose is attracting considerable interest in packaging because it provides a barrier property. Barrier properties are associated with permeable objects such as liquids, gases, water vapor, organic substances, and others. The barrier property of nanocellulose is best described through diffusion of molecules into the nanocellulose substrate or thin film. There are three stages involved to explain the diffusion mechanism of molecules through a thin film [1]. During the first stage, the surface of the thin film adsorbs the diffusing molecules. The second stage involves the movement of the molecules through the thin film. The third stage observes the desorption of the diffusing molecules out of the thin film surface. The formula to express the dissolution of gas through thin film or any other materials is shown by

$$P = D \times S \tag{11.1}$$

where P is the permeability, D is the diffusion coefficient, and S is the solubility coefficient. P can also be expressed according to Henry's law of solubility to Fick's law of diffusion:

$$P = \frac{q \times l}{A \times t \times \Delta p} \tag{11.2}$$

where q is the amount of material passing through the film, l is the flow length or thickness, A is the cross-sectional area, t is time, and Δp is the pressure between the two sides of the film.

In molecular diffusion, the most important aspect is the rate of molecule diffusion, also known as bulk flow, which can also be described as the movement of molecules from a high concentration to a low concentration area [9].

In the case of nanocellulose, the good oxygen barrier properties can be contributed to its dense structural organization. The structure network of nanofibrillated cellulose and nanocrystalline cellulose and how they affect the respective barrier properties deserves an interesting explanation. Nanofibrillated cellulose is reported to show better barrier properties than nanocrystalline cellulose film. The entanglement in nanofibrillated cellulose makes it easier to fill up the space between the fibrils at the nanoscale level. Such a non-porous film of nanofibrillated cellulose thus provides a tortuous diffusion passage to oxygen molecules [10] leading to low oxygen permeability. On the other hand, the nanocrystalline cellulose structure network in a film is more rigid and more porous than that of nanofibrillated cellulose, and thus it is less complicated.

In terms of the water vapor barrier properties, the same positive effect for the oxygen barrier discussed previously could not be observed on pristine nanocellulose if the relative humidity is high. This is due to the fact that hydroxyl groups on nanocellulose still have a high affinity to water. However, nanocellulose still shows a higher water barrier property compared to native cellulose. The film compactness from nanofibrillated cellulose reduces the penetration of water into the film. One of the most common ways to improve the water vapor properties is via surface modification of nanocellulose with hydrophobe material. This hydrophobic nanocellulose can serve both as an oxygen and water vapor barrier material for packaging. Other methods include heat treatment to improve the hydrophobicity of the surface.

11.4 Nanocellulose in Films

11.4.1 Extrusion of Nanocellulose Composite

Extrusion is performed by forcing a sample mixture at high temperature and pressure through a die, forming a sheet, film, or fiber (Figure 11.1). Nanocellulose, having outstanding physicochemical characteristics, can be used as a filler in various polymer matrices for packaging materials. Rising interest in the use of nanocellulose in the packaging field is also due to its biodegradability and renewability attributes. Incorporation of nanosized material in the polymer composite will enhance the properties, thus producing value-added or advanced end products. Many studies have been carried out to use nanocellulose as a reinforcing agent for thermoplastic resin via the melt extrusion technique [11, 12]. The resulting resin containing nanocellulose usually shows an increase in the modulus of elasticity. The reason for such improvement may be attributed to the greater interface in nanocomposites

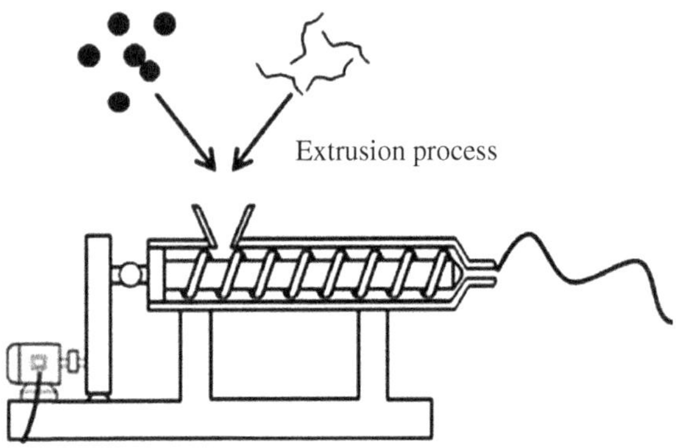

Figure 11.2 Extrusion process.

compared to a conventional composite. There are several key criteria to consider before nanocomposites can be products, namely the intrinsic properties of nanocellulose and matrix, the interfacial behavior, and the desired end product (Figure 11.2).

Blending prior to the extrusion process has become a necessary step to improve the extrusion procedure. The use of ball milling and the pre-blending process have been reported to produce a more homogeneous sample. In addition, nanocellulose is hydrophilic in nature; therefore, compatibility with nonpolar polymer is problematic. To solve this issue, nanocellulose needs to be surface modified with a hydrophobic molecule such as long chain organic acids. Low melting polymers are usually used with nanocellulose to avoid thermal degradation of cellulose during the extrusion process as nanocellulose can be thermally decomposed above 200°C [13].

11.4.2 Casting of Nanocellulose Films

The casting technique is carried out to prepare films made from suspensions of only nanocellulose or that contain a high content of nanocellulose. For this technique, good dispersibility of the mixture is important to produce film of a high quality with a good reinforcement property. Many studies have been conducted on the preparation of nanocellulose film with water as the medium [14, 15] as well as with non-aqueous solvents [16, 17].

Similar to the extrusion technique, nanocellulose needs to be modified to make it less hydrophilic if there is a need to use it in nonaqueous solvent casting. There are several studies on the use of modified nanocellulose in hydrophobic polymer matrices [17, 18]. Silylation and esterification are the most common hydrophobization techniques to make nanocellulose less hydrophilic for use with a nonaqueous solvent in nonpolar polymer matrices. Besides modification, an attempt to reduce the surface energy of nanocellulose has been carried out using surfactant by Fortunati et al. [19].

The advantage of employing the casting technique is that the film preparation can be carried out at a low temperature only for the purpose of solvent evaporation and the need for less viscous mixture for ease of film fabrication. In a different case, it is interesting to note that the use of nanocellulose helps to reduce the shrinkage problem in a film preparation, as reported by Tammelin et al. [20].

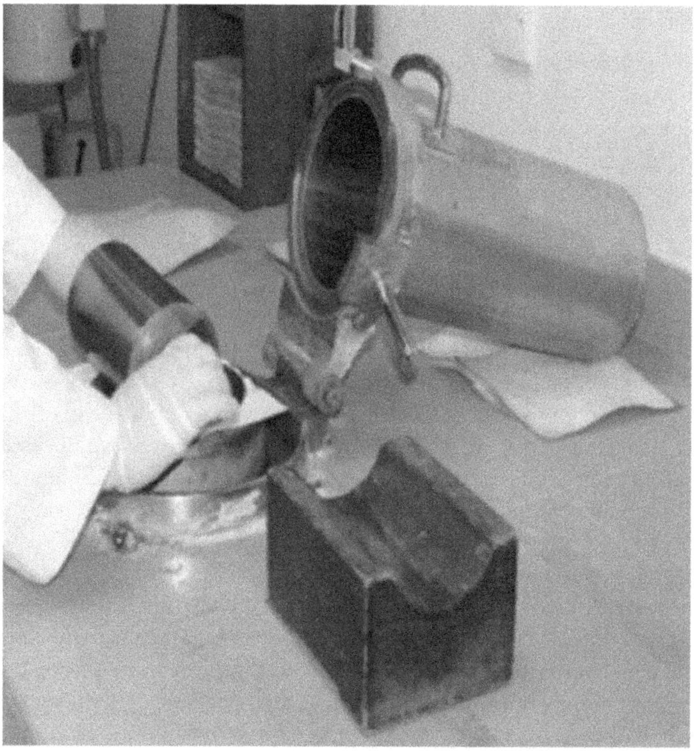

Figure 11.3 Laboratory papermaking machine for nanocellulose thin film making.

11.4.3 Filtration of Nanocellulose Composite

Preparation of a film from nanocellulose can be performed using the filtration technique. Film from nanocellulose can also be visualized as nanopaper since it can employ the same technique as in the making of paper, i.e. by using the filtration process. Filtration in the papermaking process is not new since papers are formed on a continuous paper machine called a Fourdrinier machine in which an aqueous suspension of cellulose fibers pumped into a headbox form a sheet of paper continuously on a fine woven screen by the drainage process. Nanofibrillated cellulose (NFC) can be converted into film using the same approach in the papermaking process. Studies on this have been reported by many authors [21, 22] and the films produced are of high strength, have good oxygen permeability, and are transparent. Figure 11.3 shows that nanocellulose film can be prepared using a laboratory papermaking machine.

Besides making film, nanocellulose has also been added to improve the strength of the papers. One of the challenges of forming film using this technique is the dewatering time. Nanosized cellulose decreases the dewatering time and so productivity will be affected. This is due to the high surface area of nanocellulose that makes ithold water longer than the native cellulose [23].

11.4.4 Coating

Coating can be described as a process of applying a layer of slurry on the surface of a material so that it forms a film that can act as an adhesive with the material [24]. Nanofibrillated cellulose has been used as a coating material for paper and board intended for packaging uses, as reported by many authors [25–27]. The positive effects of using nanofibrillated cellulose as a coating material for packaging materials are reported in the following studies:

1. Paper strength is increased [28]
2. Oxygen transmission is reduced [29]
3. Air permeability is reduced [29]

Nanocrystalline cellulose has also been used to coat various substrates such as paper and regenerated cellulose film and has also been found to display the same good effect as when NFC is used, especially for such properties as strength and air permeability [30].

Coating using nanocellulose has mostly been focused on aqueous solution due to the hydrophilic nature of cellulose. However, the study on surface-modified nanocellulose with polylactic acid has also been reported. It is interesting to observe that a reduction in low water permeability was reported.

Another similar technique to coating that deserves attention is impregnation of nanocellulose film. Some authors reported the impregnation of phenolic resin into pulp to produce high-strength nanocellulose composite film with an improved barrier property. This can be achieved by using the compressing procedure at high pressure, which can result in pores being coated by fillers, thus making the film more impermeable [22, 31, 32].

11.5 Nanocellulose Film in Packaging

Early development of nanocellulose-based material was focused mainly on the mechanical properties in terms of the tensile modulus [22]. However, recently the attention has been more on the barrier properties of the nanocomposite films, which can find more applications in many industrial sectors.

Nanocellulose has a high potential to be applied in various industrial sectors such as medical, electronics, membranes, packaging, paper, and coatings. It can add value to conventional products and can also lead to the development of advanced materials such as photonics, films and foams, nanocomposites, and medical devices. Since a decade ago, research and development on the use of renewable and biodegradable materials, particularly in plastic packaging technology, has been escalating. The prime objective of packaging is to prevent food and beverages, healthcare, cosmetics, and consumer goods from any disturbances in terms of physical, biochemical, and microbiological factors [1].

Two obvious advantages of nanocellulose in the film-based packaging system are its availability in providing strength to the films and also increasing the oxygen permeation resistance [13]. Film made from nanofibrillated cellulose is reported to reduce the oxygen permeability substantially, which can be competitive and comparable with the commercial synthetic polymer. Such an observation is attributed to the dense network of nanofibrils that have smaller and more uniform dimensions that lead to higher entanglements within the film [1].

Besides that, nanocellulose-based materials are available as edible, biodegradable, nontoxic and low-cost materials [33]. Nanocellulose-based film can be generally divided into food or non-food products. For instance, the industry sectors that can use nanocellulose-based film include food and beverage, medicine and pharmaceuticals, industrial packaging, and many others.

11.5.1 Food and Beverage Industry

Food can be classified as solid, liquid, dry, frozen, and fresh foods, whereas a beverage can be categorized into liquid and fresh beverage. Thus the aim of packaging is to protect the content of foods or beverages in terms of quality and safety. Important characteristics of certain nanocomposite films are its barrier against oxygen, oils and grease, water vapor, and aqueous liquid as well.

By limiting the permeation of oxygen, gases, and volatile compounds, nanocellulose film has a high ability to increase the shelf life of foods or beverages. In addition, accumulation of unwanted odors, contamination of foods, and development of food odors can be prevented. Oil and grease resistance is a very important element in the food packaging industry [34, 35]. The hold-out of oils and grease is related to nonpolar gases like oxygen. Nanocomposite film is found to prolong food shelf life and hence preserves food quality. A research by Sirviö et al. [36] found that mechanical strength of nanocellulose and alginate-based composite film increased from 70.02 MPa to 97.97 MPa and its grease barrier properties were also excellent.

Prevention of permeation of water vapor has become a challenge in cellulose-based film packaging. Typical cellulose-based films lose the ability to prevent the permeation of oxygen [14]. Various techniques can be applied to modify the film so that it can reduce the film's sensitivity against moisture [37–39]. Resistance to penetration of packaging material by an aqueous solution is important in many applications, especially for cellulose-based packaging [34, 40, 41].

11.5.2 Medicine and Pharmaceuticals

Nanocellulose-based films appear to function as a barrier to contained pharmaceutical compounds [44]. The rate of release by the confined materials was observed to be slowed by the presence of the film. Such films act as antioxidants and antimicrobials by serving as carriers for active substances [42]. Toivonen et al. [43] successfully achieved good transparency properties of the film. Deciding which nanocomposite film material has a high transparency is very difficult as the cellulosic materials need to be tiny and when wetted by matrix material in the layer there must be no detection of clumps or entangled regions.

11.6 Conclusion

Nanocellulose has a lot to offer for use in packaging applications as it can be used as fillers, strength additives in composites, a coating material as well as a self-standing thin film. Because of its attractive physicochemical properties coupled with its renewability and biodegradability characteristics, it is deemed suitable for many sustainable packaging purposes.

References

1. Nair SS, Zhu JY, Deng Y, Ragauskas AJ. High performance green barriers based on nanocellulose. *Sustainable Chem. Process*. 2014; **2**(1).
2. Habibi Y. Key advances in the chemical modification of nanocelluloses. *Chem. Soc. Rev.* 2013.
3. Turbak AF, Snyder FW, Sandberg KR. Microfibrillated cellulose, a new cellulose product: properties, uses, and commercial potential. *J. Appl. Polym. Sci. Appl. Polym. Symp.* 1983; **37**: 815–27.
4. Klemm D, Cranston ED, Fischer D, Gama M, Kedzior SA, Kralisch D, et al. Nanocellulose as a natural source for groundbreaking applications in materials science: today's state. *Mater. Today* 2018; **21**(7): 720–48.
5. De Souza Lima MM, Borsali R. Rodlike cellulose microcrystals: structure, properties, and applications. *Macromol. Rapid Commun.* 2004; **25**(7): 771–87.
6. Roman M, Winter WT. Effect of sulfate groups from sulfuric acid hydrolysis on the thermal degradation behavior of bacterial cellulose. *Biomacromolecules* 2004; **5**(5): 1671–7.
7. Naderi A, Lindström T, Flodberg G, Sundström J, Junel K, Runebjörk A, et al. Phosphorylated nanofibrillated cellulose: production and properties. *Nord. Pulp Pap. Res. J.* 2016; **31**(1): 20–9.
8. Chen L, Zhu JY, Baez C, Kitin P, Elder T. Highly thermal-stable and functional cellulose nanocrystals and nanofibrils produced using fully recyclable organic acids. *Green Chem.* 2016; **18** (13): 3835–43.
9. Lagaron JM, Catala R, Gavara R. Structural characteristics defining high barrier properties in polymeric materials. *Mater. Sci. Technol.* 2004; **20**(1): 1–7.
10. Belbekhouche S, Bras J, Siqueira G, Chappey C, Lebrun L, Khelifi B, et al. Water sorption behavior and gas barrier properties of cellulose whiskers and microfibrils films. *Carbohydr. Polym.* 2011; **83**(4): 1740–8.
11. Lendvai L, Karger-Kocsis J, Kmetty A, Drakopoulos SX. Production and characterization of microfibrillated cellulose-reinforced thermoplastic starch composites. *J. Appl. Polym. Sci.* 2016; **133**(2).
12. Ferrer A, Hoeger IC, Lu X, Rojas OJ. Reinforcement of polypropylene with lignocellulose nanofibrils and compatibilization with biobased polymers. *J. Appl. Polym. Sci.* 2016; **133** (34).
13. Ferrer A, Pal L, Hubbe M. Nanocellulose in packaging: advances in barrier layer technologies. *Ind. Crops Prod.* 2017; **95**: 574–82.
14. Aulin C, Gällstedt M, Lindström T. Oxygen and oil barrier properties of microfibrillated cellulose films and coatings. *Cellulose* 2010; **17**(3): 559–74.
15. Lu P, Xiao H, Zhang W, Gong G. Reactive coating of soybean oil-based polymer on nanofibrillated cellulose film for water vapor barrier packaging. *Carbohydr. Polym.* 2014; **111**: 524–9.
16. Grunert M, Winter WT. Nanocomposites of cellulose acetate butyrate reinforced with cellulose nanocrystals. *J. Polym. Environ.* 2002; **10** (1–2): 27–30.
17. Hassan ML, Bras J, Hassan EA, Fadel SM, Dufresne A. Polycaprolactone/modified bagasse whisker nanocomposites with improved moisture-barrier and biodegradability properties. *J. Appl. Polym. Sci.* 2012; **125** (S2): E10-E9.
18. Song J, Rojas OJ. Approaching super-hydrophobicity from cellulosic materials: a review. *Nord. Pulp Pap. Res. J.* 2013; **28**(2): 216–38.

19. Fortunati E, Peltzer M, Armentano I, Torre L, Jiménez A, Kenny JM. Effects of modified cellulose nanocrystals on the barrier and migration properties of PLA nano-biocomposites. *Carbohydr. Polym.* 2012; **90**(2): 948–56.
20. Tammelin T, Hippi U, Salminen A, Salminen T. (inventors) Preparing a film of nanofibrillated cellulose on the surface of support material, which is useful for producing, e.g., food packaging material, comprises applying and spreading suspension of cellulose on to plastic support material, 2013.
21. Nogi M, Iwamoto S, Nakagaito AN, Yano H. Optically transparent nanofiber paper. *Adv. Mater.* 2009; **21** (16): 1595–8.
22. Nakagaito AN, Yano H. The effect of morphological changes from pulp fiber towards nano-scale fibrillated cellulose on the mechanical properties of high-strength plant fiber based composites. *Appl. Phys. A* 2004; **78**(4): 547–52.
23. Rojo E, Peresin MS, Sampson WW, Hoeger IC, Vartiainen J, Laine J, et al. Comprehensive elucidation of the effect of residual lignin on the physical, barrier, mechanical and surface properties of nanocellulose films. *Green Chem.* 2015; **17**(3): 1853–66.
24. Hubbe MA, Ferrer A, Tyagi P, Yin Y, Salas C, Pal L, et al. Nanocellulose in thin films, coatings, and plies for packaging applications: a review. *BioResources* 2017; **12**(1): 2143–233.
25. Freire CSR, Fernandes SCM, Silvestre AJD, Neto CP. Novel cellulose-based composites based on nanofibrillated plant and bacterial cellulose: recent advances at the University of Aveiro – a review. *Holzforschung* 2013; **67**(6): 603–12.
26. Hamada H, Beckvermit J, Bousfield WD. Nanofibrillated cellulose with fine clay as a coating agent to improve print quality. *PaperCon 2010 Conference* (Volume Session 201). 2010: p. 11.
27. Vartiainen J, Shen Y, Kaljunen T, Malm T, Vaha-Nissi M, Putkonen M, et al. Bio-based multilayer barrier films by extrusion, dispersion coating and atomic layer deposition. *J. Appl. Polym. Sci.* 2016; **133**(2).
28. Syverud K, Stenius P. Strength and barrier properties of MFC films. *Cellulose* 2009; **16**(1): 75–85.
29. Rezayati Charani P, Dehghani-Firouzabadi M, Afra E, Blademo A, Naderi A, Lindstrom T. Production of microfibrillated cellulose from unbleached kraft pulp of Kenaf and Scotch Pine and its effect on the properties of hardwood kraft: microfibrillated cellulose paper. *Cellulose* 2013; **20**(5): 2559–67.
30. Yang S, Tang Y, Wang J, Kong F, Zhang J. Surface treatment of cellulosic paper with starch-based composites reinforced with nanocrystalline cellulose. *Ind. Eng. Chem. Res.* 2014; **53** (36): 13980–8.
31. Nakagaito AN, Yano H. Novel high-strength biocomposites based on microfibrillated cellulose having nano-order-unit web-like network structure. *Appl. Phys. A* 2005; **80**(1): 155–9.
32. Nakagaito AN, Yano H. The effect of fiber content on the mechanical and thermal expansion properties of biocomposites based on microfibrillated cellulose. *Cellulose* 2008; **15**(4): 555–9.
33. de Moura M. R, Mattoso LHC, Zucolotto V. Development of cellulose-based bactericidal nanocomposites containing silver nanoparticles and their use as active food packaging. *J. Food Eng.* 2012; **109**(3): 520–4.
34. Kisonen V, Prakobna K, Xu C, Salminen A, Mikkonen KS, Valtakari D, et al. Composite films of nanofibrillated cellulose and O-acetyl galactoglucomannan (GGM) coated with succinic esters of GGM showing potential as barrier material in food packaging. *J. Mater. Sci.* 2015; **50**(8): 3189–99.
35. Kumar V, Bollstrom R, Yang A, Chen Q, Chen G, Salminen P, et al. Comparison of nano- and microfibrillated cellulose films. *Cellulose* 2014; **21**.
36. Sirvio JA, Kolehmainen A, Liimatainen H, Niinimaki J, Hormi OEO. Biocomposite cellulose-alginate films: promising packaging materials. *Food Chem.* 2014; **151**: 343–51.
37. Cao X, Chen Y, Chang PR, Muir AD, Falk G. Starch-based nanocomposites reinforced with flax cellulose nanocrystals. *Express Polym. Lett.* 2008; **2**(7): 502–10.
38. Savadekar NR, Karande VS, Vigneshwaran N, Bharimalla AK, Mhaske ST. Preparation of nano cellulose fibers and its application in kappa-carrageenan based film. *Int. J. Biol. Macromol.* 2012; **51**(5): 1008–13.

39. Santos TM, Souza Filho MDSM, Caceres CA, Rosa MF, Morais JPS, Pinto AMB, et al. Fish gelatin films as affected by cellulose whiskers and sonication. *Food Hydrocolloids* 2014; **41**: 113–8.
40. Choi Y, Simonsen J. Cellulose nanocrystal-filled carboxymethyl cellulose nanocomposites. *J. Nanosci. Nanotechnol.* 2006; **6**(3): 633–9.
41. Shimizu M, Saito T, Isogai A. Water-resistant and high oxygen-barrier nanocellulose films with interfibrillar cross-linkages formed through multivalent metal ions. *J. Membr. Sci.* 2016; **500**: 1–7.
42. Andresen M, Stenstad P, Møretrø, T, Langsrud S, Syverud K, Johansson LS, et al. Nonleaching antimicrobial films prepared from surface-modified microfibrillated cellulose. *Biomacromolecules* 2007; **8**(7): 2149–55.
43. Toivonen MS, Kaskela A, Rojas OJ, Kauppinen EI, Ikkala O. Ambient-dried cellulose nanofibril aerogel membranes with high tensile strength and their use for aerosol collection and templates for transparent, flexible devices, *Adv. Funct. Mater.* 2015; **25** (42): 6618–26.
44. Ilyas RA, Sapuan SM, Atikah MSN, Asyraf MRM, Ayu Rafiqah S, Aisyah HA, Mohd Nurazzi N, Norrahim MNF. Effect of Hydrolysis Time on the Morphological, Physical, Chemical, and Thermal Behavior of Sugar Palm Nanocrystalline Cellulose (Arenga Pinnata (Wurmb.) Merr). *Textile Research Journal*. 2020. In print. doi: 10.1177/0040517520932393.

12

Utilization of Rice Straw as a Raw Material for Food Packaging

Rushdan Ibrahim[1], S.M Sapuan[2,3], R.A Ilyas[4,5], and M.S.N. Atikah[6]

[1] *Pulp and Paper Laboratory, Forest Research Institute Malaysia, Selangor, Kepong, 52109, Malaysia*

[2] *Institute of Tropical Forest and Forest Products (INTROP), Universiti Putra Malaysia (UPM), Selangor, Serdang, 43400, Malaysia*

[3] *Advanced Engineering Materials and Composites Research Centre (AEMC), Department of Mechanical and Manufacturing Engineering, Faculty of Engineering, Universiti Putra Malaysia (UPM), Selangor, Serdang, 43400, Malaysia*

[4] *School of Chemical and Energy, Faculty of Engineering, Universiti Teknologi Malaysia, 81310 UTM Johor Bahru, Johor, Malaysia*

[5] *Centre for Advanced Composite Materials, Universiti Teknologi Malaysia, 81310 UTM Johor Bahru, Johor, Malaysia*

[6] *Department of Chemical and Environmental Engineering, Universiti Putra Malaysia (UPM), Selangor, Serdang, 43400, Malaysia*

12.1 Introduction

Currently, there has been a rapid development in research and innovation on converting waste natural fiber to create wealth [1–7]. Huge attention has been given due to the advantages of these waste materials when compared to others. Such materials as synthetic fiber composites have very little environmental impact and are produced at a low cost, and support has been given to their potential across a wide range of applications [8–19, 34]. There are more than 200 000 ha of rice fields in Peninsular Malaysia (Table 12.1 and Figure 12.1) generating about 2.6 million tonnes of rice straw annually. Disposing of the large amount of rice straw is difficult as the options are limited by the great bulk of material, slow degradation in the soil, harboring of rice stem disease, and a high mineral content [20]. Soil incorporation and field burning have been the common practices for removing the rice straw

Bio-Based Packaging: Material, Environmental and Economic Aspects, First Edition. Edited by S.M. Sapuan and R.A. Ilyas

Table 12.1 Acreages of rice cultivation in Peninsular Malaysia.

No.	Planting area	Acreage (ha)
1	MADA	96 558
2	KADA	31 440
3	PKSM	30 600
4	PBLS	19 600
5	IADP Pulau Pinang	11 000
6	Seberang Perak	8 982
7	KETARA	4 660
8	Kemasin/Semarak	1 645
	Total	204 485

Figure 12.1 Areas of cultivation of rice in Peninsular Malaysia (colored green).

[21, 22]. Traditionally, in Malaysia combustion of the straw is done after the harvesting season of paddy. Generally, rice straw burning is done to eliminate sources of insect pests, rat infestations, and control of rice diseases. In essence, such activity causes air pollution, affects the safety of road users, and affects the atmosphere and environment, thus leading to global climate change [23].

Utilizing rice straw would reduce its environmental impact and ease the disposal problem facing agricultural operations [22]. The potential environmental benefits of diverting rice straw from open-field burning will significantly reduce air pollutants that can be a health hazard [24].

12.2 Selling Rice Straw

Rice straw needs to be collected from the field after rice harvesting (Figure 12.2) and is collected using a baler (Figures 12.3 to 12.5). Bales are stored under a shed to make it

Figure 12.2 *Harvesting rice.*

Table 12.2 *Requirement for rice straw sales.*

No.	Facility	Description	Cost per unit (RM[a])	Workers
1	Baler	To bale rice straw	150 000	2
2	Lorry	To transfer rice straw from rice field to storage	300 000	2
3	Warehouse	To store rice straw	150 000	2
4.	Forklift	To carry bale	150 000	1
		Total cost	750 000	7

[a]Note: 1 US$ = RM 4.12, where RM stands for Malaysian Ringgit.

last longer (Figure 12.6). Table 12.2 displays the requirement for rice straw sales. The baled rice straw is currently sold for RM150 per tonne, so it can generate extra income of RM390 million to farmers annually.

12.3 Selling Pulp

One way to utilize rice straw is as a raw material for pulp, which is a lignocellulosic fibrous material prepared by chemically, mechanically, or chemimechanically separating cellulose fibers from rice straw. Chemimechanical pulp can be used to produce pulp for handcrafts, pulp molded paperboard, and newsprint industries.

In this study, we found that rice straw can be made into pulp by chemical, semi-chemical, and chemimechanical pulping processes via soda-anthraquinone (AQ), chemimechanical pulp (CMP), and cold soda refiner mechanical pulp (RMP), respectively [25]. Each pulping process produced pulp and paper with various properties [26, 27]. Cold soda RMP gave the highest pulp yields, but the semi-chemical pulping process had the highest paper properties [28]. The pulp yields for RMP, cold soda mechanical pulping, CMP, and

Figure 12.3 Baler machine with capacity of 120–240 bales per hour.

soda-anthraquinone pulping are tabulated in Table 12.3. The properties of paper made from different pulping processes are presented in Table 12.4.

The process of producing chemimechanical pulp are:

1. Raw material preparation – cutting and cleaning (Figure 12.7)
2. Pulping (Figure 12.8)
3. Cleaning (Figure 12.9)
4. Storing and packaging (Figure 12.10 and Table 12.5)

Figure 12.4 Rice straw bale weight of 500 kg.

Table 12.3 Pulping properties of rice straw.

Property	Process			
	RMP	Cold soda RMP	CMP	Soda-AQ
Total yield (%)	93.2	85.5	73.6	44.7
Screened yield (%)	77.7	75.5	69.2	44.7
Sieve (%)	15.5	10.0	4.6	0

The pulp yield is about 80%, so 2.6 million tonnes of rice straw will produce approximately about 2 million tonnes of chemimechanical pulp. The price of chemimechanical pulp is about US$400 per tonne or RM1600 per tonne. Two million tonnes of pulp will have a value of RM 3.2 billion.

A small pulp mill utilizes 10 tonnes of rice straw and needs about 1200 ha of rice fields. Two hundred thousand hectares of rice fields in Peninsular Malaysia can be used to provide167 pulp mills with 10 tonnes per day.

Figure 12.5 Rice straw bales in rice field.

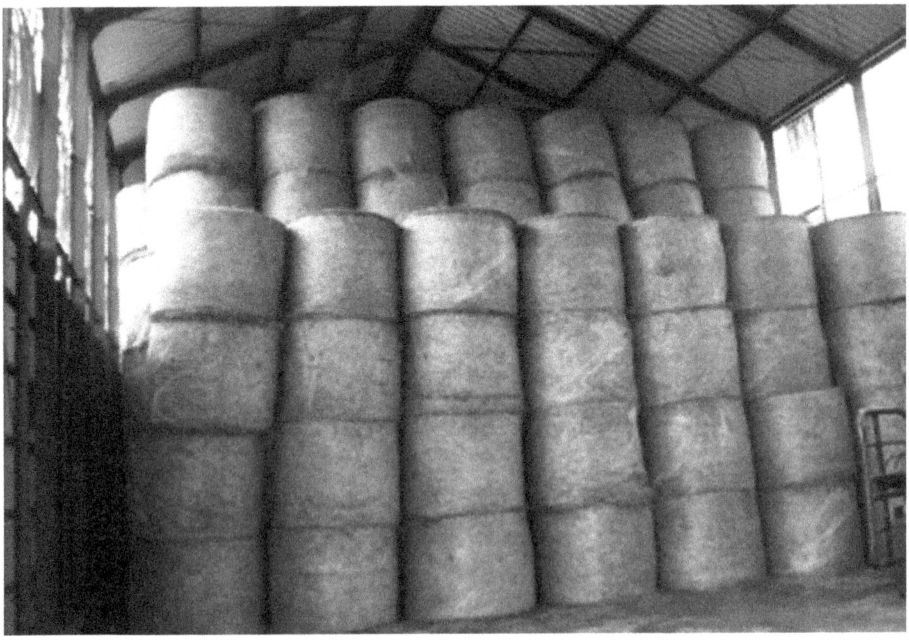

Figure 12.6 Storage of rice straw bales.

Table 12.4 *Pulp and paper properties of rice straw.*

Property	Process			
	RMP	Cold soda RMP	CMP	Soda-AQ
Density (g/cm^3)	—	0.49	0.74	0.61
Tensile index (Nm/g)	—	0.41	0.78	0.51
Tear index (mNm^2/g)	—	3.51	6.41	4.98
Burst index ($kPam^2/g$)	—	1.85	4.40	2.27
Double fold (800 K.M)	—	6.00	530.00	24.00

Figure 12.7 *Rice straw cutter – reduce rice straw to 10–30 mm in length.*

12.4 Selling Pulp Molded Products

Globally, the billion dollar packaging products industry is sourcing the raw materials from forest woods as their feedstock [29]. Forest woods are being deforested annually to cater for the needs of the packaging products industry worldwide. This has led to massive environmental problems, global warming, climate change, and a serious environmental impact to the eco-system and biodiversity [29].

Virgin pulps from cellulose fibers of rice straw have been identified and used to make biodegradable food containers, biodegradable packaging medical products, and biodegradable industrial packaging products [30, 31]. The global market is moving into sustainable and biodegradable packaging needs since 2011 and the market is growing and accelerating

Figure 12.8 Pulping machines – converting rice straw into pulp.

Table 12.5 Requirement for rice straw pulp sales.

No.	Facility	Description	Cost per unit (RM)	Worker per shift
1	Electricity sub-station	To supply electrical power to pulp mill	300 000	1
2	Pulp mill	To house pulping machinery	300 000	2
3	Pulping machinery	To convert rice straw into pulp	750 000	2
4	Forklift	To carry bale	150 000	1
		Total cost	1 500 000	6

rapidly to meet and satisfy the market requirements (Tables 12.6 and 12.7), the industry requirements and the regulation requirements. In addition to the global market needs and demands for biomass fiber pulps as an alternative material to produce packaging products, our Government is also focusing to accelerate this biomass pulp industry to generate high-impact income to the nation.

The target market for the commercialization of rice straw fibers has narrowed down to high impact industries with good financial returns. As such, the target market for our rice straw fibers is as follows:

1. Biodegradable packaging products for the food and beverages industry (e.g. food-service industry)

Figure 12.9 Washing pulp.

Table 12.6 Global packaging markets [32].

	2000	2005	Growth 2000–2005 (%)	2010	Growth 2005–2010 (%)	Growth 2000–2010 (%)
Western Europe	35 433.4	39 973.2	12.8	47 407.2	18.6	330.8
Eastern Europe	3 135.9	3 591.5	14.5	4.589.2	27.8	46.3
North America	35 930.2	42 958.4	19.6	52 482.3	22.2	46.1
Latin America	5.9115.7	7 407.8	25.2	9 447.4	27.5	59.7
Oceania	2 419.4	2 449.4	1.2	2 584.8	5.5	6.8
Asia	29 892.6	36 340.4	21.6	50 377.3	40.0	70.2
World	112 720.3	132 720.7	17.7	167 388.2	26.1	48.5

2. Biodegradable packaging products for medical industry (e.g. biodegradable bed-pans, bowls, and urinal containers)
3. Biodegradable packaging products for industrial packaging products (e.g. electronic packaging)

Processing steps for producing molded products from rice straw pulps are:

1. Stock preparation (Figure 12.11)
2. Molding (Figure 12.12)
3. Finishing and packaging (Figure 12.13 and Table 12.8)

Figure 12.10 Clean and dried pulp are kept in jumbo bag for transportation and storage.

Table 12.7 US green packaging demand [32].

US green packaging demand (million dollars)					
% Annual growth					
Item	2004	2009	2014	2004–2009	2009–2014
Green packaging demand	30 145	34 460	41 650	2.7	3.9
Recycled content packaging	27 175	31 178	37 250	2.8	3.6
Reusable packaging	2 795	2 920	3 715	0.9	4.9
Degradable packaging	175	362	685	15.6	13.6

One tonne of rice straw pulp can produce 100 000 articles of 10 g in weight. A molded product is sold for 10 cent per piece. One tonne of molded product is worth RM10 000.

12.5 Selling Paper

Paper is a thin material produced by pressing together moist fibers of cellulose pulp and drying them into flexible sheets. It is a versatile material with many uses, including writing, printing, packaging, cleaning, and a number of industrial and construction processes. Malaysia imported close to RM 6.7 billion worth of paper and paperboard in 2010 (Malaysia International Trade and Industry Report 2011 [33]. With the advent of this paper-making technology from rice straw bio-waste, the need to import pulp and paper from other countries will be minimized and may generate income from possible export in the future.

Figure 12.11 Stock preparation – pulp beating, mixing with additives, and metering.

Table 12.8 Requirement for pulp molded sales.

No.	Facility	Description	Cost per unit (RM)	Worker per shift
1	Electricity sub-station	To supply electrical power to pulp mill	300 000	1
2	Pulp molded mill	To house molding machinery	300 000	2
3	Molded machinery	To convert pulp into molded articles	750 000	2
4	Forklift	To carry pulp and products	150 000	1
		Total cost	1 500 000	6

Market risk may emerge from the use of rice straw as an unconventional raw material to produce paper. However, as the fiber characteristics and chemical content of rice straw and the characteristics of the paper after the chemical process is similar to that of wood-based raw materials like eucalyptus and it is able to meet the specification of hard wood pulp, consumers' views on unbleached paper and paperboard produced using rice straw as the raw material will be more reassuring. Also, since the use of rice straw as the raw material is also a major step toward reducing greenhouse gas emission and depletion of forests, market acceptance of this product is envisaged to be positive.

The use of rice straw for paper making reduces the dumping of bio-waste, which could result in uncontrolled release of methane into the environment if digested anaerobically and

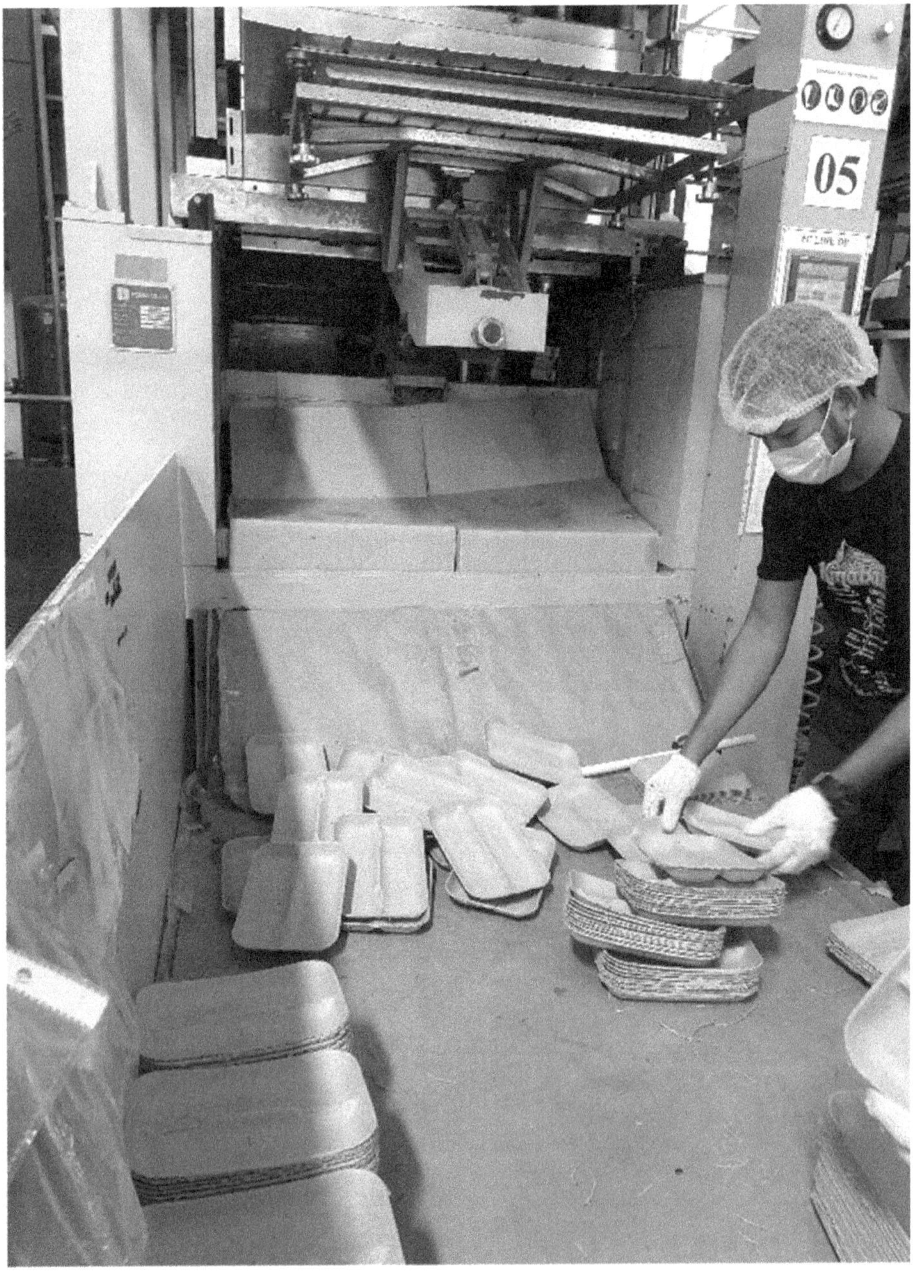

Figure 12.12 *The pulp was molded into fruit tray articles.*

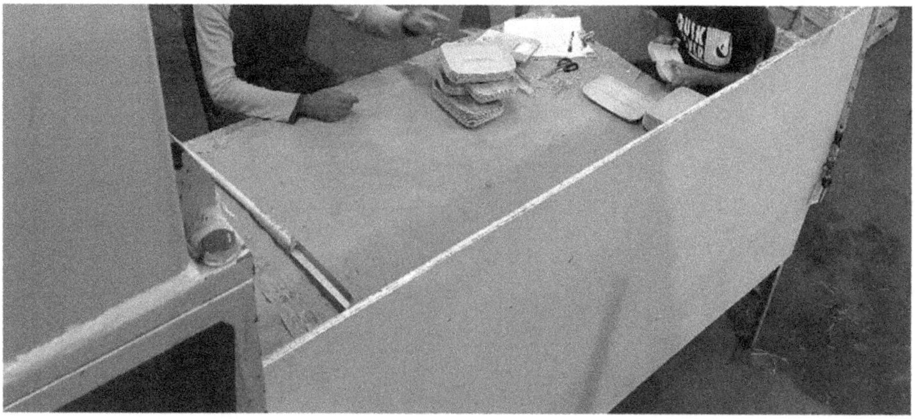

Figure 12.13 Quality control and packaging prior sales.

Table 12.9 Requirement for paper and paperboard sales.

No.	Facility	Description	Cost per unit (RM	Worke r per shift
1	Electricity sub station	To supply electrical power to pulp mill	300 000	1
2	Paper mill	To house papermaking machinery	300 000	2
3	Papermaking machinery	To convert pulp into molded articles	750 000	2
4	Forklift	To carry pulp and products	150 000	1
		Total cost	1 500 000	6

uncontrolled release of carbon dioxide if digested aerobically under natural decomposition. The release of such greenhouse gases from landfills into the environment is eliminated by using rice straw for paper making and this adds a positive impact on the environment. With the implementation of this project, the rice straw industry is safeguarded against possible smearing by NGOs.

The processing steps of producing paper and paperboard from rice straw pulps are:

1. Stock preparation (Figure 12.14)
2. Papermaking (Figure 12.15)
3. Finishing and packaging (Figure 12.16 and Table 12.9)

The economic value of 1 ton of rice straw used for unbleached paper and paperboard making is approximately RM 1600.00 assuming the selling price of USD 600 (RM 2400.00) per tonne of unbleached paper and paperboard. With such a large amount of rice straw available from the paddy field, if rice straw is used for paper making, the economic value that can be realized from this commercialization activity is very significant. Figure 12.17 shows the S.W.O.T analysis in utilization of rice straw as a raw material for pulp, molded, paper, and paperboard.

Figure 12.14 Pressure screener to purify pulp during stock preparation.

Table 12.10 Comparison of cost and selling of products from rice straw.

No.	Product	Investment cost (RM)	Start material cost (RM)	Production cost (RM)	Selling per ton (RM)
1	Rice straw	750 000	0	50	150
2	Pulp	1 500 000	150	850	1 600
3	Molded product	1 500 000	1 600	850	10 000
4	Paper and paperboard	1 500 000	1 600	850	2 400

12.6 Cost of Commercialization of Products from Rice Straw

Technology adaption of rice straw processing seems to have potential to be further commercialized. The costs of producing and the selling price of rice straw, pulp, molded product, paper, and paperboard are different (Table 12.10). A molded product has the highest selling price and its investment is comparable to paper and paperboard products (see Table 12.10).

Figure 12.15 Paper forming, drying, and calendaring in paper making.

Figure 12.16 Finishing and packaging.

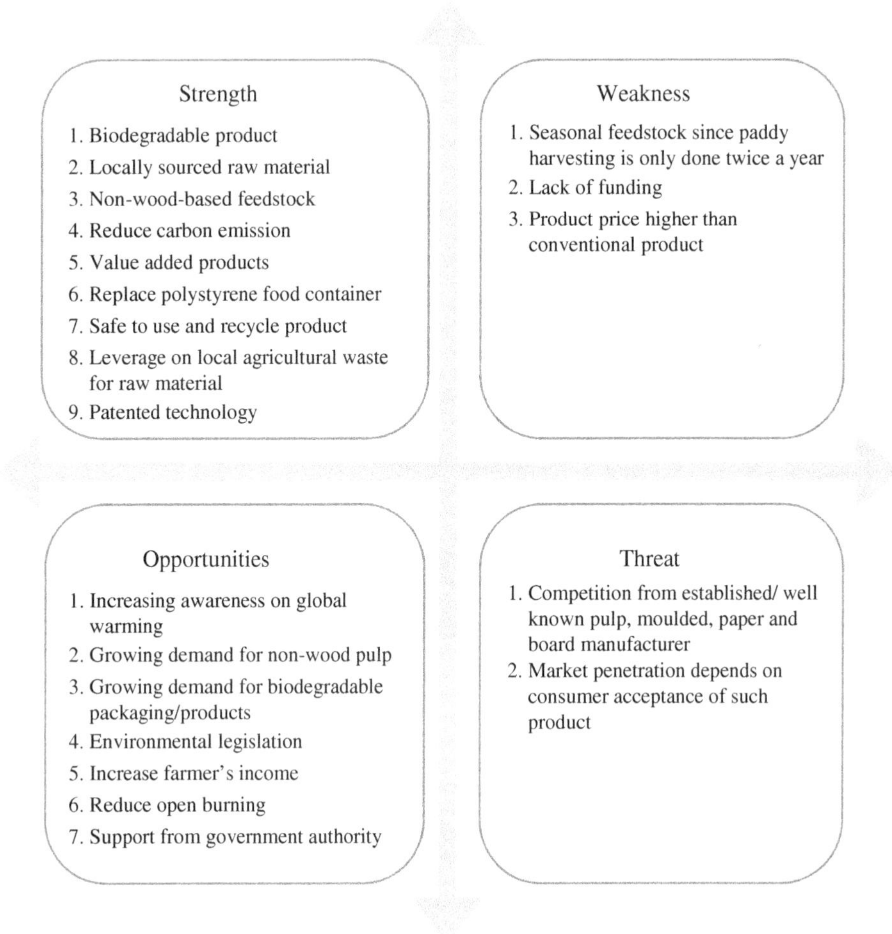

Figure 12.17 *S.W.O.T. analysis in utilization of rice straw as a raw material for pulp, molded, paper, and paperboard.*

The biodegradable packaging market has mainly been driven by the growing awareness of the sustainability of molded products and government regulations have started to focus toward the reduction of wastage and usage of biodegradable packaging materials. A business model canvas (BMC) has been made for molded products from pulp using rice straw (Table 12.11). The impact of income generation from selling the product from rice straw to the community should be deeply analyzed to achieve the outcome of FRIM's technology adaptation.

12.7 Conclusions

The byproduct of rice is 2.6 million tonnes of rice straw annually with a plantation size of more than 200 000 ha of rice fields. These byproducts can be processed into pulp by

Table 12.11 Business Model Canvas (BMC): molded products from pulp using rice straw.

7. Key partners 1. FRIM 2. FRIM Inc. 3. MESTECC 4. MADA 5. KADA 6. PKSM 7. PBLS 8. IADP Pulau Pinang 9. Seberang Perak 10. IADA KETARA 11. Kemasin/Semarak 12. Investor 13. Ministry of territories 14. Ministry of Health	**5. Key activities** Production of Biodegradable products. Supply raw Material	**1. Value proposition** **Molded biodegradable products** 1. Packaging, 2. Disposable Food 3. Containers, 4. Takeaway Containers 5. Disposable Cups and 6. Tableware 7. Antimicrobial pulp 8. Molded 9. Trash bags **"From waste to wealth"**	**2. Customer relationship** 1. Business to business 2. Supply contract 3. Distribution 4. One to one service	**3. Customer segments** 1. Federal Territory 2. (Kuala Lumpur, 3. Putrajaya, 4. Labuan) 5. Local authorities 6. International 7. Market 8. Green packaging 9. Distributor 10. Packaging 11. Manufacture 12. Hospitals
	6. Key resources 1. Raw Material Resources a. MADA b. KADA c. PKSM d. PBLS e. IADP Pulau Pinang f. Seberang Perak g. KETARA h. Kemasin/Semarak 2. Funding 3. Human Resource		**4. Channels** 1. Online: social media, website, video series, influencers 2. Offline: Event activations (collaborate/well-known brands/NGOs/Government), company visit, media gathering	
8. Cost structure 1. CAPEX (machinery, land, and building) 2. Cost of goods sold 3. Operational expenses 4. Human resource 5. Operation cost			**9. Revenue Streams** 1. Sales of biodegradable products	

FRIM – Forest Research Institute Malaysia (FRIM); FRIM Inc. – Forest Research Institute Malaysia (FRIM) Incorporation; MESTECC – Ministry of Science, Technology, Environment & Climate Change (MESTECC); MADA – Muda Agricultural Development Authority; KADA – Kemubu Agricultural Development Authority; PKSM – Pertubuhan Kemajuan Sosial Malaysia; PBLS – Projek Barat Laut Selangor; IADP Pulau Pinang – Integrated Agricultural Development Project Pulau Pinang; KETARA – North Terengganu Integrated Agricultural; NGOs – Non-Governmental Organizations; CAPEX – capital expenditure or capital expense.

using three techniques, such as mechanical, chemical, or a combination of both chemical and mechanical processes. Pulp can then be used as the raw material for fabricated, paper, paperboard, and molded products. To conclude, the investment cost and start material cost of producing these products and their selling price are different. Molded products have the highest selling price compared to others and the investment is comparable to pulp, paper, and paperboard products combined. This is because molded products, especially those of the green packaging market, are mainly driven by growing public awareness and are concerned with sustainability and government regulations that have started focusing on the reduction of wastage and usage of biodegradable packaging materials. The commercialization of rice straw can improve, which will increase the farmer's income and reduce air pollution and deforestation as well as changing waste to wealth.

References

1. Sapuan SM, Ishak MR, Leman Z, Ilyas RA, Huzaifah MRM. Development of products from sugar palm trees (Arenga pinnata Wurb. Merr): A community project. INTROPica [Internet]. 2017;12–13. Available at: http://www.introp.upm.edu.my/upload/dokumen/20180914161643l3.5_Prof._Sapuan.pdf.
2. Abral H, Ariksa J, Mahardika M, Handayani D, Aminah I, Sandrawati N, et al. Transparent and antimicrobial cellulose film from ginger nanofiber. *Food Hydrocolloids*. 2020;**98**:105266. Available at: https://doi.org/10.1016/j.foodhyd.2019.105266.
3. Sapuan SM, Ilyas RA. Sugar palm: fibers, biopolymers and biocomposites. *INTROPica* 2017;5–7.
4. Nurazzi NM, Khalina A, Sapuan SM, Ilyas RA. Mechanical properties of sugar palm yarn/woven glass fiber reinforced unsaturated polyester composites: effect of fiber loadings and alkaline treatment. *Polimery* 2019;**64**(10):12–22.
5. Ilyas RA, Sapuan SM, Ibrahim R, Abral H, Ishak MR, Zainudin ES, et al. Effect of sugar palm nanofibrillated cellulose concentrations on morphological, mechanical and physical properties of biodegradable films based on agro-waste sugar palm (*Arenga pinnata* (Wurmb.) Merr) starch. *J. Mater. Res. Technol.* 2019 ;**8**(5):4819–30. Available at: https://linkinghub.elsevier.com/retrieve/pii/S2238785419308828.
6. Ilyas RA, Sapuan SM, Atiqah A, Ibrahim R, Abral H, Ishak MR, et al. Sugar palm (*Arenga pinnata* [Wurmb .] Merr) starch films containing sugar palm nanofibrillated cellulose as reinforcement: water barrier properties. *Polym. Compos.* 2020 ;**41**(2):459–67. Available at: https://onlinelibrary.wiley.com/doi/abs/10.1002/pc.25379.
7. Hazrol MD, Sapuan SM, Ilyas RA, Othman ML, Sherwani SFK. Electrical properties of sugar palm nanocrystalline cellulose, reinforced sugar palm starch nanocomposites. *Polimery* 2020;**55**(5):33–40.
8. Mazani N, Sapuan SM, Sanyang ML, Atiqah A, Ilyas RA. Design and fabrication of a shoe shelf from kenaf fiber reinforced unsaturated polyester composites. *Lignocellulose for Future Bioeconomy* Elsevier; 2019. p. 315–32. Available at: https://linkinghub.elsevier.com/retrieve/pii/B9780128163542000177.
9. Ilyas RA, Sapuan SM, Ibrahim R, Atikah MSN, Atiqah A, Ansari MNM, et al. Production, processes and modification of nanocrystalline cellulose from agro-waste: a review. In: *Nanocrystalline Materials* . IntechOpen; 2019. p. 3–32. Available at: http://www.intechopen.com/books/trends-in-helicobacter-pylori-infection/floating-drug-delivery-systems-for-eradication-of-helicobacter-pylori-in-treatment-of-peptic-ulcer-d.
10. Ilyas RA, Sapuan SM, Ishak MR. Isolation and characterization of nanocrystalline cellulose from sugar palm fibres (*Arenga pinnata*). *Carbohydr. Polym.* 2018;**181**:1038–51. Available at: http://dx.doi.org/10.1016/j.carbpol.2017.11.045.

11. Ilyas RA, Sapuan SM, Ishak MR, Zainudin ES. Development and characterization of sugar palm nanocrystalline cellulose reinforced sugar palm starch bionanocomposites. *Carbohydr. Polym.* 2018;**202**:186–202. Available at: https://linkinghub.elsevier.com/retrieve/pii/S014486171831066X.
12. Ilyas RA, Sapuan SM, Ibrahim R, Abral H, Ishak MR, Zainudin ES, et al. Thermal, biodegradability and water barrier properties of bio-nanocomposites based on plasticised sugar palm starch and nanofibrillated celluloses from sugar palm fibres. *J. Biobased Mater. Bioenergy* 2020;**14**(2):234–48. Available at: https://www.ingentaconnect.com/content/10.1166/jbmb.2020.1951.
13. Jumaidin R, Saidi ZAS, Ilyas RA, Ahmad MN, Wahid MK, Yaakob MY, et al. Characteristics of cogon grass fibre reinforced thermoplastic cassava starch biocomposite: water absorption and physical properties. *J. Adv. Res. Fluid Mech. Therm. Sci.* 2019;**62**(1):43–52.
14. Jumaidin R, Khiruddin MAA, Asyul Sutan Saidi Z, Salit MS, Ilyas RA. Effect of cogon grass fibre on the thermal, mechanical and biodegradation properties of thermoplastic cassava starch biocomposite. *Int. J. Biol. Macromol.* 2019 ; Available at: https://doi.org/10.1016/j.ijbiomac.2019.11.011.
15. Atikah MSN, Ilyas RA, Sapuan SM, Ishak MR, Zainudin ES, Ibrahim R, et al. Degradation and physical properties of sugar palm starch/sugar palm nanofibrillated cellulose bionanocomposite. *Polimery* 2019;**64**(10):27–36.
16. Atiqah A, Jawaid M, Sapuan SM, Ishak MR, Ansari MNM, Ilyas RA. Physical and thermal properties of treated sugar palm/glass fibre reinforced thermoplastic polyurethane hybrid composites. *J. Mater. Res. Technol.* 2019 ;**8**(5):3726–32. Available at: https://doi.org/10.1016/j.jmrt.2019.06.032.
17. Jumaidin R, Ilyas RA, Saiful M, Hussin F, Mastura MT. Water transport and physical properties of sugarcane bagasse fibre reinforced thermoplastic potato starch biocomposite. *J. Adv. Res. Fluid Mech. Therm. Sci.* 2019;**61**(2):273–81.
18. Aisyah HA, Paridah MT, Sapuan SM, Khalina A, Berkalp OB, Lee SH, et al. Thermal properties of woven kenaf/carbon fibre-reinforced epoxy hybrid composite panels. *Int. J. Polym. Sci.* 2019;1–8. Available at: https://www.hindawi.com/journals/ijps/2019/5258621.
19. Ilyas RA, Sapuan SM. The preparation methods and processing of natural fibre bio-polymer composites. *Curr. Org. Synth.* 2020;**16**(8):1068–70. Available at: http://www.eurekaselect.com/178429/article.
20. Sarkar N, Aikat K. Kinetic study of acid hydrolysis of rice straw. *ISRN Biotechnol.* 2013;**2013**:1–5. Available at: https://www.hindawi.com/archive/2013/170615.
21. Shafie SM, Masjuki HH, Mahlia TMI. Life cycle assessment of rice straw-based power generation in Malaysia. *Energy*. 2014
22. Huang Y-F, Lo S-L. Utilization of rice hull and straw. *Rice*. 2019.
23. World Health Organization. *Ambient Air Pollution: A Global Assessment of Exposure and Burden of Disease*. World Health Organization. 2016
24. El Damatty AA, Hussain I. An economical solution for the environmental problem resulting from the disposal of rice straw. In: *Appropriate Technologies for Environmental Protection in the Developing World – Selected Papers from ERTEP 2007*. 2009.
25. Jani SM, Rushdan I. Mechanical properties of beating pulp and paper from rice straw. *J. Trop. Agric. Food Sci.* 2016;**44**(1):103–9.
26. Rushdan I, Nurul Husna MH, Latifah J, Ainun Zuriyati M, Sharmiza A, Salmiah U, et al. A preliminary study on the effects of biopulping on pulp property of oil palm (Elaeis guineensis). Empty fruit bunches. In: *Seventh National Conference on Oil Palm Tree Utilisation OPTUC Strategizing for Commercial Exploitation*. 2008;1–8. Available at: www.biotek.gov.my/nbs2010/program/poster/ind/abstract/C18 Rushdan Ibrahim FRIM.pdf.
27. Hassan NHM, Muhammed S, Ibrahim R. Effect of soda-anthraquinone pulping conditions and beating revolution on the mechanical properties of paper made from *Gigantochloa scortechinii* (Semantan bamboo). *Malaysian J. Anal. Sci.* 2013;**17**(1):75–84.
28. Mossello AA, Harun J, Tahir PM, Resalati H, Ibrahim R, Fallah Shamsi SR, et al. A review of literatures related of using kenaf for pulp production (beating, fractionation, and recycled fiber). *Mod. Appl. Sci.* 2010;**4**(9). Available at: http://www.ccsenet.org/journal/index.php/mas/article/view/6110.

29. Hurmekoski E, Jonsson R, Korhonen J, Jänis J, Mäkinen M, Leskinen P, et al. Diversification of the forest industries: role of new wood-based products. *Can. J. For. Res.* 2018.
30. Buranov AU, Mazza G. Lignin in straw of herbaceous crops. *Ind. Crops Prod.* 2008.
31. Kadam KL, Forrest LH, Jacobson WA. Rice straw as a lignocellulosic resource: collection, processing, transportation, and environmental aspects. *Biomass Bioenergy* 2000;
32. Tovey A. Cash is king for the future, says UK banknote printer De La Rue. *Telegraph Media.* 2015 [cited 2020 January 9]. Available at: www.telegraph.co.uk/finance/newsbysector/industry/12061880/Cash-is-king-for-the-future-says-UK-banknote-printer-De-La-Rue.html.
33. Ministry of International Trade and Industry Malaysia. Malaysia International Trade and Industry Report 2011. Kuala Lumpur; 2011.
34. Ilyas RA, Sapuan SM, Atikah MSN, Asyraf MRM, Ayu Rafiqah S, Aisyah HA, Mohd Nurazzi N, Norrahim MNF. Effect of Hydrolysis Time on the Morphological, Physical, Chemical, and Thermal Behavior of Sugar Palm Nanocrystalline Cellulose (Arenga Pinnata (Wurmb.) Merr). *Textile Research Journal.* 2020. In print. doi: 10.1177/0040517520932393.

13

Sustainable Paper-Based Packaging

Latifah Jasmani[1], Z.M.A. Ainun[2], Sharmiza Adnan[1], Rushdan Ibrahim[1], S.M. Sapuan[2,4], and R.A. Ilyas[2,4]

[1]Pulp and Paper Laboratory, Biomass Technology Programme, Forest Products Division, Forest Research Institute Malaysia, Kepong, 52109 Selangor, Malaysia

[2]Program of Pulp and Paper & Pollution Control, Laboratory of Biopolymer and Derivatives, Institute of Tropical Forest and Forest Products (INTROP), Universiti Putra Malaysia (UPM), Selangor, Serdang, 43400, Malaysia

[3]Advanced Engineering Materials and Composites Research Centre (AEMC), Department of Mechanical and Manufacturing Engineering, Faculty of Engineering, Universiti Putra Malaysia (UPM), Selangor, Serdang, 43400, Malaysia

[4]Laboratory of Biocomposite Technology, Institute of Tropical Forest and Forest Products (INTROP), Universiti Putra Malaysia (UPM), Selangor, Serdang, 43400, Malaysia

13.1 Introduction

Packaging is one of the most important and fastest growing markets, worth US$799 billion in 2012 and expected at US$1 trillion in 2018 [1], with growth at 4% per year. The prime factors that drive the global demand for packaging material are due to increasing industrial production in developing regions, increased food trading such as fruits and vegetables, increased online shopping, increased global trade of consumer goods, and increased demand for flexible packaging [2]. Petroleum-based packaging is a very convenient form of packaging and has been widely used for many areas for a long time. The popularity and versatility of this plastic material, however, cannot outweigh its disposal problem since it is not biodegradable. The increasing amount of plastic wastes in landfill and its threat to

Bio-Based Packaging: Material, Environmental and Economic Aspects, First Edition. Edited by S.M. Sapuan and R.A. Ilyas

the environment ecosystem prompts authorities to gradually reduce the use of plastic and search for alternatives. Paper-based packaging is seen as one of the alternatives to plastic as it is made from renewable and sustainable material and, most importantly, it is biodegradable. The pulp and paper industry is the downstream industry or the end user of the forest sector. Figure 13.1 shows the consumption of paper worldwide in 2016 by region. Asia was the largest paper consumer followed by Europe and North Africa. The growth on sustainable packaging is promisingly to increase due to factors from government regulation and policy and also demands from customers and consumers.

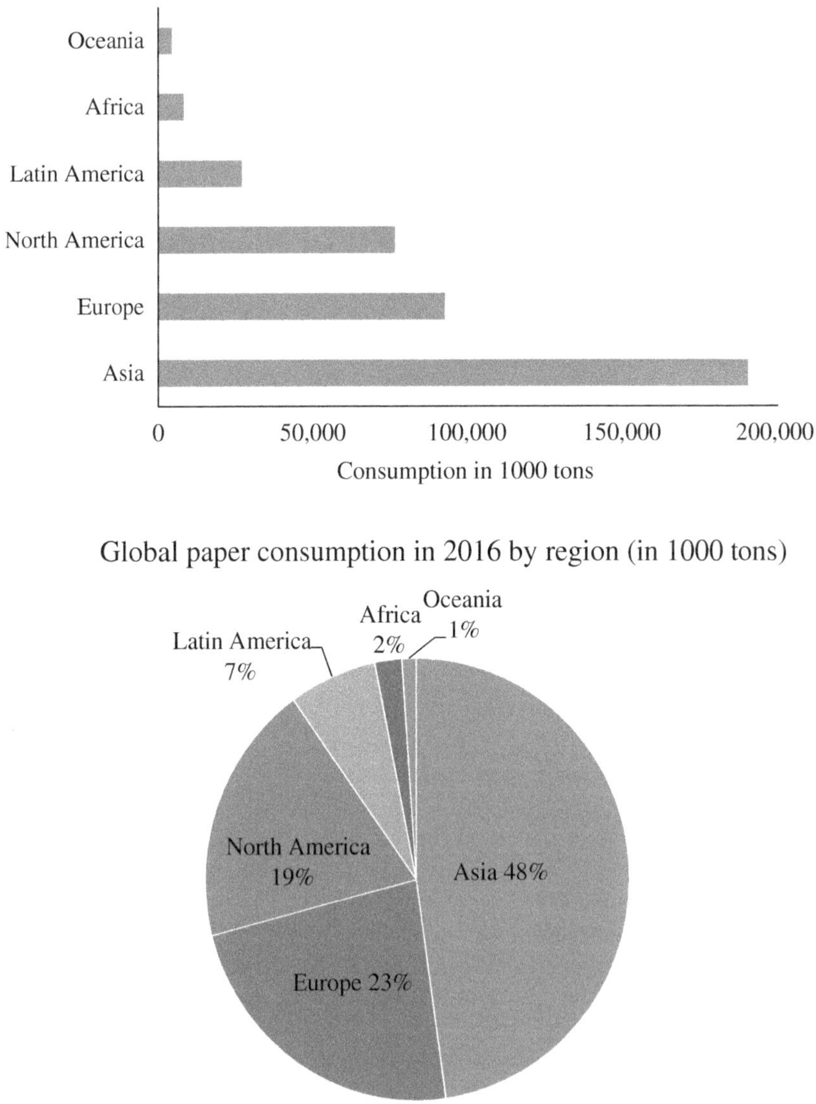

Figure 13.1 *Worldwide consumption of paper by region in 2016. Source: From Ref. [3]*

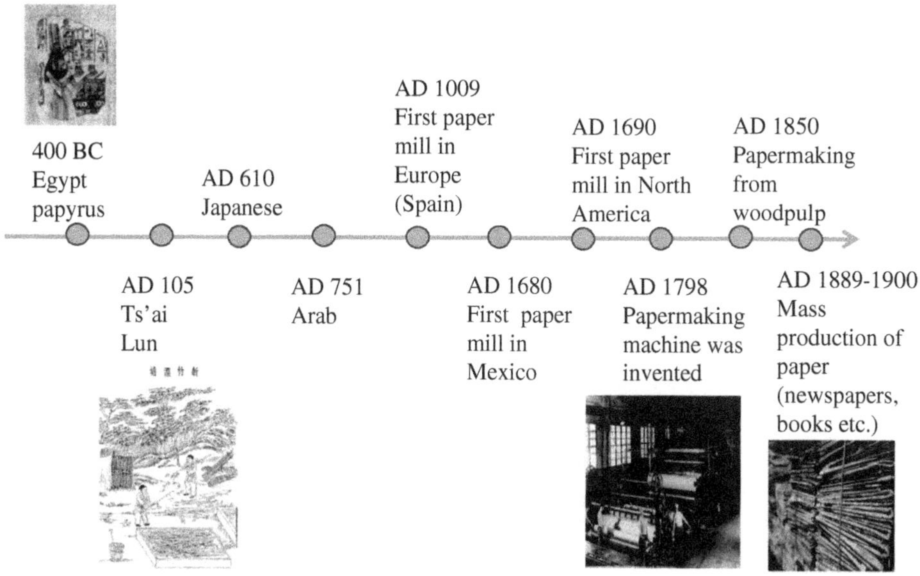

Figure 13.2 *Development of paper technology.*

Paper is described as a thin layer of cellulosic plant fibers produced on a screen by dewatering an aqueous suspension of fibers on a fine screen. The word "paper" actually originates from papyrus, the plant that ancient Egyptians used as the world's first writing object [4]. The paper they used then was produced by just beating and pressing the stem of the plant into a thin sheet. However, the paper readily available today was first developed by a Chinese dynasty, which became the most important information medium during that era. It is an indication of a positive development in education through which knowledge can be transferred to society more efficiently. Since then paper evolved to the Arab continent, followed by Europe, and has gone through a modernization phase until today (Figure 13.2).

Worldwide consumption of paper and paperboard has always been on an increasing trend. A significant amount of this is now produced for packaging (Figures 13.3 and 13.4). The total market size has been valued at about US$400 billion in 2012. Interestingly, about 30% or one-third of global packaging is led by paper and paperboard [2]. Rigid plastic is the second most used for packaging followed by glass, metal, flexible plastic, and cans. Paper-based packaging is widely used in various forms whenever products are manufactured, transferred, distributed, and consumed/used. It is usually in proportion with manufacturing industry growth.

13.2 Types of Raw Material for Paper-Based Packaging

13.2.1 Source of Fiber

There are many sources of fiber for paper manufacturing purposes, as shown in Figure 13.5. Production of paper uses mostly fibrous raw material that can be further classified into primary or secondary fiber. Primary fiber refers to virgin fiber from wood, non-wood, or

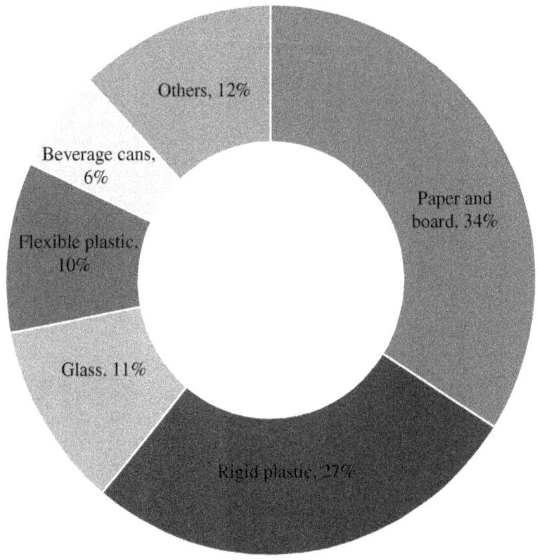

Figure 13.3 *Global consumer packaging by type. Source: Replotted from Ref. [2].*

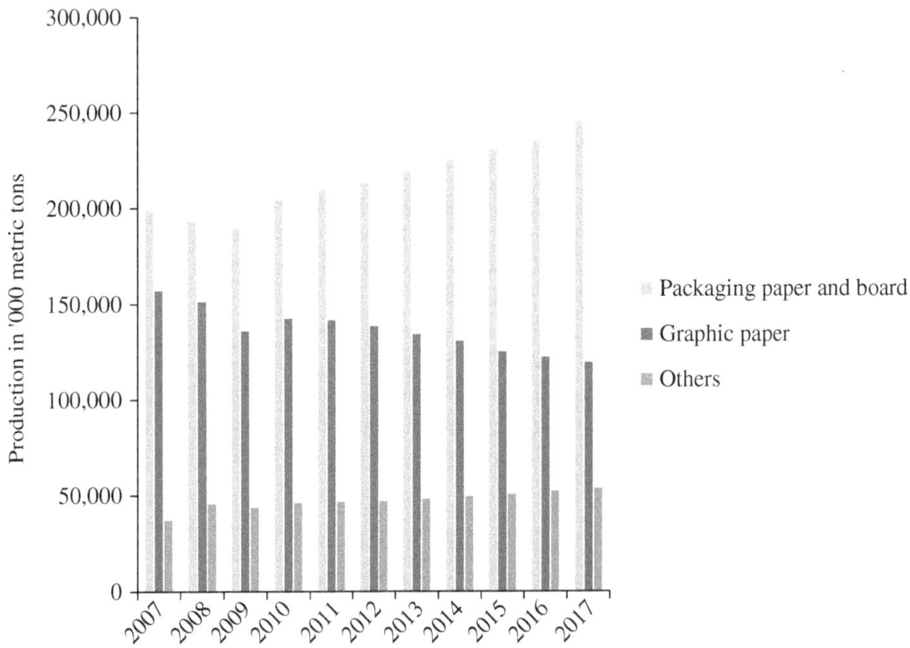

Figure 13.4 *Global paper production volume. Source: From Ref. [3]*

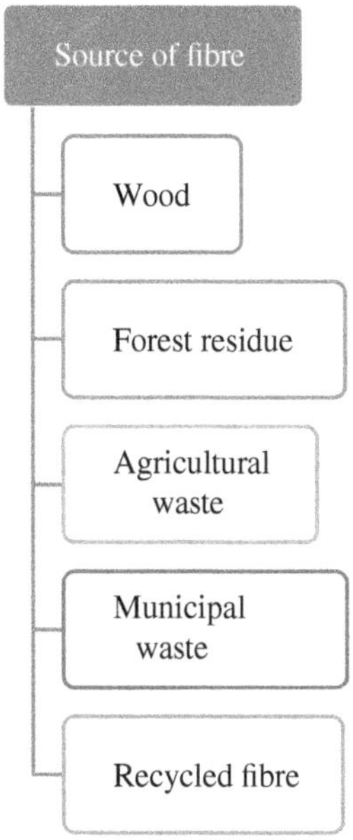

Figure 13.5 *Source of fiber.*

agricultural residues. Wood is the main raw material for the pulp and paper industry and it is globally estimated that around 97% of the world's paper and board is produced from wood pulp. Furthermore, the use of thinning material and sawmill waste is also common. Non-wood or agricultural residues are used in areas where the supply of wood is limited. On the other hand, secondary fibers are derived from recovered paper and board. In 2001, the use of primary fiber accounts for 55% whereas the remaining 45% is from secondary fibers [5].

The properties of fiber are affected by the species, age, and location of the tree. Their physical and chemical properties influence the ability to form a good bonding network during papermaking. Thus, the selection of raw materials for packaging takes into account this property to produce a good quality of paper. Long fiber provides strength and toughness to the paper whereas short fibers provide a smooth surface and low-density paper. Coniferous trees such as spruce, fir, and pine are known for their long fiber property whereas short fiber species usually come from deciduous trees such as acacia, eucalyptus, birch, and many other hardwood species. The blending of two fibers sometimes is essential for a specific end product. In the pulp and paper industry, long fiber refers to fibers having a length between 3 and 4 mm whereas short fiber refers to a length between 1 and 2 mm.

As the pulp and paper industry mainly uses wood as their raw material, their mill is usually adjacent to the source, i.e. a plantation forest. A plantation forest is a designated area of land for growing forest species that will be used for various downstream industrial needs. Common forest species used for the paper industry include spruces, firs, acacia, eucalyptus, birch, and many others. In areas or countries where wood supply is scarce and land for forest plantation is limited, agricultural residues can be considered for potential

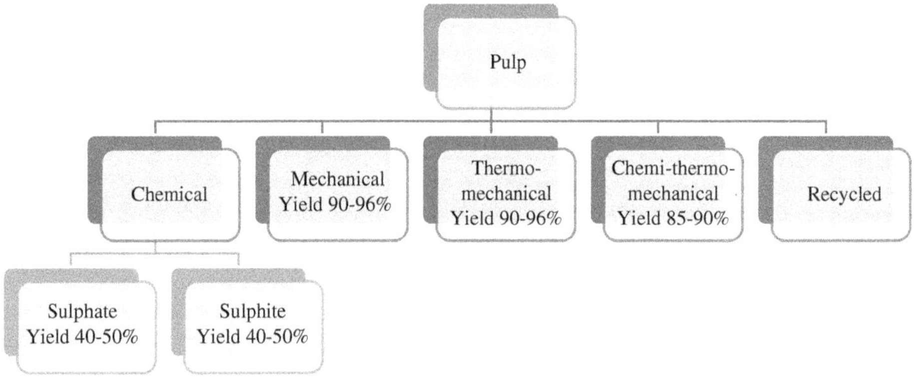

Figure 13.6 *Types of pulp and their yield.*

feedstock. For example, rice straw, bagasse, and oil palm empty fruit bunches have been used for packaging products such as molded products or liner in Malaysia.

13.2.2 Types of Pulp

There are various types of pulp that can be produced to cater for specific packaging needs. Figure 13.6 shows the different types of pulp that can be produced using different equipment and pulping conditions.

13.2.2.1 Chemical Pulp

Chemical pulp is obtained by chemical pulping of wood or non-wood. Chemical pulp is the major type of pulp produced worldwide (Figure 13.7). The removal of lignin is the main objective of pulping; the term used in pulp and paper to separate the fiber from wood. Lignin usually makes up to 30–35% of wood depending on the species. During chemical pulping, most of the lignin is removed and the pulp yield is between 45% and 55%. There are two types of chemical pulping currently in practice by the industry, namely sulfate and sulfite pulping. Sulfate pulping requires the use of sodium sulfide and sodium hydroxide whereas the sulfite process is based on the use of calcium, magnesium, and sodium bisulfide. The world production of chemical pulp, however, is dominated by the sulfate process due to its alkaline condition process and also its higher strength pulp compared to sulfite pulp. After pulping, the pulp may undergo further lignin removal via the bleaching process if pulp is to be used for high-quality printing and writing paper. Bleaching can be categorized into elemental chlorine free bleaching (ECF) and total chlorine free bleaching (TCF) processes. The development of this bleaching method is in response to the carcinogenic chemical byproducts, i.e. dioxin generated by conventional bleaching using elemental chlorine. The ECF process uses chlorine dioxide, oxygen, and peroxide whereas TCF only utilizes oxygen and hydrogen peroxide. This process produced harmless byproducts.

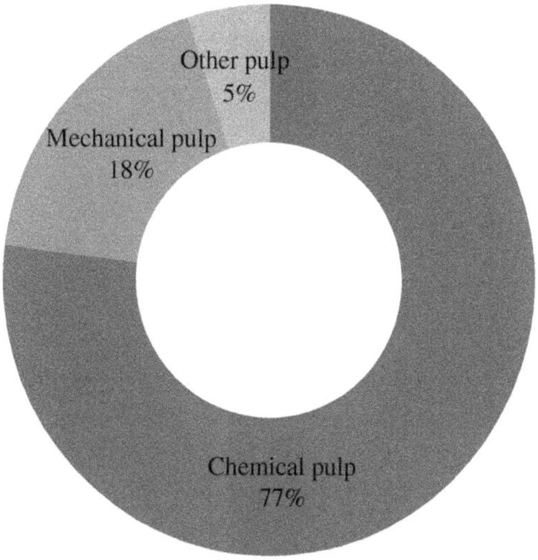

Figure 13.7 Global pulp production by grade. Source: From Ref. [3].

13.2.2.2 Mechanical Pulp

Mechanical pulping involves the use of mechanical action on wood chips via grinding or refining to defibrate the fiber from wood. The lignin is not removed from the fiber as it has been only plasticized. The lignin that remained in the fibers results in a high yield of pulp, which can go up to 97% depending on the method used. There are different processes that can be used to produce pulp mechanically. For instance, refiner mechanical pulp uses a refiner disc to crush the wood chip in the opposite direction and the refining is carried out in two stages. The stone groundwood process uses a grinding stone. Besides individual fibers, groundwood pulp contains mostly fiber bundles, broken fibers, and fines [6]. Thermomechanical pulp applies steam on wood chips prior to refining under high pressure. Pre-treatment using chemicals is commonly conducted prior to refiner mechanical pulping (RMP) or thermomechanical pulping (TMP) to improve the strength of the mechanical pulp. Mechanical pulp has lower strength properties compared to chemical pulp as most of the lignin are still present. The presence of lignin makes the fiber stiff and thus weaker paper is produced from mechanical pulp. It is usually referred to as wood-containing pulp and is used in newspaper or applications that do not requires high strength pulp. Mechanical pulp can be bleached but the lignin is modified instead of solubilized, which means that the lignin is still conserved in the pulp. The brightness, however, is not stable and therefore after some time the pulp will become yellow due to the ultraviolet (UV) exposure reaction. Table 13.1 shows the different properties of the chemical and mechanical pulp.

13.2.2.3 Recovered Paper

Recycled fiber has become an important and exclusive raw material for the pulp and paper industry as they use it the most. This is due to the introduction of a circular economy where

Table 13.1 Difference between chemical and mechanical pulp.

	Chemical pulp	Mechanical pulp
Brightness	High	Low
Yellowing	Low	High
Stiffness	High	Low
Energy	Generate energy	Energy intensive
Chemicals	High but can be recycled	Low
Strength	High	Low
Surface	Smooth	Rough

zero waste and recycling activity are practised in society nowadays. Thus, the growth of recovered paper keeps on increasing every year and even gradually surpassed the annual production of paper. Recovered paper can also be classified into different grades of chemical or mechanical pulp or usually an undefined mixture of both. For example, high grade/high price recovered pulp usually comes from white shavings, medium grade is from newspaper, whereas low grade is from paper and board. Prior to its usage in various end products, recovered paper must be subjected to sorting and repulping by agitation in water using a hydrapulper. Every repulping reduces the quality of the fibers in terms of its length and bonding ability. Thus, the introduction of primary fiber or virgin fiber must be done continuously to maintain the quality and quantity [7]. Figure 13.8 shows the recycling system in Malaysia to obtain recovered paper.

13.2.2.4 Non-fiber Material

Besides fibers, there are other materials known as non-fiber additive added into paper. This makes up to a maximum of 30% of the paper. The main component of paper, i.e. 70% to 80%, must be the fiber (reference). Non-fiber additives have functions to improve the performance of the product or increase the productivity. The examples of various non-additive fiber with their function are strength additive for strength enhancement, surface sizing for water repellant, mineral pigment for surface coating, fillers to improve surface printability or brightness or smoothness, and also chemicals to improve process productivity.

13.3 Papermaking

The principle of papermaking involves three stages, namely preparation of fiber suspension in water, formation of papersheet through the fiber network, and lastly removal of water by drainage, pressure, and drying. Further surface treatment on the paper is commonly required depending on the desired end product, such as incorporation of functional additive, surface sizing, or adhesive. In packaging applications, a sizing agent or coating material is applied to act as a barrier material to protect against water, gas, or vapor.

13.4 Types of Paper-Based Packaging

Paper-based packaging can be divided into two major components, namely box-type and paper sheet-type (Figure 13.9).

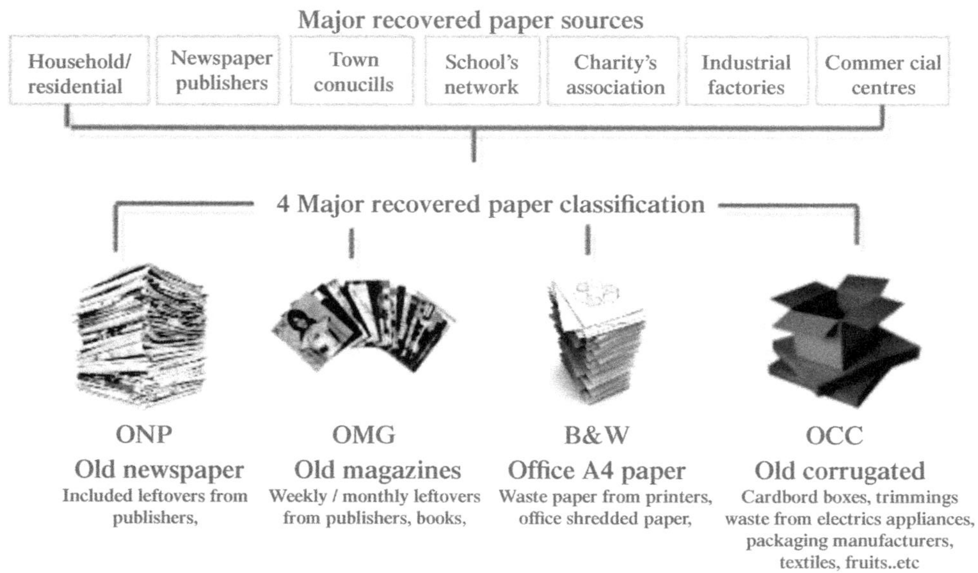

Figure 13.8 *Recycling system in Malaysia [8].*

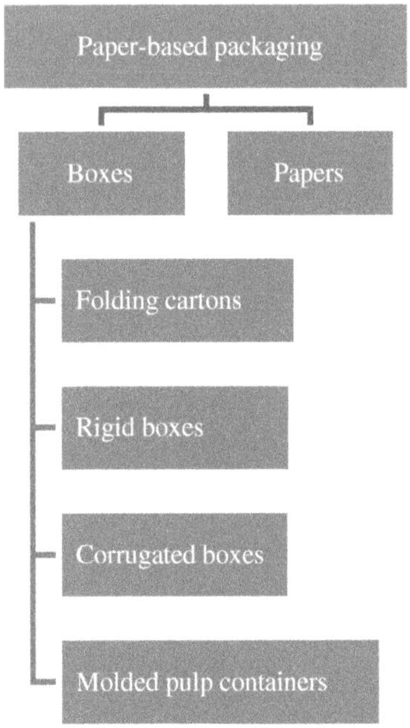

Figure 13.9 Types of paper-based packaging.

13.4.1 Boxes

In packaging, there are four common types of boxes, listed as (i) folding cartons, (ii) rigid boxes, (iii) corrugated boxes, and (iv) molded pulp containers. A packaging process of a certain product may involve the usage of all types of boxes as it travels from the manufacturer to the consumers.

13.4.1.1 Folding Cartons

Folding cartons are a type of box that is made from sheets of paperboard. The thickness is between 300 μm and 1100 μm that have been cut and counted for bending into desired shapes. In other words, folding cartons are also known as paperboard cartons or paperboard boxes. These boxes are the most popular type of paper-based packaging being used in retail stores or marketplaces such as Coated Solid Bleached Board (SBB) or Solid Bleached Sulfate (SBS), Coated Solid Unbleached Board (SUB) or Solid Unbleached Sulfate (SUS), Folding Box Board (FBB), and White Lined Chipboard (WLC). The first records of paper in terms of folding cartons being used to carry liquids on a commercial scale were found in 1908 by Dr. Winslow. He remarked on paper milk containers that were invented and sold in San Francisco and Los Angeles by G.W. Maxwell in 1906.

13.4.1.2 Rigid Boxes

A rigid box is a type of paper-based packaging that is also known as a set-up box. It has a sturdier and tougher formation compared to folding cartons. In practice, it is usually applied for high-end products that need to support heavy goods.

13.4.1.3 Corrugated Boxes

Corrugated board, corrugated fiberboard, combined board, cardboard boxes, or brown cardboard boxes are other names for corrugated boxes. In addition, transporters always refer to corrugated boxes as shipped boxes or master packs, which are boxes shipped all over the world. Kinds of boxes like corrugated fiberboard will involve layers of paper in order to deliver the flexibility and strength throughout the process of consignment. For instance, the corrugated fiberboard has three layers of paper, with a fluted layer sandwiched between two outer flat layers. The strength of the corrugated fiberboard depends to a great extent on the fluted layer. Therefore, such a corrugated fiberboard is very significant in applications such as stand-up displays.

13.4.1.4 Molded Pulp Containers

Molded pulp is a single-use packaging material manufactured from cellulosic pulp slurries and formed into discrete products using screened molds [9]. Molded pulp has found its use in the packaging field since the 1930s but the use declined in the 1970s due to the introduction of plastic foam packaging. The market has recovered recently due to environmental awareness. This kind of container is very practical in packaging bottled wines and spirits. The packaging can be packed from top to bottom with bottles in cartons. In addition, such containers have widespread use to pack eggs and fruits. Food trays of various shapes and sizes can also be formed to meet consumer needs that may substitute for ceramic plates. These types of paper-based food trays are very convenient for outdoor occasions due to their ease of handling, biodegradability, and environmental friendly characteristics. Moreover, such containers are designed to have dual ovenability of both microwave and convection ovens. To be more consumer friendly, the molded pulp container may be coated with thin thermoplastic films like polyethylene (PET) on top of the surface for safety, security, and appearance purposes. Table 13.2 shows the properties of molded products.

13.4.2 Paper Sheet

There are many types of paper sold in retail stores and the marketplace for packaging purposes. Kraft paper and bleached paper are produced before converting them into ready-to-use packaging paper for foods or goods. The converted versions of paper-based packaging under paper sheet types are listed as (i) greaseproof paper, (ii) glassine paper, (iii) vegetable parchment, (iv) waxed paper, and (v) decorative paper (Figure 13.10).

13.4.2.1 Greaseproof Paper

Greaseproof paper is a translucent paper that has been hydrated to provide oil and grease resistance as well (Figure 13.11). This kind of paper is often used for packaging butter and similar fatty foods since they resist the penetration of fat for a reasonable period. In order to produce greaseproof paper, a physical phenomenon known as hydration should be created. This can be carried out by having a prolonged beating in order to fibrillate the cellulose fibers. By applying this action, the cellulose fibers allow a large amount of water absorption. Therefore, the fibers would be gelatinized and sticky.

Table 13.2 *Properties of molded product.*

	Thick wall	Transfer	Thermoformed	Processed
Thickness	5–10 mm	3–5 mm	2–4 mm	
Source	Recycled paper	Recycled newspaper	Virgin fibers	
Molded process	One single mold	One forming mold and one transfer mold	Multiple heated molds, dried in the mold	Molded fiber products that require some type of secondary or special treatment like additional printing, coatings, or additives
Surface property	One side smooth, another side rough	Both sides smooth, better dimensional accuracy	Good dimensional accuracy, smooth rigid surface, resemble thermoformed plastic material	
Typical application	Non-fragile and heavy items	Egg trays or electronic equipment	Health and beauty, medical and scientific products	

Source: Retabulated from [10].

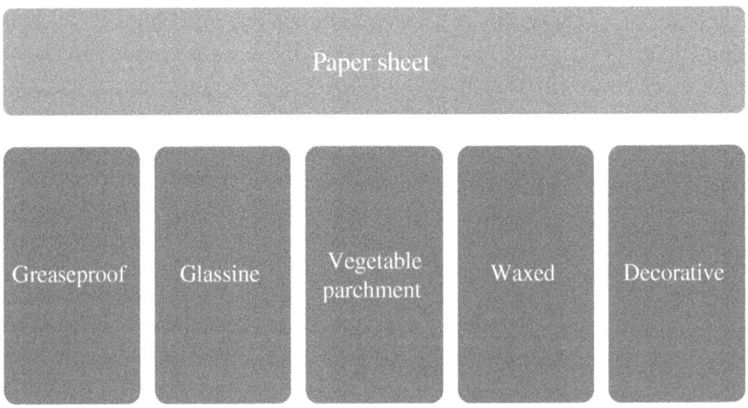

Figure 13.10 *Types of paper sheet for packaging.*

13.4.2.2 Glassine Paper

Glassine paper has a glassy, smooth surface, high density, and transparency properties (Figure 13.12). Glassine paper is prepared by further treating greaseproof paper in a super-calendar machine. This results in intimate interfiber hydrogen bonding where the refractive index approaches a value of 1.02 of amorphous cellulose. Hence, this indicates very few pores or other fiber/air interfaces that exist for scattering light or allowing liquid penetration.

Figure 13.11 Greaseproof paper.

Figure 13.12 Glassine paper.

13.4.2.3 Vegetable Parchment

Vegetable parchment is a type of paper-based packaging dedicated to pack vegetables (Figure 13.13). It obtains the name vegetable parchment from its physical similarity to animal parchment (made from animal skins). Due to its grease resistance and wet strength, vegetable parchment can easily strip from food or materials. Therefore, it is applied as an interleaver between slices of food or materials like meat or pastry. Labels and inserts in paper-based packaging products equipped with high oil or grease are regularly made from

Figure 13.13 Vegetable parchment paper.

parchment. Additionally, it can be treated with mold inhibitors and used to wrap foods such as cheese. The early process for producing parchment paper was developed in the 1850s. The process involves passing a web of high-quality and unsized chemical pulps via a bath of concentrated sulfuric acid. The cellulosic fibers swell and partially dissolve, filling the interstices between the fibers and resulting in extensive hydrogen bonding. Thorough washing in water, followed by drying on conventional papermaking dryers, causes reprecipitation and consolidation of the network. A paper that is stronger wet (even in boiling water) than dry, free of lint, odor and taste, and resistant to grease and oils is successfully produced [9].

13.4.2.4 Waxed Paper

Waxed paper is a type of paper-based packaging that provides a barrier against penetration of liquids and vapors (Figure 13.14). Many base papers are suitable for waxing, including greaseproof and glassine papers. There are three major types of waxed paper, namely (i) wet-waxed, (ii) dry-waxed, and (iii) wax-laminated. Wax that is added at the beater during the papermaking process will have the least amount of wax and therefore give the least amount of protection compared to other coating applications. Wet-waxed papers have a

Figure 13.14 Waxed paper.

Figure 13.15 Decorative paper.

continuous surface film on one or both sides, which is achieved by shock-chilling the waxed web immediately after application of the wax. This also imparts a high degree of gloss on the coated surface. Dry-waxed papers are produced using heated rollers and do not have a continuous film on the surfaces. Consequently, exposed fibers act as wicks and transport moisture into the paper.

Wax-laminated papers are bonded with a continuous film of wax that acts as an adhesive. The main purpose of the wax is to provide a moisture barrier and a heat-sealable laminant. Frequently, special resins or plastic polymers are added to the wax to improve adhesion and low-temperature performance, and to prevent cracking as a result of folding and bending of the paper.

13.4.2.5 Decorative Paper

Decorative paper is usually used in wrapping presents, which have a high demand, especially by end users at the retail stores and marketplace (Figure 13.15). Many types of such paper can be found at different grammages and thickness ranges, depending on the consumer demand and modern taste. Some of the decorative papers are able to be printed or written as well. The papers can be produced via machine or manual approaches. The machine-made decorative papers allow various pattern and colors that are very attractive. However, due to environmental friendly awareness, decorative paper for wrapping purposes can also be produced manually by using a handsheet self-made machine that can be carried out individually. Another name for man-made decorative paper is craft paper.

13.4.3 Using Types of Paper-Based Packaging

These types of paper-based packaging are dominated by two most important industries: (i) food and beverages industries and (ii) transportation industries.

13.4.3.1 Food and Beverages Industries

In modern lifestyles, a lot of food and beverages are commercialized, which involves dry food products, frozen foods, liquids, fast foods, and fresh products as well. In the

Figure 13.16 *Paper-based packaging for food and beverage.*

food industry, paper is usually used in corrugated medium boxes, wrapping papers, small–medium sized bags and sacks, milk cartons, folding cartons, tissue paper, and paper plates and cups (Figure 13.16). These papers are included in boxes and paper types of paper-based packaging, discussed earlier.

13.4.3.2 Transportation Industries

In transportation industries, packing paper like boxes, bags, and sacks is a type of paper used for packaging goods that usually are made from unbleached or bleached kraft paper having high strength (Figure 13.17). The crucial characteristics are paper physical and mechanical properties like liquid absorption, roughness, tensile strength, burst strength, tear strength, and folding endurance. However, the most highlighted characteristic is paper mechanical properties because transportation involves long distance and countless conditions that may influence the paper itself, followed by its content. Once the packaging material is having minor or major damage, the contents may be highly interrupted in terms of condition, safety, security, etc.

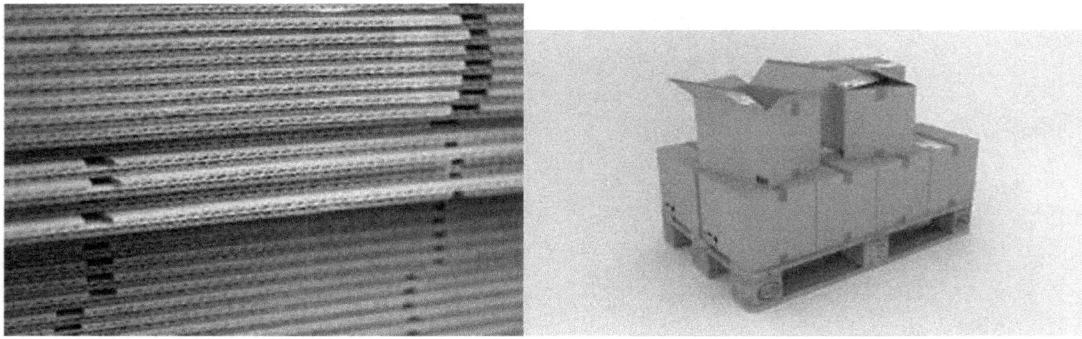

Figure 13.17 *Packaging for transportation.*

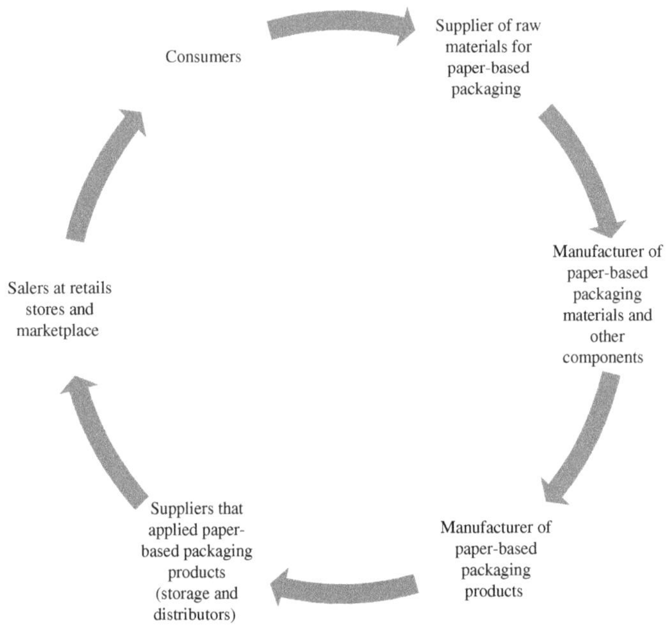

Figure 13.18 Packaging requirements for consumers.

13.5 Packaging Requirement for Paper-Based Packaging

The main purpose of packaging is to protect the product along the way before reaching the consumer. The way involves a chain of important players, as listed in Figure 13.18. Packaging is one of the important processes in food and beverages industries, as also is transportation, which should be capable of preventing deterioration of the food and beverages, or goods, from environmental influences during storage and transportation [11, 12].

13.5.1 Physical and Mechanical Characteristics of Paper

There are various specific tests that need to be done on paper in order to meet packaging requirements. Figure 13.19 shows the common test carried out on paper and paperboard. However, the list is not exhaustive as there might be other specific tests required for specific packaging needs.

Paper has a grain direction, known as machine direction (MD), and a cross machine direction (CD). In measuring paper physical and mechanical properties, these grain directions must be taken into account. In practice, MD has lesser variations compared to CD in terms of paper properties.

13.5.2 Other Requirements

Recent findings declared that the role of packaging has changed not only to defend the products but also as one of main criteria to attract and increase sales [13]. Overall, the purpose

Physical	Strength	Optical
• Grammage • Bulk and density • Thickness • Moisture content	• Tensile • Burst • Tear • Water absorption • Fold • Ring crush • Concora crush • Flat crush	• Brightness • Whiteness • Opacity • Colour

Figure 13.19 *Common tests for paper and board.*

of packaging is listed as (i) to protect products, (ii) to identify products by consumers, (iii) to ease delivery, (iv) to safeguard products, (v) to satisfy consumers, (vi) to promote, (vii) to influence buying behavior of an individual consumer, (viii) to market a communication purpose, and (ix) to catch a consumer's perception [13]. In addition, demands also come from retailers who request and require longer shelf lives in terms of food and beverage products. Hence, smart packaging technologies need to be enhanced from time to time [14, 15].

References

1. Pira S. 2017 Available at: www.smithers.com.
2. Woods T. Packaging demand surge. *Appita J.* 2016;**69**(4):284–6.
3. Statista. 2016 Available at: www.statista.com.
4. Tschudin P. (ed.) *The Invention of Paper*. IPH Congress Book; 1994.
5. Kirwan M. Paper and paperboard packaging. *Food Packaging Technology*. 2003;**241**.
6. Robertson GL. *Food Packaging: Principles and Practice*: CRC press; 2005.
7. Kirwan MJ. *Paper and Paperboard Packaging Technology*: Wiley; 2008.
8. MPPMA. Report of Malaysia Paper Industry. 2012.
9. Yam KL. *The Wiley Encyclopedia of Packaging Technology*: Wiley; 2010.
10. Su Y, Yang B, Liu J, Sun B, Cao C, Zou X, et al. Prospects for replacement of some plastics in packaging with lignocellulose materials: a brief review. *BioResources* 2018;**13**(2):4550–76.
11. Kelsey RJ. *Packaging in Today's Society*: Technomic Publishing Co., Inc., Lancaster, PA; 1989. 139 p.
12. Restuccia D, Spizzirri UG, Parisi OI, Cirillo G, Curcio M, Iemma F, et al. New EU regulation aspects and global market of active and intelligent packaging for food industry applications. *Food Control* 2010;**21**(11):1425–35.
13. Shah S, Ahmad A, Ahmad N. Role of packaging in consumer buying behavior. *Int. Rev. Basic Appl. Sci.* 2013;**1**(2):35–41.
14. Kerry JP. New packaging technologies, materials and formats for fast-moving consumer products. *Innovations in Food Packaging*: Academic Press; 2014. p. 549–84.

15. Ilyas RA, Sapuan SM, Atikah MSN, Asyraf MRM, Ayu Rafiqah S, Aisyah HA, Mohd Nurazzi N, Norrahim MNF. Effect of Hydrolysis Time on the Morphological, Physical, Chemical, and Thermal Behavior of Sugar Palm Nanocrystalline Cellulose (Arenga Pinnata (Wurmb.) Merr). *Textile Research Journal*. 2020. In print. doi: 10.1177/0040517520932393

14

Properties and Food Packaging Application of Poly-(Lactic) Acid

N.H Sari[1], S. Suteja[1], S.M Sapuan[2,3], and R.A Ilyas[4,5]

[1]*Department of Mechanical Engineering, University of Mataram, West Nusa Tenggara, Indonesia*

[2]*Advanced Engineering Materials and Composites Research Centre (AEMC), Department of Mechanical and Manufacturing Engineering, Faculty of Engineering, Universiti Putra Malaysia, Serdang, Selangor, 43400, Malaysia*

[3]*Laboratory of Biocomposite Technology, Institute of Tropical Forestry and Forest Product (INTROP), Universiti Putra Malaysia, Serdang, Selangor,43400, Malaysia, Malaysia*

[4]*School of Chemical and Energy, Faculty of Engineering, Universiti Teknologi Malaysia,81310 UTM Johor Bahru, Johor, Malaysia*

[5]*Centre for Advanced Composite Materials, Universiti Teknologi Malaysia,81310 UTM Johor Bahru, Johor, Malaysia*

14.1 Introduction: Background and Driving Forces

Poly(lactic) acid (PLA) is one kind of biodegradable thermoplastic polymer that is usually produced from renewable sources, mainly from starch [1]. Recently, PLA was mainly used in packaging as well as biomedical applications. Since PLA exhibits excellent properties, the use of PLA has been considered to extend to the sectors of agriculture, building, electronics and electrical appliances, and houseware [2].

There are three methods used to produce PLA, namely: (i) direct polycondensation of lactic acid, where the molecular weight of PLA produced is lower than 1.6 × 104, (ii) condensing azeotropic dehydration, where the PLA produced has a molecular weight reaching 15.400 and a yield of 89%, and (iii) the polymerization of nine ring openings to produce PLA with a high molecular weight from 2 × 104 to 6.8 × 105 [3].

Bio-Based Packaging: Material, Environmental and Economic Aspects, First Edition. Edited by S.M. Sapuan and R.A. Ilyas

Today, the food industry is growing rapidly and is increasingly varied, so we need a package that is able to store and maintain the quality of food. Plastic packaging is the most widely used by industry because of its low price, good mechanical properties, easy to make, transparent, and easily shaped as desired [4]. However, plastic waste is difficult to decompose by soil microorganisms and ultimately causes environmental pollution. Increased public awareness of the environment and global warming issues have demanded people to use environmentally friendly packaging. PLA has advantages compared to plastic, namely: (i) biodegradable, can decompose naturally, (ii) biocompatible, can be accepted by biological cells, (iii) produced from renewable materials, and (iv) 100% recyclable, where lactic acid hydrolysis can be reused to produce other products. In addition, this package can be decomposed by microbial enzymes and can reduce the use of petroleum as a raw material for synthetic plastics [5, 6]. Despite its several advantages, commercially available PLA has many inherent weaknesses, in particular its low toughness, low heat resistance, and brittleness that prevent it from being widely adopted for durable applications. More importantly, PLA has poor melt processability due to a narrow processing window and a very slow crystallization rate. Typically, a higher crystallinity is desirable in finished products due to its strong influence on mechanical and thermal properties. Several strategies exist to overcome the slow crystallization rate and melt processability of PLA, such as the addition of fillers [7, 8], copolymerization [9, 10], and melt-blending [10–12], all of which offer an effective medium for enhancing its overall performance. Among these, blending with inherently toughened bio-polymers is one of the most effective solutions. Examples of this are mixing PLA with poly(D, L-lactic acid) (PDLA) with a comparison of different mixed materials results in different properties [13]. PLA can form a misible mixture with polyethylene glycol (PEG) [14], polyethylene oxide (PEO) [15], and polybutylene succinate (PBS) [16], etc.

To expand the PLA application, it is very important for us to understand the various properties and applications of this food packaging application. Properties such as transition and melt temperature, thermal, mechanical, crystallinity, degradation, optical, and barrier properties must be understood in more detail. Therefore, this chapter will focus on the properties and food packaging application of PLA. This chapter also reviews the development of PLA modified films that are used for packaging applications.

14.2 Properties of PLA

In this chapter, the PLA transition and melt temperature, physical, crystallinity, mechanical, thermal, optical, degradation, grease permeability, water resistance properties, and barriers are studied.

14.2.1 Melt and Transition Temperature

The melt temperature (T_m) for PLA ranges from 150°C to 162°C (see Table 14.1). In the solid state, PLA can be either amorphous or semi-crystalline, depending on the stereochemistry and thermal history. T_g and T_m are strongly affected by the overall optical composition, thermal history, primary structure, and M_w. Above T_g, amorphous PLAs transition from glassy to rubbery and will behave as a viscous fluid upon further heating [17].

The melt enthalpy estimated for pure PLA of 100% crystallinity from 93 Jg^{-1} to 14 t Jg^{-1}) [18, 19]. The T_m and degree of crystallinity are dependent on the molar mass, thermal history, and purity of the polymer. An optical purity of at least 72–75% corresponds to

Table 14.1 Isothermal crystallization and melting behaviors of PLA composites at 105 °C [17].

Samples	Crystallization $t_{1/2}$ (min)	Melting		
		T_m (°C)	ΔH_m (J g^{-1})	X_c (%)
PLA0	6.41	161/170	36.42	38.9
PLA1	6.45	159/169	41.70	45.9
PLA2	5.89	158/168	43.28	47.2
PLA3	2.54	158/168	44.65	48.2
PLA5	1.27	156/166	44.46	50.0

about 30 isolactic–lactyl units [20]. For the formation of long isotactic sequences, a PLA crystallizes at as low as 43% optical purity. The high T_m is the result from cocrystallization of enantiomerically pure D and L PLA, leading to a stereo complex formation. The T_m of blended poly(L-lactic acid) (PLLA) and poly(D-lactic acid) (PDLA) is about 50°C higher than for the pure enantiomeric polymers.

The transition temperature (T_g) can be very important when studying mechanical properties, as seen in Table 14.1.These properties are always obtained in standard conditions (comparable to room temperature), T_g decreases at higher temperatures, and there may be a limit above which mechanical properties may degrade drastically [21]. Below T_g, PLA behaves as a glass with the ability to creep until cooled to its β-ransition temperature of approximately −45°C. Below this temperature, PLA will only behave as a brittle polymer [22]. PLA with an MW of 22 000 g mol^{-1} has a T_g of 55°C, which is only 4–5°C lower than that predicted for PLA of infinite MW [13]. For amorphous PLAs, the transition temperature (T_g) determines the upper use temperature for most commercial applications. For semi-crystalline PLAs, T_g sebesar ~58°C, and T_m sekitar 130–230°C, depending on the structure [21].

14.2.2 Crystallinity

Slow crystallization is a huge drawback of PLA which greatly limits its production efficiency. The crystallization rate of PLA can be improved by adding nucleating agent (NA), which can reduce the surface free energy barrier toward nucleation and induce PLA to crystallize at a higher temperature [23, 24]). NAs for PLA are commonly divided into three categories [25], i.e. inorganic nucleating agent (INA), organic nucleating agent (ONA), and macromolecular nucleating agent (MNA). INAs such as clay, talc, and sepiolite are rich in sources with low prices, but the incorporation of I-NAs usually reduces the mechanical properties of PLA, due to the poor compatibility with the PLA matrix [26–28]. In addition, PLA nucleation can also be increased by using POCFA (phosphorus oxychloride) because of its phosphamide structure [17]. The interaction mechanism for PLA/POCFA is shown in Figure 14.1.The isothermal crystallization and melting behaviors of PLA and PLA/POCFA at 105 °C are displayed in Table 14.1.

Table 14.1 shows that the crystallization half-time ($t_{1/2}$), noted as the time when crystallinity (X_t) reached 50%, is regarded as a very important crystallization kinetics parameter. It can be employed to characterize the crystallization rate roughly. Usually a small $t_{1/2}$ means fast crystallization. The $t_{1/2}$ of PLA is 6.41 minutes and obtains a crystallinity (χ_c) of 38.9%, meaning that PLA needs a relatively long time to complete crystallization. When POCFA is added to PLA, the $t_{1/2}$ decreases rapidly with its content increasing. The $t_{1/2}$ of PLA5 is decreased to 1.27 minutes while the χ_c is increased to 50.0%, suggesting that the

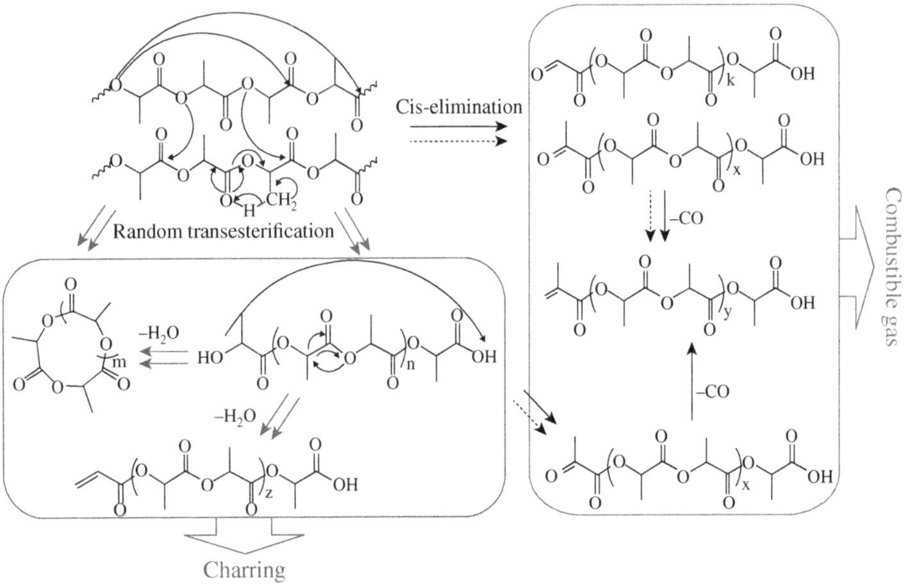

Figure 14.1 Interaction mechanism for PLA/POCFA [17].

crystallization rate and crystallinity are evidently both improved. The isothermal crystallization kinetic of PLA according to Avrani is shown in Table 14.2.

The isothermal crystallization behaviors of PLA determined by Eqs. (14.1) and (14.2) according to the Avrami model [29, 30]:

$$X_t = 1 - \exp(-kt^n) \tag{14.1}$$

$$\ln[-\ln(1 - X_t)] = \ln\, k + n\, \ln\, t \tag{14.2}$$

where k is the kinetic constant and n is the Avrami exponent, depending on the growth geometry and nucleation type of the crystals. The $t_{1/2}$ is calculated by Eq. (14.3):

$$t_{1/2} = \left(\frac{\ln 2}{k}\right)^{1/n} \tag{14.3}$$

The n value of PLA is around 4.0 and changes slightly with the temperature, which indicates that homogeneous nucleation is the main way to form crystals for PLA. The n decreases to 3.0, meaning heterogeneous nucleation becomes the main way for PLA to form three–dimensional crystals. The k value increases, indicating that the crystallization rate has increased.

The nucleation in the isothermal crystallization of PLA could be calculated by [17]

$$G = G_0 \exp[-U^*/R(T_C - T_\infty)] \exp[-K_g/T_c \Delta T f] \tag{14.4}$$

Table 14.2 *Isothermal crystallization kinetic of PLA composites from the Avrani Model [17].*

Samples	*T* (°C)	*N*	ln *k*	*k*	$t_{1/2}$
PLA0	95	3.90	−8.62	1.80×10^{-4}	8.30
	100	4.30	−8.31	2.45×10^{-4}	6.35
	105	4.23	−8.25	2.61×10^{-4}	6.41
	110	3.91	−7.88	2.80×10^{-4}	6.81
	115	3.83	−8.42	2.20×10^{-4}	8.20
PLA1	95	4.00	−8.89	1.38×10^{-4}	8.42
	100	4.12	−7,87	3.80×10^{-4}	6.19
	105	4.15	−8.10	3.03×10^{-4}	6.45
	110	4.12	−8.37	2.30×10^{-4}	7.00
	115	3.75	−8.71	1.64×10^{-4}	9.27
PLA2	95	4.16	−8.85	1.44×10^{-4}	7.68
	100	4.34	−7.92	3.64×10^{-4}	5.69
	105	4.23	−7.88	3.79×10^{-4}	5.89
	110	4.00	−8.38	2.29×10^{-4}	7.43
	115	3.85	−8.94	1.31×10^{-4}	9.27
PLA3	95	2.76	−3.44	3.19×10^{-2}	3.06
	100	2.78	−3.09	4.55×10^{-2}	2.67
	105	2.95	−3.11	4.44×10^{-2}	2.54
	110	3.08	−4.37	1.26×10^{-2}	3.67
	115	3.00	−4.86	7.74×10^{-3}	4.48
PLA5	95	2.35	−1.72	7.74×10^{-1}	1.78
	100	1.85	−0.87	7.74×10^{-1}	1.31
	105	1.71	−0.77	7.74×10^{-1}	1.27
	110	2.04	−1.06	7.74×10^{-1}	1.40
	115	2.44	−2.46	7.74×10^{-2}	2.36

where G is the spherulitic growth rate, $G_o$is a pre–exponential factor, U^* is the activation energy for the transport of segments to the crystallization site, K_g is the nucleation constant, ΔT is the undercooling T_mo $- T_c$ (where T_c is the crystallization temperature and $T_m{}^o$ is the equilibrium melting point), f is a factor given by $2T_c/(T_m{}^o + T_c)$ that accounts for the change in the heat of fusion as the temperature is decreased below $T_m{}^o$, R is the gas constant, and T_∞ is a hypothetical temperature where all motion associated with the viscous flow ceases.

The crystallinity of PLA is calculated by the formula [17]

$$\chi_c = \frac{(\Delta H_m + \Delta H_c) \times 100\%}{93.6 \times \chi_{PLA}} \tag{14.5}$$

where χ_{PLA} is the weight fraction of PLA in the composites, ΔH_m is the enthalpy of melting, ΔH_c is the crystallization enthalpy, and 93.6 J g^{-1} is the melting enthalpy of 100% PLA.

PLA composites have an obvious cold crystallization peak. The melting peak shifts to a lower temperature for PLA composites. These sharp crystallization peaks and increased χ_c indicate that the PLA molecular chains form ordered structures, leading to fast crystallization and high crystallinity.

The activation energy of a nonisothermal crystallization process for PLA and its composites was investigated by [17]

$$\frac{d\left[\ln\left(\frac{\frac{dT}{dt}}{T^2_p}\right)\right]}{d\left(\frac{1}{T_p}\right)} = -\frac{\Delta E}{R} \tag{14.6}$$

where T_p is the crystallization peak temperature and R is the gas constant. The lower activation energy values reveal the easier transformation for PLA chains from the molten state to crystallization.

Pure PDLA or PLLA has a crystalline equilibrium T_m of 207°C, but a typical T_m is in the 170–180°C range, attributed to small and imperfect crystallites, impurities, and slight racemization.

14.3 Mechanical

PLA is one of the biodegradable thermoplastic aliphatic polyesters derived from renewable resources and possesses a high glass transition temperature (T_g) and mechanical strength that is comparable to that of polystyrene [31–33]. PLA alloys usually exhibit better mechanical properties compared with pure PLA, which provides a new strategy to accelerate its commercial utilization [34].

Modification of PLA by copolymerization or physical blending is a useful way to decrease the brittleness, toughness, and heat distortion temperature. Various additives, such as plasticizers such as polyethylene glycol, polypropylene glycol, citrate esters [35], epoxidized soybean oil [36], toughening agents, reinforcing fillers, and compatibilizers, have been incorporated into PLA [7, 37, 38].

The mechanical properties of lactic acid-based polymers can be varied to a large extent, ranging from soft and elastic plastics to stiff and high-strength materials. Semi-crystalline PLA is preferred over the amorphous polymer when higher mechanical properties are desired. Semi-crystalline PLA has an approximate tensile modulus of 3 GPa, tensile strength of 50–70 MPa, flexural modulus of 5 GPa, bending strength of 100 MPa, and an elongation at break of about 4%. Different from thermal properties, the mechanical properties and crystallization behavior of PLA are very dependent on the MW and stereo-chemical makeup of the backbone. For example, the tensile modulus of PLLA increases by a factor of 2 when the MW is raised from 50 to 100 kDa with tensile strengths of 15.5–150 MPa [21]. The stereo chemical makeup is very easily controlled by the polymerization with D-lactide, L-lactide, D,L-lactide, or meso-lactide, to form random or block stereocopolymers, while the MW is directly controlled by the addition of hydroxylic compounds (i.e. lactic acid, water, and alcohols) [39].

Superior mechanical properties can be achieved by stereo-complexation of enantiomeric PLAs, which was ascribed to the formation of stereo-complex crystallites giving intermolecular cross-links. This pure stereo complex has a T_m of 230°C and mechanical properties greater than either pure polymer [40]. The impact resistance of PLA increased due to the cross-linking effects of the crystalline domains, while the tensile strength increased,

Table 14.3 *Effects of stereochemistry and crystallinity on mechanical properties [13, 39].*

Properties		PLLA	Annealed PLLA	PDLLA
Tensile strength	MPa	59	66	44
Elongation at break	%	7	4	5.4
Modulus of elasticity	MPa	3750	4150	3900
Yield strength	MPa	70	70	53
Flexural strength	MPa	106	119	88
Unnotched izod impact	Jm^{-1}	195	350	150
Notched izod impact	Jm^{-1}	26	66	18
Rockwell hardness		88	88	76
Heat deflection temperature	°C	55	61	50
Vicat penetration	°C	59	165	52

Table 14.4 *Mechanical characteristics of the NBR and its PLA blends [41].*

Properties	NBR	NP10	NP20	NP30	NP40
Tensile strength (MPa)	0.14 ± 0.04	0.16 ± 0.04	0.45 ± 0.05	0.56 ± 0.05	0.99 ± 0.06
Modulus (MPa)	0.66 ± 0.15	0.92 ± 0.20	3.79 ± 0.70	5.07 ± 0.80	16.9 ± 0.12
Elongation at break (%)	268 ± 65	246 ± 60	191 ± 57	178 ± 49	85 ± 26
Hardness (Shore A)	36 ± 2	47 ± 2	59 ± 2	65 ± 2	77 ± 2
Density (g/cc)	0.96 ± 0.01	0.98 ± 0.01	1.02 ± 0.01	1.04 ± 0.01	1.07 ± 0.01

presumably due to the stereo regularity of the chain [13, 39]. The effects of stereochemistry and crystallinity on mechanical properties are seen in Table 14.3.

Neat PLA has a higher value of tensile strength of 61.6 MPa. Fracture behavior in the tensile test changed from the brittle fracture of neat PLA to the ductile fracture of the blend. The added PLA enhances the mechanical strength of the rubber foams. The tensile strength and modulus as well as hardness of the NBR/PLA blends increased with the increase of the added PLA content [41] as seen in Table 14.4.

Conversely, Figure 14.2 discusses the addition of PBS content in PLA, which decreases the tensile strength up to 20 wt% and increases elongation, attributed to the more elastic characteristic of the neat PBS matrix. Pure PBS demonstrated an amazing value of elongation of about 224%. The impact strength of PLA/PBS blends was higher than neat PLA related to the high flexibility of PBS [42].

14.3.1 Physical

Physical properties of PLA polymers, as well as other polymers, depend on their molecular characteristics and ordered structures include crystalline thickness, crystallinity, degree of chain orientation, spherulite size, and morphology.

PLA is a biodegradable thermoplastic polymer that can be produced from lactic acid. The physical and mechanical properties of PLA can decrease if mixed with other polymers that have lower mechanical and physical properties [43]. The mechanical and physical properties of PLA and other polymers are shown in Table 14.5.

Major limitations of PLA are due to its high brittleness, low toughness, and low tensile elongation [44]. One way to improve the properties of PLA is through incorporation

Table 14.5 *Physical properties of PLA and other biopolymers [21].*

Properties	Type of biopolymer								
	Unit	PLA	PLLA	PDLA	PGA	PDLLA/PGA 50/50	PDLLA/PGA 50/50	PCL	PHB
Polymer density, ρ	G cm^{-3}	1.21–125	1.24–1.30	1.25–1.27	1.50–1.71	1.30–1.40	1.3	1.11–1.146	1.18–1.262
Tensile strength, σ	MPa	21–60	15.5–150	27.6–50	60–99.7	41.4–55.2	41.4–55.2	20.7–42	40
Tensile modulus, E	GPa	0.35–35	2.7–4.14	1–3.45	6.0–7.0	1–4.34	1.38–4.13	0.21–0.44	3.5–4
Ultimate strain, ε	%	2.5–6	3.0–10.0	2.0–10.0	1.5–20	2.0–10.0	2.5–10	300–1000	5.0–8.0
Specific tensile strength, σ^*	Nm g^{-1}	16.8–48.0	40.0–66.8	22.1–39.4	40.0–45.1	30.9–41.2	31.8–42.5	18.6–36.7	32.0–33.9
Specific tensile modulus, E^*	kNm g^{-1}	0.28–2.80	2.23–3.85	0.80–2.36	5.00–4.51	0.77–2.14	1.06–2.12	0.19–0.38	2.80–2.97
Glass transition temperature, T_g	°C	45–60	55–65	50–60	35–45	40–50	50–55	(−60)–(−65)	15.0–5.0
Melting temperature, T_m	°C	150–162	170–200	am[a]	220–233	am	am	58–65	168–182

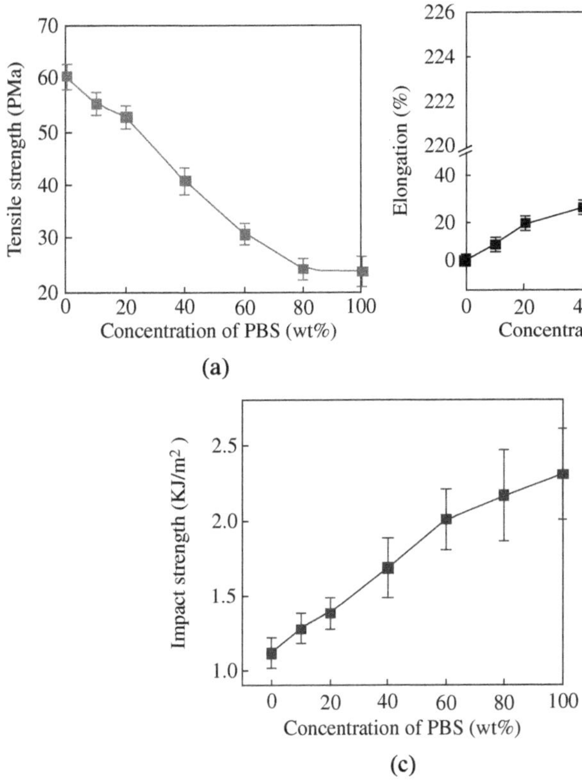

Figure 14.2 *(a) Tensile strength, (b) elongation, and (c) impact strength of PLA/PBS blends [42].*

of plasticizers. Considerable efforts have been made to enhance the mechanical properties and flexibility of PLA by blending with plasticizers such as poly(ethylene glycol), poly(propylene glycol), citrate esters [35], and epoxidized soybean oil [36].

14.3.2 Thermal Properties

Thermal properties of PLA are seen in Figure 14.3, where the initial decomposition temperature of PLA is 375°C and the temperature ascribed to the maximum rate of thermal degradation is 411°C. At temperature degradation 0–330°C, the weight loss of the specimens varied very little. This means that there was no effect of the weight loss during the manufacturing process.

The thermal stability of neat PLA is lower compared to the PLA/PBS blend shown in Table 14.6, which is attributed to the thermal stability of PBS where the T_g of the PLA and the PBS were about 63.9°C and about −44.3°C, respectively. The thermal properties of PLA are better able to be produced by adding MFC (microfibril cellulose) as the nucleating agent, which is associated with a composite crystallization time faster than that of pure PLA [45–47] and TiO_2 nanoparticles (NPs) [48].

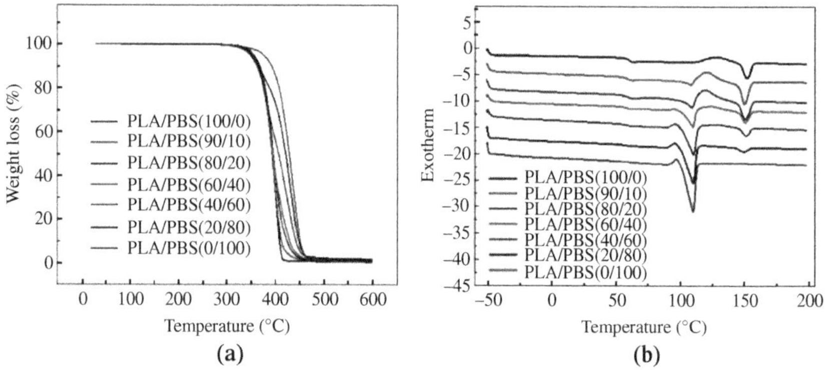

Figure 14.3 *(a) Thermogravimetric and (b) differential scanning calorimetry (DSC) curves of the neat PLA, PBS (poly(butylene succinate), and PLA/PBS blends [39].*

Table 14.6 *Thermal properties of PLA/PBs blends [13, 39].*

	PLA		PBS					
PLA/PBS (wt%)	T_g (°C)	T_m (°C)	T_g (°C)	T_m (°C)	T_{rc}	ΔH_m	X_c^a	X_c^b
100/0	63.9	152	—	—	—	14.88	15.9	—
90/10	63.2	149.5	—	108.23	120.6	22.28	23.8	20.2
80/20	63.9	150.1	—	108.58	119.5	19.00	20.3	17.2
60/40	61.2	150.3	—	109.1	124.64	17	18.0	—
40/60	60.9	150.7	—	109.5	126.86	—	—	—
20/80	—	149.6	—	109.1	—	—	—	—
0/100	—	—	−44.3	109	—	49	—	—

X_c^a and X_c^b: crystallinity for neat PLA and PBS.

14.3.3 Optical

The absorption of visible and UV light by polymers is one of the main factors affecting food quality. The transmission of visible and ultraviolet light are important parameters in the protection of products until they reach the consumer, such as juices, sports drinks, vitamins, and dairy products. Figure 14.4 shows that there is no UV transmission in the lower range of UV−C (190–220 nm) of PLA. However, at 225 nm the amount of UV light transmitted by PLA increases significantly. At 250 nm, 85% of the UV light is transmitted. At 300 nm, 95% of the UV light is transmitted. Thus, while UV−C is not transmitted, nearly all the UV−B and UV−A light passes.

14.3.4 Flame Retardancy

The flame retardancy properties of polylactide are related to the flame retardancy index (FRI). An FRI is calculated according to [17]

$$FRI = \frac{\left[THR \times \left(\frac{pHRR}{TTI}\right)\right]_{neatPLA}}{\left[THR \times \left(\frac{pHRR}{TTI}\right)\right]_{PLAcomposites}} \qquad (14.7)$$

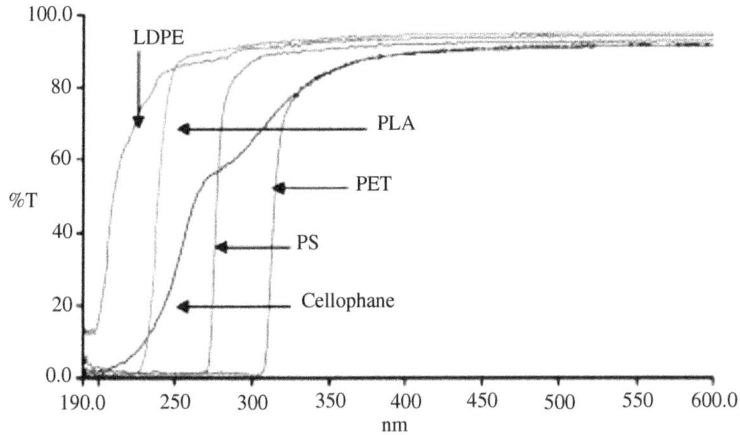

Figure 14.4 Percent transmission versus wavelength for PLA (98% L-lactide), PS, LDPE, PET, and cellophane films [3].

where HRR is the heat release rate, TTI is the time to ignition, pHRR is the peak of the heat release rate, THR is the total heat release, EHC is the average effective heat of combustion, and TSR is the total smoke release.

The pHRR and TTI of PLA are 411 kW m^{-2} and 69 seconds, respectively. The fire retardancy index (IFR) of PLA composites can be increased by adding ammonium polyphosphate (APP) [49] and POCFA [17]. The limiting oxygen index (LOI) value of PLA is 19.4 vol% and increases to 26.8 vol% with 1 wt% POCFA, and can be improved to 31.0 vol% with 5 wt% POCFA. PLA composites containing more POCFA show better flame retardancy [17], whereas with APP the reduction of flame retardancy and mechanical properties of PLA composites has a poor compatibility with the polymer matrix and is moisture sensitive [50, 51].

The addition of Si–APP and BSi–APP was found to significantly increase the LOI value of PLA composites, and the LOI values increased with the increasing loading of FR. The Si–APP and BSi–APP have better fire-retardant efficiency than APP in PLA composites, while BSi–APP is the most positive among the three FRs. The flame-retarded blend shows intermediate properties [22, 52–54] and its reaction to fire, needing the use of flame retardants [55].

14.3.5 Water Resistance

PLA demonstrated that the main drawbacks for various applications were its sensitivity to hydrolysis [52–54], the presence of water and hydrolysis processes occurring, leading to a drastic decrease in molecular weight and degradation of the mechanical properties [54, 56], and the high relative humidity when polymer is completely immersed in water [57, 58]). Four key features of the hydrolysis phenomenon are the catalysis of the ester hydrolysis by carboxylic acid end groups of PLA during the degradation (autocatalysis), accelerated chain-end scission by low pH values, decreased sensitivity of crystalline regions to hydrolysis compared to amorphous regions, and a higher rate of hydrolysis at high temperature, especially above the glass transition temperature [22, 59].

The water resistance of PLA composite was significantly improved by the addition of microencapsulated APP and polymethyl-methacrylate (PMMA). The combination of PLA with PMMA limits water diffusion and reduces the loss of phosphorus [60].

14.3.6 Grease Permeability

Grease resistance is important for packaging products containing fats or oils. Grease barrier properties are required for food packaging with a high grease content, such as fast food, and product appearance is also influenced by oil permeation through paper packaging. All PLA–based composites and mixtures show grease resistance [61]. For low PLA concentrations, PLA–AM indicates a penetration time of more than 1800. In addition, diffusivity of grease also depends on the thickness of the package. At thicker concentrations of coating, good grease resistance of kraft paper coated with double layer zein, paraffin wax, and with one single layer is well produced [62].

14.3.7 Water Vapor Permeability (WVP)

WVP is an important property because it indicates the amount of water that can pass through the packing material; for example it can determine the speed at which water molecules can diffuse through the material and its solubility. For low PLA concentrations less than 0.5%, amorphous PLA presents higher WVP than PLA/SC, as shown in Figure 14.5. The crystalline PLA presented lower WVTR than amorphous PLA [63]. Further, polymer crystallinity reduces the solubility and permeability of water. Crystallites have very low permeabilities and so reduce the cross-sectional area for diffusion, increase the diffusion path length, and impose restraints on the mobility of the amorphous phase [64]. WVP depends on hydrophobicity of the composite [65, 70].

The WVP is not affected significantly either by PLA or by its concentration, but for the paperboard, the WVP decreased significantly depending on the PLA concentration. In water vapor transport, the polymer microstructure is involved. Transport proceeds by a solution–diffusion mechanism. Typically, crystalline PLA regions are not permeable to gases [66]. Crystallites increase tortuosity, which affects the diffusion of the penetrants. The effect of a filler is similar to the crystallites regarding diffusion. The area-to-thickness ratio and orientation of the additive relative to the permeant direction affect its diffusivity.

14.3.8 Biodegradation Properties as a Packaging

Biodegradation definitions describe the action of microorganisms existing in the natural environment on the polymer and its conversion into methane and water or carbon dioxide. Two types of biodegradation are aerobic or anaerobic. No residue mean biodegradation and mineralization properties were established when the substrate was converted completely to the gaseous state. The rate of degradation was affected by the medium of biodegradation (temperature, humidity, and the chemical parameters and molecular weights and composition) of PLA.

PLA is initially degraded by hydrolysis and soluble oligomers that are formed are metabolized by cells, as seen in Figure 14.6. On the other hand, PLA degradation is largely resistant to attack by microorganisms in the soil and wastewater under ambient conditions when they are disposed of in the environment. Before biodegradation can commence, the polymer must be hydrolyzed at high temperatures of 58°C to reduce the molecular weight. The hydrolysis reaction is determined by the formula [68]

$$MW = ae^{bt} \tag{14.8}$$

where a, b, and t are constants equal to 230 kDa, 0.18 per second, and t days, respectively.

14.4 Food Packaging Application of PLA

PLA has been approved for use in food packaging. Recently, thermoform and extruded PLA packages have been developed for practical uses such as food containers (lunch boxes and fresh food packaging), drink containers such as juice and yogurt bottles, cups, blister packages, and overwrap [43]. Recently, modifiers have been investigated to increase stiffness at high temperatures and increase the degradation of PLA, or reduce costs. Some PLA modifications in the packaging application are presented in Table 14.7 [2].

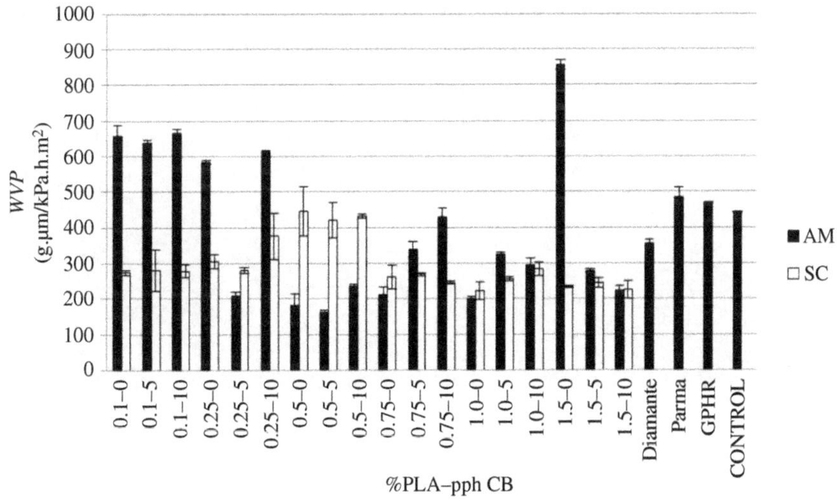

Figure 14.5 WVP (water vapor permeability) for the whole experimental design [65].

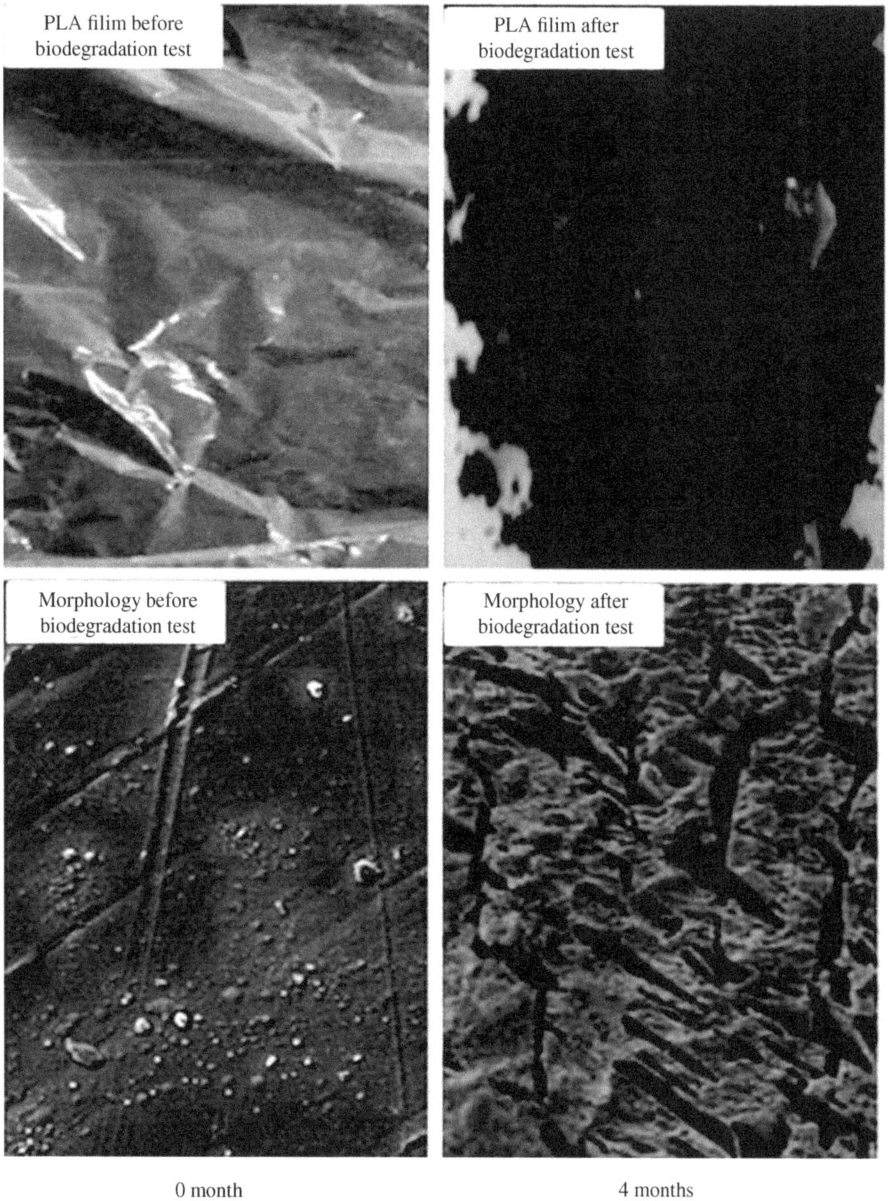

Figure 14.6 *PLA film samples and scanning electron microscopy (SEM) images before the biodegradation test and after four months of the biodegradation test [67].*

Table 14.7 PLA modifications for packaging applications [69].

Type of modification	Treatment or added material	Effect
Modifier	Citrate esters	↓T_g and improving the elongation at break
	Triacetine or tributyl citrate	↓T_g and ↑crystallinity
	Oligomeric malonate esteramides	↓T_g and improvement of the strain at break
	4,4-Methylene diphenyl diisocyanate	↓T_g to 64°C, ↑ tensile strength to 5.8 MPa good Nucleating agent for PLA crytallization
	Polyglycerol esters	Improving the elongation at break
	Polyethylene glycol and acetyl Triethyl citrate	↓T_g and crystallization rate
	Talc	↑ Ductility at more than 10%
	Bifunctional cyclic ester	Enhance PLA toughness
	Poly(1,3-butylene adipate)	↓ Storage modulus and T_g but ↑ elongation at break
	Polycabodiimide	Improve the thermal stability at 210°C for up to 30 min
Blending with:	Polyvinyl acetate	↑ Tensile strength and percent elongation
	Polyethylene oxide (POE)	Elongation at break of more than 500%
	Poly ε-caprolactone (PCL)	High improvement in mechanical properties
	Polyethylene glycol (PEG)	Enhance the crystallinity of PLA and biodegradability
	Starch with different plasticizers	Lowering the price, ↓T_g, and ↑crystallinity and ↑ biodegradability
	Polyvinyl alcohol and starch	↑ Tensile strength
	Ethylene vinyl alcohol (EVOH)	Improvement of mechanical, thermal, and biodegradability properties
	Polycarbonate	Improvement of mechanical properties and biodegradation rate
	Polyethylene glycidylmethacrylate (PEGMA)	Production of super-tough PLA materials
Copolymerization of PLA and:	DL-mandelic acid	↑T_g and improving mechanical properties
	ε-Caprolactone	Improving the decomposition temperatures and crystallinity
	Polyvinyl chloride	Improving strength and toughness
	Acrylonitrile butadiene styrene	Improved impact strength and elongation at break with a slight loss in modulus and tensile strength
Physical treatment	Vacuum compression-molding and solid-state extrusion techniques	Flexural strength and flexural modulus were improved up to 221 MPa and 8.4 GPa, respectively
	Orientation	Significant improvement in tensile and impact properties
	Annealing	↑ Toughness
	Aging	↑ T_g
	Drawing	Improvement in tensile and fracture properties

14.5 Conclusions

The properties of transition and melt temperature, mechanical, optical, grease resistance, water resistance, crystallinity, thermal, degradation, and WVP of PLA have been investigated, as well as melt and transition temperatures influenced by optical composition, thermal, and structure. The crystallization rate of PLA can be increased by adding nucleating agent and POCFA because of its phosphamide structure. Modification of PLA by copolymerization blending is a useful way to decrease the brittleness and toughness. Modifiers can increase stiffness at high temperatures and increase degradation of PLA, or reduce costs. This present study provides evidence that PLA modified films can have a packaging application.

References

1. Abral H, A Basria, F Muhammad, Y Fernandoa, F Hafizulhaq, et al. A simple method for improving the properties of the sago starch films prepared by using ultrasonication treatment. *Food Hydrocolloids*. 2019; **93**: 276–283.
2. Jamshidian M, Tehrany EA, Imran M, Jacquot M, and Desobry S. Poly-lactic acid: production, applications, nanocomposites, and release studies. *Compr. Rev. Food Sci. Food Saf.* 2010; **9**(5): 552–571.
3. Auras R, Harte B, and Susan S. An overview of polylactides as packaging materials. *Macromol. Biosci.*, 2004; **4**:835–864. doi: 10.1002/mabi.200400043.
4. Abral H, Ariksa J, Mahardika M, Handayani M, Aminahb I, et al. Transparent and antimicrobial cellulose film from ginger nanofiber. *Food Hydrocolloids*, 2020; **98**: 105266. Available at: https://doi.org/10.1016/j.foodhyd.2019.105266.
5. Urayama H, Ma HC, and Kimura Y. Mechanical and thermal properties poly(L–lactic) incorporating, various inorganic fillers with particale and whisker shapes." *Macromol. Mater. Eng.*, 2003; **287**(7): 562–568.
6. Botelho, T., Teixeira, N., and Felipe, A. Polylactic acid production from sugar molasses. International Patent WO 2004/057008 A1.
7. Pilla S, Gong S, O'Neill E, Rowell RM, and Krzysik AM. Polylactide–pine wood our composites. *Polym. Eng. Sci.*, 2008; **48**: 578–587.
8. Huda MS, Drzal LT, Mohanty AK, and Misra M. The effect of silane treated–and untreated–talc on the mechanical and physico–mechanical properties of poly (lactic acid)/newspaper fibers/talc hybrid composites. *Composites, Part B*, 2007; **38**: 367–379.
9. Grijpma DW, van Hofslot RDA, Supèr H, Nijenhuis AJ, and Pennings AJ. Rubber toughening of poly(lactide) by blending and block copolymerization. *Polym. Eng. Sci.*, 1994; **34**: 1674–1684.
10. Odent J, and Leclère P. Toughening of polylactide by tailoring phase–morphology with P[CL–co–LA] random copolyesters as biodegradable impact modifiers. *Eur. Polym. J.*, 2013; **49**: 914–922.
11. Zeng JB, Li KA, and Du AK. Compatibilization strategies in poly(lactic acid)–based blends. *RSC Adv.*, 2015; **5**: 32546–32565.
12. Kfoury G, Raquez JM, Hassouna F, Odent J, Toniazzo V, Ruch D, and Dubois P. Recent advances in high performance poly(lactide): from "green" plasticization to super–tough materials via (reactive) compounding. *Front. Chem.*, 2013; **1**, 32.
13. Perego G, Cella GD, and Bastioli C. Effect of molecular weight and crystallinity on poly(lactic acid) mechanical properties. *J. Appl. Polym. Sci.*, 1996; **59**:37–43.
14. Sheth M, Kumar RA, Dave V, Gross RA, and Carty Mc. Biodegdrable polymer blends of poly (lactic acid) and poly (ethylene glycol)." *J. Appl. Polym. Sci.* , 1997; **66**: 1495–1505.
15. Nijenhuis A, Colstee JE, and Pening AJ. High molecular weight poly (L–lactide) and poly(ethylene oxide) blends: thermal characterisation and physical properties. *J. Polym.*, 1996; **37**: 5849–5857.

16. Park JW, and Im SS. Miscibility and morphology in blends of poly (lactic acid) and poly (vinyl acetat–co–vynil alcohol). *J. Polym.*, 2003; **44**: 4341–4354.
17. Sun J, Li L, and Li J. Effects of furan–phosphamide derivative on flame retardancy and crystallization behaviors of poly(lactic acid). *Chem. Eng. J.*, 2019; **369**: 150–160
18. Fischer EW, Sterzel HJ, and Wegner G. Investigation of the structure of solution grown crystals of lactide copolymers by means of chemical reactions. *Kolloid–Z u Z Polymer*, 1973; **251**: 980–990.
19. Loomis GL, Murdoch JR, and Gardner KH. Polylactide strereocomplexes. *Polym. Prepr.*, 1990; **2**: 55.
20. Huang, J., M. S. Lisowski, J. Runt, E. S. Hall, R. T. Kean, N. Buehler, and J. S. Lin. Crystallization and microstructure of poly(l–lactide–*co–meso*–lactide). *Copolymers Macromol.*, 1998; **31**(8): 2593–2599.
21. De Velde KP, and Kiekens P. Biopolymers: overview of several properties and consequences on their applications. *Polym. Test.*, 2002; **21**: 433–442.
22. Henton DE, Gruber P, Lunt J, Randall J. in: A. K. Mohanty, M. Misra, and L.T. Drzal. *Natural Fibers, Biopolymers, and Biocomposites*. Taylor & Francis, Boca Raton, FL, 2005; 527–577.
23. Ma P, Jiang L, Xu P, Dong W, Chen M, and Lemstra PJ. Rapid stereocomplexation between enantiomeric comb–shaped cellulose–g–poly (l–lactide) nanohybrids and poly (d–lactide) from the melt. *Biomacromolecules*, 2015; **16**: 3723–3729.
24. Li H, and Huneault MA. Effect of nucleation and plasticization on the crystallization of poly (lactic acid). *Polymer*, 2007; **48**: 6855–6866.
25. Jiang L, Shen T, Xu P, Zhao X, Li L, Chen WPD, and Ma M. Crystallization modification of poly (lactide) by using nucleating agents and stereocomplexation." *e-Polymers*, 2016; **16**: 1–13.
26. Huda MS, Drzal LT, Misra M, Mohanty AK, Williams K, and Mielewski DF. A study on biocomposites from recycled newspaper fiber and poly(lactic acid)." *Ind. Eng. Chem. Res.*, 2005, **44**: 5593–5601.
27. Ayutthaya SIN, Tanpichai S, Sangkhun W, and Wootthikanokkhan J. Effect of clay content on morphology and processability of electrospun keratin/poly (lactic acid) nanofiber. *Int. J. Biol. Macromol.*, 2016; **85**: 585–595.
28. Fukushima K, D Tabuani, and Camino G. Nanocomposites of PLA and PCL based on montmorillonite and sepiolite. *Mater. Sci. Eng., C*, 2009; **29**: 1433–1441.
29. He DR, Wang YM, Shao CG, Zheng GQ, Li Q, and Shen CY. Effect of phthalimideas an efficient nucleating agent on the crystallization kinetics of poly(lactic acid). *Polym. Test.*, 2013; **32**(6): 1088–1093.
30. Kermaniyan TS, Garmabi H, and Saeb MR. Calorimetric and rheokinetic analyses merged to capture crystallization kinetics in polyamide/clay nanocomposites: revisiting predictability of models. *J. Appl. Polym. Sci.*, 2018; 46364.
31. Choi KM, Choi MC, and Han DH. Plasticization of poly(lactic acid) (PLA) through chemical grafting of poly(ethylene glycol) (PEG) via in situ reactive blending." *Eur. Polym. J.* 2013; **49**:2356–2364.
32. Xu ZR, Park HY, and Kim HY. Effects of modified MMT on mechanical properties of EVA/MMT nanocomposites and their foams. *Macromol. Symp.* 2008; **264**:18–25.
33. Choi KM, Lim SW, and Choi MC. Properties of poly(ethylene glycol)–grafted poly(lactic acid) plasticized with poly(ethylene glycol). *Macromol. Res.* 2014; **22**:1312–1319.
34. Taib RM, Hassan H, and Ishak MZ. *Polym. Plast. Technol. Eng.* 2014; **53** (2): 199–206. Tensides. U.S. Patent 20100112226 A1.
35. Emad AJA, Adeel HS, and Saadon AA. New biopolymer nanoblends based on epoxidized soybean oil plasticized poly (lactic acid)/fatty nitrogen compounds modified clay: preparation and characterization. *Ind. Crops Prod.*, 2011; **33**: 23–29.
36. Fathilah A, Chang YW, Kang SC, and Yoon JY. Thermal, mechanical and rheological properties of poly(lactic acid)/epoxidized soybean oil blends. *Polym. Bull.*, 2009; **62**: 91–98.
37. Wang H, Sun X, and Seib P. Strengthening blends of poly(lactic acid) and starch with methylenediphenyl diisocyanate. *J. Appl. Polym. Sci.*, 2001; **82**: 1761–1767.
38. Jamshidi K, Hyon SH, and Ikada Y. Thermal characterization of polylactides. *Polymer* 1988; **29**: 2229–2234.

39. Hartmann MH. in: D. L. Kaplan. *Biopolymers from Renewable Resources*." Springer–Verlag, 1998; 367–411. Berlin.
40. Yamane H, and Sasai K. Effect of the addition of poly(d–lactic acid) on the thermal property of poly(l–lactic acid). *Polymer*, 2003; **44**: 2569–2575.
41. Han DH, Choi MC, Jeong JH, Choi CM, Kim HS. Properties of acrylonitrile butadiene rubber (NBR)/poly(lactic acid) (PLA) blends and their foams. *Compos. Interfaces*, 2016. doi: 10.1080/09276440.2016.1170518.
42. Hassan E, Wei Y, Jiao H, and Muhuo Y. Dynamic mechanical properties and thermal stability of poly(lactic acid) and poly(butylene succinate) blends composites. *J. Fiber Bioeng. Inf.*, 2013; **6**(1): 85–94. doi: 10.3993/jfbi03201308.
43. Farah S, Anderson DG, and Langer R. Physical and mechanical properties of PLA, and their functions in widespread applications – a comprehensive review. *Adv. Drug Delivery Rev.*, 2016; **107**: 367–392.
44. Ljungberg N, and B Wesslén. Preparation and properties of plasticized poly(lactic acid) films. *Biomacromolecules*, 2005; **6**: 1789–1796.
45. Iwatake A, Nogi M, and Yano H. Cellulose nanofiber–reinforced polylactic acid. *Compos. Sci. Technol.*, 2008; **68**(9):2103–2106. doi: 10.1016/j.carbpol.2010.07.026.
46. Nakagaito AN, Fujimura A, Sakai T, Hama Y, and Yano H. Production of microfibrillated cellulose (MFC)–reinforced polylactic acid (PLA) nanocomposites from sheets obtained by a papermaking–like process. *Compos. Sci. Technol.*, 2009; **69**(7–8):1293–1297. doi: 10.1016/j.compscitech.2009.03.004.
47. Suryanegara L, Nakagaito AN, and Yano H. The effect of crystallization of PLA on the thermal and mechanical properties of microfibrillated cellulosereinforced PLA composites. *Compos. Sci. Technol.*, 2009. **69**(7–8):1187–1192. doi: 10.1016/j.compscitech.2009.02.022.
48. Mofokeng JP, Luyt AS, *Polym. Test.* 2015: **45**, 93.
49. Chen C, Gu X, Jin X, Sun J, and Zhang J. The effect of chitosan on the flammability and thermal stability of polylacticacid/ammonium polyphosphate biocomposites. *Carbohydr. Polym.*, 2017; **157**:1586–93.
50. Lim KS, Bee ST, Sin LT, Tee TT, Ratnam CT, Hui D, and Rahmat AR. A review of application of ammonium polyphosphate as intumescent flame retardant in thermoplastic composites. *Composites, Part B*, 2016; **84**: 155–74.
51. Wang J, Wang D, Liu Y, Ge X, and Wang Y. Polyamide–enhanced flame retardancy of ammonium polyphosphate on epoxy resin. *J. Appl. Polym. Sci.*, 2010; **108**:2644–53.
52. Helou P. Acide polylactique (PLA). *Tech. Ing.* 2017. AM 3317 V1.
53. Mohd Adnan AF, Nishida H, and Shirai Y. Evaluation of kinetics parameters for poly(L–lactic acid) hydrolysis under high–pressure steam. *Polym. Degrad. Stab.*, 2008; **93**: 1053–1058.
54. Proikakis CS, Mamouzelos NJ, and Tarantili PA, and Andreopoulos AG. Swelling and hydrolytic degradation of poly (D, L–lactic acid) in aqueous solutions. *Polym. Degrad. Stab.*, 2006; **91** (3): 614–619.
55. Gonzalez MF, Ruseckaite RA, and Cuadrado TR. Structural changes of polylacticacid (PLA) microspheres under hydrolytic degradation. *J. Appl. Polym. Sci.*, 1999; **71**(8): 1223–1230.
56. Rocca-Smith JR, Chau N, Champion D, Brachais CH, Marcuzzo E, Sensidoni A, Piasente F, Karbowiak T, and Debeaufort F. Effect of the state of water and relative humidity on ageing of PLA films. *Food Chem.*, 2017; **236**: 109–119.
57. Aso Y, Yoshioka S, Li AL, and Terao WP. Effect of temperature on mechanisms of drug release and matrix degradation of poly (D, L–lactide) microspheres. *J. Controlled Release*, 1994; **31**: 33–39.
58. Tsuji IN. Poly(L–lactide) high–temperature hydrolysis of poly(L–lactide) films with different crystallinities and crystalline thicknesses in phosphate–buffered solution. *Macromol. Mater. Eng.*, 2001; **286**: 398–406.
59. Mangin R, Vahabi H, Sonnier R, ChivasJoly C, Lopez-Cuesta JM, and Cochez M. Improving the resistance to hydrothermal ageing of flame-retarded PLA by incorporating miscible PMMA. *Polym. Degrad. Stab.*, 2018; **155**: 52–66.
60. Trezza TA, and Vergano VJ. Grease resistance of corn zein coated paper. *J. Food Sci.*, 1994; **59**: 912–915.

61. Parris N, Vergano VJ, Dickey LC, Cooke PH, and Craig JC. Enzymatic hydrolysis of Zien-wax-coated paper. *J. Agric. Food. Chem.*, 1998; **46**(10): 4056–4059. doi: 10.1021/jf980406j.
62. Shogren R. Water vapor permeability of biodegradable polymers. *J. Environ. Polym. Degrad.*, 1997; **5**: 91–95.
63. Ashley RJ. Permeability and plastics packaging. In *Polymer Permeability*; Comyn, J. Ed.; Springer , 1985; 269–308. Dordrecht, The Netherlands.
64. Aldana DS, Villa ED, Hernández MDD, Sánchez GG, Cruz QR, Gallardo SF, Castillo HP, and Casarrubias LB. Barrier properties of polylactic acid in cellulose based packages using montmorillonite as filler. *Polymers*, 2014; **6**: 2386–2403. doi: 10.3390/polym6092386.
65. Sato S, Nyuui T, Matsuba G, and Nagai K. Correlation between interlamellar amorphous structure and gas permeability in poly(lactic acid) films. *J. Appl. Polym. Sci.*, 2014; **131**. doi: 10.1002/APP.40626.
66. Kale G, Auras R, and Singh P. Comparison of the degradability of poly(lactide) packages in composting and ambient exposure conditions. *Packag. Technol. Sci.*, 2007; **20**: 49–70. doi: 10.1002/pts.742.
67. Rahul MR, Janorkar AV, and Hirt DE. Poly(lactic acid) modifications. *Prog. Polym. Sci.*, 2010; **35** (3): 338–356. Available at: https://doi.org/10.1016/j.progpolymsci.2009.12.003
68. Weng YX, Jin YJ, Meng QY, Wang L, Zhang M, and Wang YZ. Biodegradation behavior of poly(butylene adipate-*co*-terephthalate) (PBAT), poly(lactic acid) (PLA), and their blend under soil conditions. *Polym. Test.*, 2013; **32**: 918–926. doi: 10.1016/j.polymertesting.2013.05.001.
69. Wei LL, Wang DY, Chen HB, Chen L, Wang XL, and Wang YZ. Effect of a phosphorus-containing flame retardant on the thermal properties and ease of ignition of poly(lactic acid). *Polym. Degrad. Stab.*, 2011; **96** (9): 1557–1561.
70. Ilyas RA, Sapuan SM, Atikah MSN, Asyraf MRM, Ayu Rafiqah S, Aisyah HA, Mohd Nurazzi N, Norrahim MNF. Effect of hydrolysis time on the morphological, physical, chemical, and thermal behavior of sugar palm nanocrystalline cellulose (*Arenga pinnata* (Wurmb.) Merr). *Textile Research Journal*. 2020. In print. doi: 10.1177/0040517520932393.

15

Poly(Lactic) Acid Modified Films for Packaging Applications

Jissy Jacob[1], Sabu Thomas[2,3,4], and Sravanthi Loganathan[5]

[1] PG & Research Department of Chemistry, St. Thomas College, Pala, Kerala, 686574, India
[2] School of Chemical Sciences, Mahatma Gandhi University, Kottayam, Kerala, 686560, India
[3] International and Inter University Centre for Nanoscience and Nanotechnology , Mahatma Gandhi University, Kottayam, Kerala, 686560, India
[4] School of Energy Materials, Mahatma Gandhi University, Kerala, Kottayam, 686560, India
[5] Processing Engineering Division, CSIR- Central Electrochemical Research Institute(CECRI), Karaikudi, Tamil Nadu, 630003, India

15.1 Introduction

In the past two decades, the production rate of conventional plastics has increased at a rapid rate in order to meet the growing demand for new developments in the field of science and technology. It has been estimated that ~140 million tonnes of conventional plastics are produced globally for various applications like packaging, building and construction, automobile, electrical and electronics, as well as aerospace [1, 2]. Among these applications, the packaging sector occupies the principal position since ~35% of the conventional plastics are being utilized annually. The primary role of food packaging material is to provide protection to the food products stored and to prevent damage. The secondary function is to provide nutrition as well as ingredient details to the customer. At present, there has been growing interest to develop intelligent packaging material in order to indicate the freshness of the food material stored via incorporation of sensing elements. The goal of food packaging is to store food in a cost-effective way that satisfies industry requirements and consumer desires, maintains food safety, and minimizes environmental impact [3, 4].

Bio-Based Packaging: Material, Environmental and Economic Aspects, First Edition. Edited by S.M. Sapuan and R.A. Ilyas

Currently, synthetic polymers like polyethylene (PE), polyethylene terephthalate (PET), polyvinylchloride (PVC), polypropylene (PP), polyamide (PA), and polystyrene (PS) are being utilized as packaging materials. This is because these conventional plastics present cheaper cost, good mechanical (tensile as well as tear strength), barrier (oxygen, carbon dioxide, water vapor as well as aroma compounds), and heat sealability properties [5]. However, the usage of conventional plastics is being restricted as they are not biodegradable in nature, due to which environmental pollution problem has become a serious concern [6]. In reality, recycling of food packaging materials is difficult due to the fact that they are contaminated with the food product stored. Moreover, recycling of food packaging materials is not often an effective option in an economical point of view. Owing to this fact, several thousands of tonnes of plastic waste materials are dumped as landfill and left as such as municipal solid waste, which in turn leads to environmental pollution. As a result, there has been growing awareness of the need to protect the environment via utilization of eco-friendly plastic materials for food packaging applications. This conveys the fact that the food packaging material should present biodegradability in a limited time period and thus can help in the prevention of environmental pollution problem [7]. In addition to this, there prevails growing demand to search for other alternatives since petroleum resources are limited. As a consequence, utilization of bio-based polymers derived from natural resources is deemed as an appropriate option to address all of the above-mentioned drawbacks. This chapter is focused on an introduction to the basics of biopolymers, classification of biopolymers based on their source of origin. Further, the importance of poly(lactic) acid (PLA) in view of industrial packaging applications, properties and drawbacks of PLA, different routes for fabrication of PLA composites are discussed. In detail, this chapter provides information about different types of PLA composite films like PLA/layered silicate [8], PLA/carbonaceous materials [9], PLA/bio-filler [10], and PLA/mesoporous silica [11] envisaged by several research groups, and the role of such reinforcements on the food packaging properties are analyzed.

15.2 Biopolymers

According to the American Society for Testing of Materials (ASTM) and the International Standards Organization (ISO), biodegradable plastics are those that exhibit prominent change in chemical structure when subjected to definite environmental circumstances. During the degradation process, bio plastics also present loss in physical as well as mechanical characteristics, which are measured and proven by standard procedures. The degradation in the case of bio plastics proceeds via the action of natural microbes like bacteria, algae, and fungi [12, 13]. The degradation of the bio plastics can also be accelerated by breaking them into simple compounds like carbon dioxide, water, and methane. The acceleration in the degradation can be possible via the action of enzymes released by microbes such that complete degradation of bio plastics may be achieved in a confined period of time [14].

At present, several chemical industries located in the United States of America (USA) and the United Kingdom (UK) show profound interest in manufacturing biopolymer-based materials mainly due to environmental concern. For example, Imperial Chemical Industry (ICI), UK and Nature Works LLC, USA are the leading manufacturers of biodegradable polymeric materials for food packaging applications.

15.2.1 Classification of Biopolymers

The classification of biopolymers based on the method of production or their source is as follows: (i) direct extraction of biopolymers from animal or vegetal biomass like proteins and polysaccharides [15] (ii) production of biopolymers by synthetic methods via utilization of monomer from renewable resources, e.g. synthesis of PLA from bio-based monomer lactic acid obtained from fermentation of biomass [16] (iii) production of biopolymers by the microbial fermentation process, e.g. polyhydroxyalkanoates, xanthan, and pullulan [17].

15.2.2 Poly(Lactic) Acid (PLA)

As far as eco-friendly products are concerned, PLA is considered as the effective biopolymer that can act as a substitute for several petrochemical-derived polymers. This is because PLA is obtained from renewable resources and, moreover, it exhibits biocompatibility, biodegradability, and thermomechanical characteristics [18, 19]. PLA is an aliphatic polyester and can be synthesized from either D- or L-lactic acid monomers. For the production of monomers, the fermentation process is generally used. The widely used agricultural sources for the fermentation process include corn and bagasse. The U.S. Food and Drug Administration (FDA) certified various grades of PLA for commercial purposes [14]. However, the applicability of PLA with regard to food packaging is limited because of certain disadvantages like brittleness and poor gas and water vapor barrier characteristics [20, 21]. In the case of food packaging, gas and water vapor barrier properties are critical as they play an important role in preventing food deterioration by retarding the microbial growth [22].

To address the limitations associated with PLA, various attempts have been made with the aim to broaden the applicability of PLA in food packaging, which includes (i) modification of the PLA matrix with the help of plasticizers like oleic acid, fatty acid esters, and polyethylene glycol [15], (ii) co-polymerization of lactide with certain other monomers like trimethylene carbonate and ε-caprolactone [23], (iii) blending of PLA with other co-polymers like poly(butylene adipate-co-terephthalate) and poly(ε-caprolactone) [24] and (iv) incorporation of reinforcements like organoclay, graphene nanoplatelets, and cellulose nanocrystals in the PLA matrix [18]. Among all the strategies, the fabrication of PLA composites via incorporation of reinforcements is regarded as the effective route to achieve improvement in ductility, gas and water vapor barrier characteristics, as well as thermal resistance of PLA.

15.3 Modified PLA Films

15.3.1 PLA/Clay Composites

Layered silicates are known to improve the thermal, mechanical, and barrier characteristics of PLA. Hence, extensive research activities are focused to envisage the influence of layered silicates on the properties of PLA. However, enhancement in the characteristics is dependent on the dispersion level of layered silicates in the PLA matrix. The factors that influence the dispersion level include (i) interaction of layered silicates with the PLA matrix and (ii) the method used for fabrication of PLA/layered silicate composites. The

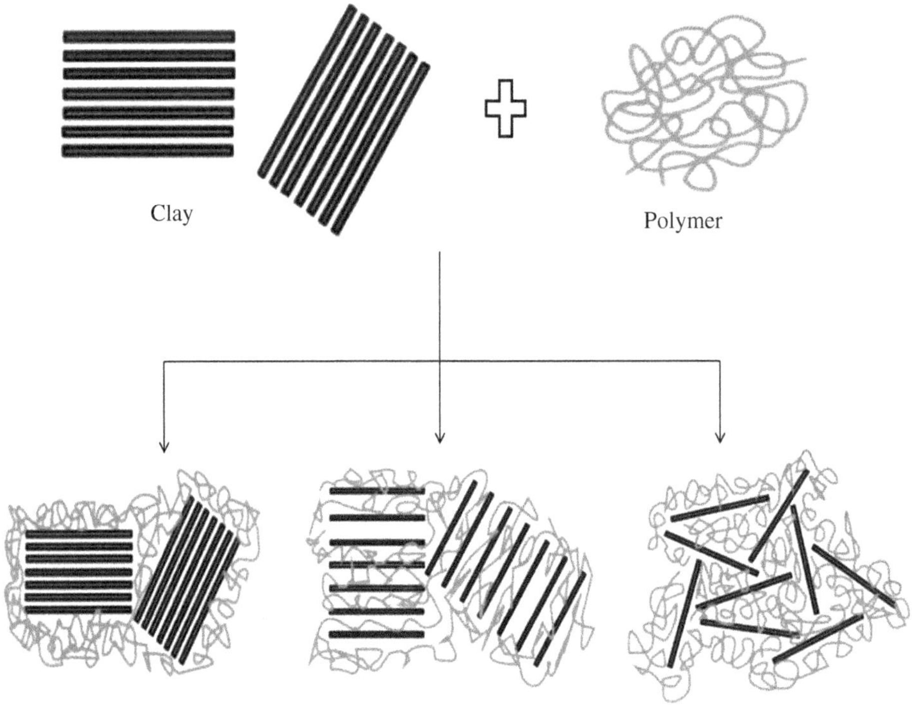

Figure 15.1 *Different structures for PLA/layered silicate composites [8].*

interaction among PLA and layered silicate usually decides the nature of final structure for the PLA composites. The three different structures that can be possible for PLA/layered silicate composites are shown in Figure 15.1.

In order to enhance the interaction ability among layered silicate and PLA, inorganic cations present in the galleries are replaced with organic cations [8]. The organic modification leads to an increase in spacing between the galleries of layered silicates such that provision will be created for distribution of polymer moieties in the interlayers. In this view, research work was reported on the effect of organically modified Cloisite 30B in enhancing the properties of PLA [25]. The PLA/organically modified Cloisite 30B composites were fabricated using the solution casting method. It was found that the PLA composite films were transparent after incorporation of organically modified Cloisite 30B in the PLA matrix. Enhancement in terms of crystallinity and storage modulus (~61%) of PLA composites was evidenced. The reason for improvement was reported to be due to the enhanced interaction achieved between diol functionalities in the organic modifier and PLA.

The influence of hexadecylamine modified montmorillonite (MMT) and mica on the thermal and mechanical properties of PLA [26]. For both PLA composites, thermal stability was found to be reduced drastically. However, the presence of clay in the PLA matrix exhibited an influence on enhancement in the tensile strength for both of the composites. For PLA/organically modified MMT, ~32% improvement in the tensile strength was evidenced. The PLA/organically modified mica showed ~57% enhancement in tensile strength property, which may be because of the high aspect ratio exhibited by the filler. For sodium MMT-reinforced PLA/starch composite, the water absorption capacity was reduced by ~12% in comparison with the neat PLA/starch system [27]. In another study, PLA/thymol/modified MMT (D43B) composites were fabricated. The presence of thymol did not affect the thermal performance of the PLA composites. However, clear improvement in terms of mechanical and oxygen barrier characteristics of PLA was evidenced in the presence of D43B. This is because the PLA matrix is entrapped in the intercalated clay or partially exfoliated clay and hence improvement in the properties was evidenced. The antioxidant activity of PLA composites was determined by the 2, 2-diphenyl-1-picrylhydrazyl (DPPH) scavenging assay test. The results indicated that the presence of thymol in the PLA composite films resulted in significant inhibition for DPPH radicals [28].

In another study, the poly(lactic acid)/layered silicate nanocomposite membrane (PLSNM) system was found to exhibit reduced gas and vapor permeabilities with the rise in clay content [29]. In the case of Cloisite 30B reinforced PLA/poly(butylene succinate) (PBS) composites, oxygen permeability was found to be reduced by ~26%, in comparison with the neat PLA/PBS system [30]. The reduction in oxygen permeability was mainly because of the tortuous path effect created by Cloisite 30B. The effect of clay (C30B) on oxygen permeability of PLA/PBS (80/20) blend and various PLA/PBS/C30B nanocomposites of different clay loadings are shown in Figure 15.2.

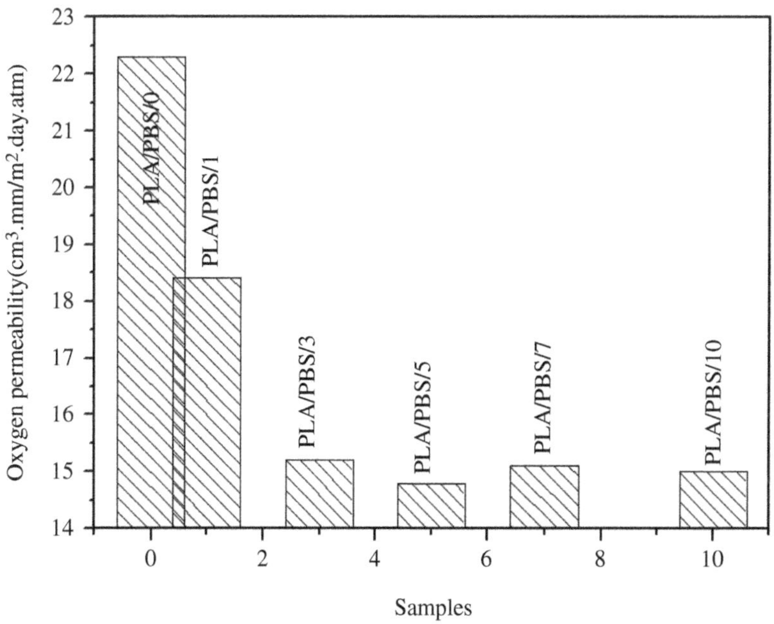

Figure 15.2 *The effect of clay (C30B) on oxygen permeability of PLA/PBS (80/20) blend and various PLA/PBS/C30B nanocomposites of different clay loadings [30].*

15.3.2 PLA/Carbonaceous Composites

Among the carbonaceous fillers, graphene (GRN) and carbon nanotubes (CNTs) are the most studied fillers for the fabrication of PLA composites. The CNTs are one dimensional in nature. When GRN sheets are rolled into a tubular shape, it results in the formation of CNTs. Both the carbonaceous fillers exhibit superior mechanical as well as thermal characteristics and hence are studied extensively as reinforcements in the PLA matrix [9]. However, achieving uniform dispersion of CNTs in the polymer matrix is a challenging task. Uniform dispersion of CNTs in the PLA matrix is a challenging task. When CNTs are directly blended in the PLA matrix, an agglomeration phenomenon was evidenced, due to which the mechanical characteristics of the composites were found to be affected [31]. Therefore, acid functionalization was carried out for CNTs and GNPs in order to enhance the dispersion of both reinforcements in the PLA matrix. In this work, melt mixing was used to blend the carbon-based reinforcements in the PLA matrix, which was then followed by the compression molding process through which PLA nanocomposite films were fabricated. The addition of reinforcement helped in improving the mechanical characteristics of PLA composites, mainly because of the strong sp^2 C–C bonding as well as the geometric arrangement of the filler in the PLA matrix [32]. The L-lactide monomer was subjected to in-situ polymerization on the surface of acid functionalized multiwalled CNTs. The interfacial interaction was found to increase due to the acid functionalization and in-situ polymerization of L-lactide. Therefore, improvement in terms of the tensile modulus for PLA composites was achieved [33]. The tensile strengths of the PLA, PLA/MWCNT-COOH, and PLA/MWCNT-g-PLAs are shown in Figure 15.3.

The blend containing PLA and poly(butylene succinate-co-L-lactate) (PBSL) was prepared and incorporated with CNTs through melt-blending. A slight enhancement in flexural strength results was obtained for PLA blend-based composites. When the CNT loadings were increased, the flexural strength was found to drop down for the composites, which might be due to agglomeration [34]. In a separate contribution, PLA/graphene composite films were fabricated in a sandwich fashion containing reduced graphene oxide (rGO)

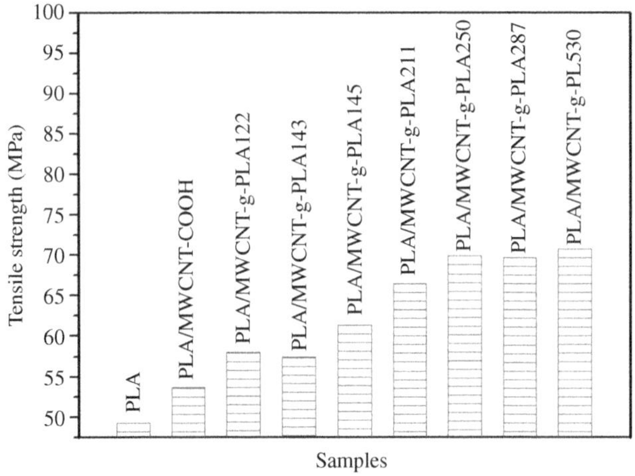

Figure 15.3 *The tensile strength of the PLA, PLA/MWCNT-COOH, and PLA/MWCNT-g-PLAs [32].*

as the core, which is then encapsulated with PLA. It was found that the PLA composite films exhibited ~87% decrement in water vapor permeation, which was due to the synergy that existed among the rGO core barrier and PLA encapsulation. In addition to this, the oxygen permeation in the PLA composite films was also reduced by ~twofold in magnitude as compared to neat PLA under both the humid and dry conditions used for the evaluation. The significant improvement achieved in the case of barrier characteristics was mainly because of the hydrophobic nature and lamellar structure possessed by the rGO platelets. The interlayer spacing between the rGO platelets and plate-like morphology led to a tortuous path and hence the improvement in terms of barrier properties for the PLA composites was evidenced [35]. The PLA/graphene oxide (GO)/clove essential oil (CLO) composite-based antimicrobial packaging films were fabricated using the solution casting approach. The content of CLO was varied between 15 wt% and 30 wt% whereas the GO content in the PLA film had been maintained at 1 wt%. It was found that the flexibility/ductility of the PLA composite films were enhanced in the presence of CLO for the PLA composite films. The improvement in flexibility for the composite films was evident from the reduction in tensile strength as well as the glass transition temperature (T_g). The presence of GO in the PLA matrix could influence the oxygen barrier properties for the PLA composites [36].

15.3.3 PLA/Bio Filler Composites

In the current scenario, the incorporation of bio fillers in the PLA matrix has caused considerable interest as they ensure biocompatibility during food storage. Therefore, PLA composites reinforced with different types of bio fillers were fabricated and the effect of such fillers on the properties of PLA has been investigated. The cellulose nanofibres (CNFs) prepared from flax fibers are reinforced in the PLA matrix via solution casting and it was evidenced that the tensile strength improved tremendously by ~59% for the composites. In addition to this, the nucleating effect provided by CNFs is confirmed by the improvement in the rate of the crystallization process [37].

The PLA/cellulose nanocrystal (CNC) composites were prepared by using the blown film technique and the influence of CNC on the oxygen as well as water vapor barrier characteristics of PLA was investigated. The PLA composite films exhibited a barrier effect to water vapor (~40%) and oxygen permeation (~75%) as compared to neat PLA, which in turn confirms the tortuous effect provided by CNC bio filler. It was discussed that the tortuous effect created by the CNC bio filler may be associated with its high crystallinity. The addition of crystalline CNC also increased the degree of crystallinity for PLA [22]. A separate study was conducted to investigate the effect of coating microfibrillated cellulose (MFC) on the PLA matrix, which was activated using plasma treatment [38]. It was found that the coating of MFC significantly improved the oxygen barrier properties for PLA.

In a separate contribution, CNC is modified with surfactant (s-CNC) in order to improve the compatibility between PLA and CNC. The neat PLA and PLA composites (PLA/CNC and PLA/sCNC) are prepared by the solution casting approach. The results showed that the PLA/s-CNC composites exhibited an improved water vapor barrier property as compared to PLA/CNC and neat PLA. This was due to the enhanced interfacial adhesion achieved among PLA and sCNC. The SEM analysis confirmed the high level dispersion of sCNC in the PLA matrix, which in turn enhanced the oxygen and water vapor barrier performance for PLA composites [39].

Similarly, the influence of addition of essential oil derived from *Tanacetum balsamita* (TBE) and ethanolic extract derived from propolis (PEE) on the shelf-life of cooked sausages stored in PLA/CNC composite films via vacuum-assisted packaging was investigated. This work demonstrated that the presence of TBE and PEE in the PLA/CNC composite films could enhance the shelf life of the sausages by up to ~50 days, when stored under refrigerated conditions, due to their antimicrobial properties. It was found that both the TBE and PEE could exhibit a synergistic antimicrobial property. The high level of antimicrobial effect was observed for PLA/CNC/TBE (2 wt%)/PEE (2 wt%) composite films used for storage of sausage [40].

The coffee silver skin (CS) was utilized as the green source for the extraction of CNC using alkali treatment, which was then followed by the hydrolysis process using sulfuric acid. It was observed that the addition of 1 wt% of CNC in the PLA matrix did not show any effect on the improvement of tensile strength as well as elongation-at-break %, in comparison with neat PLA. On the other hand, the tensile strength was found to be improvised for PLA/CNC (3 wt%) composite film. However, with further increment in the CNC loading (5 wt%), the tensile strength was found to decrease for the PLA composite. For all the loadings of CNC (1, 3, 5 wt%) in the PLA matrix, the Youngs' modulus values were found to be more enhanced than the neat PLA. The maximum improvement in the Youngs' modulus by ~24% was obtained in the case of PLA/CNC (5 wt%) composite film. In the case of water vapor permeation, the maximum reduction of ~31% was obtained for PLA/CNC (3 wt%) composite film. Similarly, reduction in the oxygen permeation by ~71% was evidenced for PLA/CNC (5 wt%) composite film. The improvement in the barrier properties was due to the enlargement in the tortuous path provided by the presence of CNCs in the PLA matrix [41]. The water vapor permeability and oxygen permeability for PLA and PLA bionanocomposite films are shown in Figure 15.4.

Other than CNCs, chitosan (CH) was also studied as a reinforcement in the PLA matrix. In this work, tributyl o-acetyl citrate (ATBC) was used as the plasticizer for the fabrication of PLA/CH composite films. It was found that the presence of ATBC could favor flexible PLA films due to its plasticizing effect. Hence, the thermal, mechanical, and transparency properties for the PLA composite films were found to be enhanced. The presence of CH in the PLA matrix favored antifungal properties for the composite films. With respect to an increase in the CH content, the films showed significant antifungal activities against *Fusarium graminearum* G87, *Aspergillus brasiliensis* ATCC 16404, and *Penicillium corylophilum* CBMF1 [42].

Likewise, nanochitosan (nCH) was also tested as a reinforcement in the PLA matrix. In this work, different weight loadings of nCH were incorporated in the PLA matrix. The composite films were fabricated using the solution casting approach. For the fabrication of PLA/nCH composites, polyethylene glycol (PEG) and polyvinyl alcohol (PVA) were used as the cross-linking agent and plasticizer, respectively. The addition of a cross-linking agent improved the heat sealing as well as tensile strength properties for PLA composites. The fabricated films could enhance the shelf-life of the packed prawn for 18 days when stored under refrigerated conditions [43].

The PLA film incorporated with α-Tocopherol (2.5 wt%) was fabricated by using the blown-film extrusion process. The releasing studies for the fabricated film were carried out

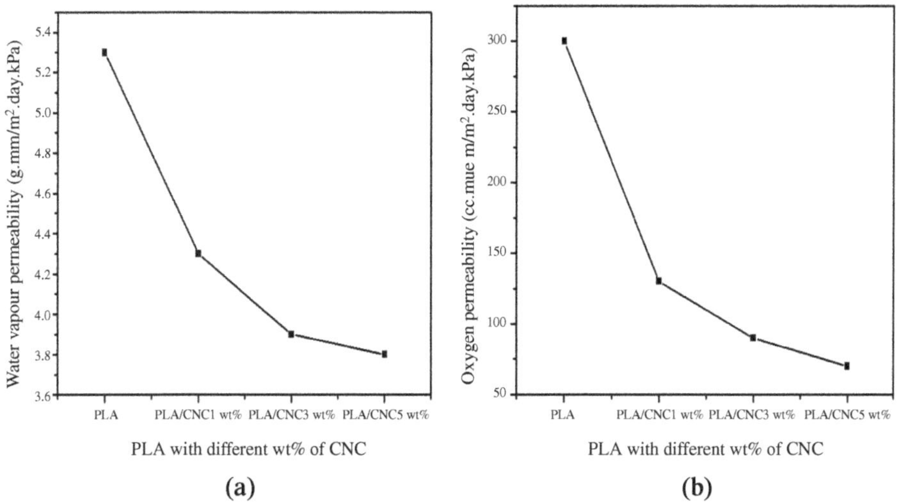

Figure 15.4 *(a) Water vapor permeability and (b) oxygen permeability for PLA and PLA bionanocomposite films [41].*

in vegetable oil and ethanol under temperature conditions ranging from 13°C to 43°C. It was found that the α-Tocopherol diffused slowly in vegetable oil as compared to ethanol. The release percentage of α-Tocopherol in vegetable oil and ethanol was calculated to be ~5.1% and ~12.9%, respectively. Hence, it was deemed that the α-Tocopherol incorporated PLA composite films could be useful for storage of oily foods at ambient temperature [44]. Similarly, the influence of bergamot essential oil (BEO), nanotitanium dioxide (n-TiO_2) and nanosilver (n-Ag) on the physicochemical as well as antimicrobial characteristics of PLA films was investigated. The formulations used for the preparation of PLA composite films are as follows: PLA/BEO, PLA/n-TiO_2/BEO, and PLA/n-TiO_2/n-Ag/BEO. The results indicated that the PLA/n-TiO_2/n-Ag/BEO composite film showed antimicrobial activity and hence could help in delaying the ripening process of the stored mangoes after post harvesting. It was reported that the quality of the mangoes was also maintained throughout the storage period. It was found that the barrier characteristics of the PLA composite films had made an influence on the physicochemical quality of the fresh mangoes during storage [45].

The cinnamon oil (CO) incorporated PLA composite films were fabricated via the solution casting method. It was found that the incorporation of CO significantly reduced the crystallization temperature (T_c) to 63°C. The decrement in T_c conveys the enhancement in the crystallization process for the PLA composite films after incorporation of CO in the PLA matrix. This was possible due to the ease in PLA chain mobility at a reduced temperature [46]. In a separate contribution, the PLA/zein composites containing different concentrations of carvacrol were fabricated. In another study, the composite fibrous films were developed from zein and PLA by incorporating carvacrol at various concentrations. The antioxidant property of the PLA composites was found to be enhanced by a maximum of ~75%, when 20% carvacrol content was loaded in the PLA/zein matrix [47].

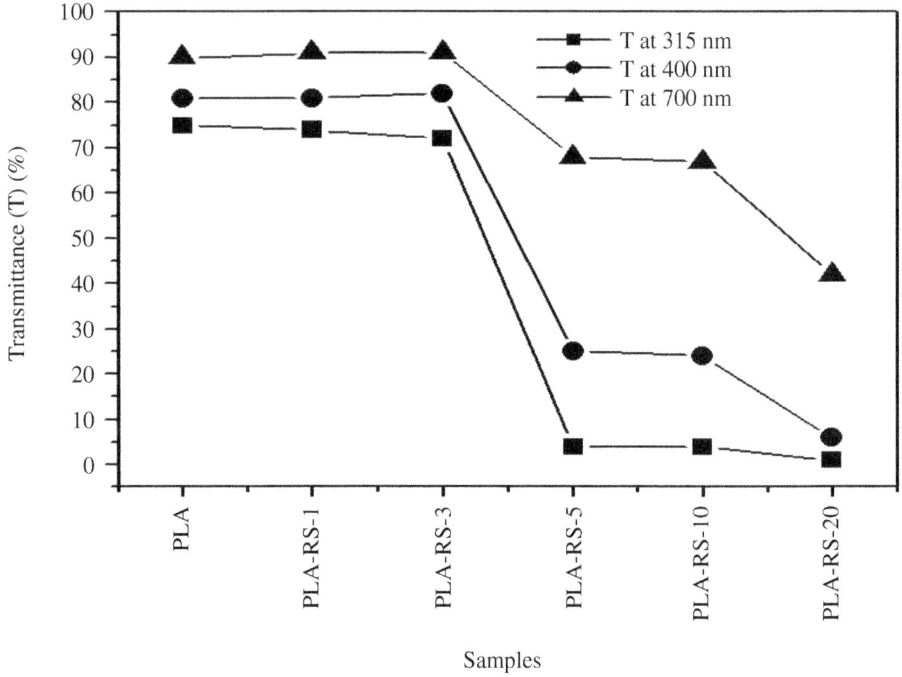

Figure 15.5 *The percentage of light transmission (%T) values for PLA and PLA-RS films [48].*

Similarly, PLA/rosin (RS)-based sustainable composite films were fabricated using the solvent-casting method. The PLA composite films were found to show enhanced UV shielding and oxygen barrier characteristics. In addition to this, the incorporation of RS improved the ductility property for the PLA composites. The results showed that the maximum of ~98% reduction in the UV light was noticed for the PLA/RS composite films in the UV-B regime. In the case of the UV-A region, ~92% reduction in the UV light was noticed for the PLA/RS composites. The most significantly complete blockage of UV light through the composite films was exhibited. In the visible region, a ~52% reduction in the transmission was noticed for the PLA/RS composite films. The percentage of light transmission (%*T*) values for PLA and PLA-RS films are shown in Figure 15.5. The oxygen permeation was found to be reduced by ~41%, indicating the tortuous path effect provided by the presence of RS in the PLA matrix. The ductility for the PLA/RS composite films were also found to be enhanced by ~67% as compared to neat PLA films [48].

15.3.4 PLA-Mesoporous Silica Composites

Recently, mesoporous silica (MS) was used as a reinforcement in the PLA matrix in limited studies. Since MS exhibits high thermal and mechanical stabilities along with a biocompatibility nature, it is being envisaged as an attractive reinforcement in the PLA matrix. The compatibility issue associated with reinforcement of MS in the PLA matrix was overcome via functionalization of MS using amine compounds. The effect of reinforcement of linear and branched amine functionalized MS on the thermal, mechanical, as well as the barrier characteristics of PLA has been studied. The linear amine compounds used in the study include tetraethylene pentamine (TEPA) and pentaethylene hexamine (PEHA). On the other hand, polyethylenimine (PEI) was used as the branched amine functionality for modification of MS.

The thermal analysis indicated that the maximum degradation temperature (T_{max}) was enhanced by ~13°C for the PEI functionalized MS reinforced PLA composites in comparison with neat PLA. The high thermal stability possessed by MS resulted in an increased degradation temperature for the PLA composites. The ductility for the PEI functionalized MS incorporated PLA films was found to be enhanced by the maximum of ~86% in comparison with the neat PLA. It was reported that the possible reason for improvement in flexibility for the PLA composite films might be due to the effective compatibility between PEI functionalized MS and PLA. The oxygen permeation for the PEI functionalized MS reinforced PLA composite films was found to be reduced by ~36% in comparison with neat PLA, which, in turn, confirms the tortuous effect created by the reinforcement [49, 51]. Likewise, PLA/MS/ascorbic acid (AA) fibrous composite films were fabricated by the electrospinning method. It was found that the films contained random orientation defect-free fibers. The PLA composite films were characterized by the presence of a porous nature. The ascorbic acid was physisorbed on the mesoporous silica surface and provided an antioxidant property for the PLA composite films [50].

15.4 Conclusions

The market for packaging materials based on biopolymers has been continuously increasing mainly due the nonbiodegradable nature of conventional plastics and depletion of fossil fuel resources. Among several biopolymers, PLA has attracted the packaging industry due to its easy production and processing ability. However, the performance of PLA-based materials for packaging applications with a view to commercialization needs to be improved as compared to petrochemical-based conventional plastics. The poor barrier properties exhibited by PLA toward oxygen as well as water vapor and its high brittleness issue have been addressed by several research groups via incorporation of nanofillers in the PLA matrix.

In this regard, inorganic fillers like layered silicates (Cloisite 30B and MMT), carbonaceous materials (CNTs and graphene), biofillers (rosin, CNC, chitosan, etc.), and porous materials (mesoporous silica) were used as reinforcements in the PLA matrix in order to enhance the performance in terms of packaging applications. The inorganic fillers mainly aided in improving the crystallinity and mechanical properties of PLA. The PLA/organically functionalized Cloisite 30B composite was found to enhance the storage modulus by ~61% as compared to neat PLA. Similarly, MMT was also found to enhance the tensile strength by ~57% for PLA. On the other hand, the carbonaceous fillers like CNTs and graphene mainly helped in improvement of thermal and barrier properties due to their high thermal stability and plate-like morphology. The presence of rGO in the PLA matrix could reduce the water vapor passage through the composite films by ~87% as compared to neat PLA. However, uniform and complete dispersion of carbon-based fillers in the PLA matrix seems to be a challenging task.

Recently, the utilization of biofiller-based reinforcements in the PLA matrix has been growing due their biocompatibility with the food during storage. The incorporation of biofillers has been shown to improve the tensile, barrier, and antimicrobial properties of PLA. The cellulose-based reinforcements have allowed for a reduction in water vapor (~40%) and oxygen permeation (~75%) for the PLA composites. The tensile strength for PLA/CNF composites was also found to be enhanced by ~59% as compared to neat PLA. In addition to this, the presence of organic plasticizers like cinnamon oil in the PLA/CNF composite was found to have a profound effect on the antimicrobial activity of the films. The incorporation of rosin in the PLA matrix presented UV shielding properties to the packaging films. The incorporation of functionalized mesoporous MS in the PLA matrix could improve the thermal, mechanical, and barrier properties for the composites. Hence, exploration of novel fillers/reinforcement procedures for fabrication of PLA composite-based packaging films with improved properties adequate for commercialization can be envisaged as an open field of research in the current trend.

References

1. Leja K, Lewandowicz G. Polymer biodegradation and biodegradable polymers–a review. *Pol. J. Environ. Stud.* 2010;**19**(2): 255-266.
2. Andrady AL, Neal MA. Applications and societal benefits of plastics. *Philos. Trans. R. Soc. B* 2009;**364**(1526):1977–84.
3. Coles R, McDowell D, Kirwan MJ. *Food Packaging Technology*: CRC Press; 2003.
4. Marsh K, Bugusu B. Food packaging–roles, materials, and environmental issues. *J. Food Sci.* 2007;**72**(3):R39-R55.
5. Siracusa V, Rocculi P, Romani S, Dalla Rosa M. Biodegradable polymers for food packaging: a review. *Trends Food Sci. Technol.* 2008;**19**(12):634–43.
6. Sorrentino A, Gorrasi G, Vittoria V. Potential perspectives of bio-nanocomposites for food packaging applications. *Trends Food Sci. Technol.* 2007;**18**(2):84–95.
7. Tharanathan R. Biodegradable films and composite coatings: past, present and future. *Trends Food Sci. Technol.* 2003;**14**(3):71–78.
8. Mittal V. Polymer layered silicate nanocomposites: a review. *Materials* 2009;**2**(3):992–1057.
9. Zhang Q, Chen G, Yoon S, Ahn J, Wang S, Zhou Q, et al. Thermal conductivity of multiwalled carbon nanotubes. *Phys. Rev. B* 2002;**66**(16):165440.
10. Bax B, Müssig J. Impact and tensile properties of PLA/Cordenka and PLA/flax composites. *Compos. Sci. Technol.* 2008;**68**(7–8):1601–7.
11. Wei L, Hu N, Zhang Y. Synthesis of polymer—mesoporous silica nanocomposites. *Materials* 2010;**3**(7):4066–79.
12. Kolybaba, M., Tabil. L., Panigrahi, S., et al. (eds) (2006) Biodegradable polymers: past, present, and future. *ASABE/CSBE North Central Intersectional Meeting*. American Society of Agricultural and Biological Engineers.
13. Armentano I, Bitinis N, Fortunati E, Mattioli S, Rescignano N, Verdejo R, et al. Multifunctional nanostructured PLA materials for packaging and tissue engineering. *Prog. Polym. Sci.* 2013;**38**(10–11):1720–47.
14. Tănase EE, Râpă M, Popa O. Biopolymers based on renewable resources-a review. *Sci. Bull. Series F Biotechnol.* 2014;**18**:188–95.
15. Vieira MGA, da Silva MA, dos Santos LO, Beppu MM. Natural-based plasticizers and biopolymer films: a review. *Eur. Polym. J.* 2011;**47**(3):254–63.

16. Basu A, Nazarkovsky M, Ghadi R, Khan W, Domb AJ. Poly(lactic acid)-based nanocomposites. *Polym. Adv. Technol.* 2017;**28**(8):919–30.
17. Gabor D, Tita O. Biopolymers used in food packaging: a review. *Acta Univ. Cinbin., Series E: Food Technol.* 2012;**16**(2).
18. Raquez J-M, Habibi Y, Murariu M, Dubois P. Polylactide (PLA)-based nanocomposites. *Prog. Polym. Sci.* 2013;**38**(10–11):1504–42.
19. Carrasco F, Pagès P, Gámez-Pérez J, Santana O, Maspoch ML. Processing of poly(lactic acid): characterization of chemical structure, thermal stability and mechanical properties. *Polym. Degrad. Stab.* 2010;**95**(2):116–25.
20. Xing C, Matuana LM. Epoxidized soybean oil-plasticized poly(lactic acid) films performance as impacted by storage. *J. Appl. Polym. Sci.* 2016;**133**(12).
21. Karkhanis SS, Stark NM, Sabo RC, Matuana LM. Blown film extrusion of poly(lactic acid) without melt strength enhancers. *J. Appl. Polym. Sci.* 2017;**134**(34):45212.
22. Karkhanis SS, Stark NM, Sabo RC, Matuana LM. Water vapor and oxygen barrier properties of extrusion-blown poly(lactic acid)/cellulose nanocrystals nanocomposite films. *Composites Part A* 2018;**114**:204–11.
23. Tempelaar S, Mespouille L, Coulembier O, Dubois P, Dove AP. Synthesis and post-polymerisation modifications of aliphatic poly(carbonate) prepared by ring-opening polymerisation. *Chem. Soc. Rev.* 2013;**42**(3):1312–36.
24. Jiang L, Wolcott MP, Zhang J. Study of biodegradable polylactide/poly(butylene adipate-co-terephthalate) blends. *Biomacromolecules* 2006;**7**(1):199–207.
25. Krikorian V, Pochan DJ. Poly(L-lactic acid)/layered silicate nanocomposite: fabrication, characterization, and properties. *Chem. Mater.* 2003;**15**(22):4317–24.
26. Chang J-H, An YU, Cho D, Giannelis EP. Poly(lactic acid) nanocomposites: comparison of their properties with montmorillonite and synthetic mica (II). *Polymer* 2003;**44**(13):3715–20.
27. Ayana B, Suin S, Khatua B. Highly exfoliated eco-friendly thermoplastic starch (TPS)/poly(lactic acid)(PLA)/clay nanocomposites using unmodified nanoclay. *Carbohydr. Polym.* 2014;**110**:430–9.
28. Ramos M, Jiménez A, Peltzer M, Garrigós MC. Development of novel nano-biocomposite antioxidant films based on poly(lactic acid) and thymol for active packaging. *Food Chem.* 2014;**162**:149–55.
29. Koh HC, Park JS, Jeong MA, Hwang HY, Hong YT, Ha SY, et al. Preparation and gas permeation properties of biodegradable polymer/layered silicate nanocomposite membranes. *Desalination* 2008;**233**(1–3):201–9.
30. Bhatia A, Gupta RK, Bhattacharya SN, Choi HJ. Analysis of gas permeability characteristics of poly(lactic acid)/poly(butylene succinate) nanocomposites. *J. Nanomater.* 2012;**2012**:6.
31. Seligra PG, Lamanna M, Famá L. PLA-fMWCNT bionanofilms with high modulus and great properties to apply in packaging and biomedicine. *Proc. Mater. Sci.* 2015;**8**:383–90.
32. Pinto V, Ramos T, Alves S, Xavier J, Tavares P, Moreira P, et al. Comparative failure analysis of PLA, PLA/GNP and PLA/CNT-COOH biodegradable nanocomposites thin films. *Proc. Eng.* 2015;**114**:635–42.
33. Yoon JT, Lee SC, Jeong YG. Effects of grafted chain length on mechanical and electrical properties of nanocomposites containing polylactide-grafted carbon nanotubes. *Compos. Sci. Technol.* 2010;**70**(5):776–82.
34. Chou P, Mariatti M, Zulkifli A, Sreekantan S. Evaluation of the flexural properties and bioactivity of bioresorbable PLLA/PBSL/CNT and PLLA/PBSL/TiO_2 nanocomposites. *Composites Part B* 2012;**43**(3):1374–81.
35. Goh K, Heising JK, Yuan Y, Karahan HE, Wei L, Zhai S, et al. Sandwich-architectured poly(lactic acid)–graphene composite food packaging films. *ACS Appl. Mater. Interfaces* 2016;**8**(15):9994–10004.
36. Arfat YA, Ahmed J, Ejaz M, Mullah M. Polylactide/graphene oxide nanosheets/clove essential oil composite films for potential food packaging applications. *Int. J. Biol. Macromol.* 2018;**107**:194–203.

37. Liu Y, Wang L, He Y, Fan Z, Li S. Non-isothermal crystallization kinetics of poly(L-lactide). *Polym. Int.* 2010;**59**(12):1616–21.
38. Meriçer Ç, Minelli M, De Angelis MG, Baschetti MG, Stancampiano A, Laurita R, et al. Atmospheric plasma assisted PLA/microfibrillated cellulose (MFC) multilayer biocomposite for sustainable barrier application. *Ind. Crops Prod.* 2016;**93**:235–43.
39. Fortunati E, Peltzer M, Armentano I, Torre L, Jiménez A, Kenny J. Effects of modified cellulose nanocrystals on the barrier and migration properties of PLA nano-biocomposites. *Carbohydr. Polym.* 2012;**90**(2):948–56.
40. Khodayari M, Basti AA, Khanjari A, Misaghi A, Kamkar A, Shotorbani PM, et al. Effect of poly(lactic acid) films incorporated with different concentrations of Tanacetum balsamita essential oil, propolis ethanolic extract and cellulose nanocrystals on shelf life extension of vacuum-packed cooked sausages. *Food Packag. Shelf Life* 2019;**19**:200–9.
41. Sung SH, Chang Y, Han J. Development of polylactic acid nanocomposite films reinforced with cellulose nanocrystals derived from coffee silverskin. *Carbohydr. Polym.* 2017;**169**:495–503.
42. Râpă M, Miteluţ AC, Tănase EE, Grosu E, Popescu P, Popa ME, et al. Influence of chitosan on mechanical, thermal, barrier and antimicrobial properties of PLA-biocomposites for food packaging. *Composites Part B* 2016;**102**:112–21.
43. Fathima P, Panda SK, Ashraf PM, Varghese T, Bindu J. Polylactic acid/chitosan films for packaging of indian white prawn (fenneropenaeus indicus). *Int. J. Biol. Macromol.* 2018;**117**:1002–10.
44. Manzanarez-López F, Soto-Valdez H, Auras R, Peralta E. Release of α-tocopherol from poly(lactic acid) films, and its effect on the oxidative stability of soybean oil. *J. Food Eng.* 2011;**104**(4):508–17.
45. Chi H, Song S, Luo M, Zhang C, Li W, Li L, et al. Effect of PLA nanocomposite films containing bergamot essential oil, TiO_2 nanoparticles, and Ag nanoparticles on shelf life of mangoes. *Sci. Hortic.* 2019;**249**:192–8.
46. Ahmed J, Mulla MZ, Arfat YA. Thermo-mechanical, structural characterization and antibacterial performance of solvent casted polylactide/cinnamon oil composite films. *Food Control* 2016;**69**:196–204.
47. Altan A, Aytac Z, Uyar T. Carvacrol loaded electrospun fibrous films from zein and poly(lactic acid) for active food packaging. *Food Hydrocolloids* 2018;**81**:48–59.
48. Narayanan M, Loganathan S, Valapa RB, Thomas S, Varghese T. UV protective poly(lactic acid)/rosin films for sustainable packaging. *Int. J. Biol. Macromol.* 2017;**99**:37–45.
49. Loganathan S, Jacob J, Valapa RB, Thomas S. Influence of linear and branched amine functionalization in mesoporous silica on the thermal, mechanical and barrier properties of sustainable poly(lactic acid) biocomposite films. *Polymer* 2018;**148**:149–57.
50. Cacciotti I, Nanni F, editors. Poly(lactic) acid fibers loaded with mesoporous silica for potential applications in the active food packaging. *AIP Conference Proceedings*; 2016: AIP Publishing.
51. Ilyas RA, Sapuan SM, Atikah MSN, Asyraf MRM, Ayu Rafiqah S, Aisyah HA, Mohd Nurazzi N, Norrahim MNF. Effect of hydrolysis time on the morphological, physical, chemical, and thermal behavior of sugar palm nanocrystalline cellulose (*Arenga pinnata* (Wurmb.) Merr). *Textile Research Journal.* 2020. In print. doi: 10.1177/0040517520932393

16

Polyhydroxyalkanoates for Packaging Application

Tengku Arisyah Tengku Yasim-Anuar[1], Mohd Nor Faiz Norrahim[2], S.M. Sapuan[3,4], R.A. Ilyas[5,6], Mohd Azwan Jenol[1], Nur Amira Mamat Razali[7], Mohd Idham Hakimi[1], Nur Farisha Abd Rahim[8], and Syed Umar Faruq Syed Najmuddin[9]

[1]Department of Bioprocess Technology, Faculty of Biotechnology and Biomolecular Sciences, Universiti Putra Malaysia, 43400 Serdang, Selangor, Malaysia

[2]Research Centre for Chemical Defence (CHEMDEF), Universiti Pertahanan Nasional Malaysia, Kem Perdana Sungai Besi, 57000 Kuala Lumpur, Malaysia

[3]Advanced Engineering Materials and Composites Research Centre (AEMC), Department of Mechanical and Manufacturing Engineering, Universiti Putra Malaysia, 43400, Serdang, Selangor, Malaysia

[4]Laboratory of Biocomposite Technology, Institute of Tropical Forestry and Forest Product (INTROP), Universiti Putra Malaysia, 43400, Serdang Selangor, Malaysia

[5]School of Chemical and Energy, Faculty of Engineering, Universiti Teknologi Malaysia,81310 UTM Johor Bahru, Johor, Malaysia

[6]Centre for Advanced Composite Materials, Universiti Teknologi Malaysia,81310 UTM Johor Bahru, Johor, Malaysia

[7]Department of Physics, Center for Defence Foundation Studies, Universiti Pertahanan Nasional Malaysia, 57000 Kuala Lumpur, Malaysia,

[8]Level 3, Putra Business School, UPM, 43400 Serdang, Selangor, Malaysia

[9]Laboratory of Vaccines and Immunotherapeutics (LIVES), Institute of Bioscience, Universiti Putra Malaysia (UPM), 43400 Serdang, Selangor, Malaysia

16.1 Introduction

Increased food and fuel demand due to growing human population has led to a search for alternative plastics. Conventional plastics have also been called synthetic plastics and are

Bio-Based Packaging: Material, Environmental and Economic Aspects, First Edition. Edited by S.M. Sapuan and R.A. Ilyas

derived from petroleum. They have been widely used for numerous applications, mainly due to their excellent properties; they are lightweight, flexible, easy to handle and store, mechanically strong as well as resistant to high and low temperatures [1]. Despite their many advantages, the use of plastics has led to severe environmental pollution, either on the land or in the ocean. It was reported that around 22–43% of plastics end up in landfills and this has led to groundwater pollution by the leaching of toxic additives [2].

In this era of modernization, numerous products rely on plastics as raw materials, including for food packaging. It is noted that the main purpose of food packaging is to maintain the quality and safety of foods during storage as well as to extend the shelf life of foods by preventing unfavorable factors such as bacterial or chemical contaminants, moisture, light and external forces that can affect the quality and safety of the foods [3, 64]. The food package has not solely been used as a container, but it should also act as a protective barrier in protecting foods. Conventionally, paper, metal, glass, and plastics are the common materials used to keep foods, although plastics are the top choice. This is mainly because they are cheap with good physicochemical properties. Despite their versatility and functionality, the use of plastics for food packaging has led to serious environmental problems since they do not easily degrade.

In line with the concern regarding environmental impact caused by plastics, several countries have launched campaigns to reduce the use of plastics. Alternatively, biopolymers from (i) natural biopolymers such as cellulose, chitosan, starch, etc., (ii) synthetic biodegradable polymers such as polylactic acid (PLA), poly(glycolic acid) (PGA), poly(butylene succinate) (PBS), etc., (iii) biopolymers produced by microbial fermentation such as polyhydroxyalkanoates (PHAs), etc., and (iv) microbial polysaccharides such as pullulan have been highlighted, as they are able to replace those synthetic plastics, especially for single-use applications such as disposable plates, cups, cutlery, diapers, trash bag, food containers, medical devices, and many more [3]. Biopolymers are generally produced from renewable biological resources, with excellent biodegradability, and possess almost similar mechanical properties to the synthetic plastics.

Nevertheless, focusing on their usage as packaging plastics, several concerns must be addressed in order to enhance the effectiveness of biopolymers in keeping the food quality with improved convenience for processing and final use. Several concerns are (i) degradation rates vary in different conditions, (ii) changes in mechanical properties during storage, (iii) potential for microbial growth, and (iv) release of harmful compounds into packed food products. Among many biopolymers, PHAs are getting the most interest, owing to their excellent biodegradability, biocompatibility, mechanical, and thermal properties [4, 5]. PHAs are also considered inert, hydrophobic, nontoxic, stable in air, and enantiomerically pure, properties that make them ideal to be used for food packaging as well as for medical purposes [6]. In fact, PHAs are now being referred as "green plastic" due to their positive environmental impact [5].

Nevertheless, PHAs are known to have a brittle characteristic, which limits their applications. Intrinsic PHAs can be considered as excellent biomaterials, but they lack proper functionalities [6]. Hence, PHAs are commonly blending, grafting, or coating, or are chemically modified to enhance their properties, with the aim to meet requirements of targeted applications. In this review, PHAs are presented here as an example of how technology and innovation can contribute to meeting societal needs for sustainable and safe packaging, especially for food.

Table 16.1 Biodegradable and nonbiodegradable biopolymers.

	Biodegradable	Nonbiodegradable
Biopolymers	Polyhydroxyalkanoates Polylactic acid Starch-based plastics	Polyamide 11 (PA11) Bio-derived polyethylene

16.2 Biopolymers

Biopolymers can be classified into two main classes, which are biodegradable and nonbiodegradable polymers. Biodegradable plastics can be defined as plastics that can be degraded biologically by microorganisms such as bacteria and fungus. According to Verhoogt et al. [7], this term can be applied if the polymer is readily degraded to carbon dioxide (CO_2) and water (H_2O). There have been many developments of biopolymers that can be degraded in the environment when discarded. Table 16.1 shows the list of biodegradable and nonbiodegradable polymers.

As seen in the table, not all bio-based polymers are biodegradable, and vice versa. In developing sustainable packaging plastics, especially for short-term use, bio-based or biodegradable plastics are preferable. Attention has been focused on biodegradable polymers that can be produced from renewable resources using various techniques and methods [8]. Several innovative techniques such as the bilayer or the use of multicomponent films with good mechanical properties have been developed to make these biopolymers applicable for packaging applications [9]. For example, a technology was developed for continuous production of laminated PVCs, with films containing 50% or more starch could reduce the water sensitivity of such films.

In practice, biopolymers need to be improved before being applied as packaging plastics, since some characteristics of these materials do not match with commercial plastics characteristics [10]. For example, most biopolymers have a thermal instability characteristic. At high temperatures at ~150°C the glucose links of the starch-based polymer start to break apart while at low temperatures retrogradation occurs. This phenomenon makes the resulting starch-based films very brittle. In addition, PHAs are known to have a narrow processing window because their melting and degradation temperatures are too close to each other. Therefore, the processability of this polymer for commercialization will be difficult. Moreover, inherent brittleness of biopolymers largely impedes their wide applications. The brittleness and low mechanical strength of biopolymers make them difficult to replace the use of polyolefins as packaging materials.

Several efforts such as plasticization, copolymerization, and blending with other polymers have been studied in order to improve their physical and mechanical properties [11]. In spite of this, copolymerization has also been extensively investigated to enhance the properties of biopolymers [11]. Properties of materials including tensile and impact strength of the copolymer can be manipulated by architecture of the molecule, sequence of monomers, and composition. Moreover, melt blending of polymers is a much more economic and convenient method than synthesizing new polymers to achieve the properties unattainable with existing polymers.

16.3 Polyhydroxyalkanoates

PHAs are polyesters of hydroxyalkanoates (HAs) synthesized by various microorganisms under limitations of nutrients and excess amounts of carbon sources [10]. PHAs have the potential for numerous applications, but due to the high cost compared to polyolefins makes it difficult to be commercialized. Several investigations such as recombinant microbial strains, cheap carbon sources, effective fermentation techniques, and mixed cultures can significantly decrease the production cost. PHAs have been the most studied among all bio-based and biodegradable polymers [8]. PHAs are thermoplastic materials that share similar properties to petroleum-based plastics, such as polypropylene (PP) and polyethylene (PE) [12]. Moreover, it has also been reported that the PHAs and synthetic plastics are thermoplastic, moldable, and could be tailor-made for numerous strengths based on the application.

In addition, PHAs are produced from renewable sources such as sugars, plant oils, and carbon dioxide, which are very useful when applied as packaging materials to replace the polyolefins from petroleum that are expected to decline [13]. Table 16.2 shows an example of bacteria strains and carbon sources used in the production of PHAs.

16.3.1 Characteristic of PHAs

PHAs are well-known microbiologically produced and biodegradable polymers. They act as carbon and energy sources for bacteria and fungus [15]. There are several types

Table 16.2 *Example of bacteria strains and carbon sources used in the production of PHAs.*

Bacteria strain(s)	Carbon source(s)	Polymer(s) produced
Microlunatus phosphovorus	Glucose, acetate	P(3HB)
Pseudomonas aerugionosa	Glucose, technical oleic acid, waste free fatty acids, waste free frying oil	Mcl-PHAs
Pseudomonas oleovorans	Octanoic acid	Mcl-PHAs
Psuedomonas putida	Glucose, octanoic acid, undecenoic acid	Mcl-PHAs
Psuedomonas putida, Psuedomonas fluorescence, Pseudomonas jessenii	Glucose, aromatic monomers	Aromatic polymers
Pseudomonas stutzeri	Glucose, soybean oil, alcohols, alkanoates	Mcl-PHAs
Rhizobium meliloti, Rhizobium viciae, Bradyhizobium japonicum	Glucose, sucrose, galactose, mannitol, trehalose, xylose, raffinose, maltose, dextrose, lactose pyruvate, sugar beet, molasses, whey	P(3HB)
Rhodopseudomonas palustris	Acetate, malate, fumarate, succinate, propionate, malonate, gluconate, butyrate, glycerol, citrate	P(3HB), P(3HB-*co*-3HV)
Spirulina platensis	Carbon dioxide	P(3HB)
Staphylococcus epidermidis	Malt, soy waste, milk waste, vinegar waste, sesame oil	P(3HB)
Cupriavidus necator	Glucose, sucrose fructose, valerate, octanoate, lactic acid, soybean oil	P(3HB), copolymers
Cupriavidus necator H16	Hydrogen, carbon dioxide	P(3HB)

Source: Mohammadi [10]; Pérez-Arauz et al. [14].

of microorganism used to produce these PHAs, such as *Bacillus megaterium*, *Rastolnia eutropha*, *Pseudomonas oleovorans*, and some other bacterial species [16]. Besides that, PHAs are aliphatic polyesters that can suffer from hydrolytic or enzymatic degradation and, by the action of microorganisms, PHAs will be degraded completely to carbon dioxide and water [17]. The high molecular weight of PHAs are synthesized and stored in the cytoplasm of a cell as water insoluble inclusions by various microorganisms [16].

This biopolymer was initially investigated by Lemoigne and was then continued by many other researchers [16]. Several researchers investigated the important application and production process of this polymer. PHAs in general are one of the biopolymers produced by microorganisms under certain unfavorable conditions such as excess carbon sources with limitations of nitrogen, oxygen, or phosphorus [4, 18]. There are several types of PHAs, namely polyhydroxy(3-hydroxybutyrate) (PHB), poly(3-hydroxybutyrate-*co*-3-hydroxyvalerate) (PHBV), poly(3-hydroxybutyrate-*co*-3-hydroxyhexanoate) (PHB-*co*-HHx). and polyhydroxyoctanoate (PHO), which can be divided into two classes, which are the short side chain and the long side chain [7].

PHB has a melting point of 180°C with a high degree of crystallization, but have been limited when applied as consumer plastics due to brittleness and narrow processing window [7]. The narrow processing window is due to the decomposition temperature, which is too close to its melting temperature. It is believed that the brittleness of this biodegradable polymer can be improved by adding an HV side chain. This is because the PHBV was found to be tougher, exhibits the isodimorphism phenomenon, and lowers the melting point of the pure PHB. Previous research done by Reis et al. [19] suggested that the copolymer with addition of 3-hydroxyvalerate (HV) units has high malleability. Due to PHAs brittleness, improvement can be achieved by the incorporation of other HA units to the polymer backbone, which results in various copolymers. Apart from that, the properties of the polymer can be improved by blending with other polymers such as PP, PE, PLA, starch, and cellulose. For example, studies of PLA/PHB blends have shown that their mechanical properties are intermediate between those of neat polymer [20].

Apart from its shortcoming in mechanical properties, PHA exhibits good barrier properties. According to previous research done by Shogren [21] on the evaluation of the water vapor permeability of biodegradable polymers, PHAs, particularly PHBV, recorded the highest permeability compared to other biopolymers tested, such as PLA, PCL, blown film of aliphatic polyester, cellulase acetate, and polyester amide. This suggest that PHAs are useful when applied as packaging materials as compared to other biopolymers.

Previous research has shown the effect on the thermal and mechanical properties of a variety of hydroxyvalerate content polymers. PHBV with different HV contents gives the different mechanical properties shown in Table 16.3.

Even though the addition of HV content can improve the mechanical properties of PHB, there is a limitation that may limit the interaction between PHB and HV. Based on the data shown in Table 16.3, it shows that 20% of HV content resulted in low mechanical properties, which was attributed to the limited interaction between PHB and HV. Besides that, too low an HV content will not improve the elongation at break of the polymer because a lower amount of HV will cause a low interaction between the polymers. Thus, it will not improve embrittlement to the PHB. Therefore, 12% of HV content is better when compared to other contents.

Table 16.3 Mechanical properties of PHBV materials.

Samples	Tensile strength (MPa)	Elongation at break (%)	Flexural strength (MPa)	Elastic modulus (GPa)
T20%	15.2	1.26	29.6	8.37
T5%	22.1	1.39	52.3	13.0
A12%	20.9	4.49	154	8.67
Biopol 12% HV	23.0	18.8	30.0	1.25
Tianan 1% HV	19.7	0.170	39.1	3.65

Source: Modi [22].

16.3.2 Biodegradability and Enzymatic Degradability of PHAs

Degradability of PHAs as a packaging material is important to be considered. As mentioned earlier, the use of plastics causes severe environmental pollution. Biodegradability is one of the unique characteristics with the ability to be degraded in various environments. Biodegradability of PHAs in natural environments such as soil, sea water, and lake water has been studied by monitoring the properties of PHAs [16]. It was also been reported that several factors can influence the biodegradability of the PHAs, such as temperature, moisture content, nutrient supply, pH level, and blending. By blending the PHAs with other polymers a better degradability of the blends is expected to be exhibited. This is an important factor that needs to be considered in order to reduce the amount of petroleum-based plastics.

Enzymatic hydrolysis can also degrade the PHAs. According to the research done by Choi et al. [23] on the study of enzymatic degradation and non-enzymatic degradation of PHBV, it was revealed that the prepared film samples were degraded by the action of enzyme PHB depolymerase. However, sample films prepared were not enzymatically hydrolyzed in 0.01 N sodium hydroxide solutions. There were no weight changes observed in this lack of enzymatic degradation over a period of 20 weeks.

16.3.3 Application of PHAs

Several interesting applications have been reported for PHAs and other biopolymers [14, 24]. PHAs are believed to have a strong potential for industrial purposes as compared to other biopolymers [2, 25]. Nowadays, there is growing concern about the current practice of plastic disposal and their impact on the environment. Several approaches have been developed to make biopolymers practicable for many applications. PHAs have been reported to be applied in many fields such as packaging, medical, tissue engineering, healthcare, and others.

Potential applications of PHAs can be divided into two groups depending on origin of the polymer [12]. The first group is where the polymer is obtained from agricultural and industrial wastes. The products derived from this polymer will be used for agricultural purposes, packaging, and low-value products. The first product derived from PHAs was Sanara shampoo (product), produced by Wella AG in 1990.

The second group is a polymer obtained from a pure substrate such as glucose, fructose, and oils from plants. The products produced are probably in small quantities but normally for high value applications such as medical applications and tissue engineering. Table 16.4 shows several applications of PHAs.

Table 16.4 *Applications of PHAs.*

Application	Description	References
Consumer products	– Food packaging, garbage bag, disposal cutlery and plates, shipping materials, blown film, bathroom accessories and others. – Applications of biopolymers are still limited due to its brittleness and thermal instability characteristic.	[10, 26]
Medicine	– Used in the medical application due to its biocompatibility characteristic. – Approved by Food and Drug Administration (FDA) as a biomaterial investigated application for a variety of medical devices. – PHAs are used for medical applications such as tissue engineering, sutures, repair patches, slings, orthopedic pins, cardiovascular patches, bone marrow scaffolds, artificial esophagus, stents, and more. – PHBV also has a potential in reconstructive surgery.	[10, 27–30]
Agriculture	– PHAs have been used in the agricultural area as mulch films. – The PHBV has good application of controlled release of insecticides which is very useful to be used in the farmer's crops.	[26]
Others	– Beauty healthcare – Construction materials – PHBV composites: material for structural application with good thermal insulation (improvement of the energy efficiency in eco-buildings).	[10, 28, 31]

PHAs possess several properties to qualify them as biomaterial for scaffolds, mainly due to their unchanging pH during degradation, thus making them well tolerated by the immune system compared to other polymers such as PLA, PGA, and polylactide-*co*-glycolid (PLGA) [32]. Zubairi et al. [33] also reported that PHAs possess five important keys that make it suitable for scaffolds; (i) biocompatible, (ii) support cell growth, (iii) organizing cells, (iv) allowing tissue ingrowth, and (v) able to degrade to non-toxic products. In another study by Evangeline and Sridharan [34], PHB synthesized from *Bacillus cereus* strain VIT-SSR1 isolated from an industrial effluent contaminated site was found to be biocompatible and can act as a potent polymer matrix for drug release.

The marine industry is also among the industries that have benefited from PHAs, as they are referred to as marine biodegradable plastics. Currently, efforts have been shown to replace plastic-based material to PHAs in the ocean, for example escape panels to trap fishes, buoys, and others. In comparison to other plastics that break down very slowly and require UV radiation, heat, and/or oxygen to degrade, PHAs can degrade in low temperatures and minimal UV exposure [35]. Hence, their lifetime is shorter compared to other plastics, which usually take hundreds to thousands of years to degrade. Dilkes-Hoffman et al. [36] and colleagues discovered that the mean rate of biodegradation of PHAs in the marine environment was approximately 0.04–0.09 mg/day, and it was also expected that a PHAs bottle can be completely degraded in 1.5 to 3.5 years. This is one of the promising solutions to encounter plastic pollution in the ocean.

Table 16.5 Tensile properties of selected PHAs/natural fiber composites.

Fiber (wt% fraction in PHAs composites)	Tensile properties		References
	Strength σ (MPa)	Modulus E (GPa)	
Wood flour (40%)	16.75	2.73	[37]
Recycled cellulose fibers (40%)	~24	3.2	[38]
Flax (40%)	40	~8.5	[39]
Pineapple fiber (30%)	~60	2.3	[40]
Rice straw (30%)	5.24	1.17	[31]
Wood fiber (40%)	16.75	2.73	[41]

16.4 Polyhydroxyalkanoate-Based Composites for Packaging Applications

The applications of polymer blending containing PHAs have been discussed in many previous researches. The blending of PHAs with other polymers can give valuable applications in food packaging. Waste lignocellulose fibers, highly available and at low cost, sourced from agricultural and industrial crops, have been investigated as fillers in PHA composites, as shown in Table 16.5. Most of the polymer blending containing PHAs and natural fibers displayed promising mechanical and physical properties [42].

Mechanical properties, low-impact resistance, and brittleness of PHAs were found to be improved when reinforced with natural fibers. Besides possessing renewability and biodegradability, natural fibers offer a number of benefits including abundant availability, low cost, low density, and high mechanical performance.

In general, PHA-reinforced natural fibers recorded a higher Young's modulus and tensile strength with increasing filler content. For example, Bhardwaj et al. [38] reported different mechanical increments for PHA reinforced with recycled cellulose fibers (RCFs) prepared by extrusion and injection molding processes. The tensile and storage moduli of the PHA composite improved by 220% and 190%, respectively, compared to neat PHA when the reinforcement was at 40 wt%.

On the previous study, Sun and co-workers investigated the effects of PHA content as a filler on the properties of starch-based films produced by extrusion [43]. The study reported that the PHA content on starch/PHA films influenced a water vapor permeability (WVP). The WVP of starch/PHA films significantly decreased as the PHA content increases. The films without PHA exhibited highest WVP (4.69×10^{-10} g m m^{-2} s^{-1} Pa^{-1}), while films containing 36% PHA recorded lowest WVP (1.93×10^{-10} g m m^{-2} s^{-1} Pa^{-1}). This indicates that the incorporation of starch to the PHA matrix improved the moisture barrier properties of the films. This behavior could be ascribed to the hydrophobicity of the PHA. Previous studies reported that the addition of a hydrophobic polymer such as PLA to starch-based films could cause a decrease in WVP when compared to the pure starch films [44].

PHAs have also been used as a filler in polymer blends. Usually, small amount of PHAs were added to improve the properties of the neat polymers. Table 16.6 shows the properties of polymer blends containing PHAs as a filler.

Table 16.6 Polymer blends of PHAs and their properties.

Polymer blends of PHAs	Properties	Reference
PHBV/PP	– Present a certain biodegradability of PP/PHBV with 10 parts per hundred of PHBV. – Reduce the accumulation of plastic waste in the environment.	[45]
PHB/PLA	– Improved the thermal distortion temperature. – Improved biodegradability of polymers. – Biodegradability increased by the addition of PHB.	[46]
PHA/PVA	– Maximum adhesion of cell culture experiments with a human keratinocyte cell line and dermal fibroblast on the electrospun. – Has a potential in the scaffold material for tissue engineering of skin.	[47]
PHBV/poly(ethylene adipate)	– Microcapsules with a microporous and smooth surface was fabricated with the incorporation of bovine serum albumin.	[48]
PHB/PHBHHX	– Scaffolds and tissue engineering materials production. – Good approach of biocompatibility.	[49]
PHB/starch	– Coating material on paper. – To reduce the production cost of biodegradable plastics. – Improved properties of material as compared to the virgin polymer.	[50]
PHBV/PCL/PS	– Increased degradation rate in the anaerobic biodegradation test.	[51]

16.5 Chemical Recycling of PHAs

Toward a sustainable development of renewable resources, attention should be given to develop a new effective recycling technique for packaging material. The carbon cycle of the polymer is also important for the sustainable polymer production and in polymer waste management [52]. This could be an effective alternative to reduce plastic wastes, saving polymer resources and reusing the recovered products obtained through pyrolysis, as shown in Figure 16.1.

The pyrolysis of polymer materials has been extensively studied over the years. Pyrolysis is also known as thermal decomposition where substances are chemically decomposed into other substances such as monomers and oligomers by the presence of heat. It can also be classified as the specific mode and controlled degradation mechanism of plastic for the chemical recycling by converting the polymer into other useful low molecular weight products [15]. A low molecular weight product obtained from the pyrolysis might be beneficial to be used in other applications.

16.5.1 Pyrolysis of PHAs

Pyrolysis of PHAs has been extensively studied by many researchers previously. PHAs are known as thermally unstable and they have the potential to be used as a source of small molecules [53] such a crotonic acid (CA) and acrylic acid [54]. The pyrolysis pathways of

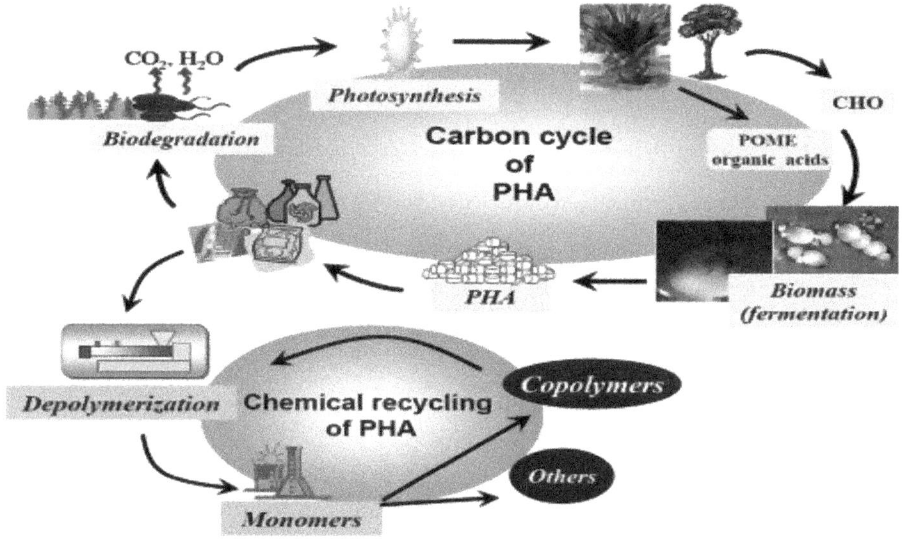

Figure 16.1 Overview of carbon cycle and chemical recycling of PHAs. Source: Ariffin et al. [52].

PHAs have been reported previously and two main kinds of monomers can be obtained, which are CA and 2-pentenoic acid (PA), depending on the chemical structure of the PHAs [53].

The random chain scission by β-elimination has been considered as the dominant mechanism for PHB pyrolysis [55]. The typical pyrolysis products of PHB pyrolysis are CA and oligomers with a crotonate end-group. The pyrolysis of PHBV showed a smooth and selective depolymerization into CA and 2-PA acids [52].

Several methods of PHAs pyrolysis were reported by Ariffin et al. [55] such as thermogravimetry (TG), differential scanning calorimetry (DSC), fast atom bombardment mass spectrometry (FAB-MS), pyrolysis-mass spectrometry (Py-MS), TG/Fourier transform infrared spectrometry (TG/FT-IR), pyrolysis-GC/FTIR, size exclusion chromatography (SEC), electrospray ionization-MS (ESI-MS), and nuclear magnetic resonance (NMR).

Norrahim et al. [56] reported that the pyrolysis of polyethylene/PHBV films at 310°C led to the degradation of PHBV into its volatile monomers and oligomers, indicating that the two polymers can be separated from each other. The ability to recycle PE and PHBV promotes recycling and cascade utilization of polymeric materials, which may contribute positively to the sustainable packaging [56].

16.5.2 Application of Crotonic Acid, 2-Pentenoic Acid, and its Derivatives

The pyrolysis products of PHAs consist of mainly CA have several interesting economical values in several applications such as cosmetic, medical application, polymer degradation, and microorganism growth inhibitor, as shown in Table 16.7.

Table 16.7 *Application of crotonic acid, 2-pentenoic acid, and their derivatives.*

Application	Description	Reference
Cosmetic products	– Copolymerization of CA produced through pyrolysis of PHBV with vinyl acetate (VA) can be used in cosmetic products such as hair spray or hair styling gel. – Grafted copolymerization of CA with AA can also improve the wettability of polyurethane.	[52, 57, 58]
Health	– Copolymerization of CA with VA has also been used to produce an interpolymer complex of poly(vinylpyrrolidone)-poly(vinylacetate-co-crotonic acid). – It was used efficiently to encapsulate bifidiobacteria in polymers to improve the shelf life and stability to simulate the gastrointestinal fluids *in vitro*.	[59]
Environment	– Poly (acrylic acid-*co*-crotonic acid) copolymers with various compositions of CA and AA are efficient as fixation agents that can be used in the recovery of heavy metals, which are very toxic in aqueous media.	[60]
Antimicrobial	– CA and oligomers also have antimicrobial characteristics that may be employed in deodorants. – CA by combination of *cis* and *trans* forms could be considered as an inhibitory compound that could prevent the growth of certain molds in food. – Besides that, the combination of CA, sorbic, salts, esters, and alpha hexanoic acids may also be employed for mold inhibiting purposes.	[61, 62]
Others	– The other copolymer produced from CA is known to be used as a component in dental applications, surface modification, plasticizers, hair styling products, herbicides, paint, textile, binder, and agrochemical industries.	[52, 63]

16.6 Future Direction and Recommendations

Interest in the use of PHAs as packaging films and composites for several applications have been growing, as discussed above. It can be concluded that PHA films could be a great alternative to reduce the environmental problem arising from plastic waste. Interestingly, PHAs can be chemically recycled, which may reduce waste generation and create new industries for economic and social development.

Several improvements can be done on the PHA composites in the future. As discussed above, several studies have been conducted by blending the biodegradable polymers with polyolefins to enhance the properties of materials. However, most of the polymer blending usually leads to an immiscible blend that could result in a marginal improvement in toughness. To overcome this problem, suitable compatibilizers can be used to increase the compatibility of PHAs in polymer blends. A good interfacial adhesion is expected to enhance the mechanical and permeability of the polymer blends. This will broaden the application of PHAs for packaging plastic in the future.

References

1. Xanthos D, Walker TR. International policies to reduce plastic marine pollution from single-use plastics (plastic bags and microbeads): a review. *Mar. Pollut. Bull.* 2017;**118**(1–2):17–26.
2. Dietrich K, Dumont MJ, Del Rio LF, Orsat V. Producing PHAs in the bioeconomy — towards a sustainable bioplastic. *Sustain. Prod. Consum.* 2017;**9**(August 2016):58–70.
3. Rhim J-W, Park H-M, Ha C-S. Bio-nanocomposites for food packaging applications. *Prog. Polym. Sci.* 2013;**38**(10–11):1629–52.
4. Yadav B, Pandey A, Kumar LR, Tyagi RD. Bioconversion of waste (water)/residues to bioplastics- a circular bioeconomy approach. *Bioresour. Technol.* 2019;122584.
5. Dwivedi R, Pandey R, Kumar S, Mehrotra D. Polyhydroxyalkanoates (PHA): role in bone scaffolds. *J. Oral Biol. Craniofacial Res.* 2020;**10**(1):389–92.
6. Butt FI, Muhammad N, Hamid A, Moniruzzaman M, Sharif F. Recent progress in the utilization of biosynthesized polyhydroxyalkanoates for biomedical applications – review. *Int. J. Biol. Macromol.* 2018;**120**:1294–305.
7. Verhoogt H, Ramsay BA, Favis BD. Polymer blends containing poly(3-hydroxyalkanoates). Polymer review. *Polymer* 1994;**35**(24):5155–69.
8. Corre YM, Bruzaud S, Audic JL, Grohens Y. Morphology and functional properties of commercial polyhydroxyalkanoates: A comprehensive and comparative study. *Polym. Test.* 2012;**31**(2):226–35. Available at: http://dx.doi.org/10.1016/j.polymertesting.2011.11.002.
9. Camann A, Dragsbeak K, Krol S, Sandgren J, Song D. Properties, recycling and alternatives to PE bags. *J. Elastomers Plast.* 2010; Worcester Polytechnic Institute. Available at: https://digitalcommons.wpi.edu/cgi/viewcontent.cgi?article=3025&context=iqp-all.
10. Mohammadi M. Alternative recovery methods of intracellular polyhydroxyalkanoates from local isolate. Commamonas sp. Eb172. Universiti Putra Malaysia; 2011.
11. Liu H, Zhang J. Research progress in toughening modification of poly(lactic acid). *J. Polym. Sci. Part B Polym. Phys.* 2011;**49**(15):1051–83.
12. Zakaria MR. Biosynthesis of poly(3-hydroxybutyrate-co-3-hydroxyvalerate). Copolymer from organic acids using Comamonas sp. EB172. Universiti Putra Malaysia; 2011.
13. Loo CY, Sudesh K. Polyhydroxyalkanoates: bio-based microbial plastics and their properties. *Malaysian Polym. J.* 2007;**2**(2):31–57.
14. Pérez-Arauz AO, Aguilar-Rabiela AE, Vargas-Torres A, Rodríguez-Hernández AI, Chavarría-Hernández N, Vergara-Porras B, et al. Production and characterization of biodegradable films of a novel polyhydroxyalkanoate (PHA) synthesized from peanut oil. *Food Packag. Shelf Life.* 2019;**20**(January):100297.
15. Ariffin H, Nishida H, Shirai Y, Hassan MA. Determination of multiple thermal degradation mechanisms of poly(3-hydroxybutyrate). *Polym. Degrad. Stab.* 2008;**93**(8):1433–9.
16. Sudesh K, Abe H, Doi Y. Synthesis, structure and properties of polyhydroxyalkanoates: biological polyesters. *Prog. Polym. Sci.* 2000;**25**(10):1503–55.
17. Silva R, Carvalho GM, Muniz EC, Rubira AF. Miscibility influence in the thermal stability and kinetic parameters of poly(3-hydroxybutyrate)/poly(ethylene terephthalate) sulphonated blends. *Polímeros* 2010;**20**(2):153–8.
18. Bhatia SK, Gurav R, Choi TR, Jung HR, Yang SY, Moon YM, et al. Bioconversion of plant biomass hydrolysate into bioplastic (polyhydroxyalkanoates) using *Ralstonia eutropha* 5119. *Bioresour. Technol.* 2019;**271**(September 2018):306–15.
19. Reis KC, Pereira J, Smith AC, Carvalho CWP, Wellner N, Yakimets I. Characterization of polyhydroxybutyrate-hydroxyvalerate (PHB-HV)/maize starch blend films. *J. Food Eng.* 2008;**89**(4):361–9.

20. Babu R, Woods T. Polymer blends with improved mechanical properties. *Soc. Plast. Eng. Plast. Res.* 2011;1–2.
21. Shogren R. Water vapor permeability of biodegradable polymers. *J. Environ. Polym. Degrad.* 1997;**5**(2):91–5.
22. Modi SJ. *Assessing the Feasibility of Poly-(3-Hydroxybutyrate-co-3-Hydroxyvalerate)(PHBV) and Poly-(Lactic Acid) for Potential Food Packaging Applications*. The Ohio State University; 2010.
23. Choi GG, Kim HW, Rhee YH. Enzymatic and non-enzymatic degradation of poly(3-hydroxybutyrate-co-3-hydroxyvalerate) copolyesters produced by *Alcaligenes* sp. MT-16. *J. Microbiol.* 2004;**42**(4): 346–52.
24. Chen GQ, Wu Q, Jung YK, Lee SY. PHA/PHB. *Compr. Biotechnol. Second Ed.* 2011;**3**:217–27.
25. Pachekoski WM, Agnelli JAM, Belem LP. Thermal, mechanical and morphological properties of poly (hydroxybutyrate) and polypropylene blends after processing. *Mater. Res.* 2009;**12**(2):159–64.
26. Philip S, Keshavarz T, Roy I. Polyhydroxyalkanoates: biodegradable polymers with a range of applications. *J. Chem. Technol. Biotechnol. Int. Res. Process Environ. Clean Technol.* 2007;**82**(3): 233–47.
27. Chan RTH, Marçal H, Russell RA, Holden PJ, Foster LJR. Application of polyethylene glycol to promote cellular biocompatibility of polyhydroxybutyrate films. *Int. J. Polym. Sci.* 2011;**1-9**.
28. Luzier WD. Materials derived from biomass/biodegradable materials. *Proc. Natl. Acad. Sci. U.S.A.* 1992;**89**(3):839–42.
29. Thwin CSU. *Fabrication, Characterization and Degradation of PHB and PHBV Microspheres for Liver Cell Growth*. National University of Singapore; 2004.
30. Biazar E, Khorasani MT, Montazeri N, Pourshamsian K, Daliri M, Rezaei M, et al. Types of neural guides and using nanotechnology for peripheral nerve reconstruction. *Int. J. Nanomed.* 2010; **5**:839.
31. Buzarovska A, Bogoeva GG, Grozdanov A, Avella M, Gentile G, Errico M. Potential use of rice straw as filler in eco-composite materials. *Aust. J. Crop Sci.* 2008;**1**(2):37–42.
32. Koller M. Biodegradable and biocompatible Polyhydroxy-alkanoates (PHA): auspicious microbial macromolecules for pharmaceutical and therapeutic applications. *Molecules* 2018;**23**(362):1–20.
33. Zubairi SI, Mantalaris A, Bismarck A, Aizad S. Polyhydroxyalkanoates (PHAs) for tissue engineering applications: biotransformation of palm oil mill effluent (POME) to value-added. *J. Teknol. UTM* 2016;**78**(1):13–29.
34. Evangeline S, Sridharan TB. Biosynthesis and statistical optimization of polyhydroxyalkanoate (PHA) produced by *Bacillus cereus* VIT-SSR1 and fabrication of biopolymer films for sustained drug release. *Int. J. Biol. Macromol.* 2019;**135**:945–58.
35. Andrady AL. Persistence of plastic litter in the oceans. In: *Marine Anthropogenic Litter*. 2015. p. 57–70.
36. Dilkes-Hoffman LS, Lant PA, Laycock B, Pratt S. The rate of biodegradation of PHA bioplastics in the marine environment: a meta-study. *Mar. Pollut. Bull.* 2019;**142**(March):15–24.
37. Qiang T, Yu D, Gao H. Wood flour/polylactide biocomposites toughened with polyhydroxyalkanoates. *J. Appl. Polym. Sci.* 2012;**124**(3):1831–9.
38. Bhardwaj R, Mohanty AK, Drzal LT, Pourboghrat F, Misra M. Renewable resource-based green composites from recycled cellulose fiber and poly(3-hydroxybutyrate-co-3-hydroxyvalerate) bioplastic. *Biomacromolecules* 2006;**7**(6):2044–51.
39. Barkoula NM, Garkhail SK, Peijs T. Biodegradable composites based on flax/polyhydroxybutyrate and its copolymer with hydroxyvalerate. *Ind. Crops Prod.* 2010;**31**(1):34–42.
40. Luo S, Netravali AN. Mechanical and thermal properties of environment-friendly "green" composites made from pineapple leaf fibers and poly(hydroxybutyrate-co-valerate) resin. *Polym. Compos.* 1999;**20**(3):367–78.

41. Singh S, Mohanty AK. Wood fiber reinforced bacterial bioplastic composites: fabrication and performance evaluation. *Compos. Sci. Technol.* 2007;**67**(9):1753–63.
42. Díez-Pascual AM. Synthesis and applications of biopolymer composites. *Int. J. Mol. Sci.* 2019;**20**(9).
43. Sun S, Liu P, Ji N, Hou H, Dong H. Effects of low polyhydroxyalkanoate content on the properties of films based on modified starch acquired by extrusion blowing. *Food Hydrocolloids* 2017;**72**:81–9.
44. Müller CMO, Pires ATN, Yamashita F. Characterization of thermoplastic starch/poly(lactic acid) blends obtained by extrusion and thermopressing. *J. Braz. Chem. Soc.* 2012;**23**(3):426–34.
45. Marcilla A, Garcia-Quesada JC, Lopez M, Gil E. Study of the behavior of blends of a poly(hydroxybutyrate-valerate) copolymer, polypropylene, and SEBS. *J. Appl. Polym. Sci.* 2009;**113**(5):3187–95.
46. Zhang M, Thomas NL. Blending polylactic acid with polyhydroxybutyrate: the effect on thermal, mechanical, and biodegradation properties. *Adv. Polym. Tech.* 2011;**30**(2):67–79.
47. Asran AS, Razghandi K, Aggarwal N, Michler GH, Groth T. Nanofibers from blends of polyvinyl alcohol and polyhydroxy butyrate as potential scaffold material for tissue engineering of skin. *Biomacromolecules* 2010;**11**(12):3413–21.
48. Atkins TW. Fabrication of microcapsules using poly(ethylene adipate) and a blend of poly(ethylene adipate) with poly(hydroxybutyrate-hydroxyvalerate): incorporation and release of bovine serum albumin. *Biomaterials* 1997;**18**(2):173–80.
49. Zhao K, Deng Y, Chen JC, Chen G-Q. Polyhydroxyalkanoate (PHA) scaffolds with good mechanical properties and biocompatibility. *Biomaterials* 2003;**24**(6):1041–5.
50. Godbole S, Gote S, Latkar M, Chakrabarti T. Preparation and characterization of biodegradable poly-3-hydroxybutyrate–starch blend films. *Bioresour. Technol.* 2003;**86**(1):33–7.
51. Lo WH, Yu J. Effects of the energy dissipation rate and surface erosion on the biodegradation of poly(hydroxybutyrate-co-hydroxyvalerate) and its blends with synthetic polymers in an aquatic medium. *J. Appl. Polym. Sci.* 2002;**83**(5):1036–45.
52. Ariffin H, Nishida H, Hassan MA, Shirai Y. Chemical recycling of polyhydroxyalkanoates as a method towards sustainable development. *Biotechnol. J.* 2010;**5**(5):484–92.
53. Morikawa H, Marchessault RH. Pyrolysis of bacterial polyalkanoates. *Can. J. Chem.* 1981;**59**(15): 2306–13.
54. Kopinke F-D, Mackenzie K. Mechanistic aspects of the thermal degradation of poly(lactic acid) and poly(β-hydroxybutyric acid). *J. Anal. Appl. Pyrolysis* 1997;**40**:43–53.
55. Ariffin H, Nishida H, Shirai Y, Hassan MA. Anhydride production as an additional mechanism of poly(3-hydroxybutyrate) pyrolysis. *J. Appl. Polym. Sci.* 2009;**111**(1):323–8.
56. Norrahim MNF, Ariffin H, Hassan MA, Ibrahim NA, Nishida H. Performance evaluation and chemical recyclability of a polyethylene/poly(3- hydroxybutyrate-co-3-hydroxyvalerate) blend for sustainable packaging. *RSC Adv.* 2013;**3**(46):24378–88.
57. Pulat M, Babayiğit D. Graft copolymerization of PU membranes with acrylic acid and crotonic acid using benzoyl peroxide initiator. *J. Appl. Polym. Sci.* 2001;**80**(14):2690–5.
58. Papantoniou C, Grognet J-C. Terpolymer of (a) crotonic acid (b) vinyl acetate and (c) allyl or methallyl esters. Google Patents; 1978.
59. Thantsha MS, Guest J, Mputle I. Comparison of different methods for release of Bifidobacterium longum Bb46 from the poly(vinylpyrrolidone)-poly (vinylacetate-co-crotonic acid) interpolymer complex matrix, and the effect of grinding on the microparticles. *World J. Microbiol. Biotechnol.* 2011;**27**(10):2443–8.
60. Bassaid S, Chaib M, Bouguelia A, Trari M. Elaboration and characterization of poly(acrylic acid-co-crotonic acid) copolymers: Application to extraction of metal cations Pb (II), Cd (II) and Hg (II) by complexation in aqueous media. *React. Funct. Polym.* 2008;**68**(2):483–91.
61. Van Walsem J, Anderson E, Licata J, Sparks KA, Mirley C, Sivasubramanian MS. Process for producing a monomer component from a genetically modified polyhydroxyalkanoate biomass. Google Patents; 2012.

62. Gooding CM. Process of inhibiting growth of molds. Google Patents; 1945.
63. Jasicka-Misiak I, Wieczorek PP, Kafarski P. Crotonic acid as a bioactive factor in carrot seeds (*Daucus carota* L.). *Phytochemistry* 2005;**66**(12):1485–91.
64. Ilyas RA, Sapuan SM, Atikah MSN, Asyraf MRM, Ayu Rafiqah S, Aisyah HA, Mohd Nurazzi N, Norrahim MNF. Effect of hydrolysis time on the morphological, physical, chemical, and thermal behavior of sugar palm nanocrystalline cellulose (*Arenga pinnata* (Wurmb.) Merr). *Textile Research Journal.* 2020. In print. doi: 10.1177/0040517520932393.

17

Manufacturing of Biobased Packaging Materials

Min Min Aung[1,2,3], Hiroshi Uyama[4], Marwah Rayung[1], Lu Lu Taung Mai[1], Moe Tin Khaing[5], S.M. Sapuan[6,7], and R.A. Ilyas[8,9]

[1]*Higher Education Centre of Excellence (HiCoE), Institute of Tropical Forestry and Forest Products, University Putra Malaysia (UPM), 43400 Serdang, Selangor, Malaysia*

[2]*Department of Chemistry, Faculty of Science, Universiti Putra Malaysia (UPM), 43400 Serdang, Selangor, Malaysia*

[3]*Chemistry Division, Centre of Foundation Studies for Agricultural Science, University Putra Malaysia (UPM), 43400 Serdang, Selangor, Malaysia*

[4]*Department of Applied Chemistry, Graduate School of Engineering , Osaka University, Suita, 565-0871, Japan*

[5]*Department Chemistry, Yadanabon University, Mandalay, Myanmar*

[6]*Advanced Engineering Materials and Composites Research Centre (AEMC), Department of Mechanical and Manufacturing Engineering, Faculty of Engineering, Universiti Putra Malaysia (UPM), Serdang, Selangor, 43400, Malaysia*

[7]*Laboratory of Biocomposite Technology, Institute of Tropical Forestry and Forest Product (INTROP), Universiti Putra Malaysia (UPM), Serdang, Selangor, 43400, Malaysia*

[8]*School of Chemical and Energy, Faculty of Engineering, Universiti Teknologi Malaysia, 81310 UTM Johor Bahru, Johor, Malaysia*

[9]*Centre for Advanced Composite Materials, Universiti Teknologi Malaysia, 81310 UTM Johor Bahru, Johor, Malaysia*

17.1 Introduction

The dynamic systems of foods have a limited shelf-life due to moisture, temperature, and hygiene condition. The quality of food packaging material will affect the food that is used or packed. Bio-based food packaging material is one of the industrial interests

Bio-Based Packaging: Material, Environmental and Economic Aspects, First Edition. Edited by S.M. Sapuan and R.A. Ilyas

Table 17.1 Classification of bioplastics durable and biodegradable.

Types	Durable	Biodegradable
Bio-based	*Cellulose acetate* Biopolyethylene Biopolyamides Biopolyurethanes, i.e.,	*Natural rubber, Linoleum,* Polyhydroxyalkanoates Polyactid acid Starch Blends, i.e.,
Fossil-based	Polyethylene Prolypropylene Polyvinylchloride	Polybutylene Adipate Co-terephthalate Polybutylene succinate Polycaprolactone

for the development of any food packaging system. There should be no compromise between food and the packaging material as far as food safety and quality are concerned [1]. Bio-based food packaging materials derived from renewable resources, biodegradable or biomass-based materials are called bioplastics. The packaging materials recognized as biodegradable according to the EU standardization committee are outlined in this chapter. Recent trends in food packaging materials are directed towards bio-based and biodegradable materials for manufacturing and industry. Bioplastic production via fermentation and chemical processes is made up of a plant base materials for the production of biomass plastic. Bioplastics can be classified into bio-based and fossil-based and further separated into biodegradable and nondegradable as listed in Table 17.1. Biodegradable plastics are currently being used as food packaging materials in global market.

17.2 Bio-Based Packaging Materials

Bio-based materials can be made from bioplastic materials and fiber-based materials that contain bio-based residues. Bio-based plastics have nonbiodegradable and biodegradable plastics and most of the bio-based food packaging uses biomass residues as raw materials. Recently, 3 million tonnes of bioplastic annually produced worldwide represented 1 % of the 300 million tonnes of plastics produced. The demand for the worldwide production capacity for bio-based plastics is expected to have been 7.85 million tonnes in 2019 [2]. The Nova Institute and the University of Applied Sciences and Arts in Hanover data show that the European bioplastics total production was 6.5 million tonnes in 2019, which is 80% of the bioplastics used in the packaging market. The increasing market demand creates not only opportunities for biomass residues producers but also for bio-based food packaging applications [3].

Bio-plastics for bio-based materials are stated under EN 13428 (prevention, reduction) by regulation and standards. The term "bio-based content" is the percentage weight of the product based on renewable raw materials. The amount of bio-based content used in bio-based materials is a minimum of 25% to establish inclusion in the many material categories. The bio-based percentage of the product is now stated on the label for packaging materials, but that depends on the country that certifies the content of bio-based materials. European regulations, Vinvotte, issues the certificates and in Germany, DIN CERTO company certifies the content in their logo [3].

The recent development in food packing materials is directed toward the biopolymer, which has ideal sources. Biopolymer is a renewable resource with excellent mechanical and barrier properties. The incorporation of additives such as antifungal, antimicrobial,

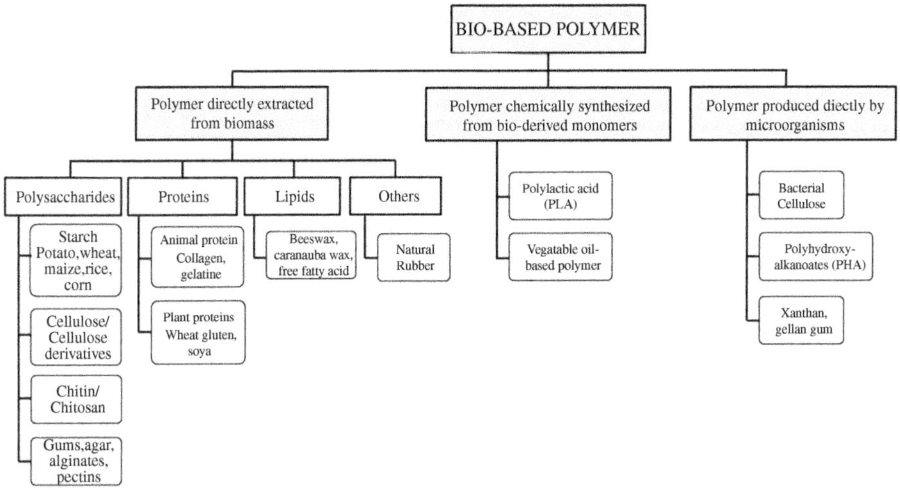

Figure 17.1 *The classification of a bio-based polymer.*

antioxidants, nutrients, and colors can be added to the biopolymer to produce safe bio-based food packaging materials [4]. The renewable polymers produce bio-based packaging materials as greener products, especially for single-use packed goods. Agricultural-derived polymers and the original bio-based polymer have three categories according to their sources. Figure 17.1 illustrates the classification of a bio-based polymer.

The most common bio-based polymers are cellulose, starch, oil, waxes, chitin, and latex. Many studies and development strategies have been discovered to optimize the potential uses of biopolymer for commercial applications, including the food packaging industry, agricultural purposes, cosmetics, the medical industry, and the pharmaceutical industry.

17.3 Food Packaging Materials

Food packaging materials have some basic needs for food quality and safety to prevent chemical contaminants, microorganisms, moisture, oxygen, light, and external forces. The food packaging material should have barrier properties to gas, vapor, and aroma permeation and must have good mechanical strength and chemical and thermal resistance properties [4]. Basic characters of food packaging materials are shown in Figure 17.2. The classification of biomass plastics and biodegradable plastics is depicted in Table 17.2.

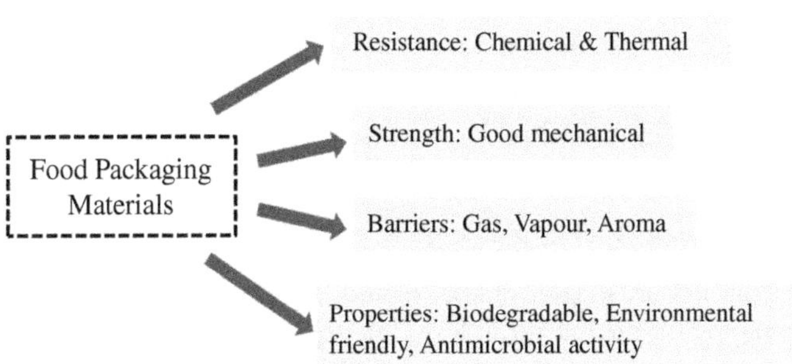

Figure 17.2 *Basic characters of food packaging materials.*

Table 17.2 Classification of biomass plastic and biodegradable plastic.

Types	Biomass	Petroleum based
Biodegradable	Biodegradable biomass plastic • Polylactic acid • Microbial polyester • Polybutlene succinate • Cellulose acetate	Biodegradable petroleum-based plastics • Polycaprolactone • Polybutylene adipate/ terephthalate • Polyvinyl alcohol
Nonbiodegradable	Nonbiodegradable biomass plastic • Bio polyethylene terephthalate • Biopolytrimethylene terephthalate • Vegetable oil-derived polyol polyurethane • Lignophenol • Bionylon 610, Bionylon 11 • Biopolyethylene, biopolypropylene	Nonbiodegradable petroleum-derived plastics • Polyethylene • Polypropylene • Polystyrene • Polyethylene terephthalate • Nylon 6, Nylon 66

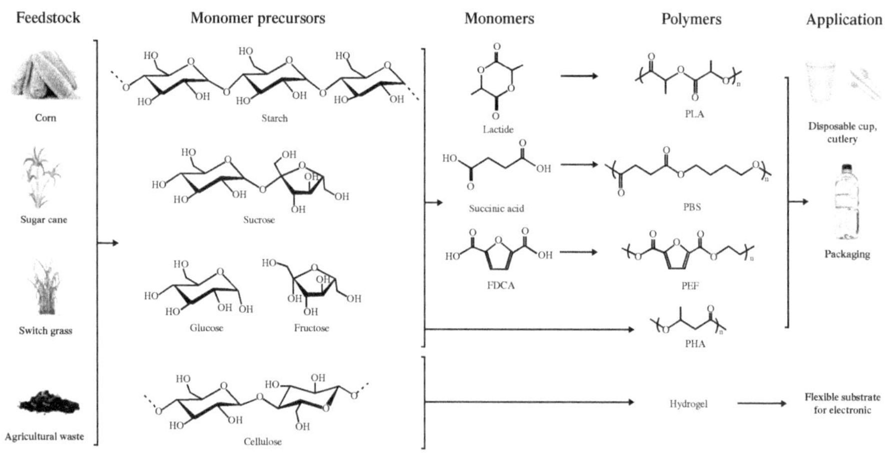

Figure 17.3 Some common sources of bio-based polymers and their usage.

Producing food packaging materials from bio-based biodegradable plastics is the current trend as there is a demand for food packaging materials that are safely and have defined properties. Bio-plastics can be classified into biomass plastics and biodegradable plastics (Table 17.2). Biodegradable plastics are extracted from renewable resources such as starch, sugar, and vegetable oil. Plastics from renewable sources (Figure 17.3) can be used to produce food packaging materials due to the availability of biomass from plants and may cost less and are environmentally friendly.

17.3.1 Biomass Plastic in Food Packaging

Most of today's plastics are made from petroleum. Some of these polymers have advanced recycling technologies at the industrial level but are eventually destroyed and incinerated, producing carbon dioxide. In order to prevent global warming, carbon-neutral plastics are being demanded from the viewpoint of materials. Therefore, "biomass plastic", which is incorporated into the natural material cycle, is attracting attention as a plastic material that is friendly to the global environment [5–7]. Interest in biodegradable plastics has been increasing due to the recent social problem of marine plastic waste. Most of the biodegradable plastics are aliphatic polyesters, and polyhydroxyalkanoic acid (PHA) and polybutylene succinate (PBS) are degraded by marine microorganisms. Bioplastic is a general term for biomass plastic and biodegradable plastic.

Since the comprehensive strategy "Biomass Nippon" was announced in 2002 in Japan, policies for the realization of a sustainable and sustainable society through the utilization of biomass have been formulated and implemented. This strategy aims to prevent global warming and achieve a recycling-based society based on the effective use of biomass and to foster strategic industries by developing unique biomass utilization methods in Japan. In addition, from the viewpoint of environmental protection on a global scale, active use of inexpensive and abundant resources of biomass raw materials, not only in Japan but also from all over the world, is required.

As an example, polylactic acid (PLA) is the most representative biomass plastic for food packaging materials. Since biomass in which carbon dioxide is immobilized by photosynthesis is produced as a raw material, the amount of carbon dioxide does not increase as a whole, even if carbon dioxide is released into the atmosphere due to combustion or biodegradation of PLA. For this reason, PLA is called a carbon-neutral material-circulating plastic. Lactic acid has an L-form and a D-form due to its chemical structure. Currently, PLA composed of L-form lactic acid is industrialized for food packaging material. In the United States, corn-based starch is used as raw materials, and production is progressing at a plant with an annual production scale of 140 000 tons. In China, thousands of tons of plants are operating and in the future, the production volume of PLA derived from biomass raw materials will increase worldwide.

In recent years, technological development related to plastic biotechnology has been rapidly progressing. Biopolyethylene (PE) made from bioethanol can be handled at the same level as petroleum-derived PE and is expected to spread. It is said that PE, a bioethanol raw material, can reduce carbon dioxide emissions by 4.3–4.9 kg per kg. In 2011, a 200 000-ton plant became operational in Brazil. In addition, industrial production of propylene and butadiene using biomass-derived raw materials has been studied and it is predicted that the development of biopolyolefins will proceed based on these. For aromatic polyesters, a technology for fermentative production of 1,3-propanediol from biomass raw materials has been developed in the United States, and biopolytrimethylene terephthalate (PTT) using 1,3-propanediol has been launched on the market. Unlike PLA and polyester produced by microorganisms, one of the monomers (terephthalic acid) is derived from petroleum, so the biomass degree is 37%. PTT is being developed for use in the automotive field for floor mats, seat skins, and interior skins. Recently, biopolyethylene terephthalate (PET) using bio-derived ethylene glycol has also been developed and is being used in various fields such as bottles packaging.

Biomass plastics are synthesized by sequential polymerization (polycondensation/polyaddition), ring-opening polymerization, etc. This chapter will introduce mainly bio-derived elastomers that have a higher added value than general-purpose plastics. Elastomers are industrially important polymers and are positioned as intermediate products between thermoplastic resins and synthetic rubbers. Thermoplastic elastomers are produced in styrene, olefin, PVC, urethane, amide, etc., and have a wide variety of uses, such as home electric appliances/office equipment parts, automobile parts, electric/electronic parts, medical parts, and sanitary parts. However, at present most of them are manufactured from petroleum-derived resins. In the first half of this chapter, as a new development of bio-based elastomers, we will introduce efforts to manufacture elastomers using polymers contained in plants. This elastomer is made by biosynthesis, but its reaction pathway is by polycondensation. In the latter half, polyols derived from vegetable oils, fats, and biopolyurethanes are described.

17.3.1.1 Eucommia Elastomer

Before the active development of biomass plastics, humans have used bio-based polymers produced by nature. A typical example is natural rubber of *Hevea brasiliensis*. Natural rubber is a *cis*-type polyisoprene contained in rubber sap. Thailand, Indonesia, and Malaysia produce 10 million tons annually in major producing countries. The *cis*-type structure is important for the development of rubber elasticity and the *trans*-type gutta percha does not show rubber elasticity. The *cis*-type polyisoprene exists as latex in the rubber sap. In order to impart practical rubber elasticity to natural rubber, it is used after being cross-linked (vulcanized) with sulfur. The elasticity is greatly improved by introducing a bridging structure with sulfur between only a part of the double bonds in the polymer. For tire applications, carbon black is further added to increase the mechanical strength of the rubber.

Rubber is classified into natural rubber and synthetic rubber mainly composed of butadiene-based rubber. However, natural rubber has been re-examined due to increasing awareness of biomass utilization for building a recycling-based society in the rubber industry, and research has been conducted on improving the productivity of natural rubber. Natural rubber contains trace amounts of proteins and lipids contained in sap, and these are known to play an important role in the physical properties of rubber. From these facts, even now, the production ratio of natural rubber and synthetic rubber has a substantially equal relationship. In addition, recently a technique for maintaining the physical properties of rubber while removing proteins has been developed and rubber gloves for reducing allergies due to proteins contained in natural rubber have been put to practical use.

Trans-type polyisoprene is produced by plants such as *Eucommia ulmoides* Oliver and balata gum tree. In recent years, research and development have been carried out to commercialize *trans*-type polyisoprene (eucommia elastomer, EuTPI) obtained from eucommia [8, 9]. Eucommia originates in China and is a deciduous tree with a height of 20 m or more and is distributed in mountainous areas below 2500 m above sea level. Gutta perkanoki and balata gum tree, which have been industrially used as *trans*-polyisoprene-producing plants, grow in the tropics, whereas eucommia grow in broad temperate zones, so they are cultivated in vast semi-arid unused areas.

Eucommia contains EuTPI in all plants such as leaves, bark, roots, and pericarp, but the content, molecular weight, and molecular weight distribution differ depending on the organ. In the case of bark and roots, it is not sustainable and is unsuitable as a raw material for biomass because logging means are required. On the other hand, using seeds that have long fruiting as biomass raw materials is effective as a sustainable means. Moreover, the EuTPI content in the seed peel exceeds 20% and its molecular weight exceeds 1 000 000, so that it is suitable for manufacturing and is an ideal nonedible biomass. Therefore, the target biomass can be stably secured without damaging the eucommia.

TPEuTPI is classified as a crystalline polymer because of its melting point, which is relatively low at about 60°C. Therefore, it is considered that processing and molding at a low temperature is easy. The glass transition point is about −60°C and the decomposition starting temperature in thermogravimetric analysis is about 320°C. The embrittlement temperature is −51°C, which is lower than that of polypropylene (PP) and can be used at low temperatures. For mechanical properties, there is a high stress with respect to the initial strain and the Young's modulus is high. The tensile properties are close to low-density polyethylene (LDPE) and PP, and the flexural modulus is close to LDPE. Since the Izod impact strength is higher than high-density polyethylene (HDPE) or PP and close to acrylonitrile-butadiene-styrene copolymer resin (ABS resin) or high-impact polystyrene, EuTPI is considered to be a material with excellent impact resistance. From the evaluation of these mechanical properties, EuTPI shows physical properties close to those of plastics such as PE and PP and is expected to be used as a modifier for plastics having a high impact resistance and poor impact resistance.

In the molding process of EuTPI, since the melting point is about 60°C, kneading with an open roll can be performed at a roll surface temperature of about 80°C, as with rubber. For this reason, studies are being made for kneading with various resins, rubbers, fillers, and various chemicals for higher performance and higher functionality. The sheet obtained by the open roll can be formed into a plate by hot pressing at about 100°C and test pieces for various physical property tests can be produced. On the other hand, since the melt flow rate indicating the fluidity of the molten plastic is low, it is considered unsuitable for twin-screw extrusion or the like.

Chemical modification using unsaturated bonds in EuTPI is a powerful approach for imparting functions and improving the performance of EuTPI, which is known to progress in cross-linking with sulfur in the same manner as natural rubber. In addition, maleic anhydride-modified EuTPI was synthesized for the purpose of constructing a reversible network structure [10]. Maleic anhydride was grafted on to EuTPI using an ene reaction. A maleated EuTPI (MTPI) was obtained by adding an aqueous sodium hydroxide solution to a THF solution of the obtained polymer (Figure 17.4). Although an excess amount of maleic anhydride was required, the graft ratio could be controlled by the charging ratio. The MTPI film swelled without dissolving in THF, which is a good solvent for EuTPI, to form an organogel. This is probably because the grafting of a carboxyl group to the TPI resulted in the formation of a polar domain in the nonpolar matrix, which acted as a cross-linking point. The organogel was decross-linked by the protonation of the carboxyl group under acidic conditions and dissolved in THF.

By reacting the maleic anhydride-modified product with amino alcohol, a hydroxyl group-grafted EuTPI derivative (OHTPI) was also synthesized. Physical cross-linking of OHTPI was investigated using the interaction of non-covalent bonds between hydroxyl groups and metal ions. When Sn (Oct) 2 (tin (II) dioctanoate) was added to a THF solution of OHTPI, an organogel was formed. This is probably because the interaction between tin and the hydroxyl group acted as a cross-linking point. When hydrochloric acid was added dropwise to this organogel, it was easily decross-linked and dissolved in THF. Thus, a reversible physical gel was developed by introducing a hydrophilic group into EuTPI.

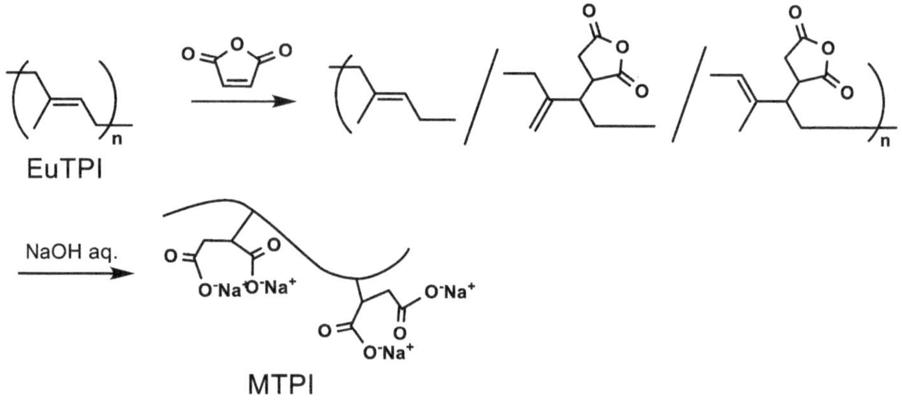

Figure 17.4 Preparation of maleated EuTPI (MTPI).

The toughness of PLLA was improved by blending poly-L-lactic acid (PLLA) and EuTPI by melt-kneading. DSC analysis showed no change in the glass transition temperature of PLLA due to the blend, indicating that PLLA and EuTPI were incompatible. Also, the crystallization rate was decreased by increasing the EuTPI content. In the tensile test of the film, the stress of the blended film was lower than that of the PLLA alone film, but the strain was about twice as large. In the flexural strength test, PLLA alone completely broke at a strain of about 3%, whereas the small amount of EuTPI reduced the bending stress.

17.3.1.2 Biopolyurethane Using Vegetable Oils

The main component of fats and oils is a triester of glycerin and fatty acid (triglyceride), but also contains a small amount of diglyceride and monoglyceride. Oils and fats are classified into vegetable oils and fats and animal fats and oils according to their raw materials and are classified into edible and industrial types according to their use. Vegetable oils and fats include soybean oil, palm oil, rapeseed oil, sunflower oil, and linseed oil, while animal oils and fats include beef tallow, lard, and fish oil. The main uses of fats and oils are foods such as margarine, shortening, dressing, and lard. In addition to being used for fuels and lubricating oils as industrial uses, fatty acids and glycerin obtained from fats and oils are used as surfactants and resin additives. Figure 17.5 shows some common examples of oil-based polymer and their usage [11].

When vegetable fats and oils are used as raw materials for polymers, carbon-carbon double bonds of fats and oils are often used for modifications [12]. Since soybean oil contains the largest amount of linoleic acid, having two unsaturated groups, it is often more suitable as a raw material for polymers than palm oil, which produces the largest amount. Alkyd resins have been used as vegetable oil-based materials that have been practically used for a long time. The alkyd resin is obtained by reacting anhydrides of phthalic acid or maleic acid with a polyhydric alcohol such as glycerin and is modified with oil or unsaturated fatty acid and is used as a paint or an adhesive. It is distinguished by excellent pigment dispersibility and paintability, as well as giving a good finish appearance and durability.

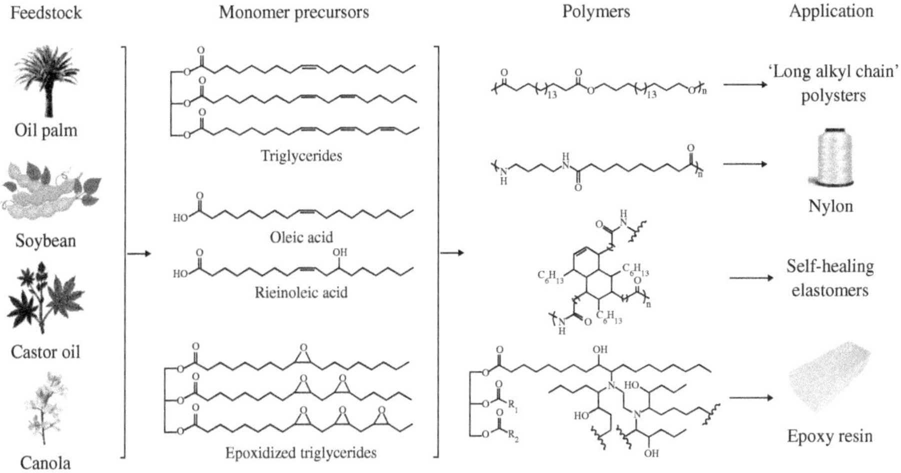

Figure 17.5 *Oil-based polymers and their usage.*

Polyurethane polyols made from vegetable oils and fats have been developed. Polyols for polyurethanes have been developed in the United States, where many soybean oils are produced, aiming at the advanced use of soybean oils. To date, several synthetic routes have been considered and some have been manufactured. Dow manufactured soybean oil-based polyols under the trademark RENUVA. Polyols have been developed that suppress the odor, a drawback of soybean oil products, and are suitable for polyurethane applications. Dow's polyols are produced by the following synthetic route. First, a fatty acid methyl ester and glycerin are synthesized by a transesterification reaction between soybean oil and methanol. Subsequently, carbon monoxide is added to the double bond of the fatty acid methyl ester to formylate, which is then converted to a primary hydroxyl group by hydrogenation. Finally, the transesterification reaction with glycerin is performed again to obtain a polyol. This method has the advantage that a polyol having a primary hydroxyl group with high reactivity can be produced, unlike the method using an epoxidized oil and fat, described below. It is also an excellent synthesis technique in that methanol and glycerin are recycled in this reaction system. The LCA of this soybean oil polyol production was studied and the energy required for production was about two-thirds that of the existing typical polyols and the carbon dioxide emission was almost zero. Therefore, the usefulness of soybean oil polyol was clarified from the viewpoint of global environmental protection. Soybean oil-based polyols have also been developed from epoxidized soybean oil. For example, when an epoxy group is hydrolyzed into a glycol, it can be used as a polyol. The exact route of production is not disclosed, but several manufacturers produce soybean oil polyols for use in flexible foams, coatings, adhesives, and elastomers.

Castor oil is a fat and oil having a specific structure in which about 90% of the constituent fatty acids are ricinoleic acid having a secondary hydroxyl group and a castor oil-based polyol for polyurethane has been developed. Castor seed contains 40–60% of oil. Castor is a member of the Euphorbiaceae family, which is native to East Africa and is now distributed worldwide. It has been used as kerosene and for constipation since ancient times and is widely used in industrial applications such as paints and printing inks. Castor oil is also important as a raw material source for bionylon. Sebacic acid, a monomer of nylon 610, having excellent flexibility and low-temperature impact resistance, is produced by oxidative hydrolysis of ricinoleic acid, a component of castor oil, by alkali treatment. 11-Aminoundecanoic acid, a monomer of nylon 11, which has excellent insulation and abrasion resistance, is also made from ricinoleic acid. Ricinoleic acid methyl ester is thermally decomposed into undecenoic acid methyl ester, which is hydrolyzed, hydrogen bromide is added, and it is then reacted with ammonia to give 11-aminoundecanoic acid.

Polyurethane foam was synthesized by mixing a soybean oil-based polyol with a petroleum-derived polyol and then the effect of the mixing ratio on various physical properties was examined. The more soybean oil polyol was used, the number of cells increased and the cell size has decreased. Soybean oil polyols had lower reactivity and higher viscosity than petroleum-derived polyols, and thus reduced thermal conductivity and compressive strength. In addition, poly(ether ester) type polyol was developed by epoxidation, hydroxylation, and transesterification of canola oil with glycol, and a bio-based polyurethane was synthesized using this polyol. The resulting polyurethane was evaluated as a wood adhesive. Diisocyanate compounds were also synthesized by chemical conversion from unsaturated fatty acids derived from canola oil and corn oil, and developed into bio-based polyurethane. Bio-based polyurethanes have been developed from 1,16-diisocyanatohexadec-8-ene, a long-chain diisocyanate derived from oleic acid.

Castor oil is used as a polyol for polyurethane because it has less than three hydroxyl groups per compound. However, castor oil is not suitable for applications such as polyurethane foam for automobiles, and therefore derivatives of castor oil have been developed. BASF has produced a polyurethane that uses a castor oil with hydroxyl and ethylene oxide and propylene oxide added to the polyol. It has a plant content of 24% and is mainly used for furniture cushions and mattresses. Mitsui Chemicals has also developed castor oil-based polyols and polyurethanes that have been put to practical use in automotive applications. The cushion feeling was greatly improved by modifying castor oil. It is possible to make 20–40% (weight) of polyurethane foam derived from plants, with CO_2 emissions reduced by about 8% compared to conventional products. As an overseas development of this technology, production of biopolyols in India, a country producing castor oil, is planned.

When designing polyurethane with a high vegetable content, the development of a high molecular weight bio-based polyol is also desired in accordance with the foam, which is the main use of polyurethane. In order to utilize the existing production process of polyurethane foam, a liquid polyol is preferable. As a bio-based polyol satisfying such requirements, a branched PLA obtained by ring-opening polymerization of lactide (or polycondensation of lactic acid) using castor oil containing ricinoleic acid and polycastor oil as a polymer as the initiator. Polyols have been developed (Figure 17.6) [13, 14]. Depending on the structure of the core fat component (number of branches, molecular weight) and the steric structure and chain length of the PLA chain, this polyol is expected to be used not only for polyols for polyurethane but also for various uses. In the polymerization of L-form lactide (LLA) using castor oil as an initiator, a good correlation is found between the charge ratio of castor oil and LLA and the molecular weight of the obtained polyol, where the molecular weight can be arbitrarily controlled. The glass transition temperature, melting point, and crystallinity of the branched PLA depends on the charge ratio, and these values were lower than those of the linear PLA.

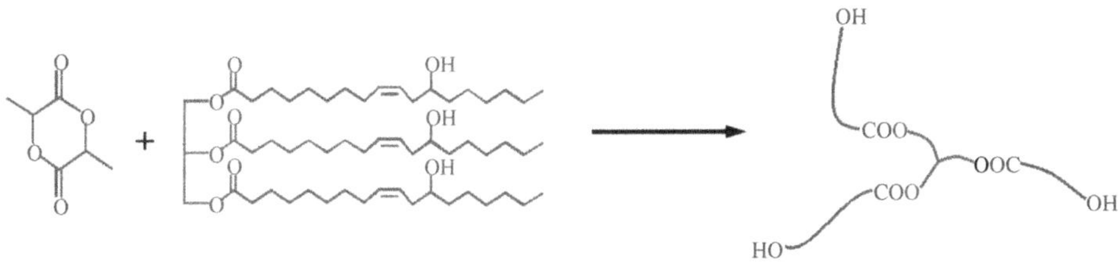

Figure 17.6 *Chemical structure representative of PLA.*

Figure 17.7 *Production of foam bio-based polyurethane using liquid branched polylactic acid.*

Synthesis of polyurethane using branched polylactic acid polyol was studied. Water, a silicone-based foam stabilizer, and 2,4-toluene diisocyanate are added to the branched PLA and the mixture is rapidly stirred at room temperature to obtain a foamed polyurethane. In addition, foamed polyurethane for car headrests was prototyped using polyol as a liquid branched PLA whose molecular weight was controlled for industrial use (Figure 17.7). For bio-based polyurethanes synthesized from polylactic acid-based polyols, the main applications are foams for cars and furniture, but important applications of polyurethane include adhesives and paints, and application to coating materials is also being studied.

A polyurethane based on dimer acid, which is a dimer of nonedible fatty acids derived from pine, has been developed. Dimer acid has a flexible structure with a long alkyl chain of C36 and is therefore used in paints, inks, adhesives, and the like. Utilizing the property that the polymer derived from dimer acid has a low glass transition temperature, utilization of polyurethane as a soft segment was studied [15]. Polyurethanes based on dimer acid showed excellent transparency and flexibility.

17.4 Properties of Bio-Based Packaging Materials

17.4.1 Biodegradable Plastic

The Japan Bioplastics Association classifies bioplastics into biomass plastics (biomass plastics) and biodegradable plastics (green plastics) [16]. Biodegradable plastics are defined worldwide as plastics that, after use, degrade into carbon dioxide and water by the action of microorganisms. The former relates to raw materials and is distinguished from petroleum-derived plastics. On the other hand, the latter concerns the properties and functions of plastics. At present, most plastics are derived from petroleum and are nonbiodegradable. In Japan, research on the development of biodegradable plastics began in 1970, due to the problem of waste plastics affecting the environment. At that time, the idea of the carbon cycle was not common and the biodegradable plastics developed and marketed are divided into those derived from petroleum and those derived from plants. Biomass plastic is a classification related to resin raw materials and the physical properties and functions of the resin are not limited. Many biomass plastics such as PLA show biodegradability, but there are also nonbiodegradable ones such as biopolyethylene (PE) and biopolytrimethylene terephthalate. On the other hand, the biodegradability relates to the function of the resin and all components of the biodegradable plastic must be biodegradable. In addition, those with a biomass plasticity of 25% or more are identified as "biomass plastic" by the Japan Bioplastics Association.

The types of biomass plastics are classified according to their structures and manufacturing methods [5, 16, 17]. From the viewpoint of the manufacturing process, (i) products produced in plants that produce biomass (one-step synthesis), (ii) products produced by microorganisms using biomass as a raw material (two-stage synthesis), and (iii) chemicals produced by fermentation using biomass as a raw material are produced as monomers (three-step synthesis) (Figure 17.8). A typical example of (i) directs plants to produce plastics, and natural rubber and eucommia elastomers are typical examples. Studies have been conducted to synthesize poly[(R)-hydroxyalkanoate] (polyhydroxyalkanoic acid, PHA) in transgenic plants, but at the present time its productivity is low and its practicality is poor. A typical example of (ii) is polyester produced by microorganisms, where plastics are produced in microorganisms using sugar and vegetable oil as raw materials. A typical example of (iii) is PLA, which produces lactic acid by fermentation of biomass and converts and polymerizes this by a chemical process to produce PLA. Recently, attention has been

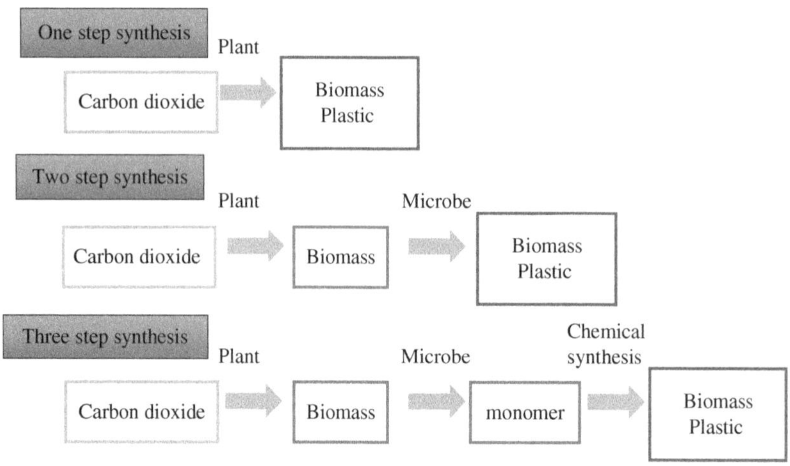

Figure 17.8 The synthesis steps of biomass plastic from plant.

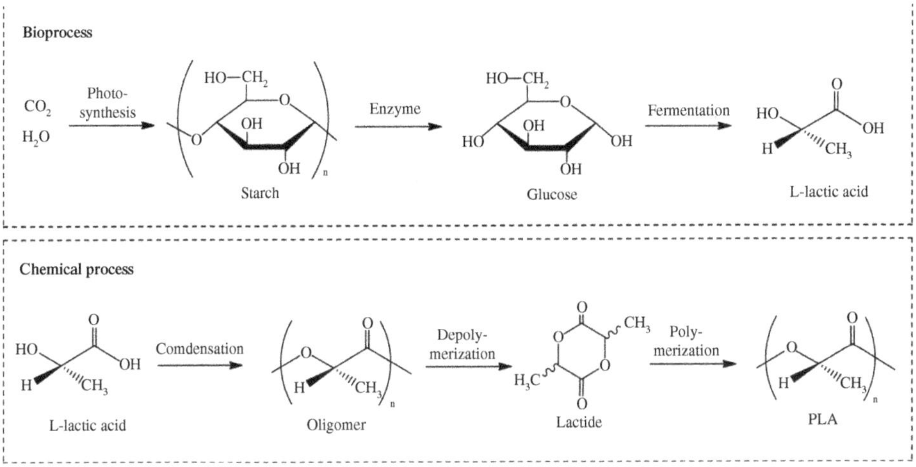

Figure 17.9 Stage production of polylactic acid.

focused on the industrial production of bio-PE produced by the polymerization of ethylene made from bioethanol.

Polylactic acid is the most typical bioplastic [18]. Since carbon dioxide is immobilized by photosynthesis and produced using biomass as a raw material, the amount of carbon dioxide does not increase even if carbon dioxide is released into the atmosphere due to combustion or biodegradation of PLA. Thus, polylactic acid is a carbon-neutral material-circulating plastic. Lactic acid has the L-form and D-form due to its chemical structure. Currently, polylactic acid composed of L-form lactic acid is industrialized. In the United States, corn-based starch is produced in a 140 000-ton plant. Thousands of tonnes of plants are operating in China and large-scale production has recently begun in Thailand. In recent years, shale gas excavation applications utilizing the biodegradability of PLA have been expanding in the United States. Figure 17.9 shows the production stage of poly(lactic acid).

Polylactic acid exhibits crystallinity, but the crystallization rate is extremely slow, which limits the molding method and is a major constraint on industrial applications. For example, it is difficult to rapidly perform injection molding, which is a typical molding of plastic. Since PLA does not crystallize under ordinary crystallization conditions and the heat resistance of PLA remains at the glass transition temperature (about 60 °C), a technique for improving heat resistance by crystallization is required. The reason that PLA is difficult to crystallize is that there is only one methine group between ester bonds per polymer unit and thus the segmental mobility of the molecular chain is poor. For crystallization of PLA, it is necessary to promote the generation of crystal nuclei and the growth of the generated crystal nuclei. A nucleating agent (crystal nucleating agent) that promotes crystallization is dispersed in a molten polymer and becomes a nucleus to promote crystallization (heterogeneous nucleation). When a crystal nucleating agent is added and polymer chains are adsorbed on the surface and crystallization proceeds epitaxially, the crystallization speed is greatly improved. Known crystallization agents include trimesinamide, melamine cyanurate, zinc phenylphosphonate, copper phthalocyanine, and the like. Recently, it has been reported that branched polylactic acid and xylan ester have excellent crystallization promoting actions.

Branched polylactic acid polyols have been developed by ring-opening polymerization of lactide (or polycondensation of lactic acid) using castor oil as an initiator [14]. It can be applied to soft segments for polyurethane and bio-based foamed polyurethane can be synthesized. In addition, branched polylactic acid has been applied as a plasticizer and crystal nucleating agent for polylactic acid [19]. The branched chain of this polymer is polylactic acid and its function can be tuned by the terminal structure. The branched polylactic acid having a terminal hydroxyl group can plasticize the polylactic acid by adding a small amount. In addition, a high-performance nucleating agent was developed by introducing a terephthalate into the terminal. Commercially available polylactic acid cannot be molded in a high-temperature mold at 90°C, even if it is held for 120 seconds. On the other hand, the compound of branched polylactic acid and polylactic acid produced using a batch-type melt kneader can be molded in a high-temperature mold at 90°C within 20 seconds, and the high crystallization-promoting ability of the branched polylactic acid enables significant productivity gains to have been achieved. The product impact resistance (Izod impact strength) of this molded product was improved by about four times that of polylactic acid and was almost equal to that of polypropylene. Regarding heat resistance (deflection temperature under load), the deflection temperature under load of polylactic acid was 60°C or less, but the molded product was improved to 130°C and showed a higher performance than polypropylene. Since the use of the branched polylactic acid as the nucleating agent greatly improves the moldability and the physical properties of the molded product as compared with the prior one, polylactic acid can be applied to the core members of automobiles and home appliances. The branched polylactic acid also acts as a nucleating agent for the microbial polyester [20]. Since the properties of polylactic acid alone are inferior to other general-purpose resins, compounding with inorganic materials and blending with other resins are also being studied. Nanocomposites of polylactic acid and clay are made using organically modified clay. The addition of clay has the effect of promoting crystallization, and the reinforcing effect also improves the mechanical properties of polylactic acid.

Microbial-produced polyester is a substance where certain microorganisms accumulate in cells as energy storage substances. This polyester poly[(R)-3-hydroxybutyrate] (P3HB) (Figure 17.10) is optically active and has a very high molecular weight. P3HB is a crystalline polymer having a glass transition temperature of 4°C and a melting point of 180°C, and its crystal structure is orthorhombic. P3HB has high crystallinity, elongation at break of 7%, Young's modulus of 3.5 GPa, and is hard and brittle. Therefore, P3HB alone is not suitable for practical use, so studies have been conducted on the biosynthesis of copolymer polyesters that lower the melting point and crystallinity. It then becomes a flexible material due to a decrease in melting point and crystallinity.

Regarding PHA, a copolymer of (R)-3-hydroxybutyrate and (R)-3-hydroxyvalerate (P(3HB-co-3HV)) was launched as a biodegradable plastic in the 1980s. Due to the high price and physical properties, sales volumes did not increase and the business was frozen in 1999. Recently, Kaneka has developed a polyester copolymer (P(3HB-co-3HH), PHBH) into which (R)-3-hydroxyhexanoate unit further increases the carbon chain length of the side chain by one. A technology to produce vegetable oil as a raw material with high efficiency has been developed and put into practical use. It has been shown to exhibit

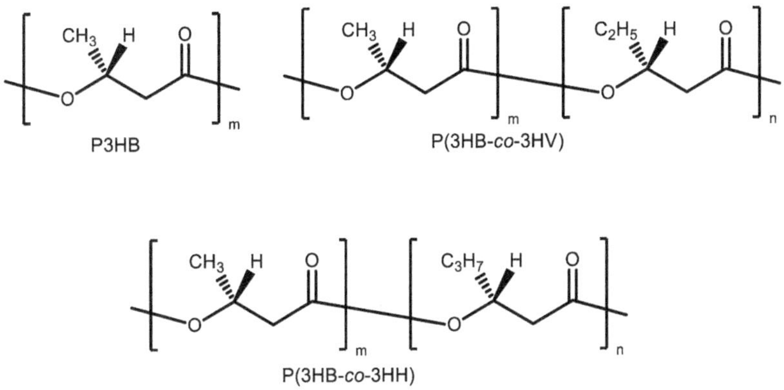

Figure 17.10 *Structure of P3HB and its derivatives.*

marine biodegradability and an increase in the number of manufacturing plants has been announced.

Polybutylene succinate (PBS) using succinic acid as a monomer has been commercialized as a biodegradable plastic, including for multifilm use in agriculture. Succinic acid, which is currently made from petroleum, can be produced by a fermentation method using biomass as a raw material. In Thailand, Mitsubishi Chemical commercialized a joint venture with a local company. PBS also biodegrades in the ocean.

17.4.2 Biodegradable Polyester Composite

Cellulose, the king of the plant kingdom, has many hydroxyl groups in its skeleton and has excellent water absorption properties. Therefore, it has applications as a raw material for fibers such as underwear, but has high moisture absorption and high crystallinity. Due to the difficulty in processing, such as molding, it has not been used for general-purpose plastics. Recently, active research has been carried out to combine cellulose nanofibers with general-purpose plastics [21–23]. Cellulose microfibrils are made of extended chain microcrystals and have high mechanical strength comparable to aramid fibers (elastic modulus: 140 GPa). The coefficient of thermal expansion is very low, equivalent to that of quartz. It has been suggested that the performance of plastics can be dramatically improved by utilizing these properties. Since the obtained composite material has the low specific gravity of cellulose, it can also achieve weight reduction and can be applied to automotive materials.

As a porous material, a monolith in which a three-dimensional network skeleton and its voids (through holes) are integrated, has attracted attention as a next-generation porous material and has been applied to high-performance materials [24]. The monolith is originally a word meaning "one lump", but the term "monolith" has been widely used in the chemical field because the term "monolith" refers to a porous material column having communication holes in the field of analytical chemistry. In a monolith having a network-like co-continuous structure, the size of the skeleton and the pores serving as the flow path can be independently controlled and the sizes can be made uniform.

In the development of a composite material using cellulose as a filler, use of a porous material (monolith) can be cited as a means for suppressing aggregation of cellulose. Cellulose is dissolved by heating the saturated calcium thiocyanate aqueous solution and a cellulose gel is formed by cooling. A cellulose monolith was synthesized by regenerating (recrystallizing) cellulose using ethanol and complexation with PHBH was studied. This monolith had a nano-order skeleton and formed a uniform network structure. A PHBH/cellulose composite sheet can be obtained by impregnating the cellulose monolith with PHBH and pressing it [25]. When the thermal properties of this composite were evaluated, the glass transition, low-temperature crystallization, and melting behavior of PHBH were observed, and the glass transition temperature of PHBH did not change. Furthermore, the low temperature crystallization shifted to the low temperature side. Polarized light microscopy observations of PHBH and the composite showed that PHBH alone formed spherulites and had a low nucleus density, but the PHBH crystals were remarkably finer in the composite. From these results, it was found that crystallization of PHBH was promoted by complexing with cellulose.

Dynamic viscoelasticity was measured to evaluate the mechanical properties of the composite. The storage modulus of the composite increased depending on the cellulose content. In the case of the composite with a high cellulose content, the decrease in storage modulus due to the melting of PHBH was suppressed and the high storage modulus was maintained, even in a high-temperature region. In the tensile test, Young's modulus and maximum stress increased in the composite as compared with PHBH alone. However, unlike the case where cellulose fibers or glass fibers were added, the use of the cellulose porous body also increased the breaking strain. This is thought to be due to the interaction between the surface of the porous body and PHBH and the miniaturization of PHBH crystals. Furthermore, the breaking energy was more than three times that of PHBH alone and the excellent toughness of the composite became clear. On the other hand, in the evaluation of the thermal expansion amount of the composite, even with the temperature higher than the glass transition temperature of PHBH, the deformation accompanying the temperature rise was suppressed in the composite and the coefficient of thermal expansion decreased. As described above, the composite material of cellulose monolith and PHBH improved with heat resistance and a composite material having excellent dimensional stability was obtained.

It is known that acetic acid bacteria (*Glconacetobacter xylinum*) produces cellulose, which is called bacterial cellulose (BC). BC is known as nata de coco and is produced cheaply in Southeast Asia from coconut water. BC is obtained as a hydrogel, of which more than 99% by weight is water. BC is a high-purity crystalline nanofiber that does not contain hemicellulose or lignin, unlike plant-derived cellulose. The BC gel is formed by fibrillating cellulose excreted from the body of bacteria and forming a ribbon-like BC nanofiber with a width of 50–100 nm to form a three-dimensional network structure. Generally, BC is synthesized by a stationary culture in a liquid medium and BC gel grows from the culture solution/air interface into the culture solution, so the BC gel has a uniform nano-sized network structure in the plane direction. It is known that it has anisotropy of having a micron-sized layered structure in the thickness direction. Further, the shape and strength of the gel can be controlled by culturing conditions. Such characteristics are not found in other polymer gels and it can be said that BC gel is a unique material, not only in vegetable cellulose but also in comparison with other polymer gels. BC is a material that has been known for a long time, but in recent years functional materials and composite materials that focus on the unique structure of BC have been actively studied [26–29]. As a method for producing the BC composite material, a dispersion method, a solution method, a culture method, and a bulk method are exemplified. The dispersion method is for a homogeneous dispersion obtained by breaking the BC gel by mechanical operation, while the solution method is for a uniform BC solution in which BC is dissolved in a cellulose solvent. This is a method of adding and compounding. These two methods are suitable for producing films and sheets, and while a uniform BC composite material can be obtained, the unique structure of BC gel is different from a plant-derived cellulose, such as a three-dimensional network structure and a layered structure, and is lost. Shortcomings have been noted. The culturing method is a method of producing a BC composite gel by culturing acetic acid bacteria in a medium in which a polymer or the like is dissolved or dispersed, but it is difficult to form a composite gel unless the material has a high affinity for the medium. The bulk method is a method in which another material is put into a BC gel and is then compounded by a chemical reaction. Although it is highly versatile and can make use of the three-dimensional structure of the BC gel, the operation is often complicated. Figure 17.11 shows the physical appearance of bacterial cellulose.

A composite monolith of BC and polylactic acid was developed. A monolith of poly(L-lactic acid) (PLLA) is obtained by a thermally induced phase separation method using a ternary solvent of 1,4-dioxane, 2-butanone, and water. The PLLA monolith exhibited a unique leaf-like morphology depending on the synthesis conditions [30]. Also, by changing the volume ratio of the solvent, especially the composition of 2-butanone, the pore diameter and skeletal system could be changed from a micro to a nano scale, and precise pore control of the monolith became possible. High heat and chemical resistance, including stereo complex crystals (sc), by dissolving the enantiomers of PLLA and poly(D-lactic acid) (PDLA) in a common high-temperature solvent, followed by mixing and cooling an sc-PLA monolith with the property was synthesized. By making the compositions of PLLA and PDLA equal, a monolith having a high crystallinity sc without any PLLA single crystal in the structure was obtained. The melting point was about 50° C higher than that of PLLA alone and the chemical resistance of polylactic acid to good solvents could be imparted. In contrast to the PLLA monolith, the sc-PLA monolith had a morphology with a connected spherical skeleton. When the PLLA monolith was synthesized in the presence of BC gel, a composite monolith (nanoplant structure) with a structure in which a leaf-like PLLA monolith was entangled with cellulose nanofibers was

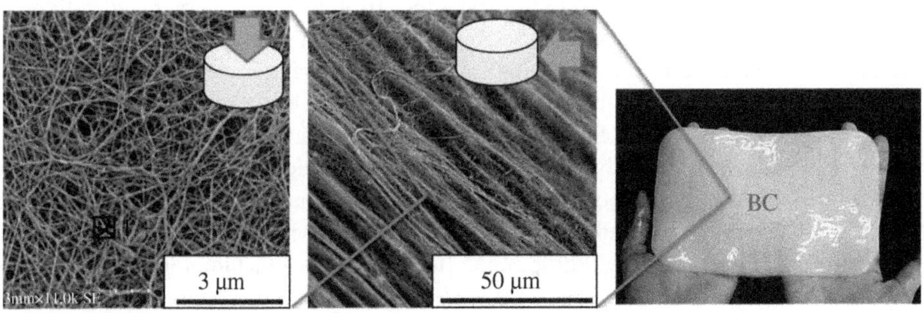

Figure 17.11 Physical appearance of bacterial cellulose.

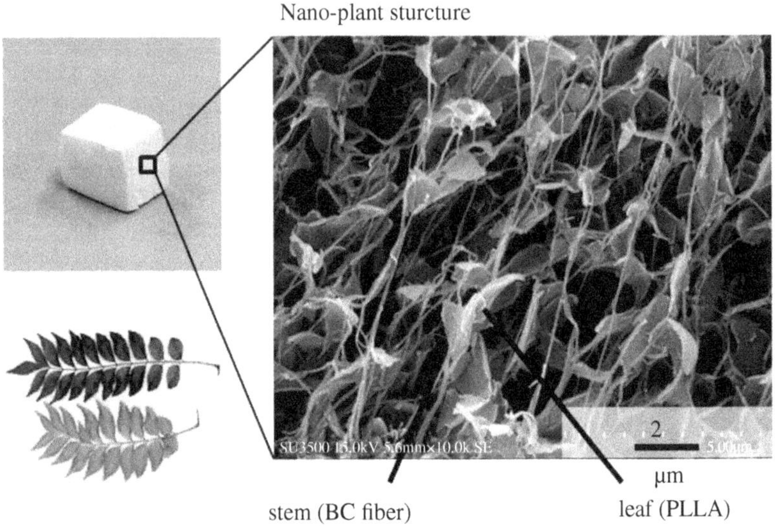

Figure 17.12 Nanoplant structure of PLLA and BC fiber.

generated (Figure 17.12) [31]. In this nanoplant, the BC fiber is the stem and the PLLA monolith is the leaf. Although the BC content of this composite monolith was less than 10%, the monolith surface quickly absorbed water.

17.5 Manufacturing Food Applications

The manufacturing of safe packaging and packaging materials is a critical part of ensuring the safety and integrity of the food supply chain. The packaging of food has essential functions for protecting from contamination, damage, and leakage. It provided the show case for marketing of the product to the end user.

The production of bio-based materials is available to a limited extent only. Manufacturing capacity data are difficult to access in order to retrieve accurate numbers of production capacities. From the Institute for Bioplastics and Biocomposites in Hannover and The Nova Institute together with European bioplastics of capacity data in 2015/2016, such data are difficult to obtain [32, 33]. Figure 17.13 illustrates the 2015/2016 figures compared to those in 2020. The increasing trend is for increased bio-based material in manufacturing. The global producers for bio-based material [34] indicated this trend in 2013, as shown in Table 17.3.

When manufacturing of bio-based food packaging materials is assessed, one of the major influencing factors is the cost of the materials. The price of bio-based food packaging materials is still considered to be higher than fossil-based plastics. However, PLA prices are now comparable for general purpose PS and PET [35]. Figure 17.14 shows the prospects for the production of eco-friendly plastics from biomass-based materials in the future. The price of bio-based materials will drop in the future with manufacturers following a structured methodology for assessing opportunities, barriers, and policy options that need to be overcome as these transitions proceed toward an economy of scale.

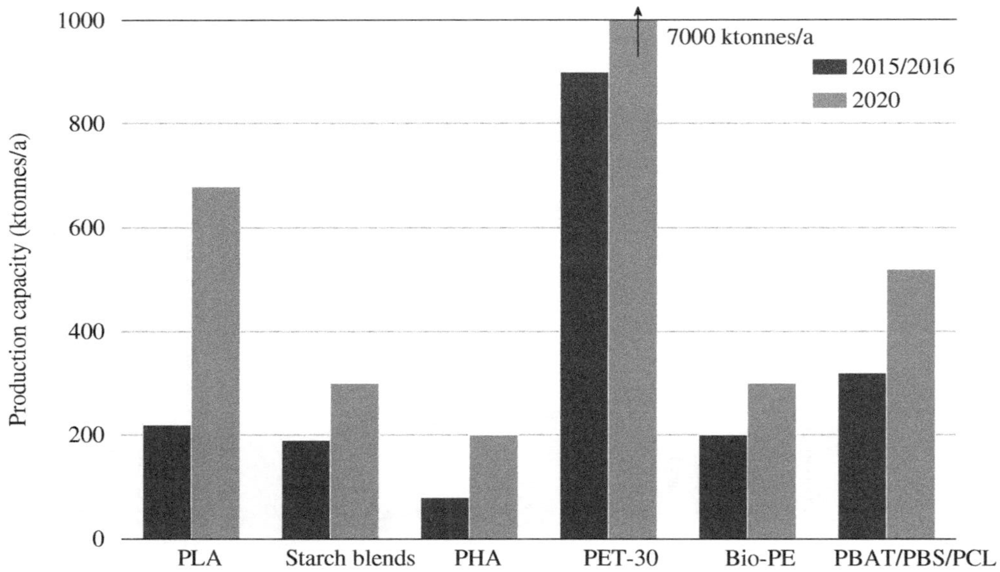

Figure 17.13 *Production capacity of bio-based polymer in 2015/2016 and 2020.*

Table 17.3 Number of producers and type of bio-based material.

Bio-based polymer type	Producers in 2013	Materials
CA	17	Food packaging (limited)
Bio-PA	9	Food packaging (limited)
Bio-PET	5	Food packaging
PBAT	4	Food packaging
PBS	10	Food packaging
PLA	28	Food packaging

Source: adapted from Nova Institute, 2015.

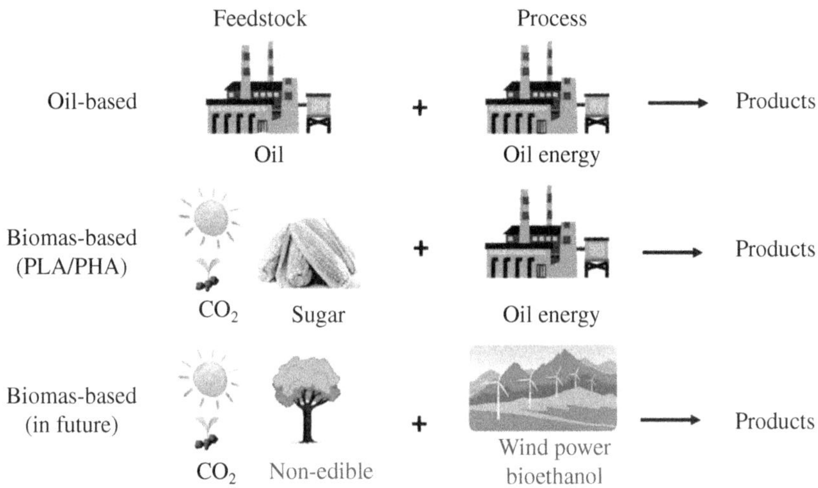

Figure 17.14 Future prospects of eco-friendly plastics process.

17.6 Food Industry and Bio-Based Materials Demand

Bio-based materials can be a substitute for fossil-based material in food packaging although some limitation will still be encountered. The factors are production cost, poor policy support, and lack of an appropriate waste management facility [36]. From biomass, bio-based material is three to five times more costly compared to fossil-based material [37–41]. The cost of producing bio-based materials include the cost of mobilizing biomass residues, the cost of technology innovations, and the lack of economies of scale. In 2011 Malaysia launched a National Biomass Strategy for 2020 for optimum utilization of biomass. The key contributor of the palm oil sector generates the largest amount of biomass. Empty fruit bunches (EFBs) converted into ethanol, which converted to bio-fuel and bio-chemicals (including bio-plastics). The cost of mobilization of palm tree residues for industrial use is the main problem. Bio-based plastics are partly or completely from biomass materials that take physical, chemical, and biological treatment to make them. Currently, a list of most bio-based food packaging materials produced by company and type of biopolymer is shown in Table 17.4.

Based on consumers strong acceptance on their part and willingness to pay higher prices, the demand for bio-based packaging products will continue to increase. The benefits will be the minimization of litter on land and sea and also helping to protect the lives of animals and marine life. The combination of these desired properties and functions of food packaging materials and commodity market prices indicate the substitution of renewable polymers in food packaging products. Figure 17.15 illustrates bioplastic use in food packaging materials, which will gradually increase demand, There is a need to change the market and people's awareness in order to lead to a high demand in usage of bio-based food packaging materials in the near future.

17.7 Conclusions and Remarks

In recent years, with the increasing of environmental awareness, demands for bio-based materials have been rapidly increasing and a remarkable increase in demand for biomass plastics and other bioplastics is expected. For the bioelastomer outlined in this chapter, technologies for converting raw materials into biomass and using plant resources directly will be developed in the future. In particular, the expected development of material design and the realization that can be made of the characteristics of biomaterials in the elasticity of materials are encouraging. Biodegradable materials that have demonstrated marine biodegradability are limited to PHBH, PBS, and the like. Although plans to increase these productions have been announced, it is considered that a considerable time may be required for widespread use as marine biodegradable plastics, due to issues such as price, physical properties, and production facilities. Therefore, it is necessary to develop a biodegradable

Table 17.4 *Food packaging production by using bio-polymer in packed materials.*

No.	Packaging product	Biopolymer	Packaged Material	Company
1	Tablewares, paper cups and gas barrier packaging	BioPBS (polybutylene succinate)		Ptt Mcc Bio-Chem Company, Thailand/Japan
2	Packaging	Bio-PE (polyethylene)	Pet food	Braskem Company, US
3	Bottle, packaged goods	BioPET (polyethylene terephthalate)		Toray Industries, The Coca-Cola Company, M&G Chemicals, Teijin Limited, and Toyota Tsusho, Japan
4	Cutlery, bag, laminated food packaging film	Polysaccharides	Fruits	Novamont Company, Italy
5	Trays, cutlery, containers, bowls, lids, portion cups	BioPLA (polylactic acid)		GreenGood USA/SMD Trading Co. Ltd

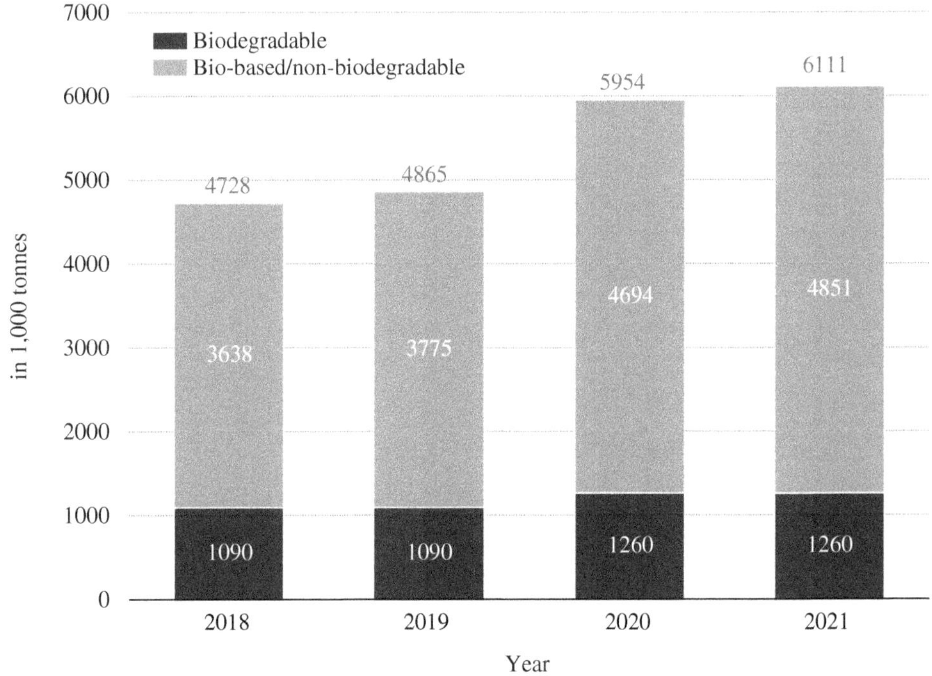

Figure 17.15 *Calculation of bio-polymers from 2015 to 2020.*

polymer other than these aliphatic polyesters. Polysaccharides such as inexpensive starch and polyvinyl alcohol that are biodegradable are considered as candidates, although they have poor water resistance. Starch-based materials that have been used for food packaging and biodegradable plastics in Europe are new. Development of a biodegradable plastic, especially for food packaging material, is being investigated. In the future, biodegradable plastics, which have been in the spotlight once again due to biodegradation, will be continuously developed and widely used as food packaging materials, contributing to global environmental conservation. It will be necessary not only to develop bio-based food packaging materials but also to build social systems that accept them.

Acknowledgments

The authors would like to acknowledge Universiti Putra Malaysia for the financial support and technical data and support from Graduate School of Engineering, Osaka University, Japan.

References

1. Semith O, Serkan O. Manufacturing of biobased packaging material for the food industry. *Acta Sci. Pol. Technol. Alimen.* 2004; **3**(2):13–17.
2. EuropaBio. 2015. *Building a Bio-based Economy for European in 2020*. Brussels: EuropaBio.

3. Rubie VC., Valeria K. Bio-based food packaging in sustatiable development. *Food Agri. Organ. United Nations* 2016; 20–37.
4. Mohanty F, Swain SK. Chapter 18, Bionanocomposites for food packaging applications. *Nanotechnol. Appl. Food* 2017; 363–379.
5. Iwata T. Biodegradable and bio-based polymers: future prospects of eco-friendly plastics. *Angew. Chem. Int. Ed.* 2015; **54**: 3210–3215.
6. Reddy M.M, Vivekanandhan S, Misra M, Bhatia S K, Mohanty A K. Biobased plastics and bionanocomposites: current status and future opportunities. *Prog. Polym. Sci.* **2013**; 10–11: 1653–1689.
7. Uyama H. Current status and future perspective of research and development on bio-based polymers. *J. Rubber Soc. Jpn* 2013; **86**: 161–168.
8. Shinya Takeno, Yoshihisa Nakazawa, Takashi Tsujimoto et al., *Biopla J.*, **55**, 18 (2014).<Please code these names: surnames are Nakazawa and Tsujimoto>
9. Shinya Takeno et al., Development and market of bio-based materials, *CMC Res.* (2015), **101**.
10. Tsujimoto T, Toshimitsu K, Uyama H, Takeno S, Nakazawa Y. Maleated trans-1,4-polyisoprene from *Eucommia ulmoides* Oliver with dynamic network structure and its shape memory property. *Polymer* 2014; **55** (25): 6488–6493.
11. Biermann U, Bornscheuer U, Meier M A, Metzger J O, Schafer H J. Oils and fats as renewable raw materials in chemistry. 2011; *Angewandte Chemie*; **50** (17): 3854–3871.
12. Llevot A. Sustainable synthetic approaches for the preparation of plant oil-based thermosets. *J. Am. Oil Chem. Soc.* 2017; **94**: 169–186.
13. Uyama H. Functional polymers from renewable plant oils. *Polymer J.* 2018; **50**: 1003–1011.
14. Tsujimoto T, Haza Y, Yin Y, Uyama H. Synthesis of branched poly(lactic acid) bearing a castor oil core and its plasticization effect on poly(lactic acid). *Polymer J.* 2011; **43**: 425–430.
15. Kasahara T, Uyama H. Elastic polyurethane ionomers from dimer acid elastic polyurethane ionomers from dimer acid. *J. Network Polymer, Japan* 2016; **37**: 175–179.
16. All Bioplastic Materials, Japan Bioplastics Association (ed.), Nikkan Kogyo Shimbun (2008).
17. Natural Plastic, Yoshiharu Kimura (ed.), Kyoritsu Shuppan (2006).
18. Polylactic Acid, Hideto Tsujizaki (ed.), Yoneda Publishing (2008).
19. Tsujimoto T, Nishio S, Uyama H. Bio-based branched polymer bearing castor oil core as a nucleating agent for poly(l-lactic acid). *J. Polym. Environ.* 2015; **23**: 559–565.
20. Hosoda N, Lee E-H, Tsujimoto T, Uyama H. Phase separation-induced crystallization of poly(3-hydroxybutyrate-co-hydroxyvalerate) by branched poly(lactic acid). *Ind. Eng. Chem. Res.* 2013; **52**: 1548–1553.
21. Cellulose with Well-Understood Names, Nanocellulose Forum (ed.), Nikkan Kogyo Shimbun (2015).
22. Cellulose Nanofiber Technical Data Book, CMC Publishing (2016).
23. Takahiro Uyama (supervised), Latest Trends in the Production and Use of Cellulose Nanofibers, CMC Publishing (2019).
24. Uyama H. Polymeric monolith – new fabrication methods and applications. *Kobunshi Ronbunshu* 2010; **67**: 489–496.
25. Hosoda N, Tsujimoto T, Uyama H. Green composite of poly(3-hydroxybutyrate-co-3-hydroxy hexanoate) reinforced with porous cellulose. *ACS Sustain. Chem. Eng.* 2014; **2**(2): 248–253.
26. Shim H, Xiang X, Karina M, Inrarti L, Uyama H. One-dimensional shrinkage and swelling of crosslinked bacterial cellulose gel. *Chem. Lett.* 206; **45**(3): 253–255.
27. Wang Q, Asoh T, Uyama H. Rapid uniaxial actuation of layered bacterial cellulose/poly (N-isopropylacrylamide) ceomposite hydrogel with high mechanical strength. *RSC Adv.* 2018; **8**: 12608–12613.
28. Bai Q, Xiong Q, Li C, Shen Y, Uyama H. Hierarchical porous carbons from poly(methyl methacrylate)/bacterial cellulose composite monolith for high-performance supercapacitor electrodes. *ACS Sustain. Chem. Eng.* 2017; **5**(10): 9390–9401.

29. Bai Q, Xiong Q, Li C, Shen Y, Uyama H, Hierarchical porous carbons from a sodium alginate/bacterial cellulose composite for high-performance supercapacitor electrodes, *Appl. Surf. Sci.*, 2018, **455**: 795–807.
30. Kanno T, Uyama H. Unique leafy morphology of poly(lactic acid) monoliths controlled via novel phase separation technology. *RSC Adv.* 2017; **7**: 33726–33732.
31. Kanno T, Uyama H. Unique ivy-like morphology composed of poly(lactic acid) and bacterial cellulose cryogel. *ACS Omega.* 2018; **3**(1): 631–635.
32. IfBB, Institute for Bioplastics and Biocomposites, European Bioplastics, Nova Institute 2015. Market statistics.
33. European Bioplastics. Bioplastics-furthering efficient waste management. Factsheet 2015.(accessed: March 16 2016)
34. Nova Institute 2015. Bio-based building blocks and polymers in the world – capacities, production and applications: status quo and trends towards 2020.
35. Verbruggen M. Bio-polymers, an industry perspective, 2013. In: *BPM Symposium*, June 27, 2013. Wageningen, Netherlands.
36. Molenveld M, Van den Oever MJA, 2015. Biobased packaging catalogue, *Wageningen Food & Biobased Research.*
37. Economic & Planning Systems, Inc. 2012. *Economic Impact Analysis of EPS Foodward Costs.* Prepared for City of San Jose. San Jose.
38. Rodenbury, T.(director of Rodenburg Polymers), Interviewed via Skype by Rubie van Crevel, February 2, 2016.
39. Van Dijk, A. (Senior Product Manager at The Greenery), interviewed via Skype by Rubie van Crevel, March 2, 2016.
40. Hairul A, Azmi B, Faris M, Yuzalmi F, Fadli H, Melbi M, Eni S, Sapuan S.M, Ilyas R.A, Ilfa S. A simple method for improving the properties of the sago starch films prepared by using ultrasonic treatment. *Food Hydrocolloids.* 201; **9**: 276–283.
41. Ilyas RA, Sapuan SM, Atikah MSN, Asyraf MRM, Ayu Rafiqah S, Aisyah HA, Mohd Nurazzi N, Norrahim MNF. Effect of Hydrolysis Time on the Morphological, Physical, Chemical, and Thermal Behavior of Sugar Palm Nanocrystalline Cellulose (Arenga Pinnata (Wurmb.) Merr). *Textile Research Journal.* 2020. In print. 10.1177/0040517520932393

18

Bioplastics: An Introduction to the Role of Eco-Friendly Alternative Plastics in Sustainable Packaging

Usman Lawal[1,2] and Ravi Babu Valapa[1]

[1]Electrochemical Power Source Division, CSIR- Central Electro Chemical Research Institute (CECRI), Karaikudi 630003, Tamil Nadu, India

[2]Academy of Science and Innovative Research (AcSIR), Ghaziabad, 201002, India

18.1 Introduction

By 2015, it is estimated that ~7.8 billion tons of plastics have been produced worldwide. The global packaging industry is worth about $851 billion with three distinct types of packaging in the supply chain: primary, secondary, and tertiary packaging [1]. The management of waste from petro-based packaging materials has become almost an impossible task due to their inherent inability to compost. High-energy consumption during processing, nonrenewable sources, and air and environmental pollution are some of the notable drawbacks associated with conventional polymer plastics. The packaging industry is a potentially important market for biopolymers because of the nonrenewable life offered by the petro-based plastics. Their nondegradable nature results in waste disposal, environmental, and ecological problems. Nondegradable plastic waste is also causing a lot of havoc to marine and wildlife. For biopolymers, the important points are those of biodegradability and compostability, antimicrobial activity, energy-efficient processes, low price, and an endless resource, environmental and ecological friendliness, among others. However, environmental friendliness and usefulness are not enough to create a market for bioplastics. They must also provide

Bio-Based Packaging: Material, Environmental and Economic Aspects, First Edition. Edited by S.M. Sapuan and R.A. Ilyas

cost-effective end products that are suitable for packaging and provide distinctive benefits in usage over regular plastics [1, 2].

Today, polymers are part of our modern life because of their desired properties, which are not limited to stability, resilience, and ease of production. Worldwide production of plastics was over 381 million tons in 2017, which is a 200% increase when compared to the 1950s when it all started. This is about the mass of two-thirds of today's world population. Packaging took the largest chunk of 141 million metric tons in 2017 and is producing about the same amount as waste [2, 3]. Europe produced 64.4 million tons, representing 18.5% of world production, while Asia produced 50.1% with China producing 29.4%, which is more than half of the Continent's production. India produced 8.3 million tons of plastic, which makes up 16.8% along with other countries in the subcontinent excluding Japan, which produced 3.9% [2, 3]. At present, about 99% of all plastic materials are manufactured by the petrochemical industries [3]. The world average for plastics utilized in the packaging industry is 39%, but this is significantly higher in developing countries. Manufacture and handling practices of plastics demand lots of energy; this increases emissions of greenhouse gases and contributes to global warming significantly. Byproducts from the burning of plastics are a health concern because they lead to poisoning and contamination [4]. An estimated 100 000 tons of plastic waste litter our water bodies worldwide. Resistance to degradation and composting, contamination of water bodies and wildlife, the significant health risk posed on nature has made plastic waste a concern to all [4].

Bioplastics have appeared as a viable alternative to solve the problems associated with petro-based plastic wastes. The Australian Bioplastics Association states that bioplastics encompass a whole family of materials, which are bio-based, biodegradable, or both [5]. These are usually derived from renewable sources, which are sustainable over time. The move to replace the conventional plastic derived from petroleum resources with biopolymers is a fair one because petroplastics consume 65% more energy during production and also emit about 80% higher greenhouse gases compared to biopolymers [5, 6]. Naturally, occurring polyesters are some of the main sources of bioplastics. These biomaterials are easily found in nature and have physicochemical properties similar to petrochemical polymers. Production of bioplastics like poly(lactic acid) (PLA), poly(hydroxybutyrate) (PHB), polybutylene succinate (PBS), polybutylene adipate terephthalate (PBAT), chitosan, poly(caprolactone) (PCL), starch, and others is estimated to be ~2.11 million tons in 2018 [2, 7]. Modified bioplastics such as PLA (polylactic acid) and PHAs (polyhydroxyalkanoates) are at the forefront of growth in the field of environmentally friendly and compostable plastics [8].PLA-derived biopolymers have been envisaged to replace PS (polystyrene), PP (polypropylene), and ABS (acrylonitrile butadiene styrene) in more demanding uses due to their excellent barrier properties and improved physical and mechanical possessions. The market share of bio-based and degradable polymers is only about 1% and is expected to grow by 60% by 2023 [2, 9]. Most bioplastics currently in use are not as proficient in some cases like the conventional plastics. However, specialized bio-based polymers are being developed for certain segment(s) of the packaging industries for fruit wrapping, shopping bags, to seafood. Polyethylene furanoate (PEF) is one such biopolymer featuring greater barrier and thermal properties, and items such as drinks, food, and non-food products are expected to be packaged ideally using this material. Among others, PBAT, chitosan, PCL, and starch composites and derivatives are also widely tested to replace petroleum-based plastics in the near future [10].

Biopolymers are a form of polymers that are obtained from renewable natural sources; they are biodegradable, mostly non-toxic, and organic in nature. These primary materials are sourced from biological systems like plants, animals, and microorganisms, and they can also be synthesized. The classes of biopolymers are starch, sugar, cellulose, and synthetic materials [11–20]. Bioplastics are biopolymers of natural sources such as carbohydrates, proteins, lipids, corn starch, chitin, etc. An agricultural feedstock is a significant source of biopolymers for food packaging applications and beyond. Petroleum and natural gas are the common sources of fossil-fuel plastics; they are otherwise called petro-based polymers. More amounts of fossil fuels are required for the production of such plastics [21].

The food processing industry has packaging as one of its essential components. Food packaging is the art, science, and technology of ultimately preparing a product for the market, as packaging essentially provides physical, emotional, chemical, and biological needs including esthetics to the food products [22]. Petrochemicals like polypropylene (PP), polyamide (PA), polystyrene (PS), polyvinyl chloride (PVC), and polyethylene terephthalate (PET) are raw materials for today's packaging materials, providing wholesome quality packaging but also generating wastes that are resistant to degradation that pollutes the environment and damages our ecology as we know it [23, 24].

This write-up explores important biopolymers behind ecologically friendly food packaging applications from the sea of unending numbers of biopolymeric materials, their sources, usefulness, varieties, and characteristics. These biopolymers are of great importance to the packaging industry as a whole because of their renewable sources, cost, and energy efficiency, eco-friendliness, diversity, among others. It is also essential to the food packaging industry as it continually seeks to reduce its carbon footprint, improve waste and pollution management, and increase its utilization of renewable resources. The aim is to demonstrate recent advances in biopolymer composites and derivatives technology and how they are helping to shape the food packaging industry of the future. Their potential for use in the sustainable and green packaging industry is also explored. More importantly, barrier and mechanical qualities are germane to the utilization of biopolymers in this chapter.

18.2 Important Biopolymers for Food Packaging

Microorganisms, plants, and animals are the main sources of biopolymers as shown in Figure 18.1. Biopolymers are renewable biomaterials with similar characteristics to petroleum-based polymers. Biopolymers are of three main classes according to the nature of their monomers and configuration of polymers formed: (i) polypeptides as in amino acids, (ii) polynucleotides as in DNA and RNA, and (iii) polysaccharides as in starch, glycogen, cellulose, etc. [25, 26]. Biopolymers with a better performance, which is important for the food packaging industries, are achieved as biopolymer nanocomposites (BNCs). Biopolymer nanocomposites are formed by adding nanofillers, plasticizers, compatibilizers, co-polymers, etc., into the polymer matrix of a biopolymer to improve its mechanical and barrier properties. Proper dispersion of the nanomaterial in the polymer matrix at about 5 wt% or lower has shown a lot of promise. Biopolymer composites result in improved functionality with enhanced mechanical, barrier, physicochemical, optical, thermal properties, etc., compared to the ordinary biopolymer or the nanofillers themselves. Several studies have shown these improvements in the properties and functionalities of composites.

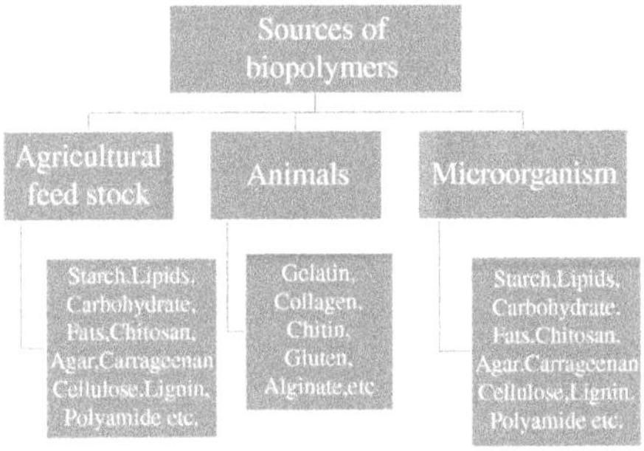

Figure 18.1 *Different sources for biopolymers [25, 26].*

Biopolymer composites have a very high potential as substitutes to the petro-based plastics because of these notable improvements in functionality and versatility [27–35].

18.2.1 Starch

Starch is the most common and cheapest biopolymer available. They are hydro-colloidal carbohydrates comprising two molecules, amylose and amylopectin. The functional properties of a starch-based bioplastic are affected by the level in which these molecules are present in it, which is also dependent on the source of the starch. A higher ratio of amylose decreases the solubility rate due to its significant linear structure. The amylopectin is highly branched and therefore increases the solubility rate of the starch. Starches from many sources like wheat, rice, and corn are used as biopolymers [27, 28].

Starch is usually used as a thermoplastic. Due to its poor resistance to the moisture and barrier property, bioplastics of starch origin have restricted uses. Improved properties are achieved by blending with other materials and additives. Modified starch bioplastics have received ample attention in the commercial food industry, mainly because of their low cost, biodegradability, and numerous applications [26, 27]. Cellulose nanofibers in conjunction with montmorillonite have been reported to drastically improve the performance of corn starch as a packaging material. Mechanical properties from tensile strength to Young's modulus are found to be enhanced while moisture vulnerability was reduced and transparency increased [28]. Starch polymer and composites find application as bioresorbable, edible coating, and films in the packaging industry.

18.2.2 Polylactic Acid (PLA)

Polylactic acid, also known as PLA, is the center of attention at the moment due to its improved food packaging qualities and mechanical features. PLA is also called polylactide and is an aliphatic polyester that is a hydrophobic thermoplastic [29]. Lactic acid is the monomer unit of PLA polymer and is produced through carbohydrate fermentation of sugar feedstock. PLA can also be chemically synthesized. The homofermentative bacterial fermentation method is the most preferred industrially due to its higher yield of lactic acid and little byproduct. PLA is made from simple sugars like maltose, glucose, sucrose, and lactose, which are often sourced from corn, potato, beet sugar, cheese, etc. PLA's popularity in the food industry as a packaging material is due to its unique qualities, which sometimes supersede those of its petroplastic counterparts [29, 30].

Among those unique qualities that made PLA as good a food packaging material as petroplastics are its high molecular weight, insolubility in water, high transparency, and significant tensile strength [30]. PLA is as flexible as PVC with the heat stability and process performance of polyethylene. PLA also demonstrates the tensile strength, flavor, and odor barrier of polyethylene. PLA film has demonstrated better mechanical properties than PS and is comparable to PET. PLA films give a good performance across a wide range of temperatures and the quantity of lactic acid that enters into food when in contact is very low compared to the amount of lactic acid in the food. This makes PLA a suitable material for food packaging applications [12, 30]. Graphene oxide (GO) and functionalized graphene nanosheets have been reported to improve the PLA matrix to give superb thermal stability and superior storage modulus [31]. The unmodified form of PLA shows poor properties like brittleness and heat instability, but this is remedied by plasticization with other materials.

18.2.3 Cellulose

Cellulose biopolymer is the most common biopolymer on earth and the most plentiful cell wall constituent in plants. Cellulose is an unbranched biopolymer of glucose linked by 1,4-β-glycosidic bonds. These linear molecules contain an average of 5000 units of glucose joined together to form fibrils linked together by hydrogen bonds. Around 50 billion tons of this renewable biodegradable biopolymer are formed year after year by plants globally. The purest form of cellulose is found in cotton seed fiber. Cellulose is often found in a mixture with other materials like lignin and hemicelluloses. Cotton, wool, linen, hemp, and straw are the main raw material sources for making cellulosic polymers and bioplastics. Pulp from wood is the main raw material for extracting cellulose, most of which is utilized for the manufacture of paper and cardboard. Films, cellulose ethers, and esters are other materials made from cellulose and are of industrial importance and made on a commercial scale for coating optical films, laminates, etc. [32].

Cellulose acetate and other esters are the focus of research in developing cellulose derivatives for use in the packaging industry. Several companies around the world already commercially produce these cellulose esters through reacting pulp with acetic anhydride. An example is EnviroPlastic from BIOCETA in the USA. Ecologically friendly packaging containers, films, tubes, etc., are commercially produced [1, 33]. Cellophane, which is modified cellulose and appears like a transparent thin film, is being used in the food packaging industry. Cellulose composites show improved water, grease, air, and bacteria barrier characteristics [33]. Cellulose derivatives are blended with other polymers such as methylcellulose to lower the water vapor permeability. Adding lipids, minerals, or fillers may also enhance the water vapor barrier property [30].

18.2.4 Chitosan

Chitosan is the second most abundant biopolymer in nature, after cellulose. It is a high molecular weight cationic natural polymer composed of β-[1–4]D-glucosamine and

N-acetyl-D-glucosamine monomers. Chitin, the main source of chitosan, is mainly isolated from crustaceans, mostly shrimp and crabs. Chitosan is also sourced from yeast, fungi, protozoa, green algae, and insects [34]. Chitosan is often produced from the N-deacetylation of chitin. Interestingly, Chitosan is a wide-spectrum antimicrobial compound and is known to show activity against a wide range of microbes, such as both gram-positive and gram-negative bacteria, fungi, and yeast [35]. The combination of its antimicrobial activity and the film-forming property has elevated chitosan as a food packaging material for the future, which will not only provide packaging for food and provide food safety but will stop microbial attacks and increase the shelf life of products. Additionally, chitosan exhibits desirable properties required for packaging materials, including good optical properties, and chitosan films are as transparent as many other petro-based plastics with high transmittance values. This ensures visibility of packaged food to the customer, which is an important tool for marketing [36].

Furthermore, these transparent chitosan films demonstrate a significant UV absorbance property, which prevents lipid oxidation in packaged foods occasioned by exposure to UV radiation. Given its morphology and variable properties, chitosan is found to be soluble in dilute acid solutions and has been widely studied due to its good film-forming properties. Chitosan can be used wholly as a film for packaging or as a coating on other packaging materials or polymers [34, 37]. Chitosan coatings enhance both the glossiness and oxygen barrier potential of paper. The application of chitosan on surfaces of the film makes it hydrophilic, which serves as a barrier for the transmission of oxygen, carbon dioxide, and ethylene [34]. The incorporation of nanoclays into chitosan increases its barrier properties across the board. Also, composite films of chitosan exhibit increased barrier levels against oxygen, grease, odor, and UV absorption, which is also true for multilayer-coated chitosan.

18.2.5 Polyhydroxyalkanoates (PHAs)

Polyhydroxyalkanoates (PHAs) are linear polyesters produced from bacterial fermentation of lipid or sugar. These are semi-crystalline, biodegradable biopolymers used in the medical industry given their inherent quality of having more ductility and less elasticity than the conventional petroplastic [38]. PHA bioplastics are products of various species of bacteria with uneven growth conditions. The PHAs formed are dependent on the choice of bacteria and are feed administered. In nature, these biopolymers are formed by bacterial breakdown of sugar and lipid, which is aided by the availability of an excess carbon source and limited by the inadequate presence of nitrogen and phosphorus. PHAs have a simple macromolecular structure of 3-hydroxy fatty acid monomers. PHAs are known to have similar thermomechanical properties as polypropylene petroplastic [39]. PHA has found usage as compostable packaging material in the manufacture of bottles, containers, laminates, films, sheets, coatings, and fibers. Potentially, hundreds of monomers and copolymers can be developed from PHAs and several are already in production with patents from around the globe. PHAs possess desirable properties like a sufficient odor and flavor barrier, heat stability, oil and grease resistance with decent tensile strength, and are also easy to dye and are printable. These attractive qualities make them attractive to the food industry as alternative packaging materials [38]. Metabolix PHA is a blend of polyhydroxy butyrate (PHB) and poly (3- hydroxyoctanoate) from Metabolix Company and is a good example of the utilization of PHAs in the packaging industry [1, 38].

18.3 Important Properties of Biopolymers for Food Packaging Applications

18.3.1 Mechanical Properties of Biopolymers

For biopolymers to effectively compete with petro-based plastics and take over, they must have at least similar mechanical properties. These important properties are tensile strength and percentage elongation among others. These parameters explain other dependencies, such as how well it will handle, transport, and store food for the packaging industry. The behavior and integrity of a polymer in the packaging industry are determined by its mechanical properties. Protuberances, puncture force, and deformation force are other properties of interest to take note of in a biopolymer for food packaging applications [39]. These values when compared to those of conventional plastics indicate the prospects of a new biopolymer for packaging utilization [40]. The mechanical properties of a biopolymer material are a function of its nature, structure, and composition elements. The tensile strength, puncture and deformation forces, etc., of a film may also be affected by the conditions under which the measurement is done as well as the measuring equipment used. Storage time is another factor affecting the mechanical performance of a biopolymer film or coating, just like a conventional petrol plastic. Standard equipment like the texture meter is used for the measurement of mechanical properties of bio-based materials.

Tensile strength is by definition the maximum load a film can carry before its breaking point. This is the total stress a film can sustain before permanent damage is done. The percentage of elongation is the film maximum elongation possible before breaking [41, 42]. These two important mechanical properties have an inverse relationship for a particular biopolymer and the elongation of a film is enhanced with reduced tensile strength and vice versa [43].

Packaging applications differ in their material needs, which means that biopolymers of different mechanical properties will be required for diverse uses. Films used for light packaging require high flexibility. Moreover, for inflexible packaging needs, such as in the fabrication of bottles and containers, higher tensile strength polymers are desired. Biopolymers made from starch provide films that are good for flexible packaging applications, but modified PLA is preferred for rigid applications given its low elongation percentage, reasonable tensile modulus, good tensile strength, and strain at break, as illustrated in Figure 18.2a, b, and c [44].

18.3.2 Barrier Property

Barrier properties of a film or coating are of immense importance to the packaging industry in general as they determine the usage of these materials accordingly. Important parameters in storage, such as the shelf life of packaged food, are dependent on the barrier property of the housing material. An important application of films used in packaging is to stop particles exchange between the environment and the packaged food. This is to ensure food quality and wholesomeness. The extent to which a biopolymeric material can prevent movements of

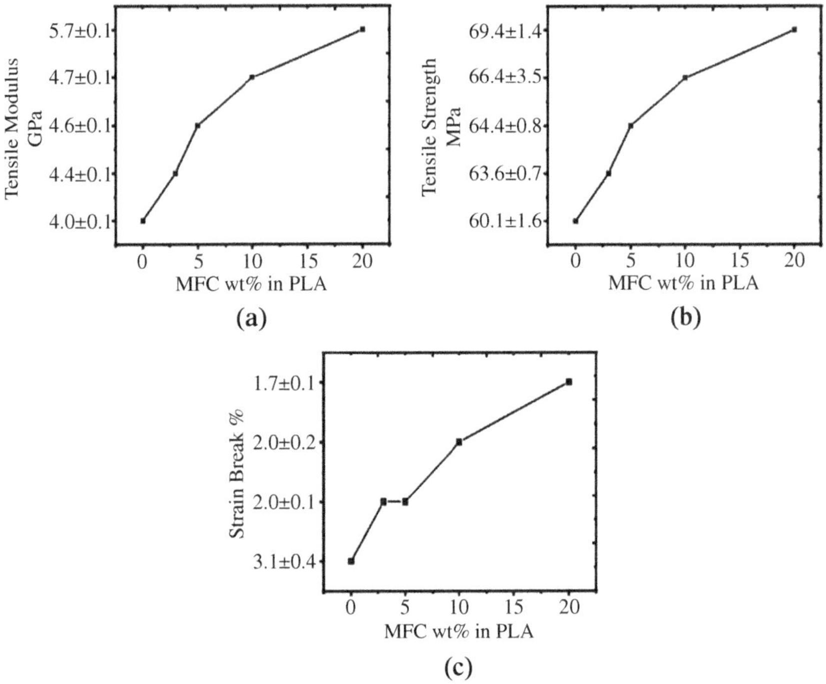

Figure 18.2 *Mechanical properties of PLA/MFC composite [44].*

certain particles like water vapor, liquids, oxygen, carbon dioxide, and organic vapors, etc., indicates its suitability for use in the food packaging industry. These particles are known to permeate through petroplastics in small amounts. This transfer of molecules happens in both directions, from the environs to the food and the other way round, bringing about a continuous decrease in food quality and shelf life [45, 46].

The degree of water permeability of a film is an important property, because of the role water plays in food safety, decay processes, wholesomeness of the packaged food, and shelf life. Water vapor permeability of biopolymers has been studied extensively because of its importance. Water vapor permeability (WVP) and oxygen permeability (OP) of selected polymers are shown in Figure 18.3a and b, where silicone elastomer (SE) gave the poorest performance in the WVP test, permeating at >3000 g μm $(m^2 \text{ day kPa})^{-1}$, while high-density polyethylene (HDPE) and polyvinylidenechloride (PVdc) gave the best performance permeating at <40 g.μm $(m^2 \text{ day kPa})^{-1}$. In OP, ethylenevinylalcohol (EVOH) gave the strongest performance, permeating at <40 cm^3 μm $(m^2 \text{ day atm})^{-1}$, with polystyrene (PS) giving the worst performance at >40 000 cm^3 μm $(m^2 \text{ day atm})^{-1}$ permeation. Importantly, these data provide critical information about shelf life sustainability and food safety of the packaging material. This is because oxygen exerts an important role in food oxidation, which brings about changes in taste, odor and color, growth of microbes, vitamin loss, enzymatic browning, and spoilage of foods. Respiration in fresh vegetables and fruits is guaranteed by suitable permeability of the packaging biomaterial. Good water

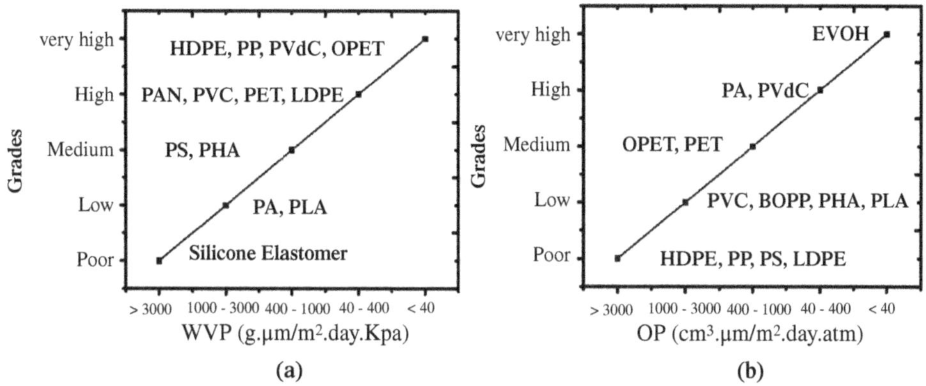

Figure 18.3 *Showing (a) oxygen and (b) water vapor permeability of selected polymers at 23–25°C. Key: ethylenevinylalcohol (EVOH), high-density polyethylene (HDPE), oriented poly ethyleneterephthalate (OPET), polyamide (PA), polyethyleneterephthalate (PET), polyhydroxyalkanoates (PHA), polylacticacid (PLA), polyvinylchloride (PVC), polyvinylidenechloride (PVdC), polystyrene (PS), polyacrylotrile (PAN), silicone elastomer (SE), biaxially oriented polypropylene (BOPP), and polypropylene (PP) [51].*

and a greater oxygen barrier are mainly shown by bioplastic composites when compared to petroplastics [47–50]. Accordingly, tailored biopolymer composites are made based on the desired permeability for different applications in the food industry.

18.3.3 Antimicrobial Properties

Consumers and scientists have been concerned about the increasing use of chemicals to extend the shelf life of food products. Increasing cases of poisoning experiences have also added to the need to have food packaging materials that are not only wholesome for packaging applications but also possess antimicrobial properties. Some biopolymers like chitosan are known to have good microbial hindering functions in addition to their ability to form packaging film. Food preservation through storage in antimicrobial films is an emerging trend. A good number of studies have been undertaken that shows the efficiency of antimicrobial coatings and films in preventing bacteria growth in food. These films have been found to prevent growth and activities of pathogens like *Staphylococcus aureus*, *Escherichia coli*, fungi, etc. [52, 53]. Antimicrobial agents from renewable natural sources are gaining traction as an added advantage to food packaging applications. Biopolymers with suitable mechanical and antimicrobial properties are a novel strategy in providing eco-friendly packaging and increased shelf life of food products [53–55].

Biopolymers from starch, cellulose and other polysaccharides, chitosan and other derivatives, especially those that are hydrophilic, show significant antimicrobial activity when used in food packaging applications. Presently numerous biodegradable biopolymers from natural sources and synthetic backgrounds are used as antimicrobial film and coating applications. Examples include proteins such as those from fish, soy, wheat gluten protein, etc., as well as hydroxypropylmethylcellulose, carboxymethyl-cellulose, or their blends [56, 57].

The knowledge of pathogen morphology and the life cycle is essential in having a targeted antimicrobial activity in food packaging. These microbes of interest have several classifications based on some parameter, such as oxygen demand, cell wall presence, temperature dependence, etc. [58]. A combination of different antimicrobial films is also used to achieve the desired shelf life based on the type of microbes to be tackled in the packaging. It is also possible to use two or more antimicrobial agents instead of only one, having complementary actions to achieve better antibacterial activity.

18.3.4 Optical Properties

The ability of packaging material to avert UV light penetration, thereby preventing food from the effect of light, is an important property that a biopolymer for packaging applications must have. The UV radiation is an oxidizing agent for oils and lipids present in our foods. The UV light emitted by the sun within the UV–visible spectrum effectively causes oxidation in exposed food and substances. These radiations cause rancidity, which results in undesirable conditions including odor, flavor, and taste changes in foods. Hence, biopolymers to be considered for food packaging applications must also possess the ability to prevent the passage of particles through it or at least minimize the access of light to the convenience of shelf life and wholesomeness of food [59].

Another criterion to meet by would be biopolymer for food application in the area of optical activity is often transparency, enough to see through. This factor is an important tool for food marketing because it increases the attractiveness of the product to the customer [60]. Biopolymers have good transparency, up to 90% have been measured for chitosan, and also PLA has excellent transparency. Additionally, these bioplastics exhibit extraordinary UV absorption properties, which is due to the presence of unsaturated bonds, carbonyl bonds, and steric effects [61–63]. However, these properties are enhanced when two or more biopolymers are combined to give a single film with more desirable features.

18.3.5 Combination with Plasticizers

Biopolymers, despite their endless and cheap renewable resources, film making ability, recommendable mechanical properties, energy efficiency, antimicrobial activity, ease of composting, and so on, do not perform wholly as packaging material like most petro-based polymers. These shortcomings in biopolymers are mostly mechanical and are augmented by the use of plasticizers. Plasticizers remedy targeted problems identified in a biopolymer for packaging application, especially their elongation abilities. This is achieved by adding plasticizers into the formulation of biopolymers to achieve the desired film [64]. Plasticizers induce plastic behavior into substances by mostly increasing their percentage of elongation and flexibility, which enhances the material's process ability, usage, and market potential [65]. The action of plasticizers is like lubricants in the biopolymer system, decreasing friction between polymer chains and thereby reducing the rigidity of the monomeric arrangements and reducing interactions between them [65, 66].

Plasticized biopolymers come up with desirable properties suitable for their use in the packaging industry. These films act with reduced tensile strength, increased elongation potential, less frictional polymer chains, etc., which means improved mechanical properties for packaging applications [65]. An ideal plasticizer is one that is compatible and miscible with the biopolymer. Mixing a biopolymer and a plasticizer can be achieved by either adding a plasticizer into a film-forming solution or by submerging the biopolymer into a solution of a desired plasticizer. Examples of plasticizers in wide use in the industry include polyethylene glycol, glycerol, and sorbitol, which are part of the family polyols; others are lipids, glucose, and sugar, depending on their suitability and need [66].

Glycerol is the most common plasticizer in use for biopolymer applications, which makes it visible in the components of many bioplastics. Glycerol eases the handling of these biodegradable films. In biopolymer films, glycerol is found to drastically reduce tensile strength while appreciably increasing the percentage of elongation. An example is in chitosan, where tensile strength was reduced from 63.1 to 22.0 MPa while the percentage of elongation was increased from 7.2 to 84.2% when plasticized with glycerol [67]. It is worth noting that the effectiveness of plasticizers on biopolymer films is affected by certain factors like the type used, the molecular weight of the biopolymer, storage time, pH of the solution, etc. The desired blend of biopolymers is made by homogenization of their solution and subsequent drying. Methods of achieving these are several, including high pressure and rotor–stator homogenization techniques, and vacuum and hot air drying methods. Homogenization is done to achieve an adequate spread of the plasticizer into the biopolymer film. The heating procedure is to induce a thermal cross-link between the polymers while disrupting the hydrogen bonds, thus giving way for more chain-to-chain interaction [68].

18.4 Biopolymers and the Future of Food Packaging

Biopolymer is a biological material that is degradable by the action of enzymes or bacteria in nature and, just like in the case of compost materials, they also possess plastic properties. These simple but important characteristics have made biopolymers desirable materials for packaging and other plastic applications, given the already ecological grade disaster being experienced around the globe with associated health and environmental consequences from petro-based plastic waste. Importantly, biopolymer with its unlimited sources has the advantage of bringing materials of usefulness not yet discovered for food packaging applications. The variety of composition, morphology, properties, and ease of mixing with nanofillers and plasticizers makes biopolymers novel for future applications [69–73].

Biopolymers have recently gained acceptance the world over including Africa, where governments are increasingly banning the use of petro-based plastic packaging to improve waste management, environmental pollution, the health of their citizens, and wildlife. Biopolymers are also welcomed by the food packaging industry, which produces the most of plastic waste for advantages offered by their variety and uniqueness, ecological and environment friendliness, low carbon footprint, low cost and renewable sources, biodegradability, etc. Edible packaging materials are an expected twist from these promising materials, which will not only provide endless resources but opportunities as well [74–80].

18.5 Conclusions

Biopolymers currently substitute about ~1% of petro-based plastics in the food packaging industry. Government policies, especially in Europe, are accelerating their use and their impact on the market. The future of packaging both fresh and processed food is one that biopolymers have a suitable application. Biopolymers are also suitable for modified atmosphere packaging and the emerging smart packaging approach, where biological sensors are used to provide information to both customers and producers. The interaction between biopolymers and food is an area that still needs to be explored and so far a healthy relationship exists. The utilization of biopolymers for food packaging will drastically increase the efficiency of agricultural processes. Agricultural waste will increasingly play an important role as a source of cheap biopolymers. These agricultural wastes and byproducts will form an important feedstock to a more efficient society of the future, where waste is converted to other uses to the benefit of humans and their environment. Finally, more research is still required in all aspects of biopolymer technology to achieve effective and sustainable growth in the use of biomaterials for food packaging applications.

Acknowledgment

The authors wish to acknowledge the CSIR-TWAS Postgraduate Fellowship Award 22/FF/CSIR-TWAS/2018.

References

1. Ajay Y, Mangaraj S, Ranjeet S., Sanjay K. D, Naveen K. M, Simran A. Biopolymers as packaging material in food and allied industry. *Inter. J. Chem. Sci.* 2018. **6**(2): 2411–2418.
2. Hannah R, Roser, M. Plastic pollution. Our world in data. June 1, 2019 (cited June 2019). Available at: https://ourworldindata.org/plastic-pollution.
3. European Bioplastics, Bioplastics facts and figures. June 1, 2019 (cited June 2019). Available at: http://www.european-bioplastics.org.
4. Clarinval A-M, Halleux J. Classification of biodegradable polymers. In: *Biodegradable Polymers for Industrial Applications*, Smith R. (Ed.), Woodhead Publishing, Limited (London). 2005. 3–29.
5. Australasian Bioplastic Association. June 2, 2019 (cited June 2019). Available at: www.bioplastics.org.au.
6. Ahvenainen R. Active and intelligent packaging: An introduction. In: *Novel Food Packaging Techniques*, R. Ahvenainen (Ed.) Woodhead Publishing, Limited, Cambridge. 2003. 45–1.
7. Loganathan S, Jacob J, Valapa R.B, Thomas S. Influence of linear and branched amine functionalization in mesoporous silica on the thermal, mechanical and barrier properties of sustainable poly (lactic acid) biocomposite films. *Polymer*. 2018. **148**, 149–157. doi: 10.1016/j.polymer.2018.06.035.
8. Halley P. Biodegradable packaging for the food industry. *Packag. Bottling Int.* 2002. **4**(4):56–57.
9. Maharana T, Mohanty B, Negi Y.S. Melt-solid polycondensation of lactic acid and its biodegradability. *Prog. Polym. Sci.* 2009. **34**:99–124.
10. Andreas K, Becker J, Börger L, Hamprecht J, Koltzenburg S, Loos R, Bernhard S.K, Schlegel K, Sinkel C, Skupin G, Yamamoto M. *Polymers, Biodegradable Ullmann's Encyclopedia of Industrial Chemistry*. 2016. Weinheim: Wiley-VCH doi:10.1002/14356007.n21_n01.pub2.
11. Rudeekit Y, Numnoi J, Tajan M, Chaiwutthinan P, Leejarkpai T. Determining biodegradability of poly(lactic acid) under different environments. 2008. *J. Mater. Mat. Min.* **18**: 83–87.

12. Auras R.S, Singh P, Singh J.J. Evaluation of oriented poly(lactide) polymers vs. existing PET and oriented PS for fresh food service containers. 2005. *Packag. Technol. Sci.* **18**:207–216.
13. Phillips G.O, Williams P. A. *Handbook of Food Proteins*. Woodhead Publishing 2011.
14. Ako K, Nicolai T., Durand D. Salt-induced gelation of globular protein aggregates: structure and kinetics. *Biomacromolecules* 2010. **11**, 864–871.
15. Plackett, D.V *Biopolymers, New Materials for Sustainable Films and Coatings*. Wiley Online Library: London. 2011
16. Babu R. P, O'Connor K, Seeram R. R. *Progress in Biomaterials*. 2013. 16–1.
17. Bin H. In *Lightweight Materials from Biopolymers and Biofibers. ACS Symposium Series*; American Chemical Society: Washington, DC. 2014.
18. Hartmann M.H. High molecular weight polylactic acid polymers . In: *Biopolymers from Renewable Resources*. Kaplan, D., Ed.; Springer: Berlin. 1998. 367–411.
19. Robertson G. L. *Food Packaging: Principles and Practice*. 3, CRC Press, Boca Raton. 2012. 35–1
20. Marra A, Silvestre C, Duraccio D, Cimmino S. Polylactic acid/zinc oxide biocomposite films for food packaging application. *Int. J. Biol. Macromol.* 2016. **88**: 254–262.
21. Vartiainen J, Tuominen M, Nättinen K. Bio-hybrid nanocomposite coatings from sonicated chitosan and nanoclay. *J. Appl. Poly. Sci.* 2010. **116**: 3638–3647.
22. Vartiainen J, Rättö M, Tapper U, Paulussen S, Hurme S. Surface modification of atmospheric plasma activated BOPP by immobilizing chitosan. *Polym. Bull.* 2005. **54**: 343–352.
23. Weber C.J, Haugaard V, Festersen R. Production and applications of bio based packaging materials for the food industry. *Food Addit. Contam.* 2002. **19**: 172–177.
24. Tripathi A.D, Srivastava S.K, Yadav A. Biopolymers potential biodegradable packaging material for food industry." In: Sajid Alavi, Sabu Thomas, K.P. Sandeep, Nandakumar Kalarikkal, Jini Varghese, and Srinivasarao Yaragall (Eds.), *Polymers for Packaging Applications*, Apple Academic Press. 2014
25. Shankar S. Rhim J.W. Preparation of sulfur nanoparticles incorporated antimicrobial chitosan films. *Food Hydrocolloids* 2018. **82**, 116–123
26. Ezeoha S. L, Ezenwanne J. N. Production of biodegradable plastic packaging film from cassava starch. *IOSR J. Eng.* 2013. **3**(10):14–20.
27. Amir H, Iran A, Manouchehr V. Functional properties of biodegradable corn starch nanocomposites for food packaging applications. *Mater. Design*. 2013. **50**. 954. Available at: https://doi.org/10.1016/j.matdes.2013.03.084.
28. Jiali L, Mi Z, Geng C, Fei C, Yi L, Pu-Xin Z. Fabrication and characterisation of starch based nanocomposites reinforced with montmorillonite and cellulose nanofibres. *Carbohydr. Polym.* 2019. (**210**) 429–436
29. Razak R.W, Aizan W. A. Overview of poly(lactic acid) links open overlay panel. In: Lee Tin Sin Abdul (Ed.), *Polylactic Acid Plastics Design Library*. 2013, 1–70. Available at: https://doi.org/10.1016/B978-1-4377-4459-0.00001-9.
30. Mensitieri G, Di Maio E, Buonocore G.G. Processing and shelf life issues of selected food packaging materials and structures from renewable resources. *Trends Food Sci. Technol.* 2011. **22**(2–3): 72–80.
31. Arriagada P, Palza H, Palma P, Flores M, Caviedes P. Poly(lactic acid) composites based on graphene oxide particles with antibacterial behavior enhanced by electrical stimulus and biocompatibility. *J. Biomed. Mater. Res. A* 2018. **106**(4):1051. doi: 10.1002/jbm.a.36307.
32. Fernanda V, Ana S.S, Fátima M.V, José P.F. Nanocellulose in green food packaging. *Crit. Rev. Food Sci. Nutr.* 2017. 1526–1537. Available at: https://doi.org/10.1080/10408398.2016.1270254.
33. Jolanta W, Tomasz R, Gabriel B, Mieczysław S, Tomasz K, & Vijay K.T. Recent progress in biodegradable polymers and nanocomposite-based packaging materials for sustainable environment. *International Journal of Polymer Analysis and Characterization*. 2018. **23**(4): 383–395. Available at: https://doi.org/10.1080/1023666X.2018.1455382.
34. Hengameh H, Mehdi B. Applications of biopolymers I: chitosan. *Monatsh. Chem.* **140**:1403.

35. Kaur S, Dhillon G.S. The versatile biopolymer chitosan: potential sources, evaluation of extraction methods and applications. *Crit. Rev. Microbiol.* 2014. **40**(2):155–75. doi: 10.3109/1040841X.2013.770385.
36. Almeida E, Frollini E, Castellan A, Coma V. Chitosan, sisal cellulose. Chitosan, sisal cellulose, and biocomposite chitosan/sisal cellulose films prepared from thiourea/NaOH aqueous solution. *Carbohydr. Polym.* 2010. **80**(3):655–664. Available at: https://doi.org/10.1016/j.carbpol.2009.10.039
37. Wang H, Qian J, Ding F. Emerging chitosan-based films for food packaging applications. *J. Agric. Food. Chem.* 2018. **66**(2):395–413. doi: 10.1021/acs.jafc.7b04528.
38. Martin K. Poly(hydroxyalkanoates) for food packaging: application and attempts towards implementation. *Appl. Food Biotech.* 2014. **1**.1 Available at: https://doi.org/10.22037/afb.v1i1.7127.
39. Amal A.A, Mahmoud M.B, Azzah A.G. Perspectives of polyhydroxyalkanoate (PHAs) biopolymer production using indigenous bacteria: screening and characterization. *J. Pure Appl. Microbiol.* 2018. **12**(4). Available at: http://dx.doi.org/10.22207/JPAM.12.4.36.
40. Arvanitoyannis I, Kolokuris I, Nakayama A, Yamamoto N, Aiba S. Physico-chemical studies of chitosan-poly(vinyl alcohol) blends plasticized with sorbitol and sucrose. *Carbohydr. Polym.* 1997. **34**:9–19. Available at: https://doi.org/10.1016/S0144-8617(97)00089-1.
41. Bonilla J, Sobral P. Investigation of the physicochemical, antimicrobial and antioxidant properties of the gelatin-chitosan edible film mixed with plant ethanolic extracts. 2016. *Food Biosci.* **16**:17–25. Available at: https://doi.org/10.1016/j.fbio.2016.07.003.
42. Dukalska Falska L, Muizniece-Brasava S, Kampuse S, Seglina D, Straumite E, Galoburda R, Levkane V. Studies of biodegradable polymer material suitability for food packaging applications. foodbalt, *Jelgava*. 2018. pp 8–64
43. Cazón P.M, Vázquez, G. Velazquez G. Novel composite films based on cellulose reinforced with chitosan and polyvinyl alcohol: effect on mechanical properties and water vapour permeability. *Polym. Test*. 2018. **69**:536544.
44. Suryanegara L, Nakagaito A.N, Yano H. The effect of crystallization of PLA on thermal and mechanical properties of microfibrillated cellulose-reinforced PLA composites. *Comp. Sci. Tech.* 2009. **69** (7–9) 11.
45. Zhong Y, Song X, Li Y. Antimicrobial, physical and mechanical properties of kudzu starch chitosan composite films as a function of acid solvent types. *Carbohydr. Polym.* 2011. **84**:335–342. Available at: https://doi.org/10.1016/j.carbpol.2010.11.041.
46. Zhang L., Mao Y, Zhou J, Cai J. Effects of coagulation conditions on the properties of regenerated cellulose films prepared in NaOH/urea aqueous solution. *Ind. Eng. Chem. Res.* 2005. **44**:522–529. Available at: https://doi.org/10.1021/ie0491802.
47. Sun Y.P, Burtrand I.L, Soon T. J, Hyun J. Park. Biopolymer composite films based on κ-carrageenan and chitosan. *Mat. Res. Bull.* (**36**) 3–4. 511–519. Available at: https://doi.org/10.1016/S0025-5408(01)00545-1.
48. Maria P, Mats S. Water vapour permeability and mechanical properties of mixed starch-monoglyceride films and effect of film forming conditions. *Food Hydrocolloids*. (**19**) 1 123–132. Available at: https://doi.org/10.1016/j.foodhyd.2004.04.021.
49. Vásconez M. Flores S, Campos C, Alvarado J, Gerschenson L. Antimicrobial activity and physical properties of chitosan-tapioca starch-based edible films and coatings. *Food Res. Int.* 2009. **42**:762–769. Available at: https://doi.org/10.1016/j.foodres.2009.02.026.
50. Vartiainen J, Motion R, Kulonen H, Rättö M, Skyttä E, Ahvenainen R. Chitosan-coated paper: effects of nisin and different acids on the antimicrobial activity. *J. Appl. Polym. Sci.* 2004. **94**:986–993.
51. Jinwu W, Douglas J.G, Nicole M.S et al. Moisture and oxygen barrier properties of cellulose nanomatrials based films. *ACS Sustainable Chem. Eng.* 2018, **6**, 49.
52. Valenzuela C, Abugoch L, Tapia C. Quinoa protein-chitosan-sunflower oil edible film: mechanical, barrier and structural properties. *LWT-Food Sci. Technol.* 2013. **50**:531–537.

53. Tharanathan R. Biodegradable films and composite coatings: past, present and future. *Trends Food Sci. Technol.* 2003. **14**(3):71–78.
54. Thakhiew W, Devahastin S, Soponronnarit S. Effects of drying methods and plasticizer concentration on some physical and mechanical properties of edible chitosan films. *J. Food Eng.* 2010. **99**:216–224. Available at: https://doi.org/10.1016/j.jfoodeng.2010.02.025.
55. Suyatma N, Tighzert L, Copinet A, Coma V. Effects of hydrophilic plasticizers on mechanical, thermal, and surface properties of chitosan films. *J. Agric. Food. Chem.* 2005. **53**:3950–3957. Available at: https://doi.org/10.1021/jf048790+.
56. Srinivasa P, Ramesh M, Tharanathan R. Effect of plasticizers and fatty acids on mechanical and permeability characteristics of chitosan films. *Food Hydrocolloids* 2007. **21**:1113–1122. doi: 10.1016/j.foodhyd.2006.08.005.
57. Srinivasa P, Baskaran R, Ramesh M, Prashanth K.H, Tharanathan R. Storage studies of mango packed using biodegradable chitosan film. *Eur. Food Res. Technol.* 2002. **215**:504–508. Available at: https://doi.org/10.1007/s00217-002-0591-1.
58. Tianqi H, Yusheng Q, Jia W, Chuncai Z. Polymeric antimicrobial food packaging and its applications. *Polymers* 2019. doi: 10.3390/polym11030560.
59. Siracusa V, Rocculi P, Romani S, Rosa M. Biodegradable polymers for food packaging: a review." *Trends Food Sci. Technol.* 2008. **19**:634–643.
60. Santacruz S., C. Rivadeneira, M. Castro. Edible films based on starch and chitosan. Effect of starch source and concentration, plasticizer, surfactant's hydrophobic tail and mechanical treatment. *Food Hydrocolloids*. 2015. **49**:89–94.
61. Ruiz-Navajas Y, Viuda-Martos M, Barber X, Sendra E, Perez-Alvarez J.A, Fernández-López J. Effect of chitosan edible films added with Thymus moroderi and Thymus piperella essential oil on shelf-life of cooked cured ham. *J. Food Sci. Technol.* 2015. **52**:6493–6501. doi: 10.1007/s13197-015-1733-3.
62. Rouhi M, Razavi S, Mousavi S. Optimization of crosslinked poly(vinyl alcohol) nanocomposite films for mechanical properties. *Mater. Sci. Eng. C.* 2017. **71**:1052–1063. doi: 10.1016/j.msec.2016.11.135.
63. Chang C. P, Wang I. C., Perng Y.S. Enhanced thermal behaiviour, mechanical properties and UV shielding of poly lactic acid (PLA) composite reinforced with nanocrystalline cellulose and filled with nanosecrite cellulose. *Cell. Chem. Technol.* 2011. **47**:1, 111–123.
64. Rong S, Mubarak N, Tanjung F. Structure-property relationship of cellulose nanowhiskers reinforced chitosan biocomposite films. *J. Environ. Chem. Eng.* 2017. **5**:6132–6136. Available at: https://doi.org/10.1016/j.jece.2017.11.054.
65. Vieira M.G.A., da Silva M.A., dos Santos L.O., Beppu M.M. Natural-based plasticizers and biopolymer films: A review. *Eur. Polymer J.* 2011. (**47**)3. 254–263. Available at: https://doi.org/10.1016/j.eurpolymj.2010.12.011.
66. Nikkola J, Vartiainen J, Vähä-Nissi M, Nättinen K. Novel packaging materials using functional thin films and coatings. *European Coatings Conference – Packaging and Can Coating*, Düsseldorf, Vincentz Network GmbH & Co. 2013.
67. Cazón P. Vázquez M. Applications of chitosan as food packaging. Materials. In: G. Crini and E. Lichtfouse (eds.), *Sustainable Agricultural Reviews*, vol. **36**, Springer, Cham. 2019. p 81–123.
68. Falguera V, Quintero J, Jiménez A, Muñoz J, Ibarz A. Edible films and coatings: structures, active functions and trends in their use. *Trends Food Sci. Technol.* 2011. **22**:292–303. Available at: https://doi.org/10.1016/j.tifs.2011.02.004.
69. Hult, E.L, Iotti M, Lenes M. Efficient approach to high barrier packaging using microfibrillar cellulose and shellac. *Cellulose* 2010. **17**: 575–586.
70. Anker M, Berntsen J, Hermansson A.M. Improved water vapour barrier of whey protein films by addition of an acetylated monoglyceride. *Innov. Food Sci. Emerg. Technol.* 2002. **3**:81–92.
71. Hairul A, Azmi B, Faris M, Yuzalmi F, Fadli F, Melbi E, et al. A simple method for improving the properties of the sago starch films prepared by using ultrasonication treatment. *Food Hydrocolloids* 2019. **93** 276–283.

72. Dipen K.R, Durgesh D. P, Ravinder K, Catalin I. P. Recent progress of reinforcement materials: a comprehensive overview of composite materials. *J. Mater. Res. Tech.* 2019. doi: 10.1016/j.jmrt.2019.09.068,
73. Halimatul M.J, Sapuan S. M, Jawaid M, Ishak M. R, Ilyas R. A. Effect of sago starch and plasticizer content on the properties of thermoplastic films: mechanical testing and cyclic soaking-drying. *Polimery J.* 2019. Available at: http://dx.doi.org/10.14314/polimery.2019.6.5.
74. Halimatul M.J, Sapuan S. M, Jawaid M, Ishak M. R, Ilyas R. A. Water absorption and water solubility properties of sago starch biopolymer composite films filled with sugar palm particles. *Polimery J.* 2019. Available at: http://dx.doi.org/10.14314/polimery.2019.9.4.
75. Sanyang M. L, Ilyas R. A, Sapuan S. M, Jumaidin R. Sugar palm starch-based composites for packaging applications. *Bionanocomp. Pack. Appl.* 2017. 125–147.
76. Adriana Z, Romina F.N, Liliana A, María S. R. Synthesis, characterization and evaluation of reactional parameters on substitution degree of N-hexyl-N-methylene phosphonic chitosan. *Carbohydr. Polym.* 2018. **202**, 1. Available at: https://doi.org/10.1016/j.carbpol.2018.08.126.
77. Joseph R. M, Alexander W, Bett J, Sadler M. Synthesis and characterization of a lignin-derived aromatic polycarbonate *J. Biobased Mater. Bioenergy*. 2018. **12**(5), 471. Available at: https://doi.org/10.1166/jbmb.2018.1793.
78. Wenbing U, Cheng P, Weijie L, Min L. et al. Graphene-based antibacterial paper. *ACS Nano*. 2010. **4**, 7, 4317. Available at: https://doi.org/10.1021/nn101097v.
79. Halimatul M.J, Sapuan S. M, Jawaid M, Ishak M. R., Ilyas R. A. Effect of sago starch and plasticizer content on the properties of thermoplastic films: mechanical testing and cyclic soaking-drying. *Polymery*. 2019. Available at:: http://dx.doi.org/10.14314/polimery.2019.6.5.
80. Ilyas RA, Sapuan SM, Atikah MSN, Asyraf MRM, Ayu Rafiqah S, Aisyah HA, Mohd Nurazzi N, Norrahim MNF. Effect of hydrolysis time on the morphological, physical, chemical, and thermal behavior of sugar palm nanocrystalline cellulose (*Arenga pinnata* (Wurmb.) Merr). *Textile Research Journal*. 2020. In print. doi: 10.1177/0040517520932393.

19

Bioplastics: The Future of Sustainable Biodegradable Food Packaging

S. Ayu Rafiqah[1], A Khalina[2,3], Khairul Zaman[4], ISMA Tawakkal[3], A.S Harmaen[1], and N Mohd Nurrazi[1]

[1]*Laboratory of Biocomposite Technology, Institute of Tropical Forestry and Forest Products (INTROP), Universiti Putra Malaysia, Serdang, 43400, Selangor, Malaysia*

[2]*Institute of Tropical Forestry and Forest Products (INTROP), Universiti Putra Malaysia, Serdang, 43400, Selangor, Malaysia*

[3]*Engineering Faculty, Universiti Putra Malaysia, Serdang, 43400, Malaysia*

[4]*Polycomposite Sdn Bhd, Taman Mutiara Galla, 70300 Seremban, Negeri Sembilan, Malaysia*

19.1 Introduction

The growing use of plastic has attracted attention, especially in the agricultural area where it enables farmers to increase their crop production [1]. The food packaging industry also contributes to the use of plastics. Generally, almost 80% of plastics in the world are thermoplastics. The most commonly used ones in the food packaging industry are usually polystyrene, polyethylene (PE), and polyvinyl chloride [2]. Nowadays, plastic waste has become a serious social problem due to disposal to landfill. The market for plastics increased exponentially to an estimated annual production of 280 million tons in 2011 [3]. Previously it was reported that an estimate of the total number of plastic particles and their weight floating in the world's oceans from 24 expeditions (2007–2013) across all five subtropical gyres, costal Australia, Bay of Bengal, and the Mediterranean was over 5 trillion and had led some researchers to claim that synthetic polymers in the ocean should be regarded as hazardous

Bio-Based Packaging: Material, Environmental and Economic Aspects, First Edition. Edited by S.M. Sapuan and R.A. Ilyas

waste [4]. Synthetic plastic in the packaging industry is the sector using the largest share of plastics, so packaging dominates in the plastic waste flow [5]. In between the years 1950 and 2007, it was reported that the production of plastic increased from 0.5 million tonnes to 260 million tonnes. This increase in usage, especially disposable items of packaging, make up 37% of all the plastic produced [6]. Plastic was invented in the 1860s but was developed for industry in the 1920s. Plastic production exploded in the 1940s, becoming one of the fastest-growing global industries. From 1950 to 2012, plastics growth averaged 8.7% per year, booming from 1.7 million tons to nearly 300 million tons today [7]. The industry has been growing for over 60 years. The biggest surge in plastics production was recorded in the 1970s as this new low-cost and disposable material began to replace more expensive materials such as glass and metal. Most plastics are nonbiodegradable and those that are biodegradable require a long time in nature before they degrade [8]. Plastics are chemically synthesized polymers produced by polymerization from petroleum products containing long chains of monomers. These chains are generally thought to be chemically inert, yet unreacted monomers and other harmful ingredients can be found in plastics. The plastics industry has existed over the last 60–70 years. Plastics have great properties, such as versatility, low weight, durability, and low costs [9, 10]. They are also safe, lightweight, easy to process, and economical [11]. The development of bioplastic is contributing greatly to the prevention of global warming [12]. By using the renewable resource as the raw material, there has been an effort to produce new biodegradable polymers. This is due to the rapid depletion of a fossil source and the high volume of nonbiodegradable plastics. In fact, in order to reduce the plastic waste material that has been accumulating daily on earth, green polymers provide an effective substitute as a new eco-friendly material [13]. However, this biopolymer has low tensile properties, low water absorption resistance, high water vapor permeability, and low thermal properties. Consequently, these weaknesses limit the applications of biopolymers. Many attempts have been made to reduce this weakness of biopolymers. In a recent study, numerous researches have been developing new thermoplastic materials that are bio-based plastics from renewable sources. Starch-based packaging material is one of the alternatives to replace the synthetic plastic market [14]. Starch is the most common chosen material for packaging films due to its good transparency, sufficient strength, and low moisture absorption, which should be able to improve the shelf life of food [15]. This present study builds on this work by further investigating potential material for food packaging that can penetrate into the global market. In this topic we will focus more on biodegradable food packaging. The choice of packaging depends on several factors, which are storage, handling, and the shelf life of food.

19.2 Types of Plastic for Food Packaging

Plastic existed in various types, which were characterized as biodegradable and synthetic plastic. Plastic derives from high molecular weight materials such as cellulose and synthetic plastic is made by chemical processing using petroleum-based [16]. Thermoplastic polyesters are derived from fossil-based resources with non-degradable properties. Their properties lead to plastic waste accumulation and petrochemical feedstock depletion [17]. A thermoplastic polymer is a type of plastic polymer that has the properties of a polymer resin. This material has unique properties that become soft when heated and can be molded

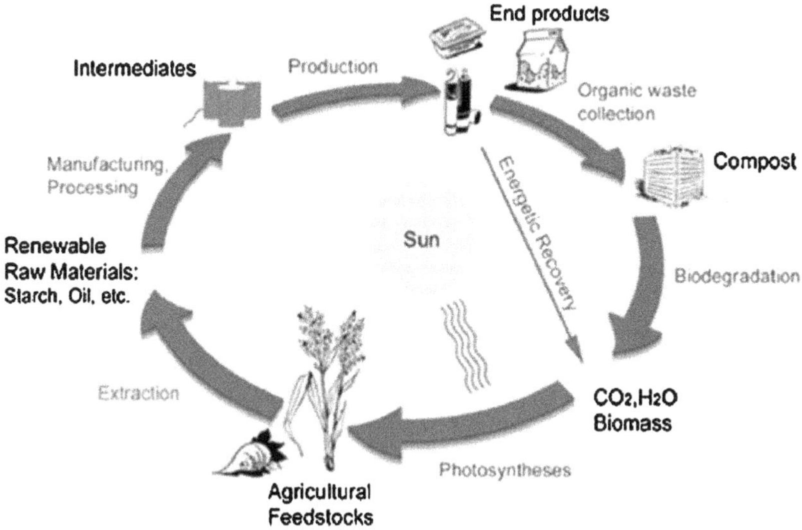

Figure 19.1 *Plastic life cycle [20].*

into any desired shape and size. Meanwhile, when it is cooled, it becomes hard and rigid and remains in the developed shape and size. Recently, biodegradable polymers have received much attention due to their wide range of applications in environmentally friendly products [18]. Polymers have started to replace natural materials in every aspect of life due to their properties. The increasing nonbiodegradable plastic issue has increased the necessity to substitute such materials with biodegradable ones and to optimize the use and disposal of conventional plastics. Biodegradable polymers such as polybutylene succinate (PBS), polycaprolactone (PCL), and polylactic acid (PLA) have attracted considerable interest due to their economic competitiveness [19]. The mechanical properties of PBS are similar to those of polyethylene and have a much more flexible structure ($Tg = -32°C$) compared to PLA ($Tg = 50°C$) with higher toughness [17]. Figure 19.1 illustrates a plastic life cycle. In view of the enormous abundance of data and knowledge about plastics, they will be presented in the following sub-topics.

19.2.1 Biopolymer

The term "bio" is often used to indicate that the materials are biodegradable in topics of polymer, composites, and packaging. The "bio" comes from Greek word "bios" means human life. However, it is also applied to indicate materials derived from renewable resources [21]. This term also enhanced awareness and concern of people regarding environmental issues driven by nonbiodegradable and nonrenewable plastics as well as the rapid depletion of fossil fuel reserves. There are also biodegradable polymers that are produced from a petroleum base and are also regarded as green polymeric matrices [22]. The European Bioplastics Association gives a wide definition of "bioplastics" that can be distinguished by three conditions: (i) bio-derived and biodegradable/compostable, (ii) fossil fuel-derived and biodegradable, or (iii) bio-derived and nonbiodegradable [21].

Examples for the first type of polymer are polylactides, polyhydroxyalkanoates (PHA), cellulose and starch, while PCL is an example of a biodegradable petroleum-based polymer. Biopolyethylene (BIO-PE) and biopolyethylene terephthalate (BIO-PET) are examples of nonbiodegradable bio-based polymers. In short, not all bio-derived polymers are biodegradable and all biodegradable polymers are not necessarily bio-based. Therefore, in this paper, "bio" in the term "biopolymers" is defined as bio-based, which means that the materials are derived from renewable resources and are biodegradable. Classification of biopolymers may be developed according to their chemical composition, the method of synthesis, the method of processing, their economic importance, or application areas where each grouping of biopolymer provides diverse and valuable data for a certain research area and application [21]. Apart from this classification, the most important attribute of biopolymers is their biodegradability or compostability. Most researches divide biopolymers into three main groups according to their production process: (i) directly extracted from natural materials like polysaccharides (e.g. cellulose, starch, chitin), lignins, proteins, and lipids; (ii) produced by chemical synthesis from renewable bio-derived monomers such as polylactide (PLA); (iii) produced by microorganisms or genetically transformed by bacteria, which includes PHA, polyhydroxybutyrate (PHB), and copolymers of hydroxybutyrate and hydroxyvalerate (PHBV) [21, 23]. However, Sharm et al. (2017) simply divided biopolymers into three main categories, which were extracted from natural, synthetic, and those produced by microorganisms. This categorization is illustrated in Figure 19.2.

19.2.2 Biodegradable Polymer – Polybutylene Succinate

PBS is one of the best biodegradable aliphatic polyesters to be considered due to its high flexibility, melt processability, good thermal stability, and chemical resistance, which are similar to those of polyethylene (PE) and polypropylene (PP) [25]. According to the American Society for Testing and Materials (ASTM Standard D-5488-94d), a biodegradable material is defined as "material capable of undergoing decomposition into carbon dioxide, methane, water, inorganic compounds, or biomass in which the predominant mechanism is the enzymatic action of microorganisms, that can be measured by standardized tests, in a

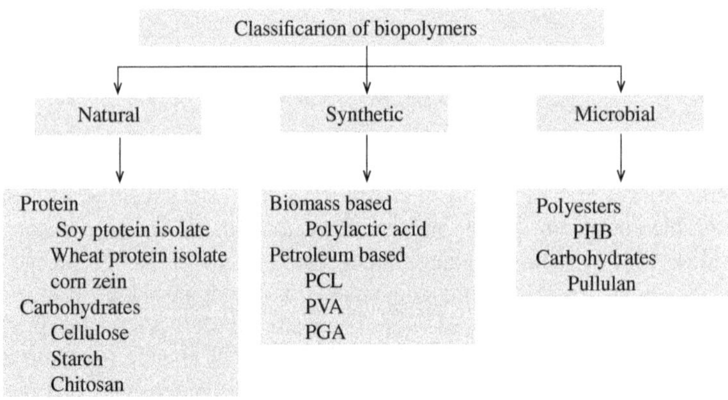

Figure 19.2 Biopolymer categorization [24].

specified period of time, reflecting available disposal condition" [26]. It was found that in the year of 2016, the manufacture of bio-based plastics was 4.2 million tons and by 2021 it has been estimated to grow by around 45% [27]. The structural features of PBS have already been analyzed by several groups and discrimination of two different crystalline forms (α and β forms) has been carried out [28]. PBS can be produced by a two-stage polycondensation starting from 1,4-butanediol and succinic acid [29]. Bio-based PBS has gained attention due to its high molecular weight and can be utilized to produce various products. However, due to its low biodegradation its utility in specific applications is limited [27]. PBS were used for production of compostable bags, mulching films, textiles, and packaging. It is targeted to contribute in a large market due to its great properties [30]. Awareness of environmental pollution has triggered people to find polymers that are able to replace petroleum-derived plastics. The structure of PBS is that of a semi-crystalline polymer and is flexible. Due to its performance, PBS has a great demand in industries. PBS has similar physical properties compared to those of polyethylene terephthalate (PET) [31]. PBS also has good elasticity properties which is useful in various applications [32]. The structure of PBS consists of an ester group that degrades into low molecular weight polymers in the presence of water. The PBS degradation rate increases with increases in temperature [33]. The chemical structure of the repeat unit is $–[O–(CH_2)m–O–CO–(CH_2)n–CO]N$, shown in Figure 19.3. The values of m and n were found to be 4 and 2, respectively. Meanwhile, its molecular weight was Mw 545,000 [31].

Figure 19.4 shows that PBS can be synthesized in various ways, such as polycondensation of succinic acid (or dimethyl succinate) and 1,4-butanediol, whereby the monomers can be obtained from fossil-based or renewable resources. The advantage of this kind of synthesis is that it is able to improve the performance in thermal and mechanical properties, including thermoplastic processability [35]. The catalytic hydrogenation of maleic acid or its anhydrite is the common method of synthesizing petrochemical succinic acid.

By using the renewable resources to produce succinic acid, it can be costly compared to the petroleum-based processes. Recent studies used different types of microorganisms for

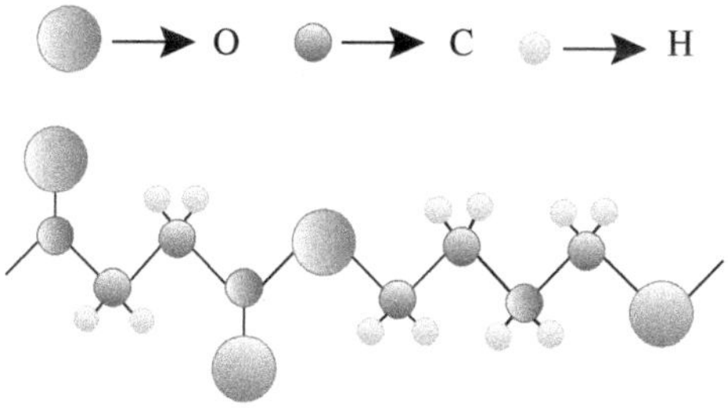

Figure 19.3 *Chemical structure of polybutylene succinate (PBS) [34].*

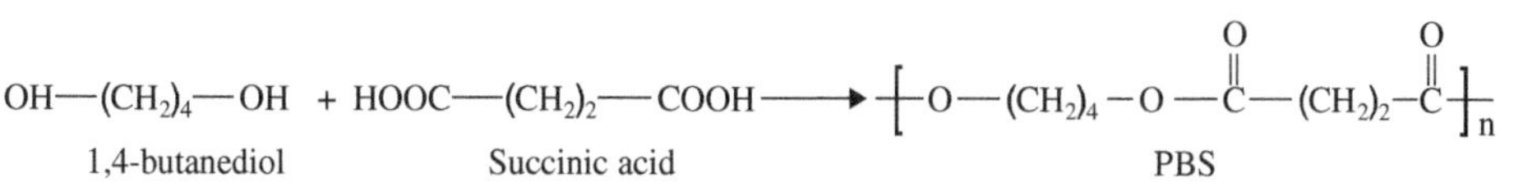

Figure 19.4 *Synthesis of PBS [36].*

succinic acid production. Among them are *Anaerobiospirillum succiniciproducens*, *Actinobacillus succinogenes*, and *Mannheimia succiniciproducens* [37]. Nevertheless, due to high glucose concentrations and exposure to air, the organism experienced instability and deterioration [38].

19.2.3 Biodegradable Polymer – Polylactic Acid

PLA or polylactic acid was discovered in 1932 by Carothers (at DuPont). He was only able to produce a low molecular weight PLA by heating lactic acid under vacuum while removing the condensed water. However, the main problem at that time was to increase the molecular weight of the products and, finally, by ring-opening polymerization of the lactide, succeed to produce high-molecular weight PLA [39]. PLA is an emerging thermoplastic aliphatic polyester produced from sustainable and renewable resources. In the sense of the European standard (EN 13432) [40], PLA is an aliphatic polyester made from lactic acid (2-hydroxy propionic acid) building blocks. PLA was also categorized as a biodegradable and compostable thermoplastic. Generally, PLA was derived from renewable plant sources, such as starch and sugar [41]. Recently, researchers have extensively studied PLA due to its great potential to replace conventional petrochemical-based polymers. Current research has found that PLA has achieved chemical, mechanical, and biological properties equivalent or superior to current conventional polymers [42]. Polylactide (PLA) is one of the most promising biodegradable polymers due to its mechanical properties, thermoplastic processability, and biological properties, such as biocompatibility and biodegradability. Lactic acid is the main component in PLA, which contributes high mechanical strength toward the PLA structure [43]. PLA was produced by condensation polymerization from lactic acid via ring-opening polymerization. Lactic acid was produced from the fermentation process of sugar, which comes from carbohydrate sources such as corn and tapioca [44]. Other researchers have found that PLA can be synthesized by three main synthetic methods, which are from direct polycondensation, shown in Figure 19.5, ring opening

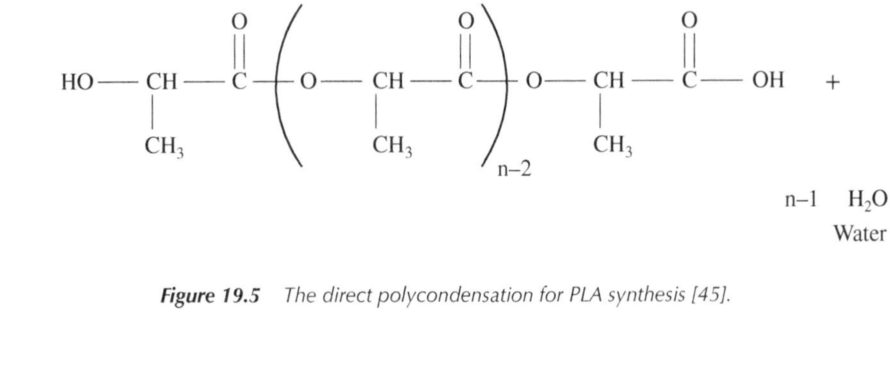

Figure 19.5 The direct polycondensation for PLA synthesis [45].

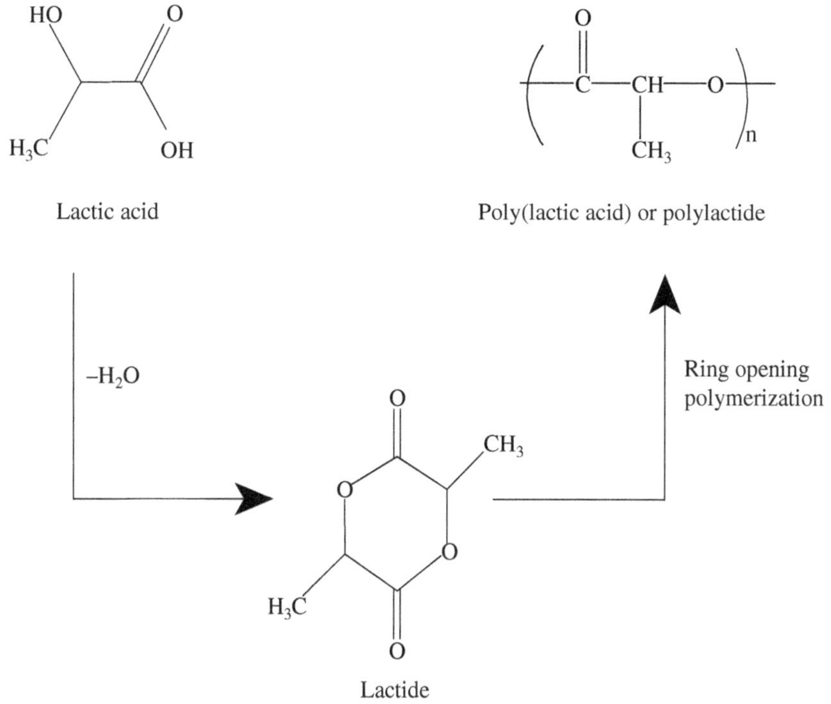

Figure 19.6 The ring opening polymerization mechanism for PLA synthesis [45].

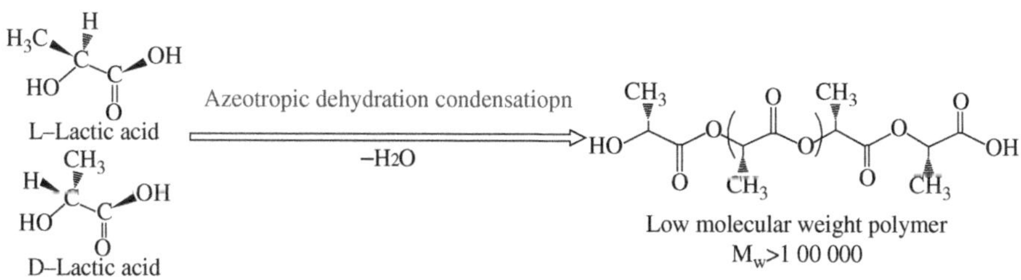

Figure 19.7 Illustrating the azeotropic dehydrative condensation for PLA synthesis [45].

polymerization, shown in Figure 19.6, and azeotropic dehydrative condensation, as illustrated in Figure 19.7 [45].

19.3 Food Packaging

Food packaging is essential in order to protect and preserve the food inside it. In the packaging industry, polyethylene (PE), polypropylene (PP), and PET are the most used due to their good properties, which are biological resistance and excellent water barriers. However, these types of plastic give a negative impact on the environment

[26]. Biodegradable packaging is an alternative to replace synthetic packaging and non-biodegradable petroleum-based packaging. The demand for biodegradable packaging has increased because it can reduce the impact of packaging on the environment [46]. The compostability attribute is very important for biopolymer materials because although recycling is energy expensive, composting allows disposal of the packages in the soil. Biological degradation produces only water, carbon dioxide, and inorganic compounds without toxic residues. According to European Bioplastics, biopolymer manufactured using renewable resources have to be biodegradable and especially compostable, so they can act as fertilizers and soil conditioners [20]. Recently, in this global economy packaging not only is essential to enable effective distribution and preservation of food and other consumer products, but must also facilitate their end-use convenience and communication at consumer levels. With these important functions, packaging has become the third largest industry in the world and represents about 2% of gross national product (GNP) in developed countries [47]. An effective food packaging has the ability to provide containment and protects food products during distribution and storage from external and internal unfavorable conditions, such as water vapor, microorganisms, gases, orders, dust, and mechanical shock and vibrations [48]. Packaging that is classified as active packaging can be defined as a system in which the product, package, and the environment interact in a positive way to extend the shelf life or to achieve some desired characteristics. Active packaging is also able to extend the shelf life of food and improve safety or sensory properties while maintaining the quality of the packaged food. Figure 19.8 shows types of active food packaging that are commercially in

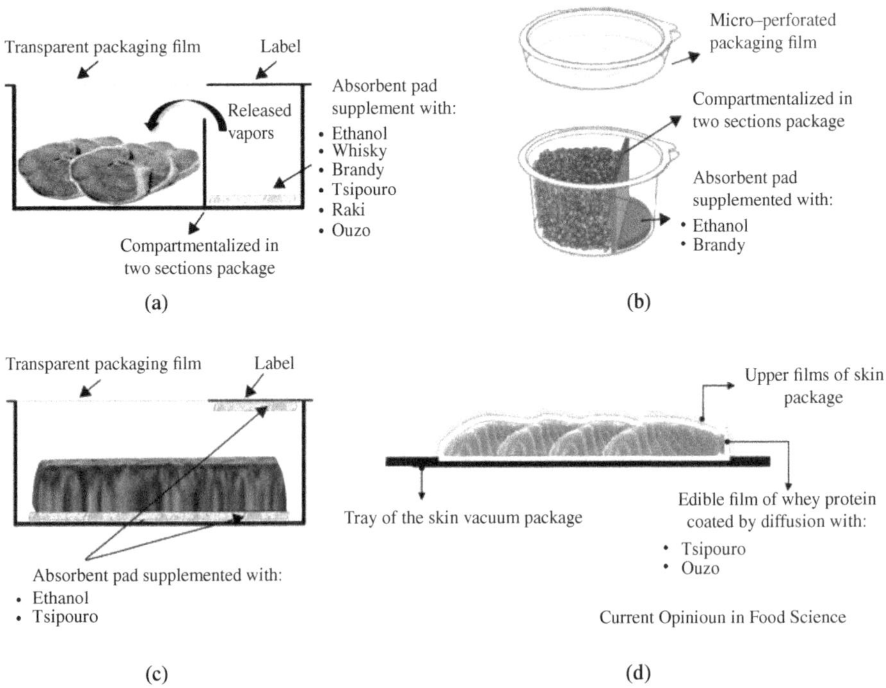

Figure 19.8 Commercially available types of active packaging [49].

the global market [49]. Active packaging had attracted consumer attention, especially in European countries, as it naturally preserved the food. Due to this, the food industry was eager to invest more on packaging quality and safety [50].

19.3.1 Starch-Based Bioplastic Packaging

Starch is one of the most abundant natural resources and also an important biopolymer and has been widely used in the food industry as paper, thickener, and gelling agent. Tapioca starch is made from the root of the cassava plant through a process of washing and pulping. The production of modified starch is known as "environmentally friendly," producing materials that are widely used as stabilizers, viscosifying agent, coatings, and thickeners [51]. Starch is made from two homopolymers of D-glucose and amylase, which consist of α-D(1,4)-glucan and an amylopectin branch. It has the same backbone structure as amylose but with many α-1,6 branch points, as illustrated in Figure 19.9 [52]. The relative amounts of amylose and amylopectin depend on the plant source. Corn starch granules typically contain approximately 70% of amylopectin and 30% of amylose [53]. Enriching starch with natural antioxidant or antimicrobial substances will extend its functional properties and provide bioactive biomaterials [54]. The structure of starch can be divided into four types, which are molecules, blocklets, shells, and granules. Blocklets are the building blocks of

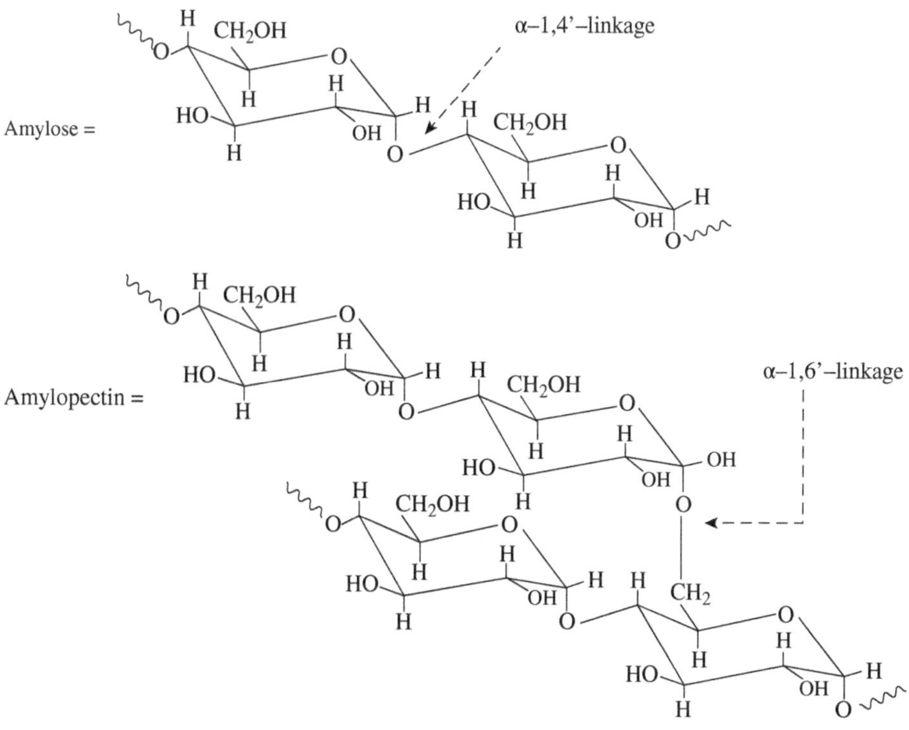

Figure 19.9 Molecular structure of starch [52].

shells and granules [55]. Starches from various plant sources, such as wheat, corn, rice, and potato, have received extensive attention in relation to structural and physicochemical properties. Starch is widely used in food, pharmaceutical, and biomedical applications because of its biocompatibility, biodegradability, nontoxicity, and abundant sources [54]. Starch was widely used in the food packaging industry due to its properties, which are low cost and abundance [53]. Thermoplastic starch (TPS) was obtained from granular disruption by shear forces applied at a high temperature in the presence of plasticizer. Unfortunately, TPS film has disadvantages from moisture absorption and poor mechanical performance [56]. From previous research it was stated that starch has great potential to become an alternative for the food packaging industry to overcome disadvantages of current food packaging in the market. However, the components in conventional starch-based polymer material are not completely inert and can cause migration of other substances into the food [52]. Plastic packaging film is usually produced by only a few methods, namely extruder, casting, extrusion coating, extrusion lamination, and metallization. All these processes have their own advantages and disadvantages depending on the polymer and thickness required [26].

19.3.2 Oxygen Transmission Rate

The oxygen barrier property for a food packaging container for fresh products plays an important role in food preservation. The oxygen barrier is quantified by the oxygen permeability coefficient (OPC), which indicates the amount of oxygen that permeates per unit of area and time in packaging materials [20]. Food products must have an adequate shelf life in order to maintain food quality in terms of nutrition and safety. An inadequate shelf life not only effect sales and results but potentially lead to food-borne disease and damage to the food brand name of the product [48]. Generally, a plastic surface has small holes that enable small gaseous and water vapors to transmit into the polymer matrix. Biodegradable polymer was presented a value one or more order of magnitude below the synthetic polymer used, such as PET and OPC. Apart from this, by increasing the temperature, the permeability of oxygen also will increase [57]. The oxygen permeability expressed in cc $(m^2\ s)^{-1}$ is given. The OPC is correlated with the oxygen transition rate (OTR) by the following equation: $\mathbf{OPC} = \mathbf{OTR} \times \iota/\Delta\mathbf{P}$, where ι is the thickness of the film (m) and $\Delta\mathbf{P}$ is the difference between the oxygen partial pressure across the film [20]. Development in food packaging with the concept of incorporating active compounds into packaging materials or its surrounding conditions is referred to as "active packaging," which is evolving progressively today [58]. This innovation has gained interest among researchers due to the demand for prolonging food shelf life, maintaining food quality and food safety, and enhancing food organoleptic properties. Moreover, environmentally friendly active packaging and natural preservatives could be better options to overcome health concerns and environmental issues [59]. Gas permeability through a polymeric material proceeds mostly in the amorphous phase and is determined by two basic physical processes: solubility of a gas at a gas/polymer interface and diffusion of this gas within the polymer. Using Henry's law describing gas solubility in a polymer and Fick's laws that characterize gas diffusion in that material, gas permeability may be expressed as a function of both solubility and diffusion coefficients. Various methods for investigation of gas permeability through polymeric films and membranes are the subject of many

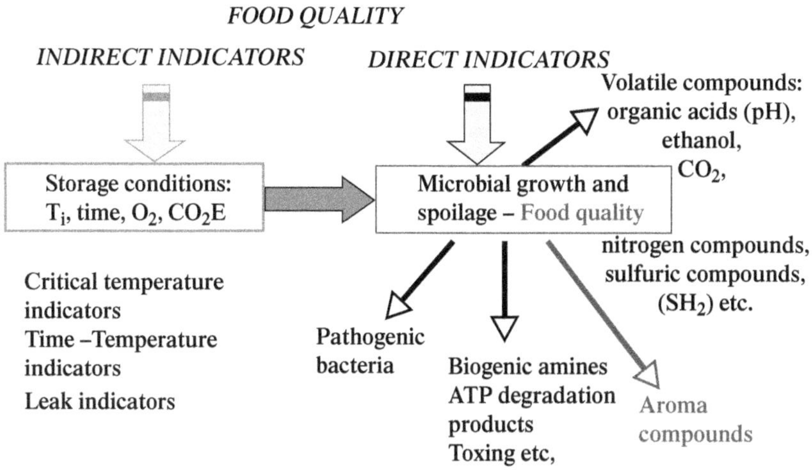

Figure 19.10 Food quality indicators: main indirect indicators and the trend toward direct indicators of microbial growth and spoilage and other food qualities attributes [61].

scientific studies and practical applications [60]. Differences in the solubility of specific gases may influence the diffusivity of gases across the film. The second step in the permeability phenomenon, the diffusion, depends on the size, shape, and polarity of the penetrating molecule of the permeant and on the crystallinity, degree of cross-linking, and polymer chain segmental motion of the polymer matrix. The general food indicators or food storage conditions, such as temperature, time, oxygen, or carbon dioxide content, and thus, indirectly, information on food quality, are illustrated in Figure 19.10 [61].

19.3.3 Water Vapor Transmission Rate (WVTR)

The water vapor barrier properties are also an important factor to consider for food packaging applications. The packaged product whose physical or chemical deterioration is related to its equilibrium moisture content is of great importance for maintaining or extending its shelf life. The water vapor barrier is quantified by the water vapor permeability coefficients (WVPCs), which indicate the amount of water vapor that permeates per unit of area and time in a packaging material (kg m $(m^2\ s\ Pa)^{-1}$) [20, 62]. Water vapor permeability (WVP) is one of the crucial parameters in film production that greatly affects the shelf life of a product. It measures the movement of water vapor molecules on the film matrix [59]. A previous research study on WVP shows that the presence of different hydrophilic and hydrophobic groups due to the presence of phenolic compounds can cause some interactions with the hydrophobic region of protein molecules in gelatin [63]. The water vapor transmission testing was carried out using ion experiments and were conducted by employing the standard gravimetric method (ASTM, 1993). For constant conditions, a temperature adjustable heater–cooler oven (Nuve ES500, Turkey) was used. Humidity in the oven was adjusted by using standard saturated solutions of LiCl and NaCl (Merck) and was continuously monitored using two calibrated humidity sensors [64].

19.4 Active Food Packaging

Active packaging is defined as an intelligent or smart system that involves interactions between package systems that deliberately incorporates components that release or absorb substances into or from the packaged food gas atmosphere and complies with consumer demands for high-quality, fresh-like, and safe products [65]. Therefore, active packaging should manage to provide a barrier to external detrimental factors, as the packaging system plays an active role in food preservation and quality during the marketing process [66]. Other researches define active packaging as a concept of incorporating active compounds such as antioxidants into packaging materials or the surrounding conditions [59]. There are basically two methodologies for producing antioxidant packaging systems, which are independent devices and antioxidant packaging materials that are used in the manufacture of the package, that is, the active agent is incorporated in the walls of the package exerting its action by absorbing undesirable compounds from the headspace or by releasing antioxidant compounds to the food [67]. Active packaging can be classified into two main types: nonmigratory active packaging acting without intentional migration and active releasing packaging allowing a controlled migration of nonvolatile agents or an emission of volatile compounds in the atmosphere surrounding the food [61] The inclusion of natural compounds in biopolymer formulations had been suggested to improve the functionality of the packaging in terms of mechanical, barrier, antioxidant, and antimicrobial properties [68]. Figure 19.11 shows the concept of packaging functions, including its advanced packaging

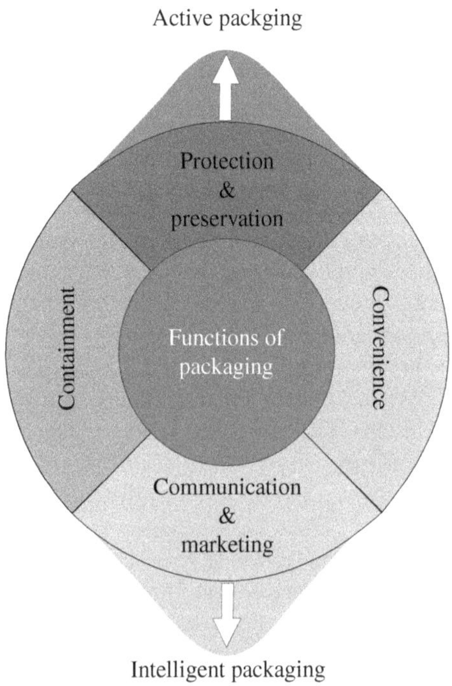

Figure 19.11 Concept of packaging functions including its advanced packaging systems, active and intelligent packaging [48].

systems and active and intelligent packaging [48]. The nature of the active agents that can be added is very diverse (organic acids, enzymes, bacteriocins, fungicides, natural extracts, ions, ethanol, etc.) as well as the nature of the materials into which they are included, such as papers, plastics, metals, or combinations of these materials [61].

19.4.1 Antimicrobial Food Packaging

Demand for safe, minimally processed, "fresh" food products presents major challenges to the food-packaging industry to develop packaging concepts for maintaining the safety and quality of packaged food. Antimicrobial packaging also categorizes as active packaging by adding active ingredients in the packaging system. Antimicrobial packaging can take several forms including [69]:

1. Addition of sachets pads containing volatile antimicrobial agents into packages.
2. Incorporation of volatile and nonvolatile antimicrobial agents directly into polymers.
3. Coating or adsorbing antimicrobials on to polymer surfaces.
4. Immobilization of antimicrobials to polymers by ion or covalent linkages.
5. Use of polymers that are inherently antimicrobial.

Antimicrobial packaging is able to kill or inhibit spoilage and pathogenic microorganisms that contaminate food. By insertion of antimicrobial agents also prevents microbial growth and extend shelf life of food [50]. Study from other researchers found that antimicrobial action reduces the growth rate of microorganisms in the packaging and also acts as a barrier for moisture and oxygen in the packaging materials [70]. The active functions may also include scavenging of ethylene, emission of ethanol and flavors, and antimicrobial activity. Microbial contamination has the potential to increase the risk of food-borne illness. The method of preserving foods by the traditional method from the effect of microbial growth includes thermal processing, drying, freezing, refrigeration, irradiation, modified atmosphere packaging, and adding antimicrobial agents or salts. Unfortunately, some of these techniques cannot be applied to some food products, such as fresh meats and ready-to-eat products [71]. Several conditions should be considered when designing an antimicrobial packaging system. First, the regulatory status of the antimicrobial agent is important. Some antimicrobial systems rely on diffusion or release of the antimicrobial agent in order for it to be effective. This would place the packaging material under the classification of an indirect food additive and the material would require review by the US Food and Drug Administration (FDA) along with similar scrutiny in other countries [72]. Antimicrobials can be added to food formulations directly or through a slow release process from packaging materials. The application of antimicrobial films allows for the migration of the antimicrobial to the film surface and provides a continuous antimicrobial effect on the food during extended exposure [73]. A few natural spices have been used as antimicrobial agents in food packaging. One of these is basil leaves from *Ocimum* genus. Basil is a popular culinary herb and its essential oils have been used extensively for many years in the flavoring of confectionary and baked goods and condiments. Naturally derived antimicrobial agents such as basil are becoming increasingly more important in antimicrobial packaging as they present a perceived lower risk to consumers [74]. From a previous study, a transparent cellulose film was prepared from ginger nanofiber with chemicals and ultrasonication for resistance against microbial activity. This purification method used acid hydrolysis of ginger fiber, which has a high cellulose content (88%) [75, 76]. The antimicrobial action can be obtained by two types of mechanisms. Preservative releasing or migrating concepts contain a preservative within their mass or on their surface, the purpose of which is to migrate partly or completely into the food or in the space surrounding the food and from there to exercise its preservative action. For both concepts, migrating and nonmigrating, intensive contact between the active material and the food product is required [77]. Lastly, antimicrobial food packaging developed with nanotechnology represents an impact on consumers. However, the toxicological impact of nanoparticles must be evaluated to determine the positive or negative effects on food safety [59, 78, 79].

References

1. Picuno P. Innovative material and improved technical design for a sustainable exploitation of agricultural plastic film. *Polym. Plast. Technol. Eng.* 2014;**53**(10):1000–11.
2. Lahimer MC, Ayed N, Horriche J, Belgaied S. Characterization of plastic packaging additives: food contact, stability and toxicity. *Arabian J. Chem.* 2017;**10**:S1938-S54.
3. Luijsterburg B, Goossens H. Assessment of plastic packaging waste: material origin, methods, properties. *Resour. Conserv. Recycl.* 2014;**85**:88–97.
4. Eriksen M, Lebreton LC, Carson HS, Thiel M, Moore CJ, Borerro JC, et al. Plastic pollution in the world's oceans: more than 5 trillion plastic pieces weighing over 250,000 tons afloat at sea. *PLoS One* 2014;**9**(12):e111913.
5. Dahlbo H, Poliakova V, Mylläri V, Sahimaa O, Anderson R. Recycling potential of post-consumer plastic packaging waste in Finland. *Waste Manage. (Oxford)* 2018;**71**:52–61.
6. O'Brine T, Thompson RC. Degradation of plastic carrier bags in the marine environment. *Mar. Pollut. Bull.* 2010;**60**(12):2279–83.
7. Gourmelon G. Global plastic production rises, recycling lags. New Worldwatch Institute analysis explores trends in global plastic consumption and recycling. Available at: http://www worldwatch org. 2015.
8. Jalal N. G Binging on plastics containing bisphenol A. *J. Microbiol. Biotechnol. Rep.* 2017;**1**(1).
9. Chidambarampadmavathy K, Karthikeyan OP, Heimann K. Sustainable bio-plastic production through landfill methane recycling. *Renewable Sustainable Energy Rev.* 2017;**71**:555–62.
10. Rochman CM, Browne MA, Halpern BS, Hentschel BT, Hoh E, Karapanagioti HK, et al. Policy: classify plastic waste as hazardous. *Nature* 2013;**494**(7436):169.
11. Paraskevopoulou D, Achilias DS, Paraskevopoulou A. Migration of styrene from plastic packaging based on polystyrene into food simulants. *Polym. Int.* 2012;**61**(1):141–8.
12. Kurokawa N, Kimura S, Hotta A. Mechanical properties of poly (butylene succinate) composites with aligned cellulose-acetate nanofibers. *J. Appl. Polym. Sci.* 2018;**135**(24):45429.
13. Luzi F, Fortunati E, Jiménez A, Puglia D, Pezzolla D, Gigliotti G, et al. Production and characterization of PLA_PBS biodegradable blends reinforced with cellulose nanocrystals extracted from hemp fibres. *Ind. Crops Prod.* 2016;**93**:276–89.
14. Sanyang M, Ilyas R, Sapuan S, Jumaidin R. Sugar palm starch-based composites for packaging applications. *Bionanocomposites for Packaging Applications*: Springer; 2018. p. 125–47.
15. Abral H, Basri A, Muhammad F, Fernando Y, Hafizulhaq F, Mahardika M, et al. A simple method for improving the properties of the sago starch films prepared by using ultrasonication treatment. *Food Hydrocolloids* 2019;**93**:276–83.
16. Piringer OG, Baner AL. *Plastic Packaging Materials for Food: Barrier Function, Mass Transport, Quality Assurance, and Legislation*: Wiley; 2008.

17. Georgousopoulou I-N, Vouyiouka S, Dole P, Papaspyrides CD. Thermo-mechanical degradation and stabilization of poly(butylene succinate). *Polym. Degrad. Stab.* 2016;**128**:182–92.
18. Tsao C-W, DeVoe DL. Bonding of thermoplastic polymer microfluidics. *Microfluid. Nanofluid.* 2009;**6**(1):1–16.
19. Negrin M, Macerata E, Consolati G, Quasso F, Genovese L, Soccio M, et al. Gamma radiation effects on random copolymers based on poly(butylene succinate) for packaging applications. *Radiat. Phys. Chem.* 2018;**142**:34–43.
20. Siracusa V, Rocculi P, Romani S, Dalla Rosa M. Biodegradable polymers for food packaging: a review. *Trends Food Sci. Technol.* 2008;**19**(12):634–43.
21. Johansson C, Bras J, Mondragon I, Nechita P, Plackett D, Simon P, et al. Renewable fibers and bio-based materials for packaging applications–a review of recent developments. *BioResources* 2012;**7**(2):2506–52.
22. AL-Oqla FM, Omari MA. Sustainable biocomposites: challenges, potential and barriers for development. *Green Biocomposites*: Springer; 2017. p. 13–29.
23. Tang X, Kumar P, Alavi S, Sandeep K. Recent advances in biopolymers and biopolymer-based nanocomposites for food packaging materials. *Crit. Rev. Food Sci. Nutr.* 2012;**52**(5):426–42.
24. Talegaonkar S, Sharma H, Pandey S, Mishra PK, Wimmer R. Bionanocomposites: smart biodegradable packaging material for food preservation. *Food Packaging*: Elsevier; 2017. p. 79–110.
25. Supthanyakul R, Kaabbuathong N, Chirachanchai S. Random poly(butylene succinate-co-lactic acid) as a multi-functional additive for miscibility, toughness, and clarity of PLA/PBS blends. *Polymer* 2016;**105**:1–9.
26. Ferreira A, Alves V, Coelhoso I. Polysaccharide-based membranes in food packaging applications. *Membranes* 2016;**6**(2):22.
27. Bi S, Tan B, Soule JL, Sobkowicz MJ. Enzymatic degradation of poly(butylene succinate-co-hexa methylene succinate). *Polym. Degrad. Stab.* 2018;**155**:9–14.
28. Papageorgiou DG, Zhuravlev E, Papageorgiou GZ, Bikiaris D, Chrissafis K, Schick C. Kinetics of nucleation and crystallization in poly(butylene succinate) nanocomposites. *Polymer* 2014;**55**(26):6725–34.
29. Frollini E, Bartolucci N, Sisti L, Celli A. Biocomposites based on poly(butylene succinate) and curaua: mechanical and morphological properties. *Polym. Test.* 2015;**45**:168–73.
30. Di Lorenzo ML, Androsch R, Righetti MC. Low-temperature crystallization of poly (butylene succinate). *Eur. Polym. J.* 2017;**94**:384–91.
31. Uesaka T, Ogata N, Nakane K, Shimizu K, Ogihara T. Structure and physical properties of cellulose acetate/poly(butylene succinate) blends containing a transition metal alkoxide. *J. Appl. Polym. Sci.* 2002;**83**(8):1750–8.
32. Li Y-D, Zeng J-B, Wang X-L, Yang K-K, Wang Y-Z. Structure and properties of soy protein/poly(butylene succinate) blends with improved compatibility. *Biomacromolecules* 2008;**9**(11): 3157–64.
33. Kanemura C, Nakashima S, Hotta A. Mechanical properties and chemical structures of biodegradable poly(butylene-succinate) for material reprocessing. *Polym. Degrad. Stab.* 2012;**97**(6):972–80.
34. Yu L, Ke S, Zhang Y, Shen B, Zhang A, Huang H. Dielectric relaxations of high-k poly(butylene succinate) based all-organic nanocomposite films for capacitor applications. *J. Mater. Res.* 2011;**26**(19):2493–502.
35. Xu J, Guo BH. Poly(butylene succinate) and its copolymers: research, development and industrialization. *Biotechnol. J.* 2010;**5**(11):1149–63.
36. Jamaluddin N, Razaina M, Ishak ZM. Mechanical and morphology behaviours of polybutylene (succinate)/thermoplastic polyurethaneblend. *Proc. Chem.* 2016;**19**:426–32.
37. Song H, Lee SY. Production of succinic acid by bacterial fermentation. *Enzyme Microb. Technol.* 2006;**39**(3):352–61.

38. Nghiem N, Kleff S, Schwegmann S. Succinic acid: technology development and commercialization. *Fermentation* 2017;**3**(2):26.
39. Jamshidian M, Tehrany EA, Imran M, Jacquot M, Desobry S. Poly-lactic acid: production, applications, nanocomposites, and release studies. *Compr. Rev. Food Sci. Food Saf.* 2010;**9**(5):552–71.
40. Bendahou D, Bendahou A, Grohens Y, Kaddami H. New nanocomposite design from zeolite and poly(lactic acid). *Ind. Crops Prod.* 2015;**72**:107–18.
41. Lim L-T, Auras R, Rubino M. Processing technologies for poly(lactic acid). *Prog. Polym. Sci.* 2008;**33**(8):820–52.
42. Rasal RM, Janorkar AV, Hirt DE. Poly(lactic acid) modifications. *Prog. Polym. Sci.* 2010;**35**(3):338–56.
43. Gupta B, Revagade N, Hilborn J. Poly(lactic acid) fiber: an overview. *Prog. Polym. Sci.* 2007;**32**(4):455–82.
44. Auras RA, Lim L-T, Selke SE, Tsuji H. *Poly(Lactic Acid): Synthesis, Structures, Properties, Processing, and Applications*: Wiley; 2011.
45. Saba N, Jawaid M, Al-Othman O. An overview on polylactic acid, its cellulosic composites and applications. *Curr. Org. Synth.* 2017;**14**(2):156–70.
46. de Moraes Crizel T, Costa TMH, de Oliveira Rios A, Flôres SH. Valorization of food-grade industrial waste in the obtaining active biodegradable films for packaging. *Ind. Crops Prod.* 2016;**87**:218–28.
47. Han JH. *Innovations in Food Packaging*: Elsevier; 2005.
48. Mihindukulasuriya S, Lim L-T. Nanotechnology development in food packaging: a review. *Trends Food Sci. Technol.* 2014;**40**(2):149–67.
49. Kapetanakou AE, Skandamis PN. Applications of active packaging for increasing microbial stability in foods: natural volatile antimicrobial compounds. *Curr. Opin. Food Sci.* 2016;**12**:1–12.
50. Ahvenainen R. *Novel Food Packaging Techniques*: Elsevier; 2003.
51. Makmoon T, Foungfuchat A, Jiratumnukul N. Modified tapioca starch as a rheology modifier in acrylic dispersion system. *Prog. Org. Coat.* 2013;**76**(6):959–62.
52. Lu D, Xiao C, Xu S. Starch-based completely biodegradable polymer materials. *Express Polym. Lett.* 2009;**3**(6):366–75.
53. Avella M, De Vlieger JJ, Errico ME, Fischer S, Vacca P, Volpe MG. Biodegradable starch/clay nanocomposite films for food packaging applications. *Food Chem.* 2005;**93**(3):467–74.
54. Kaur L, Singh J, Liu Q. Starch–a potential biomaterial for biomedical applications. *Nanomaterials and Nanosystems for Biomedical Applications*: Springer; 2007. p. 83–98.
55. Bertoft E. Understanding starch structure: recent progress. *Agronomy* 2017;**7**(3):56.
56. López OV, Castillo LA, Garcia MA, Villar MA, Barbosa SE. Food packaging bags based on thermoplastic corn starch reinforced with talc nanoparticles. *Food Hydrocolloids* 2015;**43**:18–24.
57. Auras R, Harte B, Selke S. Effect of water on the oxygen barrier properties of poly (ethylene terephthalate) and polylactide films. *J. Appl. Polym. Sci.* 2004;**92**(3):1790–803.
58. Realini CE, Marcos B. Active and intelligent packaging systems for a modern society. *Meat Sci.* 2014;**98**(3):404–19.
59. Adilah AN, Jamilah B, Noranizan M, Hanani ZN. Utilization of mango peel extracts on the biodegradable films for active packaging. *Food Packag. Shelf Life* 2018;**16**:1–7.
60. Żenkiewicz M, Richert J. Permeability of polylactide nanocomposite films for water vapour, oxygen and carbon dioxide. *Polym. Test.* 2008;**27**(7):835–40.
61. Dainelli D, Gontard N, Spyropoulos D, Zondervan-van den Beuken E, Tobback P. Active and intelligent food packaging: legal aspects and safety concerns. *Trends Food Sci. Technol.* 2008;**19**:S103-S12.
62. Siracusa V. Food packaging permeability behaviour: a report. *Int. J. Polym. Sci.* 2012;**2012**.
63. Wu J, Chen S, Ge S, Miao J, Li J, Zhang Q. Preparation, properties and antioxidant activity of an active film from silver carp (*Hypophthalmichthys molitrix*) skin gelatin incorporated with green tea extract. *Food Hydrocolloids* 2013;**32**(1):42–51.
64. Dirim SN, Özden HÖ, Bayındırlı A, Esin A. Modification of water vapour transfer rate of low density polyethylene films for food packaging. *J. Food Eng.* 2004;**63**(1):9–13.

65. Ozdemir M, Floros JD. Active food packaging technologies. *Crit. Rev. Food Sci. Nutr.* 2004;**44**(3):185–93.
66. Gómez-Estaca J, López-de-Dicastillo C, Hernández-Muñoz P, Catalá R, Gavara R. Advances in antioxidant active food packaging. *Trends Food Sci. Technol.* 2014;**35**(1):42–51.
67. Brody AL, Bugusu B, Han JH, Sand CK, Mchugh TH. Innovative food packaging solutions. *J. Food Sci.* 2008;**73**(8):107–16.
68. Valdés A, Mellinas AC, Ramos M, Garrigós MC, Jiménez A. Natural additives and agricultural wastes in biopolymer formulations for food packaging. *Front. Chem.* 2014;**2**:6.
69. Appendini P, Hotchkiss JH. Review of antimicrobial food packaging. *Innovative Food Sci. Emerg. Technol.* 2002;**3**(2):113–26.
70. Han JH. Antimicrobial food packaging. *Novel Food Packag. Tech.* 2003;**8**:50–70.
71. Quintavalla S, Vicini L. Antimicrobial food packaging in meat industry. *Meat Sci.* 2002;**62**(3):373–80.
72. Cooksey K. Effectiveness of antimicrobial food packaging materials. *Food Addit. Contam.* 2005;**22**(10):980–7.
73. Jin T, Zhang H. Biodegradable polylactic acid polymer with nisin for use in antimicrobial food packaging. *J. Food Sci.* 2008;**73**(3):M127-M34.
74. Suppakul P, Miltz J, Sonneveld K, Bigger SW. Antimicrobial properties of basil and its possible application in food packaging. *J. Agric. Food. Chem.* 2003;**51**(11):3197–207.
75. Abral H, Ariksa J, Mahardika M, Handayani D, Aminah I, Sandrawati N, et al. Transparent and antimicrobial cellulose film from ginger nanofiber. *Food Hydrocolloids* 2020;**98**:105266.
76. Abral H, Ariksa J, Mahardika M, Handayani D, Aminah I, Sandrawati N, et al. Highly transparent and antimicrobial PVA based bionanocomposites reinforced by ginger nanofiber. *Polym. Test.* 2019:106186.
77. Vermeiren L, Devlieghere F, Debevere J. Effectiveness of some recent antimicrobial packaging concepts. *Food Addit. Contam.* 2002;**19**(S1):163–71.
78. Espitia PJP, Soares NdFF, dos Reis Coimbra JS, de Andrade NJ, Cruz RS, Medeiros EAA. Zinc oxide nanoparticles: synthesis, antimicrobial activity and food packaging applications. *Food Bioprocess Technol.* 2012;**5**(5):1447–64.
79. Ilyas RA, Sapuan SM, Atikah MSN, Asyraf MRM, Ayu Rafiqah S, Aisyah HA, Mohd Nurazzi N, Norrahim MNF. Effect of hydrolysis time on the morphological, physical, chemical, and thermal behavior of sugar palm nanocrystalline cellulose (*Arenga pinnata* (Wurmb.) Merr). *Textile Research Journal.* 2020. In print. doi: 10.1177/0040517520932393.

20

Renewable Sources for Packaging Materials

R.A Ilyas[1,2], S.M Sapuan[3,4], H.A Aisyah[5], Rushdan Ibrahim[6], M.S.N. Atikah[7], H.N. Salwa[4], Min Min Aung[8,9,10], S.O.A. SaifulAzry[4], L.N. Megashah[11], and Z.M.A. Ainun[12]

[1]*School of Chemical and Energy, Faculty of Engineering, Universiti Teknologi Malaysia,81310 UTM Johor Bahru, Johor, Malaysia*

[2]*Centre for Advanced Composite Materials, Universiti Teknologi Malaysia,81310 UTM Johor Bahru, Johor, Malaysia*

[3]*Advanced Engineering Materials and Composites Research Centre (AEMC), Department of Mechanical and Manufacturing Engineering, Faculty of Engineering, Universiti Putra Malaysia (UPM), Serdang, Selangor, 43400, Malaysia*

[4]*Laboratory of Biocomposite Technology, Institute of Tropical Forestry and Forest Product (INTROP), Universiti Putra Malaysia, Serdang, Selangor, 43400, Malaysia*

[5]*Institute of Tropical Forest and Forest Products (INTROP), Universiti Putra Malaysia (UPM), Serdang, Selangor, 43400, Malaysia*

[6]*Pulp and Paper Laboratory, Forest Research Institute Malaysia, Kepong, Selangor, 52109, Malaysia*

[7]*Department of Chemical and Environmental Engineering Faculty of Engineering, Universiti Putra Malaysia (UPM), Serdang, Selangor, 43400, Malaysia*

[8]*Higher Education Centre of Excellence (HiCoE), Institute of Tropical Forestry and Forest Products, University Putra Malaysia (UPM),43400 Serdang, Selangor, Malaysia*

[9]*Department of Chemistry, Faculty of Science, Universiti Putra Malaysia (UPM),43400 Serdang, Selangor, Malaysia*

[10]*Chemistry Division, Centre of Foundation Studies for Agricultural Science, University Putra Malaysia (UPM),43400 Serdang, Selangor, Malaysia*

[11]*Department of Bioprocess Technology, Faculty of Biotechnology and Biomolecular Sciences, Universiti Putra Malaysia (UPM), Serdang, 43400, Selangor, Malaysia*

[12]*Program of Pulp and Paper & Pollution Control, Laboratory of Biopolymer and Derivatives, Institute of Tropical Forestry and Forest Products, Universiti Putra Malaysia (UPM), Serdang, 43400, Selangor, Malaysia*

Bio-Based Packaging: Material, Environmental and Economic Aspects, First Edition. Edited by S.M. Sapuan and R.A. Ilyas

20.1 Introduction

Bio-based packaging has been a move toward more sustainable materials. Industries and universities have been working on bio-based materials as an alternative to conventional petrochemical-derived polymers [1–7]. Plastics, for example, have been commonly used for packaging as they are light, cheap, and flexible when it comes to the final usage. However, ecological problems have been increasing due to the weakness of these petroleum-based plastics in terms of biodegradability [8–13]. According to Thakur et al. [14], 7.8 to 8.2 million tonnes of poorly used plastics are estimated to enter the sea every year.

New bio-based polymers and bio-based conventional polymers are expected to drive the growth of research and development rapidly in the next decade due to the high demand in industries [15–23]. According to Emmambux et al. [24], global dependence on petrochemical-based materials and environmental effects has raised pressure on depleting nonrenewable resources. Hence, sustainable alternatives are more preferable due to being much greener, cleaner, post-use biodegradable, compostable, and efficient [25–38]. Hao et al. [39] conducted a survey to analyze factors influencing consumers' willingness to buy green packaging. The survey uncovered the fact that the environmental effect, quality of green packaging, commodity, and packaging cost are the main factors driving the willingness of consumers to buy items. In addition, the survey showed that there are also challenges for bio-based packages to fulfill apart from their advantages.

Enactment is another viewpoint that cannot be disregarded. It is important to distinguish a standard method for identifying a bio-based material. This makes it simpler for a buyer to separate between the items and it encourages arranging of waste. European standards, such as the European Directive for Packaging and Packaging Waste 94/62/EC, were established in 1994 to have standardized packaging and packaging waste rules among the EU members. This is displayed in the umbrella standard EN13427, as shown in Figure 20.1 [40].

Some of the polymers from renewable sources are attractive from an environmental point of view. However, some of the property limitations, such as mechanical and water

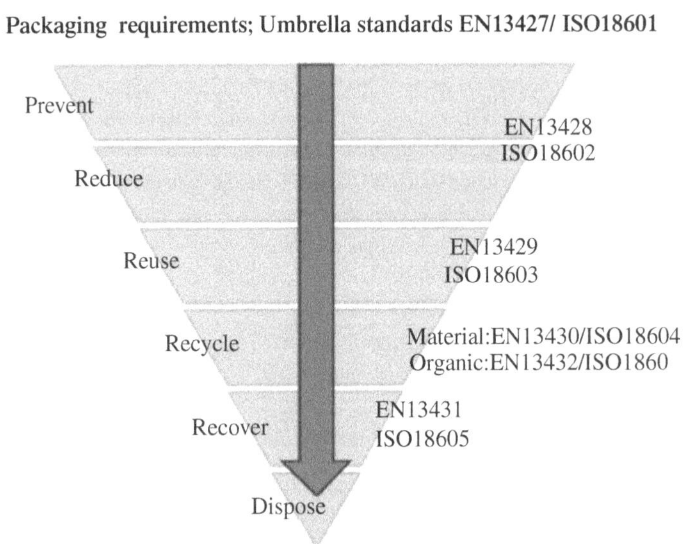

Figure 20.1 Representation of packaging and packaging waste standards [40].

barrier properties, as well as the high cost of bio-based materials, pose limitations for their extensive applications, especially for commercial applications. Therefore, these aspects have opened the door for mass research activities in the development of polymer blends, biocomposites, hybrid biocomposites, and interlaminate composites containing polymers from renewable sources as one of the components. In this chapter, the aim is to review recent research activities related to various renewable sources for the application in packaging materials.

20.2 Packaging Materials from Bio-based Materials

Bio-based plastics are eco-friendly compared to conventional plastics. In addition, bio-based plastics can be defined as plastics made from natural resources and are renewable, whereas biodegradable plastics are referred to by how the plastics behave at the end of their lives. Biodegradable plastic is a material that decomposes naturally when introduced in the environment by the action of living organisms, usually microorganisms. Robertson et al. [41] defined biobased packaging as the products made up of renewable sources, for instance from agricultural and marine renewable sources, and has the potential to biodegrade by composting. Illustrations of bioplastics are displayed in Figure 20.2. From the figure, it can be summarized that plastics can be classified into four categories: (i) conventional plastics (i.e. polyethylene terephthalate [PET], polypropylene [PP], and polyethylene [PE]); (ii) biodegradable and fossil-based plastics (i.e. polycaprolactone [PCL], poly(butylene adipate-co-terephthalate [PBAT], and poly(vinyl alcohol) [PVOH]); (iii) nonbiodegradable and biobased plastics (i.e. biobased-PE, PET, polyamides [PA] and polytrimethylene terephthalate [PTT]); and (iv) biodegradable and biobased plastics (i.e. polyhydroxyalkanoates [PHAs], polylactic acid [PLA], starch, protein, chitin, chitosan,

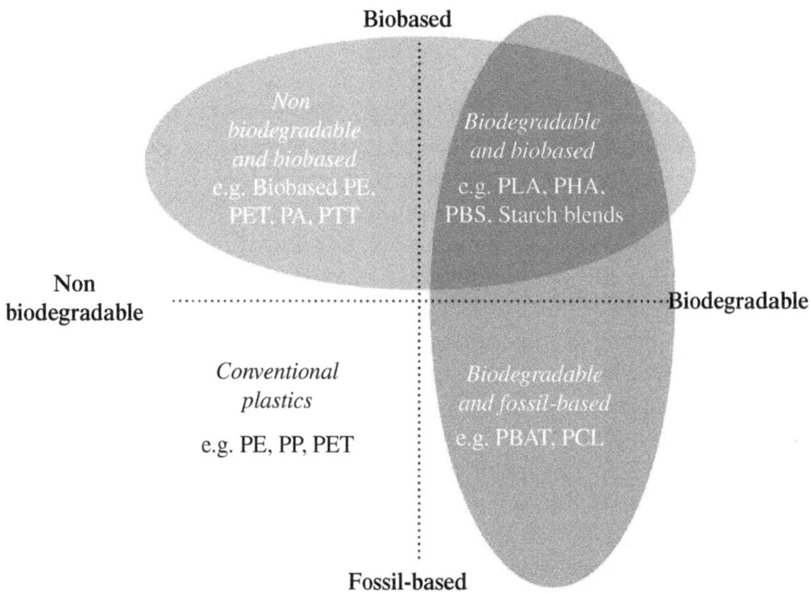

Figure 20.2 Conventional petroleum-based plastics and bioplastics that made up of biodegradable polymers [44].

and polybutylene succinate [PBS]). According to several researchers [41–43], most products in the twentieth century were made from chemical-based and petroleum-derived materials. In the twenty-first century, due to the development of research, science, and technology, natural-derived materials are regaining their popularity. Naturally-derived products will contribute a huge part in the industry, as mentioned by Robertson et al. [41]. Table 20.1 shows the overview of the currently most important groups and types of bio-based plastics.

One of the difficulties confronting the bio-based material industry in its endeavors to deliver bio-based materials is to coordinate the strength of the materials with an item time-frame of realistic usability [45, 46]. The most significant parameters for controlling the strength of bio-based materials are water barrier, pH, oxygen, and temperature. In addition, dry items might be securely put away for extended periods, though wet food would have constrained capacity periods. Prior to utilizing bio-based materials for food stuffs, the consequences for sustenance quality and sanitation must be inspected.

20.3 Development of Bio-based Packages

Environmental factors caused by conventional plastics have led to the development of biobased packages. Biodegradable packages originate from renewable sources available in nature, such as cellulose, starch, and sugar. Hence, natural microorganisms such as bacteria, algae, and fungi are able to decompose these packages [14]. Thakur et al. [14] portrayed bioplastic degradation by enzymes, as shown in Figure 20.3.

20.3.1 Polycarbonates from Sugars and Carbon Dioxide

Recently, the world has produced more plastic waste than ever before. Much of it flows into the oceans and is turning them into a plastic soup [48]. Drinking water can also be contaminated by this waste. Scientists estimated that by 2050, there could be more waste plastics in the ocean than fish, with up to 80% at that time. Researchers at the Centre for Sustainable Chemical Technologies (CSCT) of the University of Bath, UK, have developed a technology to produce biodegradable and biocompatible plastics using sugar and carbon dioxide as the main ingredients of the new biodegradable plastic [49–51]. These scientists have produced alternative polycarbonates from sugars and carbon dioxide in a new process that also uses low pressures and room temperature, making the plastic cheaper and safer to produce. In addition, this new type of polycarbonate can be biodegraded back into carbon dioxide and sugar using enzymes from soil bacteria. Packaging like these can break down naturally, producing only the amount of gas that is first used to spread back into the environment. The resulting plastic has similar physical properties to those derived from petrochemicals, which are strong, transparent, and scratch-resistant [49–51].

When biodegradable plastics are used rather than items made from the traditional process of refinement, there is less emission of greenhouse gasses that can flee into the atmosphere. Every year, more than 100 million tonnes of plastics [52] are consumed, which means that the standard 5 : 1 production ratio indicates that this industry produces 500 million tons of carbon dioxide spreading to our atmosphere every year. This number is equivalent to 19 million vehicles' annual emissions.

Table 20.1 *Overview of currently most important groups and types of bio-based plastics [47].*

Bio-based plastics (group)	Type of polymer	Types/Structure/Production
Cellulose polymers	Polysaccharides	Organic cellulose esters; Regenerated cellulose
Polyamides (PA)		
a. PA11		Bio-based monomer 11 aminoundecanoic acid from castor oil
b. PA610	Polyamide	Monomer sebacic acid from castor oil
c. PA6		Bio-based monomer caprolactam by fermentation of sugar
d. PA66		Bio-based adipic acid by fermentation
e. PA69		Bio-based monomer obtained from oleic acid via azelaic (di)acid
Polyethylene (PE)	Polyolefin	Bio-based monomer ethylene obtained from ethanol; ethanol is produced by fermentation of sugar
Polyhydroxyalkanoates (PHAs) Polyester	Polyester	Direct production of PHA by fermentation
Polylactide (PLA)	Polyester	Bio-based monomer (lactide) by fermentation, followed by polymerization
Polytrimethylene terephthalate (PTT)	Polyester	Bio-based 1,3-propanediol (1,3-PDO) by fermentation plus petrochemical terephthalic acid (or DMT)
Polyvinylchloride (PVC)	Polyvinyls	Monomer vinyl chloride can be obtained from bio-based ethylene (from ethanol)
Polyurethanes (PUR)	Polyurethanes	React polyol with isocyanate. Bio-based polyol can be produced from vegetable oils.
Starch plastics	Polysaccharides	Partially fermented starch; Thermoplastic starch (TPS); Chemically modified starch Starch blends; Starch composites
Thermosets		
a. Epoxy resins	Epoxy resin	Diglycidyl ether of bisphenol A (DGEBA) derived from bisphenol A and epichlorohydrin (ECH). ECH can be produced by glycerine- to epichlorohydrin (GTE) process; glycerine is a by-product of bio-diesel production.
b. Epoxidised vegetable oils	Epoxide	Addition of oxygen to alkenes
c. Thermosets based on 1,2- PDO and 1,3-PDO	Unsaturated polyester	Polycondensation of unsaturated and saturated dicarboxylic acids with diols.
d. Alkyd resins	Alkyd resin	Condensation polymerization of polyols, organic acids and fatty acids or triglyceride oils.
Other thermoplastics[a]		
a. Other polyesters (PBT, PBS, PBSL, PBSA, PBST, PBAT, PET, PEIT PVAc, Polyacrylates, PTN, PTI, thermoplastic elastomoers)	Polyester	Various carboxylic acids, various alcohols

(continued)

Table 20.1 *(continued)*

Bio-based plastics (group)	Type of polymer	Types/Structure/Production
b. Other ethylene-based compounds (e.g. polystyrene and EPDM rubber)	Various	Ethylene by dehydration of bio-ethanol, reacted with other compounds
c. Methanol-based compounds (e.g. phenolic resins, urea formaldehyde resins, melamine formaldehyde resins)	Various	Syngas by gasification of biomass, and synthesis of methanol, reacted with other compounds
d. Propylene-based compounds (e.g. PP, polyacrylates, PUR, PA)	Various	Thermochemical propylene production via bionaphtha plus steamcracking or via biomethanol, followed by Lurgi's methanol-to- propylene (MTP) process or UOP's methanol-to-olefins process.

[a]Abbreviations: PBT = polybutylene terephthalate; PBS = polybutylene succinate; PBSL = polybutylene succinate-co-lactate; PBAT = polybutylene adipate-co-butylene terephthalate; PET = polyethylene terephthalate; PEIT = polyethylene-co-isosorbite terephthalate; PVAc = polyvinyl acetate; PTN = polytrimethylene naphthalate; PTI = polytrimethylene isophthalate; EPDM = ethylene propylene diene M-class rubber; PP = polypropylene; UOP = Universal Oil Products LLC.

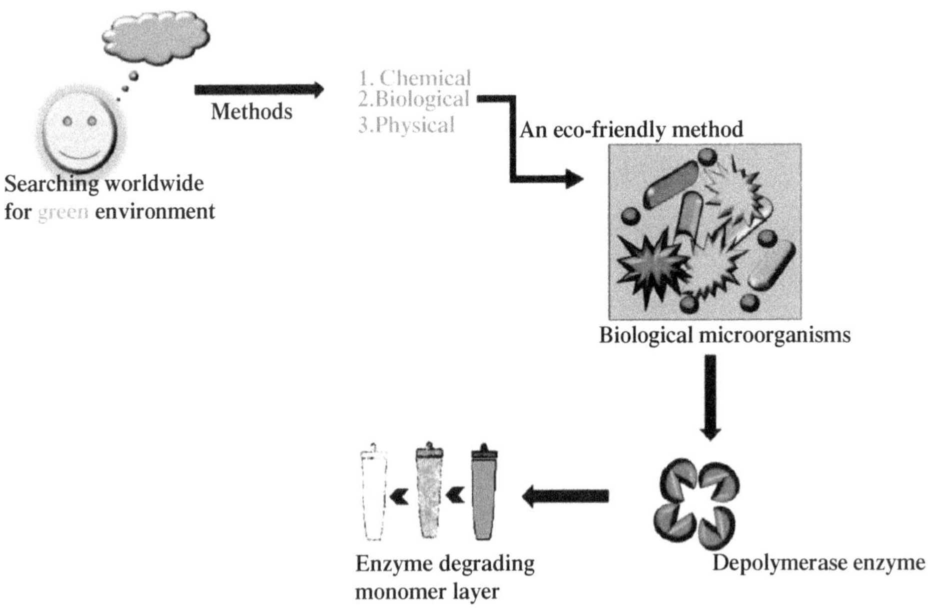

Figure 20.3 *Diagram of biodegradation made with enzymes [14].*

20.3.2 Chitosan

According to Ritchie and Roser [48], implementing a thin-layer biobased packaging gives significant energy and surrounding advantages in the drying process. In addition, it can decrease the use of coating materials by around 30%. Next, biomaterials can also be used to make environmental friendly packaging, such as resorbable chitosan packing. The packaging is not harmful to the environment as its material is soft, flexible, and non-toxic. The layer of the material also uses a multilayer coating so that the structure is strong [19, 29, 53, 54]. The presence of biodegradable packaging can benefit the environment as the packaging can decompose efficiently without causing pollution. This advantage can only occur if the items are properly disposed of, which means that these items must be treated in a compost-like manner.

20.3.3 Plant Cell Wall Biopolymers

Research and study on bio-based materials have opened opportunities in gaining knowledge about plant cell wall biopolymers. Obtaining functional and chemical information from plant cell wall biopolymers has further allowed the production of important tools to be invented. The study of such biological molecules inspired and widened the possibilities to create more novel nanoscale materials. In addition, efficient transformation of bioresources could be made as it is of great interest to the engineering of materials. Enzymatic susceptibility of biological materials is also illustrated as an advantage to the newly nanostructured materials [55].

20.3.4 Polyhydroxyalkanoate

Polyhydroxyalkanoate (PHA) is a bacterial biopolyester with growing industrial interest but with still limited commercial availability [56]. PHA is a biopolymer material that was developed rapidly in the last 20 years. PHA is a polyester synthesized in nature by various microorganisms, including through bacterial fermentation of renewable feedstocks, such as lipids and sugars. Unlike PHA, PLA production involves a two-stage process (fermentation to monomer, followed by a conventional polymerization process). Meanwhile, PHA is produced directly via fermentation of carbon substrates within the microorganism. PHA accumulates as granules within the cytoplasm of cells. It serves as a microbial energy reserve material. PHA has a semi-crystalline structure, with the degree of crystallinity ranging from about 40% to around 80% [47]. Biosynthesis of specific monomers and PHA polymers depends on many factors, including the types of carbon sources on which biological growth depends, the types of metabolic pathways through which organisms convert these carbon sources into PHA monomers, and the substrate specificity of enzymes involved in PHA synthesis.

20.3.5 Polylactic Acid

Polylactic acid (PLA) is derived from renewable sources such as sugar cane, sugar palm starch, or corn starch. PLA is mainly produced through condensation and polymerization processes. The ring-opening polymerization process is a procedure that uses a metal catalyst in a mix with lactide to create bigger PLA molecules. Similarly, the condensation process goes through the same procedure with different temperatures and produces different byproducts [57]. There are few types of PLA, for instance, racemic PLLA, regular PPLA, PDLA, and PDLLA. All of them have different yet similar characteristics because they are derived from renewable sources. PLA is a thermoplastic polyester, which means that it can be heated to its melting point at around 150–160°C, cooled, and heated again without any degradation.

20.3.6 Starch

Starch, a completely biodegradable polysaccharide that is biosynthesized by numerous plants, is one of the most abundant renewable resources. Starch as an agro-sourced polymer has received much attention recently due to its various advantages, such as low cost, wide availability, and total compostability without toxic residues. Common commercial starches are typically derived from stems of palm (sugar palm, sago palm), tubers (sweet potato, potato, bengkuang), cereals (wheat, corn, sorghum, rice), legumes (green pea, mung bean), and roots (cassava) [44–55]. Starch is mostly made up of highly branched amylopectin and linear amylose in its granular form. Therefore, starch can be considered as a crystalline material. Moreover, starch would undergo spontaneous destructurization when mixed with a limited amount of water and is subjected to heat and shear. This would result in a homogeneous melt known as thermoplastic starch (TPS), which possesses thermoplastic characteristics [36, 58]. TPS is a renewable and flexible material that can be easily utilized in different thermoplastification processes with a standard equipment used in manufacturing synthetic polymers. These processes include solution casting, injection molding, extrusion blow molding, injection compression molding, and extrusion [5, 54]59-68].

20.3.7 Protein

Proteins are distinguished from other macromolecules due to their structure, which is based on approximately 20 amino acid monomers, rather than just a few or even one monomer, such as glucose in the case of cellulose and starch. Most proteins contain 100–500 amino acid residues [69, 70]. The structures of proteins can be altered or modified by various chemical and physical processes. These include lipid interfaces, acids and alkalis, and metal ions, as well as mechanical treatment, heat, irradiation, and pressure [70]. In the development of film bioplastics, these modifications are usually used in the formation process to optimize the configurations and interactions of proteins. These modifications would also result in better bioplastic film properties.

20.3.8 Chitin and Chitosan

According to Ma et al. [71], chitin is a polysaccharide with linear chains composed of 2–acetamide–2–deoxy–β–D–glucopyranose units that are linked by glycosidic bonds $\beta(1 \rightarrow 4)$. Furthermore, chitin is the second most abundant organic substance in the biosphere, surpassed only by cellulose. Meanwhile, chitosan is often produced by alkaline

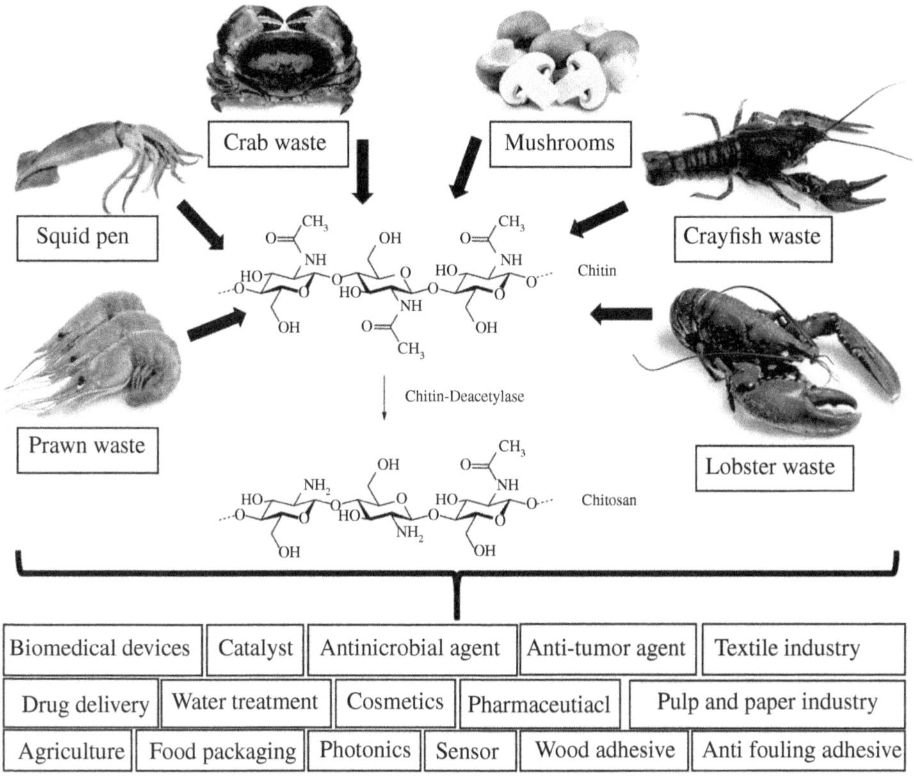

Figure 20.4 *Molecular structures of chitin and chitosan, and their sources and applications.*

deacetylation of α–chitin. Chitin and chitosan are important linear polysaccharides consisting, respectively, of 2–acetamide–2–deoxy–β–D–glucopyranose (GlcNAc) and 2–amino–2–deoxy–β–D–glucopyranose (GlcN) linked by β(1 → 4) [72]. According to Nataraj et al. [72], chitin can be found in the skeletal structure of invertebrates, such as annelids, arthropods, mollusks, and coelenterates, and in the cell walls of diatoms and some fungi. Figure 20.4 shows the molecular structures of chitin and chitosan, and also their sources and applications.

20.4 Decomposition of Biodegradable Plastics

A study conducted by the Biodegradable Products Institute [73] in the 2000s found that the decomposition in biodegradable plastics of these natural materials is incredibly dependent on the climate conditions in the area. In the process, temperature and humidity play important roles. When the weather becomes colder, composting occurs more slowly. The process almost comes to a complete stop during high humidity. This means that in the equatorial and far northern climates, most of these possible advantages with the product are unlikely to be seen.

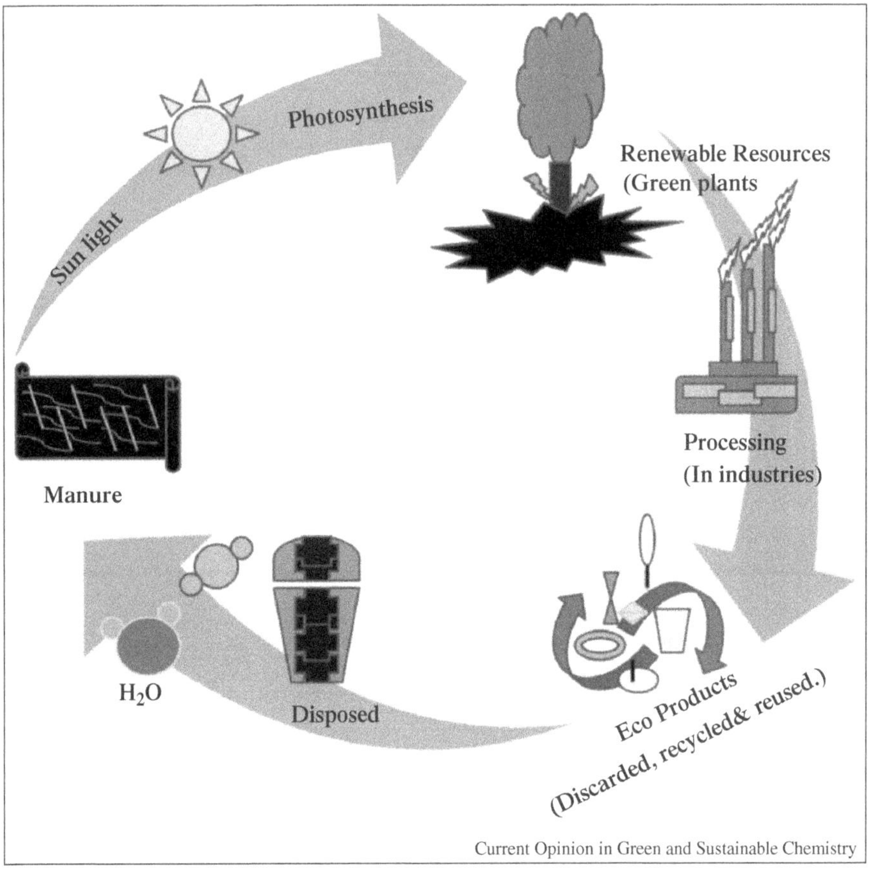

Figure 20.5 *Biobased package life cycle [14].*

Bio-based packaging could also reduce the dependency on fossil resources [74]. Crude oil or gas, the starting point for any traditional plastics, has fewer resources and is expected to last for only a few more decades and become substantially more expensive. Almost all of these fossil materials are used to serve the globe population's energy demand. For a while, in the energy industry, the search for alternative sources, such as wind or solar power, has been an important topic. For the plastics industry, an early switch to renewable sources is equally important. Although only 4% of global oil consumption is used to produce plastics and another 4% is used to generate plastic energy, it takes enough time to develop new technologies needed to process renewable sources and prepare for the "post-oil era" [75].

The degradation processes of bio-based packaging are being widely studied, as mentioned by numerous researchers [14, 73, 76, 77]. As biopolymers fulfill the basic needs of life cycle environmental effects and assessment for proper disposal, they tend to gain more attention in the market. Thakur et al. [14] described the reuse of bio-based packages via photosynthesis, as shown in Figure 20.5.

20.5 Renewable Energy Production Using Biobased Packaging Waste

Renewable energy can be created by using the waste of bio-based packaging. End-of-life decomposition of bio-based packaging can also provide the environment with benefits if recycling is no longer possible. Bio-based packaging contains valuable energy that can be recovered in combined heat and power plants. The renewable part of the material releases the same amount of carbon dioxide that was originally removed from the atmosphere by plants [75].

Bio-based materials are receiving attention in all areas due to their increased environmental footprint and the freedom from the petroleum resources they use. Today, renewable materials can be used to deliver a huge range of technological adhesive innovations that can require equal or greater performance than their petroleum-based marketing counterparts.

Usually, an adhesive joint provides a polymeric factual basis that is connected to the substratum via chemical bonds, physiochemical attractions, or even physical interlinking. The style in which this polymer is applied is as relevant as its chemical composition because it shows the circumstances under which it must be added and therefore the potential end uses. It also directly impacts considerations such as the adhesive spread on the surface and the contact area, which in turn greatly affects the adhesive forces that can be developed. Also, often used in medicinal applications such as plasters are bio-based polysaccharides. Their advantage over synthetic adhesives, in this case, is their nontoxicity and their ability to absorb moisture without losing adhesion, as well as enhancing some strength properties [78, 79].

20.6 Cost of Bio-based Materials

Petroleum-based materials have been widely used as packaging materials because the materials are cheap, have favorable properties, and are easy to be manufactured as desired [80]. Regardless of these advantages, petroleum-based materials have the downside of being harmful to the environment. In addition, environmental awareness among consumers has begun to drive toward a more sustainable approach. As a solution to the environmental problems, new bio-based packaging has been introduced. The bio-based materials are making an impact in a positive approach as the materials are derived from renewable feedstock. Among the advantages of bio-based packaging is that it produces a lower carbon footprint, has a higher energy-efficient process, produces a reduction in raw material use, availability from renewable sources, and improves some functions of the packaging properties [80, 81].

Next, bio-based materials packaging usually requires more than the usual cost to make it. As reported by Ezgi et al. [82], these types of materials cost the production twice that of a typical plastic. This is because it is still new to the industry and not many people are aware of green technology. Therefore, people usually tend to use cheap packaging that is harmful to living things. There are still some companies that prefer to use a conventional plastic bag as the initial cost of the raw material is cheap and can give a lot of profit to them.

Besides, the absence of regulations in some countries to control this issue is also a challenge. As stated by Ezgi [82], the production of biodegradable packaging was expected to boost to around 6.7 million tonnes by the year 2018. Nevertheless, a lot of countries have yet to provide any legislation about their local industry in the production of biodegradable packaging, usage, or waste management. In addition, although many experiments have been conducted to determine the degradation of biodegradable packaging, a standard that needs to be followed by the industry is a must so that it will allow a comparison between a local standard and an international standard. This is to ensure that all the materials and steps used from the beginning until the finished products are in a good condition and can give positive impacts. Some recommendations, such as labeling regulations, can improve the system based on the product's raw material usage, energy use, and energy discharged from manufacture.

Although bio-based materials promise a much brighter future, there are still challenges to be faced in the current era. Since bio-based packaging is not widely spread, conventional processing of these bio-based packages tends to generate a huge amount of waste [31]. In addition, tailoring the biobased material to the current industry's needs is also a challenge to be faced. Furthermore, manufacturing bio-based packaging is a much more holistic approach, for reflecting the target products, variable environment, and minimizing cost and energy is a lot easier said than done. Furthermore, developing such bio-based material requires the use of novel materials, reducing wastage, and employing a standardized procedure [31].

20.7 Life Cycle Assessment

By considering green products as a move toward environmental sustainability, bio-based products have become a subject of interest to many and are being widely studied. Within this perspective, producing packages that are made of renewable materials will have a potential solution to the environment from constant degrading. The food industry is an example of adapting bio-based materials and a start to move toward a much cleaner environment. Life cycle assessment (LCA) is a method usually carried out to measure the environmental effect of food packages. Based on a defined functional unit, LCA aids by quantifying the environmental effects of a product or service. The functional unit, in this case, serves the purpose as a quantitative representation of the function given by a product or service. Thus, this becomes the basis of an environmental impact investigation. Regardless of each industry having its own customized standards and practices related specifically to its industry, the LCA approach was scientifically established and tailored for the food packaging industry and is still in its early stages, providing less force to drive more industries toward using the bio-based packaging [83, 84].

20.8 Social Consumption Behavior

A basic representation of social consumption behavior postulated by Philip Kotler is shown in Figure 20.6. In the beginning, stimuli on information, such as marketing and external sources, will be obtained by consumers. Marketing stimuli in this context are pricing, places, and promotion. External stimuli, on the other hand, are psychological effects due to sociocultural surroundings. Upon receiving such information, different consumers process the issue of cognition, information gathering, and decision making differently when it comes to purchasing. Thus, various ways of consumption are formed among consumers at the end [39].

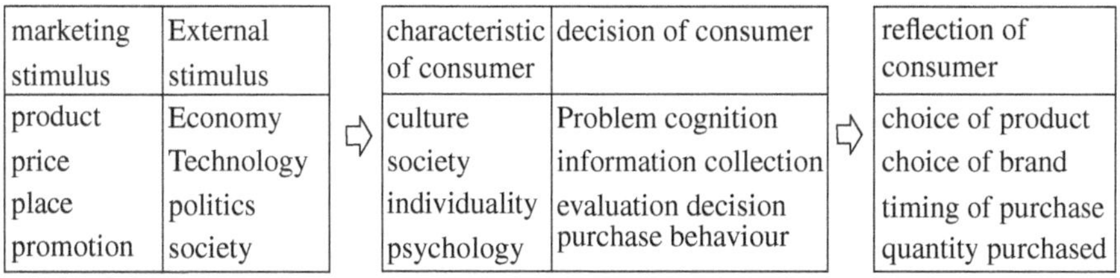

Figure 20.6 *Basic model to a consumer's behavior [39].*

20.9 Conclusions

In the last few decades, petroleum-based plastics have dominated the food packaging industry. Although these materials have brought a lot of attraction and convenience to food, agro, and packaging industries, they have also brought along with them problems related to the renewability and safe disposal of these materials. Due to growing concern over environmental and wildlife problems of these nondegradable materials, interest has shifted toward research and development, and promoting the use of bio-based plastics. Bio-based plastics is a term used for sustainable packaging materials that are not only derived from renewable resources (i.e. produced from agro/food sources, materials such as starch, cellulose, etc.), but are also biodegradable in nature and considered safe to be used in food applications. The current chapter reviews the progress of research in renewable sources for bio-based sustainable packaging materials. To sum up, bio-based food packaging leads to a positive impact on the environment, wildlife, and society.

Acknowledgment

Acknowledgment is addressed to the Ministry of Education, Malaysia, for supporting research funding with the project names "Universiti Putra Malaysia Grant Scheme Hi-CoE (6369107)" and "Fundamental Research Grant Scheme FRGS/1/2017/TK05/UPM/01/1 (5540048)."

References

1. Sapuan SM, Ilyas RA. Sugar palm: fibers, biopolymers and biocomposites. *INTROPica* 2017;5–7.
2. Sapuan SM, Ishak MR, Leman Z, Ilyas RA, Huzaifah MRM. Development of products from sugar palm trees (*Arenga pinnata* Wurb. Merr): A community project. *INTROPica*. 2017;12–3. Available at: http://www.introp.upm.edu.my/upload/dokumen/2018091416164313.5_Prof._Sapuan.pdf.
3. Ilyas RA, Sapuan SM, Sanyang ML, Ishak MR, Zainudin ES. Nanocrystalline cellulose as reinforcement for polymeric matrix nanocomposites and its potential applications: a review. *Curr. Anal. Chem.* 2018;**14**(3):203–25. Available at: http://www.eurekaselect.com/156148/article.
4. Sanyang ML, Ilyas RA, Sapuan SM, Jumaidin R. Sugar palm starch-based composites for packaging applications. In: M.S. Jawaid and S. Kumar (Eds.), *Bionanocomposites for Packaging Applications*. Cham: Springer International Publishing; 2018. p. 125–47. Available at: http://link.springer.com/10.1007/978-3-319-67319-6_7.
5. Ilyas RA, Sapuan SM, Ibrahim R, Atikah MSN, Atiqah A, Ansari MNM, et al. Production, processes and modification of nanocrystalline cellulose from agro-waste: A review. In: *Nanocrystalline*

Materials. IntechOpen; 2019. p. 3–32. Available at: http://www.intechopen.com/books/trends-in-helicobacter-pylori-infection/floating-drug-delivery-systems-for-eradication-of-helicobacter-pylori-in-treatment-of-peptic-ulcer-d.

6. Ilyas RA, Sapuan SM, Ibrahim R, Abral H, Ishak MR, Zainudin ES, et al. Thermal, biodegradability and water barrier properties of bio-nanocomposites based on plasticised sugar palm starch and nanofibrillated celluloses from sugar palm fibres. *J. Biobased Mater. Bioenergy* 2020;**14**:1–13.
7. Goh GD, Yap YL, Agarwala S, Yeong WY. Recent Progress in additive manufacturing of fiber reinforced polymer composite. *Adv. Mater. Technol.* 2019 ;**4**(1):1800271 Available at: http://doi.wiley.com/10.1002/admt.201800271.
8. Zinoviadou KG, Gougouli M, Biliaderis CG. Innovative biobased materials for packaging sustainability. In: *Innovation Strategies in the Food Industry*. Elsevier; 2016. p. 167–89. Available at: https://linkinghub.elsevier.com/retrieve/pii/B978012803751500009X.
9. Hamad K, Kaseem M, Ko YG, Deri F. Biodegradable polymer blends and composites: an overview. *Polym. Sci. Ser. A.* 2014;**56**(6):812–29.
10. Lavoine N, Givord C, Tabary N, Desloges I, Martel B, Bras J. Elaboration of a new antibacterial bio-nano-material for food-packaging by synergistic action of cyclodextrin and microfibrillated cellulose. *Innovative Food Sci. Emerg. Technol.* 2014;**26**(December 2014):330–40. Available at: http://dx.doi.org/10.1016/j.ifset.2014.06.006.
11. Nurazzi NM, Khalina A, Sapuan SM, Ilyas RA, Rafiqah SA, Hanafee ZM. Thermal properties of treated sugar palm yarn/glass fiber reinforced unsaturated polyester hybrid composites. *J. Mater. Res. Technol.* 2019 . Available at: https://linkinghub.elsevier.com/retrieve/pii/S2238785419308312.
12. Norizan MN, Abdan K, Ilyas RA, Biofibers SP. Effect of fiber orientation and fiber loading on the mechanical and thermal properties of sugar palm yarn fiber reinforced unsaturated polyester resin composites. *Polimery* 2020;**65**(2):34–43.
13. Nurazzi NM, Khalina A, Sapuan SM, Ilyas RA. Mechanical properties of sugar palm yarn / woven glass fiber reinforced unsaturated polyester composites: Effect of fiber loadings and alkaline treatment. *Polimery* 2019;**64**(10):12–22.
14. Thakur S, Chaudhary J, Sharma B, Verma A, Tamulevicius S, Thakur VK. Sustainability of bioplastics: Opportunities and challenges. *Curr. Opin. Green Sustainable Chem.* 2018 Oct;**13**:68–75. Available at: https://linkinghub.elsevier.com/retrieve/pii/S2452223618300233.
15. DeGruson ML. Biobased packaging. *Reference Module in Food Science*. Elsevier; Amsterdam, The Netherlands, 2016. Available at: https://linkinghub.elsevier.com/retrieve/pii/B9780081005965032066.
16. Cano A, Cháfer M, Chiralt A, González-Martínez C. Physical and antimicrobial properties of starch-PVA blend films as affected by the incorporation of natural antimicrobial agents. *Foods* 2015;**5**(1):3.
17. Jumaidin R, Khiruddin MAA, Asyul Sutan Saidi Z, Salit MS, Ilyas RA. Effect of cogon grass fibre on the thermal, mechanical and biodegradation properties of thermoplastic cassava starch biocomposite. *Int. J. Biol. Macromol.* 2019.Available at: https://doi.org/10.1016/j.ijbiomac.2019.11.011.
18. Jumaidin R, Saidi ZAS, Ilyas RA, Ahmad MN, Wahid MK, Yaakob MY, et al. Characteristics of cogon grass fibre reinforced thermoplastic cassava starch biocomposite: water absorption and physical properties. *J. Adv. Res. Fluid Mech. Therm. Sci.* **62**, 2019;(1):43–52.
19. Ilyas RA, Sapuan SM, Ibrahim R, Abral H, Ishak MR, Zainudin ES, et al. Effect of sugar palm nanofibrillated cellulose concentrations on morphological, mechanical and physical properties of biodegradable films based on agro-waste sugar palm (*Arenga pinnata* (Wurmb.) Merr) starch. *J. Mater. Res. Technol.* 2019 ;**8**(5):4819–30. Available at: https://doi.org/10.1016/j.jmrt.2019.08.028.
20. Jumaidin R, Ilyas RA, Saiful M, Hussin F, Mastura MT. Water transport and physical properties of sugarcane bagasse fibre reinforced thermoplastic potato starch biocomposite. *J. Adv. Res. Fluid Mech. Therm. Sci.* 2019;**61**(2):273–81.

21. Ilyas RA, Sapuan SM, Ishak MR, Zainudin ES. Water transport properties of bio-nanocomposites reinforced by sugar palm (*Arenga pinnata*) nanofibrillated cellulose. *J. Adv. Res. Fluid Mech. Therm. Sci.* 2018;**51**(2):234–46.
22. Aisyah HA, Paridah MT, Sapuan SM, Khalina A, Berkalp OB, Lee SH, et al. Thermal properties of Woven Kenaf/carbon fibre-reinforced epoxy hybrid composite panels. *Int. J. Polym. Sci.* 2019;**2019**(December):1–8. Available at: https://www.hindawi.com/journals/ijps/2019/5258621.
23. Asyraf MRM, Ishak MR, Sapuan SM, Yidris N. Woods and composites cantilever beam: a comprehensive review of experimental and numerical creep methodologies. *J. Mater. Res. Technol.* 2020;
24. Emmambux MN, Stading M, Taylor JRN. Sorghum kafirin film property modification with hydrolysable and condensed tannins. *J. Cereal Sci.* 2004 Sep;**40**(2):127–35. Available at: https://linkinghub.elsevier.com/retrieve/pii/S0733521004000876.
25. Ilyas RA, Sapuan SM, Ishak MR, Zainudin ES. Sugar palm nanofibrillated cellulose (*Arenga pinnata* (Wurmb.) Merr): effect of cycles on their yield, physic-chemical, morphological and thermal behavior. *Int. J. Biol. Macromol.* 2019 ;**123**:379–88. Available at: https://linkinghub.elsevier.com/retrieve/pii/S0141813018331477.
26. Ilyas RA, Sapuan SM, Ishak MR. Isolation and characterization of nanocrystalline cellulose from sugar palm fibres (*Arenga pinnata*). *Carbohydr. Polym.* 2018;**181**:1038–51. Available at: http://dx.doi.org/10.1016/j.carbpol.2017.11.045.
27. Abral H, Ariksa J, Mahardika M, Handayani D, Aminah I, Sandrawati N, et al. Highly transparent and antimicrobial PVA based bionanocomposites reinforced by ginger nanofiber. *Polym. Test.* 2019 106186. Available at: https://doi.org/10.1016/j.polymertesting.2019.106186.
28. Mazani N, Sapuan SM, Sanyang ML, Atiqah A, Ilyas RA. Design and fabrication of a shoe shelf from Kenaf fiber reinforced unsaturated polyester composites. In: *Lignocellulose for Future Bioeconomy*. Elsevier; 2019. p. 315–32. Available at: https://linkinghub.elsevier.com/retrieve/pii/B9780128163542000177.
29. Ilyas RA, Sapuan SM, Ibrahim R, Abral H, Ishak MR, Zainudin ES, et al. Sugar palm (*Arenga pinnata* (Wurmb.) Merr) cellulosic fibre hierarchy: a comprehensive approach from macro to nano scale. *J. Mater. Res. Technol.* 2019 ;**8**(3):2753–66. Available at: https://doi.org/10.1016/j.jmrt.2019.04.011.
30. Maisara AMN, Ilyas RA, Sapuan SM, Huzaifah MRM, Nurazzi NM, Saifulazry SOA. Effect of fibre length and sea water treatment on mechanical properties of sugar palm fibre reinforced unsaturated polyester composites. *Int. J. Recent Technol. Eng.* 2019;**8**(2S4):510–4.
31. Tumwesigye KS, Oliveira JC, Gallagher MJS. Integrated sustainable process design framework for cassava biobased packaging materials: critical review of current challenges, emerging trends and prospects. *Trends Food Sci. Technol.* 2016 Oct;**56**:103–14. Available at: https://linkinghub.elsevier.com/retrieve/pii/S0924224416300577.
32. Xie F, Pollet E, Halley PJ, Avérous L. Starch-based nano-biocomposites. *Prog. Polym. Sci.* 2013;**38**(10–11):1590–628. Available at: http://dx.doi.org/10.1016/j.progpolymsci.2013.05.002.
33. Zini E, Scandola M. Green composites an overview. *Polym. Compos.* 2011;**32**(12):1905–15.
34. Halimatul MJ, Sapuan SM, Jawaid M, Ishak MR, Ilyas RA. Water absorption and water solubility properties of sago starch biopolymer composite films filled with sugar palm particles. *Polimery* 2019;**64**(9):27–35.
35. Halimatul MJ, Sapuan SM, Jawaid M, Ishak MR, Ilyas RA. Effect of sago starch and plasticizer content on the properties of thermoplastic films: mechanical testing and cyclic soaking-drying. *Polimery* 2019;**64**(6):32–41.
36. Syafri E, Sudirman, Mashadi, Yulianti E, Deswita, Asrofi M, et al. Effect of sonication time on the thermal stability, moisture absorption, and biodegradation of water hyacinth (*Eichhornia crassipes*) nanocellulose-filled bengkuang (*Pachyrhizus erosus*) starch biocomposites. *J. Mater. Res. Technol.* 2019 ;**8**(6):6223–31. Available at: https://doi.org/10.1016/j.jmrt.2019.10.016.
37. Azammi AMN, Ilyas RA, Sapuan SM, Ibrahim R, Atikah MSN, Asrofi M, et al. Characterization studies of biopolymeric matrix and cellulose fibres based composites related to functionalized fibre-matrix

interface. In: Kheng Lim Goh Sabah, Thomas Rangika, Thilan De Silva, and M.K. Aswathi (Eds.), *Interfaces in Particle and Fibre Reinforced Composites*. 1. London: Elsevier; 2020. p. 29–93. Available at: https://linkinghub.elsevier.com/retrieve/pii/B9780081026656000030.

38. Kathirselvam M, Kumaravel A, Arthanarieswaran VP, Saravanakumar SS. Isolation and characterization of cellulose fibers from *Thespesia populnea* barks: a study on physicochemical and structural properties. *Int. J. Biol. Macromol.* 2019;**129**:396–406. Available at: https://linkinghub.elsevier.com/retrieve/pii/S014181301836656X.
39. Hao Y, Liu H, Chen H, Sha Y, Ji H, Fan J. What affect consumers' willingness to pay for green packaging? Evidence from China. *Resour. Conserv. Recycl.* 2019 ;**141**:21–9. Available at: https://linkinghub.elsevier.com/retrieve/pii/S0921344918303665.
40. Molenveld M, Van den Oever M. Biobased Packaging Catalogue, Wageningen Research., Food & Biobased. Wageningen Research, Food & Biobased. 2015;
41. Robertson, G. L. (2011).Packaging and food and beverage shelf life. In: David Kilcast and Persis Subramaniam (Eds.), *Food and Beverage Stability and Shelf Life* (pp. 244–272), Oxford, United Kingdom: Woodhead Publishing. doi: 10.1016/B978-1-84569-701-3.50007-4. ,
42. Robertson G. L. State-of-the-art biobased food packaging materials. In: Emo Chiellini (Ed.), *Environmentally Compatible Food Packaging* (pp.3–28), Oxford, United Kingdom, Woodhead Publishing. 2008. doi: 10.1533/9781845694784.1.3.
43. Ballesteros LF, Michelin M, Vicente AA, Teixeira JA, Cerqueira MÂ. *Lignocellulosic Materials and their Use in Bio-Based Packaging*. Cham: Springer International Publishing; 2018. (Springer Briefs in Molecular Science). Available at: http://link.springer.com/10.1007/978-3-319-92940-8.
44. Ilyas RA, Sapuan SM. The preparation methods and processing of natural fibre bio-polymer composites. *Curr. Org. Synth.* 2020;**16**(8):1068–70. Available at: http://www.eurekaselect.com/178429/article.
45. Peças P, Carvalho H, Salman H, Leite M. Natural fibre composites and their applications: a review. *J. Comp. Sci.* 2018; **2**, 66.
46. Abral H, Ariksa J, Mahardika M, Handayani D, Aminah I, Sandrawati N, et al. Transparent and antimicrobial cellulose film from ginger nanofiber. *Food Hydrocolloids*. 2020;**98**(August 2019):105266. Available at: https://doi.org/10.1016/j.foodhyd.2019.105266.
47. Shen L, Haufe J, Patel MK. Product overview and market projection of emerging bio-based plastics. *Group Science, Technology and Society*. 2009 (June):41.
48. Ritchie H, Roser M.Plastic Pollution [Internet]. OurWorldInData.org. 2019. Available at: https://ourworldindata.org/plastic-pollution.
49. Gregory GL, Hierons EM, Kociok-Köhn G, Sharma RI, Buchard A. CO_2 -driven stereochemical inversion of sugars to create thymidine-based polycarbonates by ring-opening polymerisation. *Polym. Chem.* 2017;**8**(10):1714–21. Available at: http://xlink.rsc.org/?DOI=C7PY00118E.
50. Gregory GL, Kociok-Köhn G, Buchard A. Polymers from sugars and CO_2: ring-opening polymerisation and copolymerisation of cyclic carbonates derived from 2-deoxy- d -ribose. *Polym. Chem.* 2017;**8**(13):2093–104 Available at: http://xlink.rsc.org/?DOI=C7PY00236J.
51. Gregory GL, Jenisch LM, Charles B, Kociok-Köhn G, Buchard A. Polymers from sugars and CO_2 : synthesis and polymerization of a d -mannose-based cyclic carbonate. *Macromolecules* 2016;**49**(19):7165–9. Available at: https://pubs.acs.org/doi/10.1021/acs.macromol.6b01492.
52. Geyer R, Jambeck JR, Law KL. Production, use, and fate of all plastics ever made. *Sci.Adv.* 2017 ;**3**(7):e1700782. Available at: http://advances.sciencemag.org/lookup/doi/10.1126/sciadv.1700782.
53. Tang XZ, Kumar P, Alavi S, Sandeep KP. Recent advances in biopolymers and biopolymer-based nanocomposites for food packaging materials. *Crit. Rev. Food Sci. Nutr.* 2012;**52**(5):426–42. Available at: http://www.tandfonline.com/doi/abs/10.1080/10408398.2010.500508.
54. Ilyas RA, Sapuan SM, Ishak MR, Zainudin ES. Development and characterization of sugar palm nanocrystalline cellulose reinforced sugar palm starch bionanocomposites. *Carbohydr. Polym.* 2018 ;**202**:186–202. Available at: https://linkinghub.elsevier.com/retrieve/pii/S014486171831066X.

55. Dudefoi W, Villares A, Peyron S, Moreau C, Ropers M-H, Gontard N, et al. Nanoscience and nanotechnologies for biobased materials, packaging and food applications: new opportunities and concerns. *Innovative Food Sci. Emerg. Technol.* 2018 ;**46**:107–21. Available at: https://linkinghub.elsevier.com/retrieve/pii/S1466856417302497.
56. Corre YM, Bruzaud S, Audic JL, Grohens Y. Morphology and functional properties of commercial polyhydroxyalkanoates: a comprehensive and comparative study. *Polym. Test.* 2012;**31**(2):226–35.
57. Nazrin, A., Sapuan, S. M., Zuhri, M. Y. M., Ilyas, R. A., Syafiq, R., Sherwani, S. F. K., Nanocellulose reinforced thermoplastic starch (TPS), polylactic acid (PLA), and polybutylene succinate (PBS) for food packaging applications. *Frontiers in Chemistry* 2020; **8**: 213; 1–12.
58. Nafchi AM, Moradpour M, Saeidi M, Alias AK. Thermoplastic starches: properties, challenges, and prospects. *Starch/Staerke* 2013;**65**(1–2):61–72.
59. Ilyas RA, Sapuan SM, Ishak MR, Zainudin ES. Sugar palm nanocrystalline cellulose reinforced sugar palm starch composite: Degradation and water-barrier properties. In: IOP Conference Series: Materials Science and Engineering. 2018.
60. Abral H, Basri A, Muhammad F, Fernando Y, Hafizulhaq F, Mahardika M, et al. A simple method for improving the properties of the sago starch films prepared by using ultrasonication treatment. *Food Hydrocolloids* 2019 ;**93**:276–83. Available at: https://linkinghub.elsevier.com/retrieve/pii/S0268005X1831539X.
61. Ilyas RA, Sapuan SM, Ishak MR, Zainudin ES, Atikah MSN. Nanocellulose reinforced starch polymer composites : a review of preparation, properties and application. In: *Proceeding: 5th International Conference on Applied Sciences and Engineering (ICASEA, 2018).* Capthorne Hotel, Cameron Highlands, Malaysia: Global Academic Excellance (M) SDN BHD; 2018. p. 325–41.
62. Ilyas RA, Sapuan SM, Atiqah A, Ibrahim R, Abral H, Ishak MR, et al. Sugar palm (*Arenga pinnata* [Wurmb.] Merr) starch films containing sugar palm nanofibrillated cellulose as reinforcement: water barrier properties. *Polym. Compos.* 2020 ;**41**(2):459–67. Available at: https://onlinelibrary.wiley.com/doi/abs/10.1002/pc.25379.
63. Nazrin A, Sapuan SM, Ilyas RA. Thermoplastic starch blended poly(lactic) acid for food packaging application : Mechanical properties. In: *6th Postgraduate Seminar on Natural Fiber Reinforced Polymer Composites* 2018. Serdang; p. 79–84.
64. Ilyas RA, Sapuan SM, Ishak MR, Zainudin ES, Atikah MSN. Characterization of sugar palm nanocellulose and its potential for reinforcement with a starch-based composite. In: *Sugar Palm Biofibers, Biopolymers, and Biocomposites* 1. | Boca Raton, FL : CRC Press/Taylor & Francis Group, 2018. CRC Press; 2018. p. 189–220. Available at: https://www.taylorfrancis.com/books/9780429813542/chapters/10.1201/9780429443923-10.
65. Atikah MSN, Ilyas RA, Sapuan SM, Ishak MR, Zainudin ES, Ibrahim R, et al. Degradation and physical properties of sugar palm starch / sugar palm nanofibrillated cellulose bionanocomposite. *Polimery.* 2019;**64**(10):27–36.
66. Jumaidin R, Afif M, Khiruddin A, Asyul Z, Saidi S, Sapuan M, et al. Effect of cogon grass fibre on the thermal, mechanical and biodegradation properties of thermoplastic cassava starch biocomposite. *Int. J. Biol. Macromol.* 2019; Available at: https://doi.org/10.1016/j.ijbiomac.2019.11.011.
67. Sapuan SM, Ilyas RA, Ishak MR, Leman Z, Huzaifah MRM, Ammar IM, et al. Development of sugar palm–based products: a community project. In: S.M. Sapuan, J. Sahari, M.R. Ishak, and M.L. Sanyang (Eds.), *Sugar Palm Biofibers, Biopolymers, and Biocomposites.* 1 Boca Raton, FL : CRC Press/Taylor & Francis Group, 2018. CRC Press; 2018. p. 245–66. Available at: https://www.taylorfrancis.com/books/9780429813542/chapters/10.1201/9780429443923-12.
68. Ilyas RA, Sapuan SM, Ishak MR, Zainudin ES. Effect of delignification on the physical, thermal, chemical, and structural properties of sugar palm fibre. *BioResources* 2017;**12**(4):8734–54.
69. Haghpanah JS, Yuvienco C, Civay DE, Barra H, Baker PJ, Khapli S, et al. Artificial protein block copolymers blocks comprising two distinct self-assembling domains. *ChemBioChem* 2009;**10**(17):2733–5 Available at: http://doi.wiley.com/10.1002/cbic.200900539.

70. Fennema O. Chemical changes in food during processing—an overview. In: *Chemical Changes in Food during Processing*. Dordrecht: Springer Netherlands; 1985. p. 1–16. Available at: http://link.springer.com/10.1007/978-94-017-1016-9_1.
71. Ma X, Qiao C, Wang X, Yao J, Xu J. Structural characterization and properties of polyols plasticized chitosan films. *Int. J. Biol. Macromol.* 2019 **135**:240–5. Available at: https://linkinghub.elsevier.com/retrieve/pii/S0141813019316885.
72. Nataraj D, Sakkara S, Meghwal M, Reddy N. Crosslinked chitosan films with controllable properties for commercial applications. *Int. J. Biol. Macromol.* 2018 ;**120**:1256–64. Available at: https://linkinghub.elsevier.com/retrieve/pii/S014181301832316X.
73. Yepsen R. *Zero Waste Requires Visionary Thinking*. Biodegradable Products Institute, Inc. 2019.
74. Jagadeesh D, Kanny K, Prashantha K. A review on research and development of green composites from plant protein based polymers. *Polym. Compos.* 2015;
75. Charlottenburg A, Rosenheim H. Benefits of biobased rigid packaging. *European Bioplastic*. 2015;
76. Adamcová D, Radziemska M, Zloch J, Dvořáčková H, Elbl J, Kynický J, et al. SEM analysis and degradation behavior of conventional and bio-based plastics during composting. *Acta Univ. Agri. Silvi. Mendel. Brun.* 2018;**66**(2):349–56.
77. Jurczyk S, Musioł M, Sobota M, Klim M, Hercog A, Kurcok P, et al. (Bio)degradable polymeric materials for sustainable future—Part 2: Degradation studies of P(3HB-co-4HB)/Cork composites in different environments. *Polymers* . 2019 ;**11**(3):547. Available at: https://www.mdpi.com/2073-4360/11/3/547.
78. Noreen A, Zia KM, Zuber M, Tabasum S, Zahoor AF. Bio-based polyurethane: an efficient and environment friendly coating systems: a review. *Prog. Org. Coat.* 2016 ;**91**:25–32. Available at: https://linkinghub.elsevier.com/retrieve/pii/S0300944015303192.
79. Latif FEA, Abidin ZZ, Cardona F, Biak DRA, Tahir PM, Abdan K, et al. Physicochemical, structural and mechanical evaluation of bio-based epoxidized jatropha oil blended with amine-cured epoxy resin as new hybrid matrix. *J. Eng. Sci. Technol.* 2019;
80. Nešić A, Ružić J, Gordić M, Ostojić S, Micić D, Onjia A. Pectin-polyvinylpyrrolidone films: a sustainable approach to the development of biobased packaging materials. *Composites Part B*. 2017 Feb;**110**:56–61. Available at: https://linkinghub.elsevier.com/retrieve/pii/S135983681631294X.
81. Johansson C, Bras J, Mondragon I, Nechita P, Plackett D, Simon P, et al. Packaging applications – A review of recent developments. *Bioresources* 2012;**7**(2):1–47.
82. Ezgi Bezirhan Arikan, Havva Duygu Ozsoy. A review: Investigation of bioplastics. *J. Civil Eng. Archit.* 2015 **28**;**9**(2). Available at: http://www.davidpublisher.org/index.php/Home/Article/index?id=116.html.
83. Aggarwal A, Schmid M, Patel MK, Langowski H-C. Function-driven investigation of non-renewable energy use and greenhouse gas emissions for material selection in food packaging applications: case study of yoghurt packaging. *Procedia CIRP* 2018;**69**:728–33. Available at: https://linkinghub.elsevier.com/retrieve/pii/S2212827117309186.
84. Ilyas RA, Sapuan SM, Atikah MSN, Asyraf MRM, Ayu Rafiqah S, Aisyah HA, Mohd Nurazzi N, Norrahim MNF. Effect of hydrolysis time on the morphological, physical, chemical, and thermal behavior of sugar palm nanocrystalline cellulose (*Arenga pinnata* (Wurmb.) Merr). *Textile Research Journal*. 2020. In print. doi: 10.1177/0040517520932393.

21

Environmental Advantages and Challenges of Bio-Based Packaging Materials

R.A Ilyas[1,2], S.M. Sapuan[3,4], Rushdan Ibrahim[5], M.S.N. Atikah[6], M.R.M. Asyraf[7], Mohd Nor Faiz Norrahim[8], S.O.A. SaifulAzry[4], and Z.M.A. Ainun[9]

[1]*School of Chemical and Energy, Faculty of Engineering, Universiti Teknologi Malaysia,81310 UTM Johor Bahru, Johor, Malaysia*

[2]*Centre for Advanced Composite Materials, Universiti Teknologi Malaysia,81310 UTM Johor Bahru, Johor, Malaysia*

[3]*Advanced Engineering Materials and Composites Research Centre (AEMC), Department of Mechanical and Manufacturing Engineering, Faculty of Engineering, Universiti Putra Malaysia (UPM), Serdang, Selangor, 43400, Malaysia*

[4]*Laboratory of Biocomposite Technology, Institute of Tropical Forest and Forest Products (INTROP), Universiti Putra Malaysia, Serdang, Selangor, 43400, Malaysia*

[5]*Pulp and Paper Laboratory, Forest Research Institute Malaysia, Kepong, Selangor, 52109, Malaysia*

[6]*Department of Chemical and Environmental Engineering, Faculty of Engineering, Universiti Putra Malaysia (UPM), Serdang, Selangor, 43400, Malaysia*

[7]*Department of Aerospace Engineering, Faculty of Engineering, Universiti Putra Malaysia, 43400, Serdang, Selangor, Malaysia*

[8]*Research Centre for Chemical Defence (CHEMDEF), Universiti Pertahanan Nasional Malaysia, 57000, Kuala Lumpur, Malaysia*

[9]*Program of Pulp and Paper & Pollution Control, Laboratory of Biopolymer and Derivatives, Institute of Tropical Forestry and Forest Products, Universiti Putra Malaysia (UPM),Serdang, 43400, Selangor, Malaysia*

Bio-Based Packaging: Material, Environmental and Economic Aspects, First Edition. Edited by S.M. Sapuan and R.A. Ilyas

21.1 Introduction

In this era of globalization, issues of improvement or development of current technologies will lead to improvements in life. One of the prevailing improvements is in the food processing industry, which consists of various stages: production, trade, packaging, transportation, storage, marketing, and consumption [1]. Currently, issues on food packaging are a top issue around the world. Almost all countries in the world use plastic for food packaging. The awareness on the usage of bioplastics is still not comprehensive as many still use fossil-based plastics for packaging. This type of packaging can affect consumers in an unsustainable manner [1].

High consumption of plastics leads to severe impacts on climate change and the environment [2–5]. For example, greenhouse gas emissions (GHGs) can cause health problems to humans and wildlife. Therefore, the necessary actions need to be taken in order to prevent GHG. One of the actions that should be carried out is to search and develop more specialized bio-based plastics for large-scale applications [6–11]. Hence, necessary actions should be taken in order to replace the conventional plastics with agricultural-based plastics. Bioplastic is basically a plastic that is partially or completely made from natural resources. European bioplastics is defined as bio-based plastic products that can be decomposed through the *natural degradation process* [6].

Generally, the problem occurs due to the introduction and production of traditional petrochemical plastics such as polyethylene (PE) and polyethylene terephthalate (PET) before the development of renewable raw materials for plastic production. As time passed, the innovation of plastics was only focusing on the production and manufacturability process of the conventional material without making any effort to search or develope an alternative material. In the end, recent studies have established numerous scientific findings that allow bio-based plastics to replace conventional plastics [7–26]. The findings show that bio-based plastic is friendlier due to its natural origin, which usually comes from plants. In the end, bio-plastics are described as innovative materials, since they are made from bio-based resources and exhibit biodegradable properties [14, 27–30].

The term "bio-based" means that the product is partly derived from biomass (plants). The term *biodegradable* is a chemical process carried out by microorganisms that causes the material to decompose into carbon dioxide, water, and other metabolic compounds such

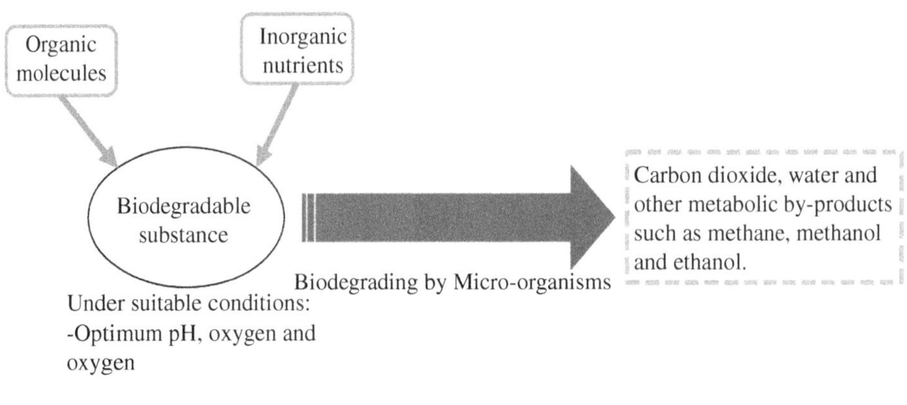

Figure 21.1 Biodegradable process of bio-based material.

as methanol and ethanol. Later, the byproduct can act as fertilizer for plants. Figure 21.1 shows the biodegradation process of bio-based materials. The process depends on several factors: material type, surrounding environmental conditions, and applications [6].

21.2 Advantages of Bio-Based Packaging Materials

21.2.1 Reduction of Waste

One of the benefits of using bio-based packaging material is the reduction of waste. A recent study calculated that the packaging waste obtained was around 78.81 million tons or 31.6% of the metropolitan solid waste (MSW) in 2003 in the United States of America (USA), 56.3 million tons or 25% of the MSW in 2005 in Europe, and 3.3 million tons or 10% of the MSW in 2004 in Australia. Moreover, the USA is implementing a strategy for packaging waste disposal, majority by landfill, followed by recycling, burning, and composting. Since landfill requires large spaces and will result in the age of GHG and contaminants [31].

Plastics make up around 14% of the world's waste stream. Around 32 million tons of waste are produced every year, with only 9% from the total waste being recycled. The remaining waste is dumped into landfills that will lead to more occupation of space in the future. This causes a stream of waste that will smear across a huge portion of land areas without any specific purpose, except to be contained. With decreased available land and the high price of land in the city, a waste stream can cause problematic issues to the lives of urban societies [32].

21.2.2 Reduction in Greenhouse Gas Emission

Utilizing biodegradable plastics aids in reducing the amount of substance outflows that cause ozone depletion. Roughly more than 100 million tons of plastic are produced every year. Considering the standard 5 : 1 ratio of carbon dioxide generation to plastic production shows that this plastic industry produces 500 million tons of carbon dioxide every year. This number is equivalent to the yearly discharges of 19 million vehicles.

In order to reduce the effect of GHG, a culture of reusing plastic material has to be nourished. Every year, the net carbon reserve funds are up to 30%, with a few specialists proposing that the investment funds should be as high as 80%. Changing to biodegradable plastics can considerably diminish the quantity of ozone-depleting substance emanations created by the industry, regardless of the underlying budgetary costs that are required to make progress.

21.2.3 Rapid Decomposition

Bioplastic is a compostable material, which implies that it degrades along with time and inevitably mixes innocuously into the dirt [15, 29, 33–35]. The chemical properties of some biodegradable polymeric materials can be altered in order to ensure the degradation time, either in weeks, months, or years depending on their applications. For instance, when corn starch is immersed in water, its atoms will gradually retain and subsequently swell up to degrade into little pieces. Later, microscopic organisms will act on the material and lead to

Table 21.1 Comparing bio-based packaging with petroplastics.

	Bio-plastics	Petroplastics
Sustainable	Yes	No
Renewable	Yes or moderately	No
Compost in the environment	Compostable and/or biodegradable	Some ecological by polymer oxidation
Polymer range	Little but growing	Large
GHG emissions	Commonly low	Comparatively high
Fossil fuel usage	Commonly low	Comparatively high
Suitable land use	Directly low	None

decomposition. Subsequently, the byproduct of the material will then treat the soil and later can be beneficial to plants.

21.2.4 Sustainability

An environmentally friendly product is a product that contains natural materials or bio-based materials. Several environmental benefits can be seen when bio-based packaging materials are used. Firstly, bio-based packaging does not contain toxins and can yield a lower percentage of GHG. For example, bio-based packaging materials reduce GHG emissions by releasing only 0.49 kg of carbon dioxide (CO_2) from 1 kg of resin produced [36]. However, when compared with conventional plastic, the process induces around 2–3 kg of CO_2, which is higher than that of bio-plastic. A transition from the actual non-biodegradable plastics to the bio-based plastic materials is one of the approaches that can be taken to help the planet reduce the high rate of environmental pollution [37]. Hence, it is important that bio-based packaging materials are produced using biopolymers as an alternative solution to petroleum-based plastics. Table 21.1 shows the comparison of the functional properties between bio-based and petro-plastic packaging during the manufacturing process.

21.2.5 New Marketing Opportunities and Export Industries

Using bio-based materials as food packaging is not only advantageous to the environment but may also bring benefits to organizations that use them in their businesses. Both non-government organizations (NGOs) and private companies can implement the application of bio-based materials in their product packaging. They are considered as stakeholders in contributing to the environment, which makes them responsible and will subsequently gain trust among customers. In this case, their products will be preferable in the markets and customers are likely to purchase their products compared to other products that do not use bio-based materials for their food packaging. Indirectly, this can be a form of marketing strategy to improve profit gains. Examples of companies that have applied the usage of bio-based material are the Coca Cola Company and Poland Spring, who have reduced the usage of plastic components in their packaging products [32].

Besides acting as a new marketing strategy, the usage of bio-based material in food packaging may create new export commodities. Approximately 290 000 tons of biodegradable plastics were produced by China in 2016. About 130 000 tons were consumed domestically and the remainder were exported. This contributed to a recorded 13% of increase of sales as compared to the previous year. This proves that bio-based materials may act as a new export commodity and help in increasing sales activities. This may occur due to the high demand for bioplastics in the market as more people begin to acknowledge the benefits of using bioplastics for the good of the environment. Switching to this type of product may help in eliminating the potential impacts of pollution on the environment over time. In conclusion, bio-based plastics are not only beneficial to the environment but they also contribute to the increase in market demand and supply for plastic-producing countries. Subsequently, the demand for bio-based plastic is able to create many employment opportunities for the society later on [32].

21.3 Challenges of Bio-Based Packaging Materials

21.3.1 Inappropriate Regulations

The Organisation for Economic Co-operation and Development (OECD) reports that the lack of policy support in establishing the bio-plastics industry has led to a disadvantage in the competition for biomass. Poor policy support for bio-based packaging of food is demonstrated by:

- Poorly managed regulations for bio-based packaging and approval of food contact
- Perplexing labels on food packaging based on bio and compostable specifications
- Restricted facts on the renewable use of biomass

There have been circumstances where standards are used for unrelated products or manufacturing processes. A related challenge for bio-based economy participants is that the current standards in the industry are not in line with actual facts. An example of the standard EN 13432 that defines the technical specification for the compostability of bioplastic products is given where the composting cycle described in the standard does not correspond to the way in which the composters run their processes [1]. Therefore, it is not practical to comply with this standard.

In the market where standards are lacking in the bio-based industry, there are some identified gaps and needs. There is a lack of standardization of recycling methods and symbols. Hence, a standard should be established in order to ensure that bio-based plastic is handled properly.

21.3.2 Lack of Composting Facilities

Waste collection and composting facilities must be in place to reap the benefits of flexible end-of-life scenarios of bio-based and compostable food packages. Most developing countries do not have industrial compost facilities or landfills and equipment for energy recovery. Even though the facility is available, there can also be very limited capacity available in the composting facilities for bio-based products. The diagrammatic sketch of the waste composting system that collect, sort, and compost bio-based materials is illustrated in Figure 12.2. This system is called composting *in situ* with complete equipment

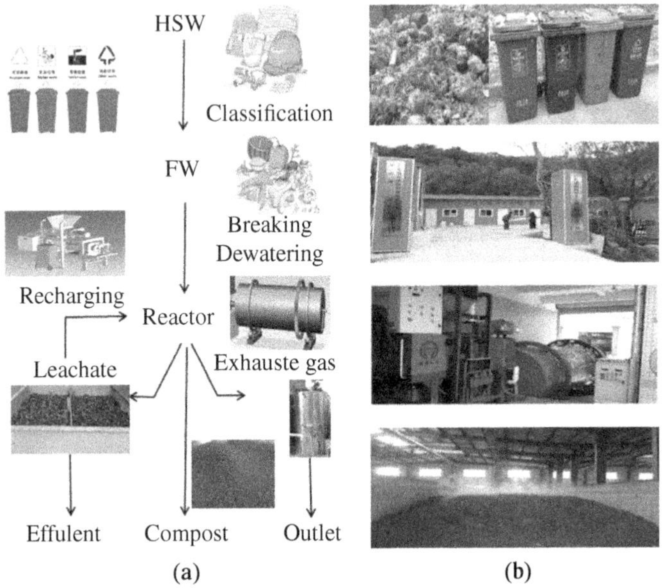

Figure 21.2 The flow chart (a) and schematic diagram (b) of the composting system. Source: Reproduced with permission from Guo et al. [38], Elsevier. Note: household solid waste (HSW) and food waste (FW).

(CPEM) [38] and is operated in a continuously horizontal-flow mode. Firstly, the source separation (SS) of food waste (FW) is sorted again by the composting staff and recycled useful resources at the same time. Next, the FW is shattered and compressed before sending it to the continuously dynamic aerobic composting reactor by a spiral feeder. Each device is individually operated by the controller. From a composting reactor, three types of material will be produced: (i) leachate, (ii) compost, and (iii) exhaust gas.

Roughly 1900 industrial composting plants in Europe were accounted for an annual capacity of more than 19 million tons of waste in 2006, which was around 0.79% of the total waste treated in Europe [39].

A study by the Institute for Local Self-Reliance (ILSR) in the USA also shows that composting facilities in the nation only allow garden waste composting. The study shows that 71% of the 4914 composting operations are compost yard trimmings based on 44 reported states [40]. It is estimated that approximately 100 000 million tons of bio waste are generated each year [41].

21.3.3 Manufacturing Costs

Most of the products in bio-based food packaging are quite expensive. Currently, no accurate and defined cost comparison statistics have been done between fossil-based and organic food packaging materials. However, some studies have deduced that residual biological materials are three to five times more costly than fossil-based packaging materials [1]. The main causes of the higher production costs are:

- Mobilization cost of residues from biomass sites
- Cost required for technological innovations
- Failure to economize

Only in certain periods of the year is biomass product dispersed. Bioplastic involves problems related to the storage and transportation of biomass residues [42]. It is a difficult and expensive process to preserve a lasting biologically based production process from the leftover biomass. The real trick is to mobilize biomass in a sustainable manner that economically meets industrial demands.

Technological breakthroughs in biotechnology and biomass conversion technologies have resulted in a larger scale of demand for bio-based packaging. Nevertheless, the innovative research and development (R&D) investment in bio-based material-related technologies based on biomass residues used in food packaging is relatively small [43, 44]. In particular, biomass residues are often more heterogeneous than virgin foodstuffs. Therefore, innovation is essential for adjusting the present bio-conversion systems to fresh feedstock kinds or for technology development. Investment in the production and development of bio-based packaging demands a strategy structure that seeks to promote feasible use, recycling, and reusability [41].

Today's carbon industry relies heavily on cheap fossil feedstocks. Large-sized petrochemical facilities, extensive capital for carbon, oil and natural gas, and issues economically related to the mining of these fossil assets have needed to be noticed and to be creatively handled with proper enforcement and techniques. Unlike fossil fuels, the economic benefits of cost-effective scales are absent from the use of most organic materials. Because of the size, performance, or size of the businesses during use, the cost potential benefits are economically cost efficient, with costs per output unit usually declining as fixed expenses are distributed by several output units [42].

21.4 Conclusions

The demand of bio-based material is growing rapidly in the world. The need to push for wide use of bio-based materials can directly contribute to the care of the earth. This change can contribute to peaceful lives as there will be no longer any need to think further about methods to eliminate old nonbiodegradable plastics from the world. This is something that is worth investing in as it concerns the longevity of the beloved earth, instead of investing in projects that may only give short-term benefits. All the challenges will be worth the effort as they will be able to improve the future and will benefit the current generation and the generations to come.

References

1. Van Crevel R. Bio-based food packaging in sustainable development. Challenges and opportunities to utilize biomass residues from agriculture and forestry as a feedstock for bio-based food packaging. Food and Agriculture Organization of the United Nations. 2016. Available at: http://www.fao.org/forestry/45849-023667e93ce5f79f4df3c74688c2067cc.pdf.

2. Sapuan SM, Ishak MR, Leman Z, Ilyas RA, Huzaifah MRM. Development of products from sugar palm trees (Arenga pinnata Wurb. Merr): A community project. INTROPica. 2017;12–13. Available at: http://www.introp.upm.edu.my/upload/dokumen/2018091416164313.5_Prof._Sapuan.pdf.
3. Sapuan SM, Ilyas RA. Sugar palm: fibers, biopolymers and biocomposites. *INTROPica* 2017;5–7.
4. Aisyah HA, Paridah MT, Sapuan SM, Khalina A, Berkalp OB, Lee SH, et al. Thermal properties of woven kenaf/carbon fibre-reinforced epoxy hybrid composite panels. *Int. J. Polym. Sci.* 2019;**2019**(December):1–8. Available at: https://www.hindawi.com/journals/ijps/2019/5258621/ .
5. Asyraf MRM, Ishak MR, Sapuan SM, Yidris N, Ilyas RA. Woods and composites cantilever beam: a comprehensive review of experimental and numerical creep methodologies. *J. Mater. Res. Technol.*. 2020. doi: 10.1016/j.jmrt.2020.01.013.
6. Chen YJ. Bioplastics and their role in achieving global sustainability. *J. Chem. Pharm. Res.* 2014;**6**(1):226–31.
7. Ilyas RA, Sapuan SM, Atiqah A, Ibrahim R, Abral H, Ishak MR, et al. Sugar palm (*Arenga pinnata* [Wurmb .] Merr) starch films containing sugar palm nanofibrillated cellulose as reinforcement: water barrier properties. *Polym. Compos.*. 2020;**41**(2):459–67. doi: 10.1002/pc.25379.
8. Jumaidin R, Khiruddin MAA, Asyul Sutan Saidi Z, Salit MS, Ilyas RA. Effect of cogon grass fibre on the thermal, mechanical and biodegradation properties of thermoplastic cassava starch biocomposite. *Int. J. Biol. Macromol.*. 2019. Available at: http://link.springer.com/10.1016/j.ijbiomac.2019.11.011.
9. Sanyang ML, Ilyas RA, Sapuan SM, Jumaidin R. Sugar palm starch-based composites for packaging applications. In: *Bionanocomposites for Packaging Applications*. Cham: Springer International Publishing; 2018. p. 125–47. Available at: http://link.springer.com/10.1007/978-3-319-67319-6_7.
10. Jumaidin R, Saidi ZAS, Ilyas RA, Ahmad MN, Wahid MK, Yaakob MY, et al. Characteristics of cogon grass fibre reinforced thermoplastic cassava starch biocomposite: water absorption and physical properties. *J. Adv. Res. Fluid Mech. Thermal Sci.* **62**,. 2019;1):43–52.
11. Jumaidin R, Ilyas RA, Saiful M, Hussin F, Mastura MT. Water transport and physical properties of sugarcane bagasse fibre reinforced thermoplastic potato starch biocomposite. *J. Adv. Res. Fluid Mech. Thermal Sci.* 2019;**61**(2):273–81.
12. Atiqah A, Jawaid M, Sapuan SM, Ishak MR, Ansari MNM, Ilyas RA. Physical and thermal properties of treated sugar palm/glass fibre reinforced thermoplastic polyurethane hybrid composites. *J. Mater. Res. Technol.*. 2019**8**(5):3726–32. doi: 10.1016/j.jmrt.2019.06.032.
13. Halimatul MJ, Sapuan SM, Jawaid M, Ishak MR, Ilyas RA. Effect of sago starch and plasticizer content on the properties of thermoplastic films: mechanical testing and cyclic soaking-drying. *Polimery* 2019;**64**(6):32–41.
14. Abral H, Ariksa J, Mahardika M, Handayani D, Aminah I, Sandrawati N, et al. Highly transparent and antimicrobial PVA based bionanocomposites reinforced by ginger nanofiber. *Polym. Test.*. 2019 106186. doi: 10.1016/j.polymertesting.2019.106186.
15. Ilyas RA, Sapuan SM, Ibrahim R, Abral H, Ishak MR, Zainudin ES, et al. Effect of sugar palm nanofibrillated cellulose concentrations on morphological, mechanical and physical properties of biodegradable films based on agro-waste sugar palm (*Arenga pinnata* (Wurmb.) Merr) starch. *J. Mater. Res. Technol.*. 2019 **8**(5):4819–30. Available at: https://linkinghub.elsevier.com/retrieve/pii/S2238785419308828 .
16. Syafri E, Sudirman, Mashadi, Yulianti E, Deswita, Asrofi M, et al. Effect of sonication time on the thermal stability, moisture absorption, and biodegradation of water hyacinth (Eichhornia crassipes) nanocellulose-filled bengkuang (*Pachyrhizus erosus*) starch biocomposites. *J. Mater. Res. Technol.*. 2019**8**(6):6223–31. Available at: https://doi.org/10.1016/j.jmrt.2019.10.016.
17. Ilyas RA, Sapuan SM. The preparation methods and processing of natural fibre bio-polymer composites. *Curr. Org. Synth.* 2020;**16**(8):1068–70. Available at: http://www.eurekaselect.com/178429/article .

18. Atikah MSN, Ilyas RA, Sapuan SM, Ishak MR, Zainudin ES, Ibrahim R, et al. Degradation and physical properties of sugar palm starch / sugar palm nanofibrillated cellulose bionanocomposite. *Polimery* 2019;**64**(10):27–36.
19. Ilyas RA, Sapuan SM, Ishak MR. Isolation and characterization of nanocrystalline cellulose from sugar palm fibres (*Arenga pinnata*). *Carbohydr. Polym.* 2018;**181**:1038–51 . doi: 10.1016/j.carbpol.2017.11.045.
20. Mazani N, Sapuan SM, Sanyang ML, Atiqah A, Ilyas RA. Design and fabrication of a shoe shelf from kenaf fiber reinforced unsaturated polyester composites. In: *Lignocellulose for Future Bioeconomy*. Elsevier; 2019. p. 315–32. Available at: https://linkinghub.elsevier.com/retrieve/pii/B9780128163542000177.
21. Nurazzi NM, Khalina A, Sapuan SM, Ilyas RA, Rafiqah SA, Hanafee ZM. Thermal properties of treated sugar palm yarn/glass fiber reinforced unsaturated polyester hybrid composites. *J. Mater. Res. Technol.*. 2019 **9**(2) 1606–1618. Available at: https://linkinghub.elsevier.com/retrieve/pii/S2238785419308312 .
22. Ilyas RA, Sapuan SM, Ishak MR, Zainudin ES. Effect of delignification on the physical, thermal, chemical, and structural properties of sugar palm fibre. *BioResources* 2017;**12**(4):8734–54.
23. Norizan MN, Abdan K, Ilyas RA, Biofibers SP. Effect of fiber orientation and fiber loading on the mechanical and thermal properties of sugar palm yarn fiber reinforced unsaturated polyester resin composites. *Polimery* 2020;**65**(2):34–43.
24. Nurazzi NM, Khalina A, Sapuan SM, Ilyas RA. Mechanical properties of sugar palm yarn / woven glass fiber reinforced unsaturated polyester composites: Effect of fiber loadings and alkaline treatment. *Polimery* 2019;**64**(10):12–22.
25. Azammi AMN, Ilyas RA, Sapuan SM, Ibrahim R, Atikah MSN, Asrofi M, et al. Characterization studies of biopolymeric matrix and cellulose fibres based composites related to functionalized fibre-matrix interface. In: *Interfaces in Particle and Fibre Reinforced Composites [Internet]*. 1. London: Elsevier; 2020. p. 29–93. Available at: https://linkinghub.elsevier.com/retrieve/pii/B9780081026656000030 .
26. Ilyas RA, Sapuan SM, Ibrahim R, Abral H, Ishak MR, Zainudin ES, et al. Sugar palm (*Arenga pinnata* (Wurmb.) Merr) cellulosic fibre hierarchy: A comprehensive approach from macro to nano scale. *J. Mater. Res. Technol.* 2019 ;**8**(3):2753–66. doi: 10.1016/j.jmrt.2019.04.011.
27. Abral H, Basri A, Muhammad F, Fernando Y, Hafizulhaq F, Mahardika M, et al. A simple method for improving the properties of the sago starch films prepared by using ultrasonication treatment. *Food Hydrocolloids* . 2019;**93**:276–83. Available at: https://linkinghub.elsevier.com/retrieve/pii/S0268005X1831539X .
28. Abral H, Ariksa J, Mahardika M, Handayani D, Aminah I, Sandrawati N, et al. Transparent and antimicrobial cellulose film from ginger nanofiber. *Food Hydrocolloids*. 2020;**98**(August 2019):105266. doi: 10.1016/j.foodhyd.2019.105266.
29. Ilyas RA, Sapuan SM, Ibrahim R, Abral H, Ishak MR, Zainudin ES, et al. Thermal, biodegradability and water barrier properties of bio-nanocomposites based on plasticised sugar palm starch and nanofibrillated celluloses from sugar palm fibres. *J. Biobased Mater. Bioenergy* 2020;**14**(2):234–48. Available at: https://www.ingentaconnect.com/content/10.1166/jbmb.2020.1951 .
30. Abral H, Atmajaya A, Mahardika M, Hafizulhaq F, Kadriadi, Handayani D, et al. Effect of ultrasonication duration of polyvinyl alcohol (PVA) gel on characterizations of PVA film. *J. Mater. Res. Technol.*. 2020:1–10. Available at: https://doi.org/10.1016/j.jmrt.2019.12.078.
31. Kale G, Kijchavengkul T, Auras R, Rubino M, Selke SE, Singh SP. Compostability of bioplastic packaging materials: an overview. *Macromol. Biosci.* 2007.
32. Regoli N. 22 Advantages and Disadvantages of Biodegradable Plastics. ConnectUS. 2019. Available at: https://connectusfund.org/7-advantages-and-disadvantages-of-biodegradable-plastics.
33. Ilyas RA, Sapuan SM, Ishak MR, Zainudin ES. Sugar palm nanocrystalline cellulose reinforced sugar palm starch composite: Degradation and water-barrier properties. In: *IOP Conference Series: Materials Science and Engineering*. 2018.

34. Ilyas RA, Sapuan SM, Ishak MR, Zainudin ES. Development and characterization of sugar palm nanocrystalline cellulose reinforced sugar palm starch bionanocomposites. *Carbohydr. Polym.*. 2018;**202**:186–202. Available at: https://linkinghub.elsevier.com/retrieve/pii/S014486171831066X .
35. Ilyas RA, Sapuan SM, Ishak MR, Zainudin ES. Water transport properties of bio-nanocomposites reinforced by sugar palm (*Arenga pinnata*) nanofibrillated cellulose. *J. Adv. Res. Fluid Mech. Thermal Sci.* 2018;**51**(2):234–46.
36. Yu J, Chen LXL. The greenhouse gas emissions and fossil energy requirement of bioplastics from cradle to gate of a biomass refinery. *Environ. Sci. Technol.*. 2008;**42** (18):6961–6966.
37. Costa SS, Druzian JI, Machado BAS, de Souza CO, Guimarães AG. Bi-functional biobased packing of the cassava starch, glycerol, licuri nanocellulose and red Propolis. *PLoS One*. 2014;**9**(11):e112554. Available at: 10.1371/journal.pone.0112554.
38. Guo W, Zhou Y, Zhu N, Hu H, Shen W, Huang X, et al. On site composting of food waste: a pilot scale case study in China. *Resour. Conserv. Recycl.*. 2018;**132**:130–8. Available at: https://linkinghub.elsevier.com/retrieve/pii/S0921344918300338 .
39. Gilbert J. *The European Compost Network 2002–2012- Ten Years of Successful Engagement in Biowaste Management*. Rennes: ORBIT. 2007. Available at: http://www.compostnetwork.info/wordpress/wp-content/uploads/2012/06/1_Gilbert_ECN-workshop_ECN-development.pdf.
40. Platt B, Goldstein N, Coker C. *State of Compost in the US: What, Why, Where & How*. Washingtion: Institute of Local self-Reliance (ILSR). 2014. Available at: http://ilsr.org/wp-content/uploads/2014/07/state-of-composting-%09in-us.pdf.
41. EuropiaBio. Building a bio-based economy for Europe in 2020. Brussels: EuropaBio. 2015. Available at: http://www.bio-economy.net/reports/files/building-%09a-bio-based-economy-for-Europe-in-2020.pdf.
42. Wageningen U. Setting up international biobased commodity trade chains. A guide and 5 examples in Ukraine. *Food and Biobased Research*. 2014.
43. OECD. Policies for bioplastics in the context of a bioeconomy, OECD Science, Technology and Industry Policy Papers, No.10. OECD Publishing. 2013.
44. Ilyas RA, Sapuan SM, Atikah MSN, Asyraf MRM, Ayu Rafiqah S, Aisyah HA, Mohd Nurazzi N, Norrahim MNF. Effect of hydrolysis time on the morphological, physical, chemical, and thermal behavior of sugar palm nanocrystalline cellulose (*Arenga pinnata* (Wurmb.) Merr). *Textile Research Journal*. 2020. In print. doi: 10.1177/0040517520932393.

22

Life Cycle Assessment of Bio-Based Packaging Products

H.N. Salwa[1], S.M. Sapuan[1,2], M.T. Mastura[3], M.Y.M Zuhri[4], and R.A. Ilyas[5,6]

[1]Laboratory of Biocomposite Technology, Institute of Tropical Forest and Forest Products (INTROP), Universiti Putra Malaysia (UPM), Serdang, Selangor, 43400, Malaysia
[2]Advanced Engineering Materials and Composites Research Centre (AEMC), Department of Mechanical and Manufacturing Engineering, Faculty of Engineering, Universiti Putra Malaysia (UPM), Serdang, Selangor, 43400, Malaysia
[3]Faculty of Mechanical and Manufacturing Engineering Technology, Universiti Teknikal Malaysia Melaka (UTeM), Hang Tuah Jaya, Durian Tunggal, Melaka, 76100, Malaysia
[4]Department of Mechanical and Manufacturing Engineering, Faculty of Engineering, Universiti Putra Malaysia (UPM), Serdang, Selangor, 43400, Malaysia
[5]School of Chemical and Energy, Faculty of Engineering, Universiti Teknologi Malaysia,81310 UTM Johor Bahru, Johor, Malaysia
[6]Centre for Advanced Composite Materials, Universiti Teknologi Malaysia,81310 UTM Johor Bahru, Johor, Malaysia

22.1 Packaging: Function and Materials

The consumer packaged goods (CPG) markets greatly rely on packaging. Packaging allows products to be delivered to consumers in excellent condition and spoilage can be avoided. Packaging also plays an important role in extending shelf life and minimizing wastage for unpreserved foods such as fresh meat and fish, ready-prepared meals, and salads. Other products, including electronic equipment such as a mobile phone, as well as other domestic appliances such as vacuum cleaners, kettles, and toasters, all rely on packaging to protect them against the potential damage encountered in the distribution chain.

Bio-Based Packaging: Material, Environmental and Economic Aspects, First Edition. Edited by S.M. Sapuan and R.A. Ilyas

In principle, packaging functions as containment, preservation, protection, and identification [1]. It also needs to ensure that products are dispensed conveniently and safely as well as helping to improve sales where it is the key component that can provide commercial advantages [2]. This applies to all "products" including raw materials and ingredients for pharmaceuticals and cosmetics, components for the automotive and aeronautical industries, and materials for packaging while being shipped to their point of use. Without a product, packaging has no reason to exist and does not have a "life" of its own.

Packaging also prevents or reduces the potential of physical damage to the product during operations, storage and handling in a warehouse, transportation for distribution for sale, display, and finally the usage phase. Without proper design and construction of packaging, a product could be damaged during the handling stages and which could also be caused by environmental factors such as dust, dirt, birds, and insects. Meanwhile, preservation using packaging could avoid or reduce changes due to biological and chemical hazards, which would initiate product spoilage. The main objective of preservation is to extend the shelf life of a product when stored at the recommended conditions, will remain safe, and retain desired sensory, chemical, physical, and microbiological characteristics [1]. Another function of packaging is to provide convenience [1–3]. The design of a pack determines the way a pack is handled and must be constructed for easy handling, convenience, and safety. The pack should be able to be picked up, opened, and unpacked, and potential damage to the contents and personnel should be minimized. Packaging also functions as the source of product information where it provides the best medium for delivering valuable information to the people handling the product, aside from the final consumer. This includes information on the identity of the product, its weight/volume, handling, and possibly unpacking/repacking instructions.

Packaging materials selection is an important process to ensure product quality and freshness during handling, distribution, and storage. Traditionally, food packaging materials used are glass, metals (aluminum, foils and laminates, tinplate, and tin-free steel), and paper and paperboards. Today several materials are often combined to form food packages that take advantage of each material's functional or aesthetic properties [4]. Petroleum-based materials used for single-use food packaging have given a serious global environmental problem due to their non-degradability [5]. Key properties for packaging applications are tensile strength, tear strength, and elongation, which are mostly relevant practically in flexible packaging during form, fill, and seal machine operations [1]. A crucial property of materials for food packaging applications are moisture and gas barrier properties because they determine the material's suitability to provide the required shelf life of the product. Cost consideration is also a major factor in manufacturing decision making and thus density of polymeric materials is vital because plastic resins are traded by weight [1].

Verghese et al. [6] highlighted the fact that the most widely used materials for packaging are paper and board materials, although they are mostly used for secondary packaging. For package with direct containment of a product, plastic materials are still popular because of their flexibility in a wide range of applications and forms, such as rigid containers (bottles, jars), film wrapping, and adhesives. A report by the Ellen MacArthur Foundation [7] described how plastic packaging has not only impacted the environment but has also incurred an economic cost. Additionally, productivity of the ocean has been reduced and urban infrastructure has become clogged with the associated emissions from its production. The cost of after-use and management of plastic packaging waste is much higher than the profit pool of the plastic packaging industry itself. The environmental impact of packaging is an important aspect of consumers' product perception where different packaging materials are perceived to have different environmental impacts [3].

Life cycle assessment (LCA) can be utilized to evaluate the technical solutions employed in the production process to identify impacts originating not only from the production itself but also from the phases of use and end-of-life (EOL) issues. This chapter will discuss thoroughly each phase of the LCA outline and recent different LCA studies conducted on packaging products constructed from bio-based materials will also be reviewed.

22.1.1 Bio-Based Materials for Packaging Applications

Bio-based materials denote that they are derived from plants, although "biomass" can also mean animal-derived material. Bio-based polymers have been defined by DeGruson [5] as polymers produced from renewable resources usually categorized in to three, as shown in Figure 22.1.

Emblem and Emblem [1] and Nazrin et al. [8] found that there are two main groups of bio-based polymers based on the production techniques from biomass, i.e. extraction of natural polymers such as polysaccharides starch or cellulose available in all plants or through methods to produce monomers, for example by hydrolysis, fermentation, or other microorganism attacks, which are then synthesized into polymers, such as lactic acid to polylactic acid (PLA). This is aligned with Saba et al.'s [9] description of biopolymers, which is divided into two main categories, i.e. those derived from living organisms and those that need to be polymerized. Despite this, Verghese [10] outlined bio-based polymer materials that were designed particularly for packaging applications, such as (i) starch-based polymers, (ii) cellulose-based polymers, and (iii) polylactic acid. These bio-based polymer properties and characteristics are summarized in Table 22.1.

The prime environmental attributes of biopolymers are their biodegradability and compostability. Biodegradable materials imply that they can be completely degraded when they encounter natural microorganisms and will produce carbon dioxide and water. Bio-based materials are derived from renewable carbon resources, with a closed-loop carbon dioxide cycle. After disposal and going through incineration, the generated carbon dioxide is converted again into biomass through the photosynthesis process [12]. However, bio-based polymer materials have inadequacies in their properties, such as lower water permeability, poor oxygen barrier characteristics, and relatively poor thermal and mechanical properties. Many studies have been conducted on various methods to overcome these limitations, including blending with other biopolymers, combining a bio-based polymer with reinforcing minerals or fibers, and from recent significant studies on nanotechnology to form nanocomposites by embedding nanoparticles or particulates in biopolymer. Fortunati et al. [13] stated that nanocomposites would be an effective method to enhance the physical properties of bio-based polymers, without affecting their transparency characteristic. Cellulose nanocrystals are among the most studied nanoparticles to be adopted as reinforcements for biobased polymers and Abral et al. [14] recently developed a transparent nanocellulose film with good antimicrobial properties from ginger nanofiber with chemicals and ultrasonication. Previously, they realized that simple ultrasonication treatment could improve the tensile and physical properties of starch based film [15] and also discovered the suitability of poly(vinyl alcohol) (PVA)/ginger fiber bionanocomposite film for packaging applications

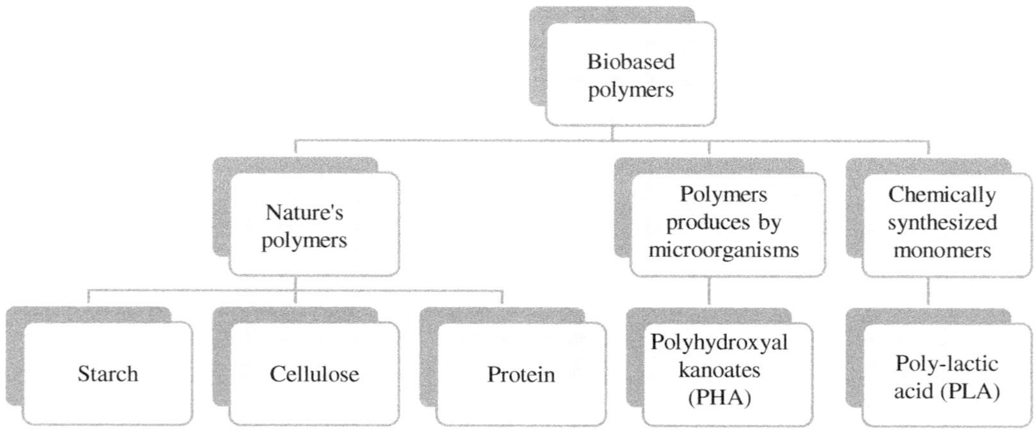

Figure 22.1 *Bio-based polymer categories [5].*

Table 22.1 Bio-based polymers for packaging applications [10, 11].

No.	Biobased polymers	Starch-based polymers	Cellulose-based polymers	Polylactic acid (PLA)
1	**Characteristics**	Clear to translucent	Transparent and permeable to water and oxygen, strong, and stiff with moderate impact resistance	Transparent, has good gas-barrier properties and is resistant to oils; strong and stiff with moderate impact resistance
2	**Manufacturing method**	Injection molding and film applications	Extrusion and injection-mold	Fabricated by injection molding, sheet
3	**Packaging applications**	Loose fill, cutlery, cups, foamed trays, shopping bags, mulch film, compost bags, and film	Film wrap for bakery products, confectionary and fresh produce, twist wrap, and glued bags	Food containers and bottles
4	**Density**	Pure thermo-plastic starch (TPS) is 1.2–1.25 g cm^{-3}	Approximately 0.9 g cm^{-3}	1.24–1.30 g cm^{-3}
5	**Sample product in market**	 Starch t-shirt bag (www.paardekooper.nl)	 Candy packs in compostable cellulose (https://www.greenerpackage.com/compost_biodegrade)	 PLA-based food container produced by Floreon Transforming Packaging (https://www.cosmeticsdesign-europe.com/article/2015/04/22/floreon-tech-targets-pla-bioplastic-improvement)

[16]. A few of the other recent studies on bio-based polymer materials for packaging purposes are summarized in Table 22.2.

22.1.2 Packaging Product Life Cycle

Throughout the product life cycle, the ability to reduce the environmental impact of each stage lessens as the product progresses through the life cycle. The life cycle begins with a conceptual design, where it influences the production and must be studied extensively as making the finest decisions is crucial. Furthermore, the total cost of a product is committed with an approximation of 70–80% during the design stage as stated by Brundage et al. [23]. According to Grant [24], a product life cycle begins with extraction of raw materials, manufacturing phase, use, and EOL stage with transportation and distribution activities going on between many of these phases. At the EOL, there is a potential for materials and products to return to the raw materials production or the manufacturing phase. Figure 22.2 illustrates the life cycle of products [24].

Each packaging material, such as paper, fossil-derived polymer, and bio-based polymer, exhibits its own unique properties and life cycle that generate environmental impacts at

Table 22.2 Selected recent studies on bio-based polymer materials for packaging application.

Biobased polymer materials	Scope of study	Study findings	Author/year
Starch-based	Three types of starch were used to produce the bioplastic sheets (potato, tapioca, and colocasia) and their properties of tensile properties (tensile strength and % of elongation at break), degradation rate, and water absorption were studied.	• Starch content was greatest in tapioca tubers than potato and colocasia. • Tapioca starch films absorb more water. • Tapioca starch film with 0.27 ml glycerol showed much higher tensile strength. • All three bioplastic films were completely degradable in the soil.	Kuruvila et al. (2018) [17]
	To characterized in terms of morphology, footprint, crystallinity (XRD), light transmittance, biodegradability, physical, water barrier, thermal (TGA, DSC, and DMA), and mechanical properties of bionanocomposites in the form of films made from sugar palm starch (SPS) and sorbitol/glycerol with different nanofiller sugar palm nanocrystalline cellulose (SPNCC) compositions (0–1.0 wt%) using solution casting method.	• The SPS/SPNCCs nanocomposite films exhibited higher crystallinity, tensile strength, Young's modulus, thermal, and water-resistance compared to the neat SPS film. • The tensile strength and moduli of the bionanocomposites increased after being reinforced with SPNCCs and the optimum nanofiller content was 0.5%.	Sapuan and Ilyas (2018) [18]
	Characterization on composites based on natural cellulose fibers (CF) and hydroxyethyl cellulose (HEC) in the form of green packaging films.	• Deep eutectic solvents (DESs) are a highly promising plasticizer for the cellulose-based biocomposites with similar or even better plasticizing effect compared to conventional plasticizer. • The DES based on tetrabutylammonium bromide and propylene carbonate was most efficient at providing thermoformability.	Sirviö et al. (2018) [19]

Table 22.2 *(continued)*

Biobased polymer materials	Scope of study	Study findings	Author/year
Cellulose-based	Mechanical properties evaluation (toughness, burst strength, and distance to burst), water adsorption, light-barrier properties, and transparency of composite films based on cellulose, glycerol, and polyvinyl alcohol.	• Discovered that polyvinyl alcohol enhances the toughness but decreases when glycerol concentration is increased (from 23.41 to 10.55 MJ m^{-3}) • Polyvinyl alcohol increased the film thickness. The films showed higher burst strength (up to 12 014 g) than other biodegradable films. • The films have optimal values of transparency like those values of synthetic polymers. • Glycerol produced a UV protective effect in the films, an important effect for food packaging to prevent lipid oxidative deterioration. • It is feasible to obtain cellulose-glycerol-polyvinyl alcohol composite films with improved properties.	Cazón et al. (2018) [20]
	To improve the mechanical and thermal properties of composites, bamboo cellulose poly (lactic acid) (PLA), and nanowhiskers (BCNW) were extracted and introduced into PLA composites as fillers. PLA/BCNW biofilms were fabricated by solution casting with different BCNW contents (0, 0.5, 1.0, 1.5, 2.0, 2.5, 3.0, 3.5, and 4.0 wt%). The characteristics of the biofilms were investigated.	• The results showed that the maximum tensile modulus and elongation at break of 427.72 ± 19.32 MPa and 22.27 ± 3.50% were reached at 2.5 wt% and 1.0 wt% loadings, respectively. Both homogeneity and stereocomplexed crystallites were observed, and hetero-geneous nucleation effect was confirmed. With the addition of BCNW, the crystallite size of PLA/BCNW composites increased remarkably, and the largest crystallinity was 30.7 ± 0.9% with 2.5 wt% BCNW. These results provided data support for enlarging the application of PLA.	Qian et al.(2018) [21]

Table 22.2 *(continued)*

Biobased polymer materials	Scope of study	Study findings	Author/year
Polylactic acid (PLA)	To investigate mechanical properties of PLA biocomposites incorporating *Aloe Vera* fibers of varied treatment time (24, 48, 72, 120, and 168 h).	• Fiber treatment time of 72 h exhibited highest improvement in tensile, flexural, and compressive behavior of the developed biocomposites. The biocomposites incorporating treated (72 h) Aloe Vera fibers exhibited 104.9% higher impact strength as compared to neat PLA. Hence, the use of sodium bicarbonate treatment for lignocellulosic-fiber surface modification ensures greener and safer processing of biocomposites without compromising their mechanical properties.	Chaitanya and Singh (2018) [22]

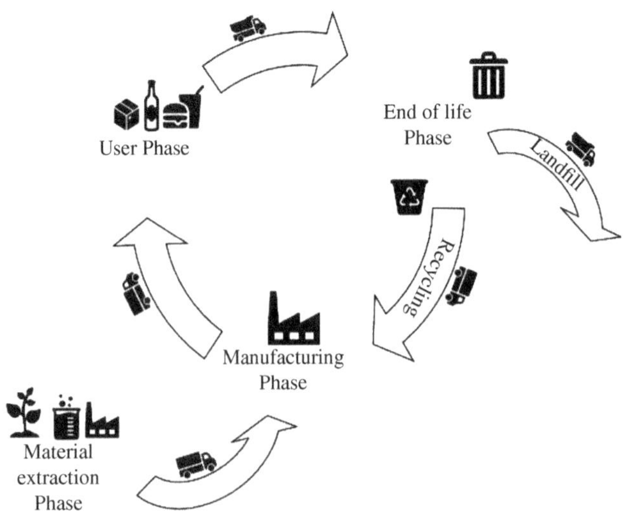

Figure 22.2 *Packaging product life cycle [24].*

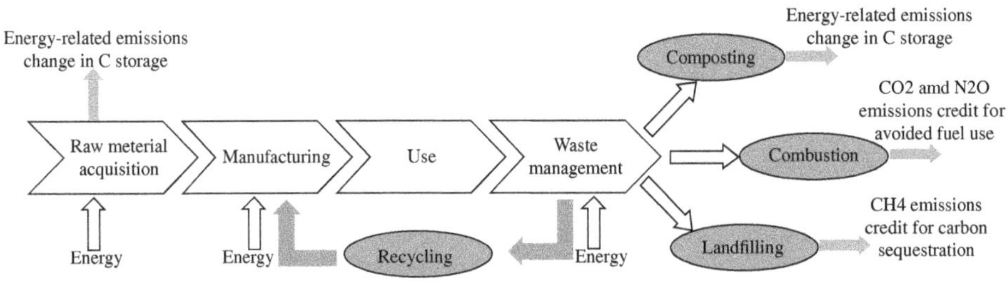

Figure 22.3 *Municipal solid-waste life cycle with end of life options [25].*

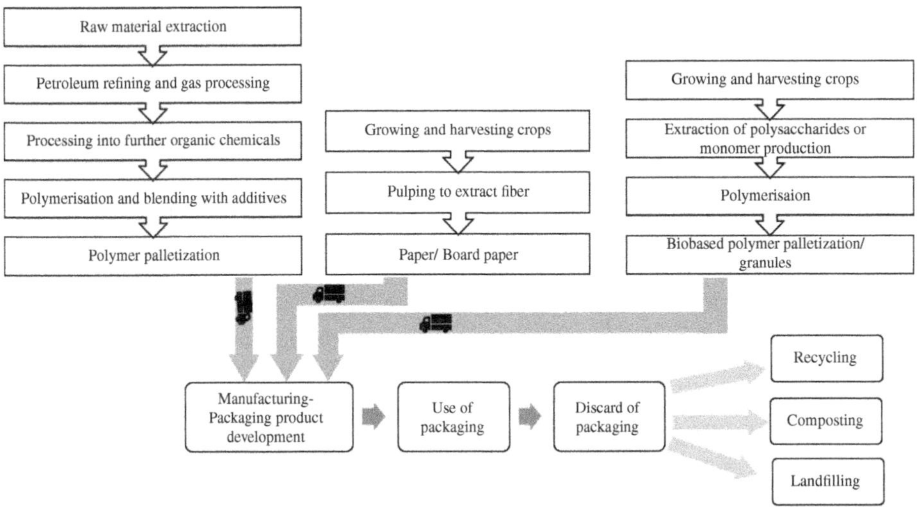

Figure 22.4 *Life cycle of packaging made from petroleum-based materials, paper, and bio-based materials with different processes before arrival at the manufacturing gate and moving on until the end of life options.*

different stages of the supply chain and for different environmental indicators. Plastic packaging products had the largest increase for all materials in municipal solid waste for the past decades. The entire life cycle of packaging waste must be evaluated when considering alternatives to land-filling options, which include a combination of source reduction, recycling, incineration, and composting [25]. This life cycle of municipal solid waste is illustrated in Figure 22.3 [19]. The significant differences of life cycles of packaging made from paper, petroleum-based, and bio-based materials are mainly during raw material acquisition and their EOL alternatives. These individually complex processes and life cycles for each material are important to be understood and must regularly be reviewed in order to have an informed packaging design decision and strategy. Figure 22.4 explains the life cycle of packaging with different sources of raw materials.

22.2 Life Cycle Assessment (LCA)

22.2.1 Background of LCA

Around the 1970s, packaging had been of great consumer concern and many LCA studies were on process-based analysis [26]. Studies were mostly to compare the environmental impacts of different materials of similar packaging products, such as molded pulp trays and plastic trays and plastic bottles and refillable glass bottles. Those studies were commonly appointed by companies producing or using the packaging and Coca Cola was the pioneer in 1969, even though the results were not disclosed directly to consumers but were mainly used for internal purposes as guidance to reduce life cycle impacts. In the 1990s, the International Organization of Standardization (ISO) later supported the initiative of a formal process of standardization to develop a global standard for LCA. The lack of a standardized methodology where different studies of the same product could produce opposite results with different methodological choices created concerns from industry. LCA was increasingly used for product development and marketing of greener products. The standard developed gave attention to the principles and framework, the goal and scope definition, the life cycle impact assessment, and the life cycle interpretation, which coded ISO 14040, ISO 14041, ISO 14042, and ISO 14043, respectively. A revision in 2006 introduced ISO 14044, which compiled ISO 14041, ISO 14042, and ISO 14043. ISO 14040:2006 defines LCA as the "compilation and evaluation of the inputs, outputs and of the potential environmental impacts due to a product-system throughout its life-cycle" [27]. The LCA method can estimate the impact of a final product by accounting for the impacts across all its full life cycle from resource extraction to EOL [23]. Figure 22.5 depicts the LCA framework according to ISO 14040:2006.

LCA can be utilized as a supporting tool for design to learn and assess the technical solutions to be employed in the production process to minimize the impacts originating not only from the production itself but also from the phases of use and EOL. LCA could as well be the comparative assessment tool that has been applied in numerous researches [28–33]. LCA is also being used for product or process selection, design and optimization, and it is possible to attach it with simulation techniques and design tools to help organizations learn of the environmental consequences of their decisions and actions both on and off site [23]. The LCA method can be applied to estimate the impact of a final product by conducting a full analysis according to a detailed understanding of how and where the product was produced and how it will be disposed of. The process comprises information collection on the inputs and outputs, such as emissions, waste, resources, and processes (life cycle inventory) and translating these to environmental consequences such as the contribution to climate change, smog creation, eutrophication, acidification, and human and ecosystem toxicity [34]. The cost incurred to complete the information of every input and output at each stage of life can be very high and requires significant investment in both time and data collection. Therefore, software vendors and government agencies developed databases, such as the European Reference Lifecycle Database and the United States Lifecycle Inventory Database (ASTME2986-15, 2015), of reusable "unit processes" that can be linked together [23].

As explained earlier, the process of manufacturing a packaging product covers the activities of acquisition of stock materials, equipment, energy use, and tooling to calculate the

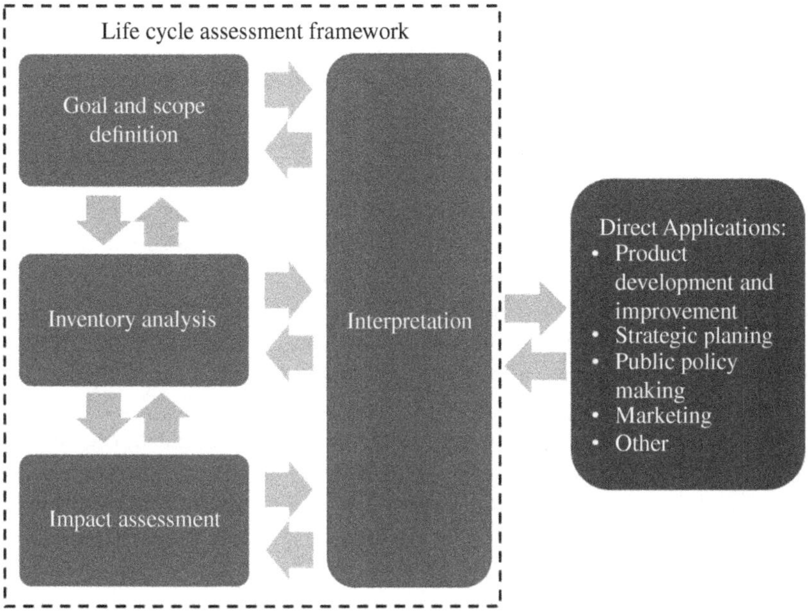

Figure 22.5 *Phases of an LCA framework according to ISO 14040:2006.*

amount of emissions and waste created throughout the life cycle of a product [23, 35]. Whereas LCA in principle includes six important phases in a product life cycle, i.e. materials extraction > manufacturing and waste production > packaging > transportation > product use > product disposal, on the other hand, the four main steps in performing the LCA technique are (i) goal and purpose, (ii) life cycle inventory (LCI), (iii) life cycle impact assessment (LCIA), and (iv) interpretation of results.

22.2.2 LCA Approaches

There are two approaches to LCA, namely the attributional and consequential. The aims of the development of these approaches are to provide answers to questions relating to different systems of modeling [23, 36]. The attributional-LCA (ALCA) provides information about the impacts of the processes employed to produce, consume, and dispose of a product. On the other hand, consequential-LCA (CLCA) provides information about the consequences of changes in the level of output, consumption, and disposal of a product, including the effects both inside and outside the life cycle of the product [36]. Most of the LCA studies employ the attributional approach (ALCA), which involves accounting for the impacts of a product system without considering any external influences. This approach attempts to provide information on the global burdens portion that can be directly associated with a product throughout its life cycle. Some LCA studies on bio-based products or processes use the consequential approach (CLCA) to avoid the problem of allocation in multifunctional bio-based systems and include the activities that have indirect implications on the product system [37]. CLCA is also known as a change-oriented approach with an attempt made to provide information on the environmental impacts that occur, directly or indirectly, because of a decision. CLCAs analyze the product system but also consider the market mechanisms, state of advancement of technologies, and the co-products indirectly affecting of the utilization considered product in case of a multifunctional product system. This approach would be very useful for policy and decision making. An example of a CLCA study by Alvarenga et al. [38] was based on the assessment of the effects on the indirect land use change (iLUC) caused by expansion of a sugarcane plantation. They emphasized that the effects of iLUC should be considered when assessing new bio-based products, as some projected environmental benefits could have been null and void if there is a low control on deforestation caused by iLUC.

LCA has expanded into a wide range of applications, both at downstream and upstream activities, such as eco-labeling, energy systems, product designs, transportation, and food production, which further highlight its functionality and effectiveness [34]. Each of ALCA and CLCA studies has its own goal and system boundaries assigned. The choice to use which LCA approach, system boundary, methodology, and others solely depends on the researchers and there must be reasons to justify the decisions before taking these results into the eco-design framework [37]. Several software items are available in the market to assist LCA implementation:

I. SimaPro, Pre-Product Ecology Consultants, Amersfoort, NL, http://www.pre.nl/simapro
II. Boustead, Boustead Consulting Ltd., West Grinstead, UK, http://www.boustead-consulting.co.uk
III. Gabi, Polymer Institute, University of Stuttgart, Germanyhttp://www.peproduct.de/GABI/htdoc/home_ englisch.htm
IV. Umberto, Institute for Environmental Informatics, Hamburg, Germanyhttp://www.umberto.de/english

22.3 LCA Goal and Scope (Definition of a Functional Unit and System Boundary)

The main elements in the first phase of LCA operation can be illustrated as in Figure 22.6, based on the description of Parameswaranpillai and Vijayan [39]. LCA begins with a detailed definition of the reasons for the study implementation and the intended application. The goal of the study and the scope of work are defined with respect to the functional unit (FU), system boundaries, assumptions, and limitations of the study.

22.3.1 Functional Unit (FU)

An FU is described as the functional requirements of a product system for a certain period, which enables different systems to be regarded as functionally equivalent for direct comparison analysis [25]. According to Lorite et al. [40], FU can also be understood as a reference unit that represents the level of performance of the product system. FU is different in every study associated with a packaging product because of a different goal and scope of the study. For example, Gubkaya and Banar [30] considered 1 m^2 of packaging film as the FU in their

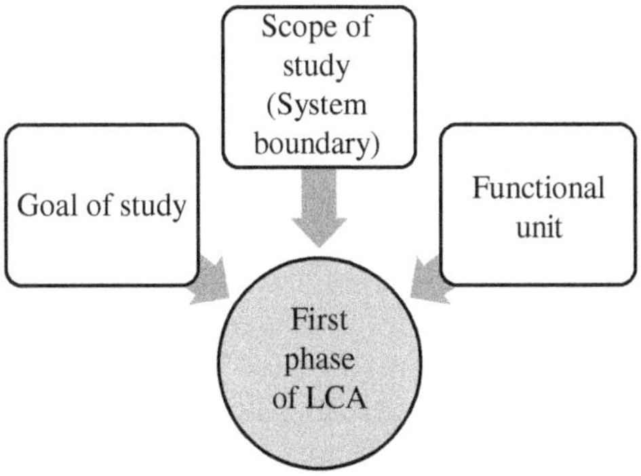

Figure 22.6 *Main elements of the first phase of LCA.*

study of a new biodegradable film based on orange peel-derived pectin jelly and corn starch to compare environmentally with a low-density polyethylene (LDPE) film. In the study by Lorite et al. [40] to evaluate the environmental impact of PLA-based active packaging compared to commonly used polyethylene terephthalate (PET) for a shelf life comparison, the FU selected was "providing customers with 100 000 kg of fresh fruits for one year," which enables a comparison between shelf lives. Bernstad and Saraiva [29] selected "transportation of ten mangos per packaging from the area of production to the end-consumer" as their FU to compare its environmental performance of an identical packaging produced from a recycled high-density polyethylene (HDPE) frame with a high-impact polystyrene (HIPS) tray, natural fibers of a reinforced/recycled HDPE composite frame with an HIPS tray, and commercial cardboard, whereas the FU defined by Nikoliae et al. [33] was 1000 units of 500 ml PLA or PET bottles that are intended to be filled with drinking water. The same thickness and forms of bottles assumed in this study but with different weights is due to the different densities of PLA and PET (1.23 and 1.33 g cm^{-3}, respectively). An earlier study done by Bohimann [25] on LCA of two different polymers, i.e. PLA derived from corn and polypropylene (PP), utilized food packaging applications for one metric ton (1000 kg) of yogurt purchased by the consumer. This means the quantity of packaging exploited for one metric ton of yogurt purchased and consumed.

22.3.2 System Boundary

System boundaries are described whether an LCA study represents a complete analysis (cradle-to-grave) or a partial analysis (cradle-to-gate or gate-to-grave) [34, 39]. In a studied product system, boundaries must be defined clearly to separate the system from the environment. Cradle-to-grave includes all life cycle phases beginning with the production phase that comprises material and component productions, a use phase, and an up to EOL processing of the product. On the other hand, cradle-to-gate is a partial product life cycle involving resource extraction and a production phase to the factory gate before it is sent to the manufacturer for application or consumers [34]. In the LCA ISO standards, i.e. ISO 14040 (principles and framework) and ISO 14044 (requirements and guidelines), it is highlighted that the cradle-to-grave approach is not the only approach to perform

LCA. Other approaches like cradle-to-gate studies, gate-to-gate studies, and specific parts of the phase in the life cycle may also be operated in LCA studies with appropriate rationalization.

A complete and detailed LCA system boundary draws a perimeter around the activities that are included within its assessment. This usually starts with raw material extraction and considers each step of material transformation, their intermediate states, and finally the synthesis of the final product. Processing stages start and end at this boundary and within this defined periphery, material flow and transformation take place to produce the final product of interest. During each and every transformation stage, other chemicals and energy are required as input [41]. Manufacturing activities usually generate hazardous effects to the general society and also cause alarming deprivation of the natural resources. High energy consumption, waste generation, and greenhouse gas emissions are among the most bothering impacts to the environment. Consequently, industries are demanded to evaluate their performances toward realizing the three sustainability dimensions, i.e. economic development, social development, and environmental protection [42]. In the design phase of a product, a product designer may consider an alternative attribute with a less energy intensive process but does not compromise on the specification's requirement.

The cradle-to-gate system boundary for biocomposite film made from orange peel-derived pectin jelly and corn starch by Gunkaya and Banar [30] starts from raw material extraction to disposal, including chemical materials production, starch production, electricity production, and biocomposite film production. It was compared with petroleum-based polymer, the LDPE whole life cycle from generation of the raw material to the final polymer product. In this system, they did not include transportation of individual materials from production plants to the laboratory and the use stage, which is the point when a consumer purchases the product until disposal, because they decided that this phase is relatively short to compare with the other stages in the study. Therefore, the emissions at use phase were considered as zero. The system boundary developed for this study is shown in Figure 22.7.

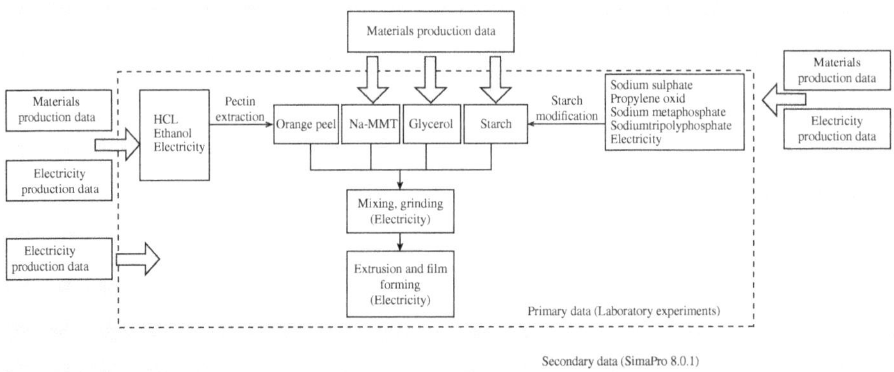

Figure 22.7 Biocomposite film production based on orange peel-derived pectin jelly–corn starch to compare the LDPE film. Source: From Gunkaya and Banar [30].

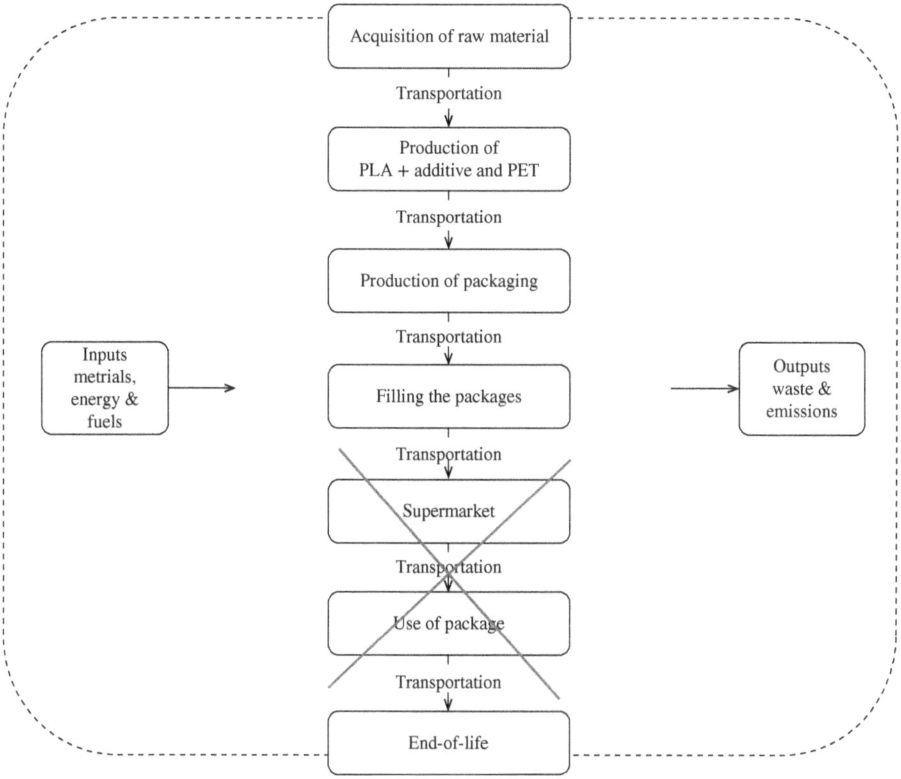

Figure 22.8 *System boundary of LCA for PLA-based and PET packages from the cradle to the grave. Source: From Lorite et al. [40].*

Lorite et al. [40] studied PLA-based packaging and PET throughout their life cycles from the cradle to the grave, i.e. they started from the extraction of raw materials and ended with the disposal of the packaging. For the PLA packaging, in addition to PLA/nanoclay and PLA/nanoclay/surfactant, a third formulation of PLA/nanowhiskers/surfactant was also included in the LCA. They excluded selling and using these packages and transportation between the two phases. Their assumption was that both packages of PLA and PET are similar; i.e. take-away packages of the same shape and size with fresh-cut fruit. They also assumed, based on physicochemical and microbiological results, that the shelf-life of the food product of PLA/nanoclay/surfactant was expected to be longer compared to that of the PET packaging (15 days). Figure 22.8 is the system boundary used in the Ingrao et al. study [36].

Ferreira Saraiva et al. [43] did an LCA study of a Brazilian mango packaging consisting of a reusable frame made from a high-density polyethylene reinforced with natural sponge fiber residue and a high-impact polystyrene recyclable tray. They compared the environmental performance with those of an identical packaging produced without natural fibers and a commercial cardboard packaging. The system boundaries of both the composite and the cardboard packaging used in this study and composite packaging production is shown in Figure 22.9.

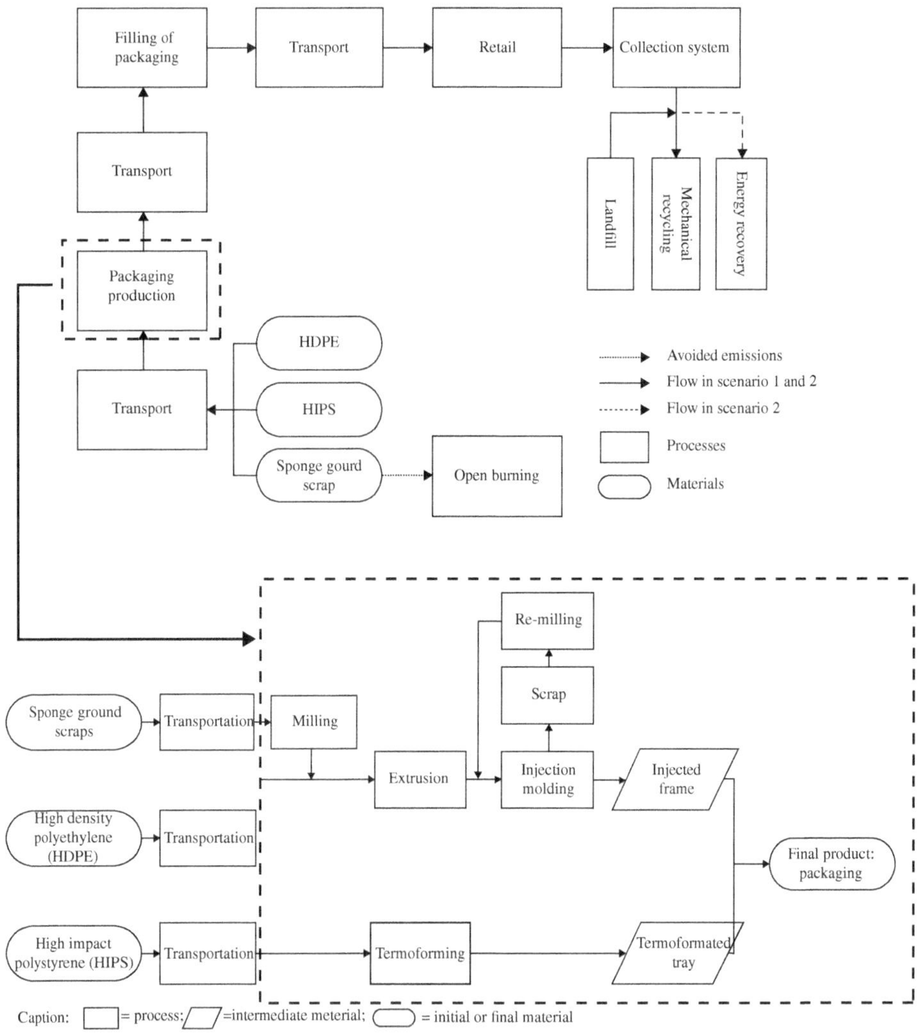

Figure 22.9 System boundary used for mango packaging and composite packaging production scheme as used in Saraiva et al. (2016) [29].

22.4 Life Cycle Inventory (LCI)

Life cycle inventory (LCI) is the second phase of LCA where usage of resources and materials and the consumption of fuels and energy need to be quantified. Transportation associated with a product in its life cycle of the analyzed FU must also be recorded [41, 44]. It is the crucial phase in LCA where data are collected to calculate material use, energy input, and pollutant emissions during the entire life cycle of a product or process. These data can be obtained from companies engaged in product fabrication and processing activities and from published databases [39]. Compiled data on emissions and resources consumed attributed to a specific product are recorded. There are several databases that have been developed

Table 22.3 *LCI developed based on system boundary of biodegradable composite films (Figure 22.7), based on orange peel-derived pectin jelly and corn starch.*

Product name	Sub-product name	Primary data (material/electricity consumption data based on experiments)	Secondary data (antibutional data EcoInvent v.3)
Biocomposite film	Pectin jelly	Orange peel	–
		HCl	Hydrogen and chloride reaction
		Ethanol	99.7% ethanol production data through ethylene hydration
		Water	Ultra-pure water
	Modified starch	Starch	Corn starch production
		Propylene oxide	90–95% propylene oxide production through ethylene hydration
		Sodium sulfate	Mannheim process of NaCl and H_2SO_4
		Sodium tripolyphosphate	Dehydrate process
		Sodium metaphosphate	Sodium tripolyphosphate data was used because of lack of data
		Water	Ultra-pure water
	LS	Na-MMT	Activated bentonite
	Glycerol	Glycerol	Glycerine
	Water	Water	Ultra-pure water

Source: From Gunkaya and Banar [30].

that allow users to insert new information or data specific to their products and processes. The most comprehensive international LCI database is Ecoinvent v3 (http://www.ecoinvent.org/database) [35]. From the list of LCI compiled, contributions to different environmental impact categories, such as global warming potential (GWP), acidification, eutrophication, or human toxicity, etc., will be generated. Other processes or activities outside the LCA boundary are ignored and thus environmental damage caused by them will not be counted. A set of LCI consists of inputs and outputs that represent the flow of material and energy used within the technical structure of the LCA, as well as emissions that are generated from the processes. All inputs and emissions of recorded data in LCI will to be translated into environmental impacts in the next phase. The LCI developed by Gunkaya and Banar [30] was built using primary data gathered from laboratory experiments supplemented by secondary data from the EcoInvent database (v3.0) of SimaPro 8.0.1 software. Table 22.3 illustrates these LCI data. They measured directly the electricity consumption data in the laboratory, which was normalized to the FU in the production of the biocomposites resin and the shaping of the resin into film. The electricity mix profile of Turkey, based on electricity generation resources of Turkey in 2012, was generated using EcoInvent v.2 data, i.e. 43.6% natural gas, 28.1% lignite coal, 24.2% hydraulic, 2.5% wind, and 0.5% geothermal; others were ignored.

It is interesting to mention that Bernstad and Saraiva's study [29] on the environmental performance of Brazilian mango fruits from producer to end-consumer originally consisted of a reusable frame, made from HDPE reinforced with natural sponge fiber residue and a high-impact polystyrene recyclable tray, and compared it with identical packaging constructed without natural fibers and a commercial cardboard packaging. The data gathered in the study from two packaging systems and the transportation involved are exhibited in Table 22.4.

All of the data or information access to process (and chemistry) details, data selection, and compilation demands an extensive amount of time and effort. The reliability of any LCA investigation and its outcomes are entirely dependent on the quality and adequacy of data used in constructing LCA models. Therefore, precaution and extra steps in data selection are important. This was supported by the work of Khoo et al. [41] who declared that selectivity, quality, and adequacy is the first challenge of applying LCA. Acknowledging these challenges, they proposed eight criteria to serve as guidance in data selection for the development of LCI (Figure 22.10). This is expected to help maintain confidence in LCA applications and its associated results.

22.5 Life Cycle Impact Assessment (LCIA)

Packaging materials go through different life cycles, which then generate certain environmental impacts at different stages of the supply chain and for different environmental indicators. Verghese et al. [10] summarized these impacts for different packaging materials, shown in Table 22.5.

The third phase of the LCA framework is the life cycle impact assessment, which focuses on evaluating and understanding the environmental impacts established by the LCI analysis. ISO 14040 outlines the impact assessment phase with the following components: (i) classification: assigning inventory results to impact categories; (ii) characterization: modeling inventory data within impact categories; the characterization process involves defining characterization factors to convert each pollutant emission into equivalent potentials represented by a reference substance (e.g. the CO_2 equivalent); (iii) weighting: aggregating inventory data in very specific cases to combine the impact categories into a single score [39]. Impact categories represent the negative effects to the environment due to substances emitted and resources used that caused the occurring damage. The impact categories are grouped into major ones and are termed "damage categories" as they represent the environmental sectors suffering the damage. Table 22.6 recaps the commonly used impact categories in LCA and their damage categories accordingly.

Inventory data exhibit what has changed environmentally caused from the production of a given product, but it must relate to the specific environmental objectives of LCA so that a correct interpretation can be made. Grant [24] explains this through Figure 22.11 on how inventory flows undergo a series of chemical, physical, and biological processes that link the original emission or resource use to the eventual environmental damage endpoint category. Indicators in LCA can be defined at the endpoint, but often indicators are defined at a midpoint somewhere along this environmental mechanism.

An extensive variety of LCIA methodologies has been developed, based on midpoint and/or endpoint indicators such as CML 2002, Eco-Indicator 99, EDIP (1997e2003), EPS2000, Impact 2002þ, LIME, LUCAS, ReCiPe, Swiss Ecoscarcity or Ecological scarcity, and TRACI MEEuP methodology [35]. Output emissions are organized into classes according to the effect they have on the environment and each emission might belong to more than one classification. In the case of chemicals emitted that contribute to the greenhouse effect or to ozone layer depletion, they are divided between the two classes,

Table 22.4 *Key data in Bernstad and Saraiva's [29] packaging system and transportation distances used in modeling the production of the composite box in internal consumption scenario (1) and export scenario (2).*

	Process	Value	Unit
Composites	Milling of luffa scrap	0.24	kWh kg^{-1}
	Extrusion of composite materials	4.88	kWh kg^{-1}
	Injection of composites material/HDPE	13.64	kWh kg^{-1}
	Thermoforming of HIPS tray	78.18	kWh kg^{-1}
	Energy content (HIPS)	36.2	MJ kg^{-1}
	Energy content (HDPE)	36.2	MJ kg^{-1}
	Energy content (luffa scrap)	15.9	MJ kg^{-1}
	Ash generation (secondary waste)	22.2	kg kg^{-1}
	Diesel use in plastic recycling	0.011	L kg^{-1}
	Electricity use in plastic recycling	0.398	kWh kg^{-1}
Cardboard	Pulp wood input	1.19	kg kg^{-1}
	Energy content (cardboard)	14.7	MJ kg^{-1}
	Ash generation (secondary waste)	130.2	kg kg^{-1}

Product	Transport process	km	Scenario 1	Scenario 2	Routes
Sponge-gourd	Production → Transformation	422	•	•	Minas Gerais →Rio De Janeiro
HDPE and HIPS	Production → Transformation	1632	•	•	Triunfo →Rio De Janeiro
Frame + tray	Transformation → Mango producer	1865	•	•	Rio De Janeiro → Juazeiro
Frame + tray + mangos	Mango production → Retail/Harbor (BR)	506	•		Juazeiro → Rio De Janeiro
Frame + tray + mangos	Harbor (BR) → Harbor (Europe)	8598		•	Rio De Janeiro → Rotterdam
Frame + tray + mangos	Harbor (Europe) → Retail (Europe)	1000		•	Assumed
Frame	Retail (Europe) → Harbor (Europe)	1000		•	Assumed
Frame	Harbor (Europe) → Harbor (BR)	8598		•	Rotterdam → Rio De Janeiro
Tray	Retail → Incineration/Landfill	100	•	•	Assumed
Tray	Retail → Recycler	100	•	•	Assumed
Frame	Harbor (BR) → Mango producer	506		•	Rio De Janeiro → Juazeiro
Pulp	Production → Transformation	1574	•		Porto Seguro → Pacajus
Cardboard box	Transformation → Mango producer	777	•		Pacajus → Juazeiro
Cardboard box + mangos	Mango production → Retail/Harbor (BR)	506	•		Juazeiro → Rio De Janeiro
Cardboard box + mangos	Harbor (BR) → Harbor (Europe)	8598		•	Rio De Janeiro → Rotterdam
Cardboard box + mangos	Harbor (Europe) → Retail (Europe)	1000		•	Assumed
Cardboard box	Retail → Incineration/Landfill	100	•	•	Assumed
Cardboard box	Retail → Recycler	100	•	•	Assumed

Criterion 1 : Information Source

- Inventory gathering and data compilation where primary or first-hand industrial data would serve as the most reliable source and quality of information to support any LCA objectives and its outcomes.

Criterion 2 : System boundary

- System boundary defined including the scope of data gathered will influence its overall results. Inventory data is only considered for processes that are within the technical periphery of the LCA study or system boundary

Criterion 3 : Process data

- In the case where industrial or process-based data is unavailable from LCA datasets, technical information is needed to generate the input–output details required for a good, high quality LCA investigation. Technological variations can alter the input values used in LCA.

Criterion 4 : Data completeness/ Comparability

- In preparation of LCI for each different synthesis route, the same types of pollutants should be presented for comparison.

Criterion 5 : Representativeness

- For the case where an exclusively designed – or proprietary – process is employed to produce a product/chemical, the "import" of LCI from other types of methods or processes for making the same chemical should not be applied. The outcome of making use of unrelated, irrelevant data for any LCA might lead to erroneous conclusions.

Criterion 6 : Scale

- Data generated from experimental records should not be directly applied in the LCA of industrial scale up considerations. The reason lies in un-optimized processes carried out at laboratory scales (discovery stage) where manufacturing factors such as environmental regulations, energy use, waste management, safety, scalability are not taken into consideration

Criterion 7 : Location

- Location-dependent LCI are highly recommended for such cases where: biomass resources are utilized in the LCA system, and these resources are produced from different a geographical area or region

Criterion 8 : Transparency, Traceability, Reproducibility

- "Transparency" of data requires assumptions or calculations used in obtaining any values to be properly stated and documented. Recorded details of how data is obtained, methods of calculations, and any set of assumptions are all documented for enough data transparency. All information should be "traceable" to its source; and "reproducibility" of data is highly dependent on appropriate documentation procedures.

Figure 22.10 *Eight criteria of data selection for the development of LCI as proposed by Khoo et al. [41].*

Table 22.5 *Different packaging materials and triggering factors of environmental impacts [10].*

Packaging material	Environmental impact categories				
	Climate change	Energy	Land use	Eutrophication	Water use
Aluminum	Electricity for smelting	Electricity for smelting	Mining	Electricity for smelting	Electricity for smelting
Steel	Pig iron production	Coking coal in pig iron production	Coking and storage	Pig iron production	Pig iron production
Glass	Electricity for melting	Natural gas in melting	Secondary and tertiary packaging	Electricity for melting	Electricity for melting
Paper and board	Electricity for pulping	Growing and harvesting of trees	Growing and harvesting of trees	Electricity for pulping	Pulping/Electricity for pulping
Non-renewable thermoplastics	Polymerization	Raw material extraction	Electricity	Polymerization	Electricity
Renewable thermoplastics	Polymer processing/growing and harvesting crop	Growing and harvesting crop	Growing and harvesting of crop	Growing and harvesting crop	Growing and harvesting crop

Table 22.6 LCA common impact and damage categories.

Damage category	Impact category	Unit	Description of impact
Human health	Carcinogen/ non-carcinogen	kg1,4-DCB-eq	Human toxicity (contributes to toxic conditions to human)
	Respiratory organ-ics/inorganics	kg1,4-DCB-eq	Human toxicity (contributes to toxic conditions to human)
	Ozone layer depletion	kg CFC-11-eq	Contributes to depletion of stratospheric ozone
Eco-system quality	Aquatic eco-toxicity	kg1,4-DCB-eq	Contributes to toxic conditions to plants and animals
	Terrestrial eco-toxicity/ acidification/ nitrification	kg1,4-DCB-eq	Contributes to toxic conditions to plants and animals
	Aquatic acidification	kg SO_2-eq	Contributes to acid deposition in water
	Aquatic eutrophication	kg N-eq	Provision of nutrients contributes to biological oxygen demand
Climate change	Global warming	Kg CO_2 eq	Contributes to atmospheric absorption of infrared radiation
Resources use	Abiotic/ non-renewable energy	Water – m^3; land – m^2a	Contributes to depletion of non-renewable resources
	Biotic/mineral extraction	MJ surplus, kg oil-eq	Contributes to depletion of renewable resources

Source: Extrapolated from references [34] and [45].

whereas nitrogen oxides emissions may simultaneously belong to several classes, such as aquatic toxicity, acid rain, and eutrophication. Assigning and converting LCI results into numerical indicator results is called characterization [36]. ISO/TR 14047 (2003) defines these environmental impact classification factors and each is briefly explained in Table 22.7.

22.6 Life Cycle Results Interpretation

Interpretation of results is the fourth phase of LCA. In this phase, the findings from inventory analysis and impact assessment are combined to reach conclusions and formulate recommendations. A sensitivity analysis can be carried out to understand how model parameters influence the LCA results or how critical uncertain parameters help reduce environmental impacts [39]. The results can be presented in impact categories or grouped together into damage categories. Normalization can be performed by dividing the impact category

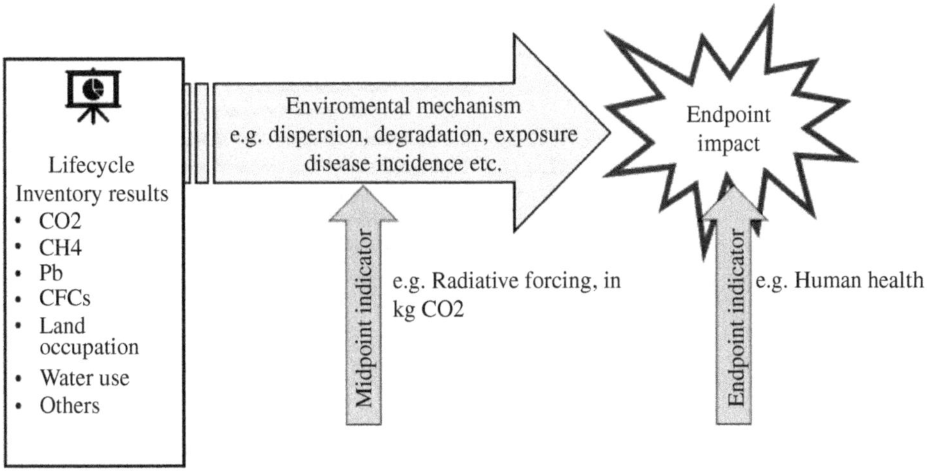

Figure 22.11 *Relationship of LCI to midpoint and endpoint indicators [24].*

Table 22.7 *ISO/TR 14047 (2003) environmental impact classification factors [46].*

Environmental impact classification factor	Description
Acidification potential (AP)	Consequence of acids being emitted to the atmosphere and subsequently deposited in surface soils and waters. AP classification factors are mainly based on the contributions of SO_2, NO_x, HCl, NH_3, and HF and expressed as SO_2 equivalent.
Aquatic toxicity potential	Calculated based on the maximum tolerable concentrations of different toxic substances in water by aquatic organisms.
Human toxicity potential	Calculated in kg and considers the releases of materials toxic to humans in air (A), water (W), and soil (S) based on human toxicological factors.
Eutrophication potential (EP)	Pollution state of aquatic ecosystems where the overfertilization of water and soil has turned into an increased growth of biomass. EP is calculated in kg based on a weighted sum of the emissions of nitrogen and phosphorus derivatives. The classification factors for EP are expressed as phosphates equivalents.
Global warming potential (GWP)	Calculated for each different greenhouse gas, including CO_2, N_2O, CH_4, and volatile organic compounds (VOCs). GWP is expressed as a CO_2 equivalent.
Nonrenewable/abiotic resource depletion potential	Calculated for fossil fuels, metals, and minerals by dividing the quantity of resource used by the estimated total world reserves of that resource.
Ozone depletion potential	The potential of depletion of the ozone layer due to the emissions of chlorofluorocarbon compounds and other halogenated hydrocarbons.
Photochemical oxidants creation potential	Photochemical smog is caused by the degradation of VOCs and nitrogen and is expressed in kg of ethylene.

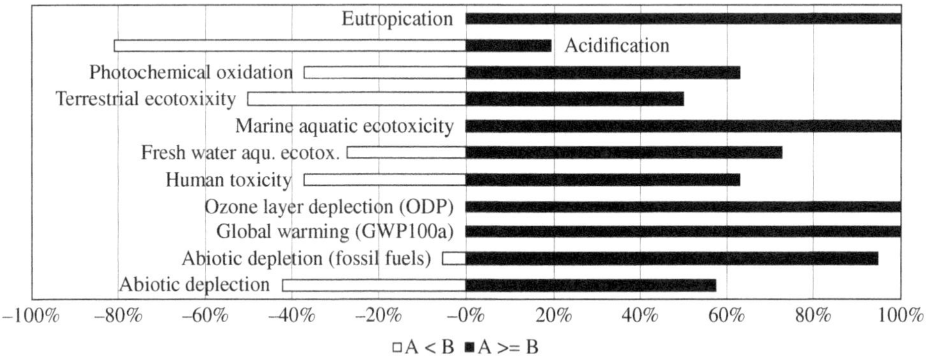

Figure 22.12 *Monte-Carlo characterized LCIA comparison between the biocomposite film and LDPE film [30].*

scores by the average person's annual contribution, allowing for the combination of categories, whose scores now have no units. Since some environmental impacts can be considered more important than others, a weight can also be assigned to the normalized impact score [34].

Ingrao et al. [36] performed an ALCA to assess both environmental impacts and improvement potentials in the life cycle of PLA-based food packaging products. The FU chosen was 1 kg of trays where the maximum capacity of each tray was 800 cm^3 (11.36 g). The study highlighted that the total damage is equal to 1.85 mpt and is mainly from the production (for almost 49.7%) and transport (for 25.43%) of the PLA granules, the electricity consumption for their processing (for 12.2%), and for 5.94% to the delivery of the produced trays. From the findings, the studies concluded that the most impacting phase from production of the required amount of PLA granules was largely due to the corn-cultivation phase. Other significant impacts come from the energy consumed in the processing of those granules, but transportation was the major issue because of the long distances traveled, which is considered in the study from America to Italy.

Gunkaya and Banar [30] applied the Monte-Carlo method, a built-in function in SimaPro 8.0.1 software, and simulated results of a characterized LCIA comparison between the biocomposite film and the LDPE film. The results were obtained on the uncertainty assessment in terms of the probability of the biocomposite film (A) having a lower impact than the LDPE film (B) on the left and on the right, and the probability of the LDPE film having a lower impact than the biocomposite film, are depicted in Figure 22.12. The percentile distribution of the impacts into processes in the life cycle of the orange peel-derived pectin jelly–corn starch biocomposite film exhibited in Figure 22.13 shows the production processes (modified starch, pectin jelly, glycerol production) and shaping processes (extrusion and film forming). The impact of the biocomposite film was mainly contributed by the electricity consumption for shaping, modified starch production, and pectin jelly production [50]. They concluded that revision is needed for the biocomposite resin and shaping stages so that less usage of additives and energy could be achieved.

Extraction of key findings on the impacts from other recent LCA studies on bio-based packaging are organized in Table 22.8.

Table 22.8 Key impact findings in recent LCA studies on biobased packaging.

Biobased materials	System boundary	Packaging application	Key findings on impacts	Reference
PLA/cassava starch blend	Cradle-to-production gate	Single-use packaging box	• Almost all impact categories of PLA/starch box are 0.14–3.01 times **lower** than PET box, except EU impact.	[47]
Starch conversion degradable plastics, corrugated paper, and kraft paper	Cradle-to-grave	Cushion packaging material	• Biomass degradable packaging material has the **smallest impact** on human health, ecological environmental quality, and resource consumption.	[48]
Polyethylene/natural fiber-composite and cardboard material		Mango box packaging	• Incineration with energy recovery makes the 100% biocardboard box preferable to the mango box. • Environmental impacts from cardboard packaging were between 2% and 14% of the corresponding ones from the composite packaging. • The higher **electricity input**, in transformation, and higher **fuel consumption in transportation** produce higher contributions to the impact categories of *Particulate 1 matter, Acidification, Eutrophication, and Photochemical ozone formation*.	[49]
Orange peel-derived pectin jelly and corn starch	Cradle-to-gate	Packaging film	• 63% probability that the biocomposite film has a **higher impact than the LDPE film.** • 50, 58, and 73% for terrestrial ecotoxicity, element-based abiotic depletion, and freshwater aquatic ecotoxicity, respectively. Other indicators show no/little uncertainty and are **higher for the biocomposite** film except for acidification.	[30]

(*continued*)

Table 22.8 *(continued)*

Biobased materials	System boundary	Packaging application	Key findings on impacts	Reference
PLA/nanoclay-based active packaging	Cradle-to-grave	Ready-to-eat fresh cut fruits packaging	• PLA-based packaging has a **higher energy demand** due to the manufacturing and mixing of the additives. • The PLA-based packaging with nanowhisker additives production requires **more energy and materials** and so possess the largest impacts in all categories compared to the other PLA-based solutions. • Although PLA-based packaging with additives require more energy for their production, LCA results clearly showed that PLA-based packaging with nanoclay and surfactant is very competitive with PET for environmental impact.	[40]
Foamy polylactic acid (PLA)	Cradle-to-grave	Trays for fresh-food packaging	• The most impacting phase is represented by the **production of the required amount of PLA granules**, mainly due to the corn-cultivation phase. • Other significant impacts come from the **energy consumed in the processing of those granules** but, mostly, from their **transport** to the tray manufacturing factory due to the long distances traveled (from America to Italy) and the means utilized. • The **transport issue causes high impacts** in ways to worsen the life-cycle environmental sustainability of the PLA-trays	[36]

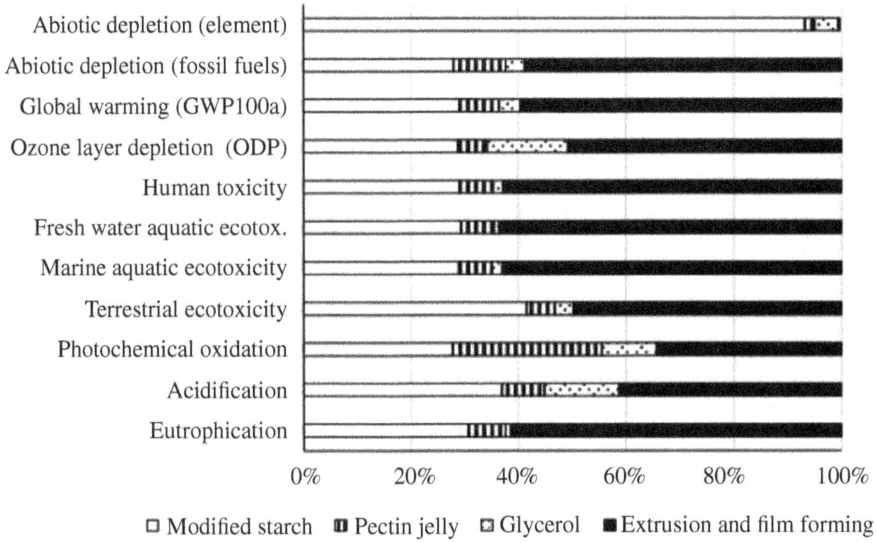

Figure 22.13 *Percentile distribution of the impacts into processes in the life cycle of the orange peel-derived pectin jelly–corn starch biocomposite film [30].*

22.7 Conclusions

Bio-based compostable packaging produced from renewable natural resources where its raw materials are from crops has always been expected to have less environmental impacts and could resolve the waste management issues. Some studies have proven that the whole life cycle of bio-based packaging material has a lower impact on the environment and possesses a better environmentally friendly property. Studies that are employing LCA used to compare environmental performance of newly developed bio-based composites materials with conventional plastic materials. However, general comparisons and conclusions of bio-derived materials are always better over the full life cycle, which may be misleading.

The bio-based composites manufacturing stage in studies shows that the highest impact comes from the process basis because processing of biocomposite film is at present not optimized from an industrial perspective. Nevertheless, there is currently continuous development to improve technologies and EOL treatment processes, including recycling that would permit bio-based materials to become the perfect alternative materials in the packaging industry. A packaging sustainability business strategy may be driven by corporate and brand positioning, supply chain requirements, solid waste management, resource efficiency, and climate, which motivate the selection of certain materials. It is important to note that each bio-based material used for packaging application possesses its individual properties and life cycle, which influence the environmental impacts generated. Different stages of the supply chain may produce different environmental impact indicators. Transport activities at every stage of the product are also understood to contribute a high impact toward the environment. A packaging design strategist must understand and always be informed on the latest establishment of the whole complex activities throughout the life cycle. In a nutshell, embedding an assessment of the life cycle of materials during the packaging product design process is helpful in estimating the environmental impacts and could identify specific activities during the whole life cycle of the packaging product that contribute more impact than others.

Acknowledgments

The authors wish to express the highest appreciation to the Ministry of Education Malaysia through Universiti Putra Malaysia Grant Scheme HiCOE (6369107) for the financial support. The first author is very grateful for the study sponsorship from Public Service Department (JPA), Malaysia.

References

1. Emblem A, Emblem H. *Packaging Technology*. Woodhead Publishing Limited; 2012 [cited 2017 Nov. 1]. 411–440 p. Available at: http://www.sciencedirect.com/science/book/9781845696658.
2. van Rompay TJL, Deterink F, Fenko A. Healthy package, healthy product? Effects of packaging design as a function of purchase setting. *Food Qual. Prefer.* 2016; **53**: 84–9. Available at: http://dx.doi.org/10.1016/j.foodqual.2016.06.001.
3. Fernqvist F, Olsson A, Spendrup S. What's in it for me? Food packaging and consumer responses, a focus group study. *Br. Food J.* 2015; **117**(3): 1122–35. Available at: http://www.emeraldinsight.com/doi/10.1108/BFJ-08-2013-0224.
4. Marsh K, Bugusu B. Food packaging? Roles, materials, and environmental issues. *J. Food Sci.* 2007; **72**(3): R39–55. Available at: http://doi.wiley.com/10.1111/j.1750-3841.2007.00301.x.
5. DeGruson ML. Biobased polymer packaging. *Reference Module in Food Science*. Elsevier. 2016. Available at: http://linkinghub.elsevier.com/retrieve/pii/B9780081005965033734.
6. Verghese K, Crossin E, Jollands M. Packaging materials. In: *Packaging for Sustainability*. London: Springer London; 2012. p. 211–50. Available from: http://link.springer.com/10.1007/978-0-85729-988-8_6
7. World Economic Forum, Ellen MacArthur Foundation and McKinsey & Company, The New Plastics Economy – Rethinking the Future of Plastics (2016), Available at: http://www.ellenmacarthurfoundation.org/publications).
8. Nazrin A, Sapuan SM, Zuhri MYM, Ilyas RA, Syafiq R, Sherwani SFK. Nanocellulose reinforced thermoplastic starch (TPS), polylactic acid (PLA), and polybutylene succinate (PBS) for food packaging applications. *Front. Chem. Front. Chem.* **8**:213. Available at: https://doi.org/10.3389/fchem.2020.00213.
9. Saba N, Jawaid M, Al-Othman O. An overview on Polylactic acid, its cellulosic composites and applications. *Curr. Org. Synth.* 2017 [cited 2017 Oct. 12]; **14**(2): 156–70. Available at: http://www.eurekaselect.com/openurl/content.php?genre=article&issn=1570-1794&volume=14&issue=2&spage=156.
10. Verghese K, Lewis H, Fitzpatrick L. editors. Packaging for sustainability. *Packaging for Sustainability*. London: Springer, London; 2012.1–384 p. Available at: http://link.springer.com/10.1007/978-0-85729-988-8.
11. Emblem A. Packaging and society. *Packag. Technol.* 2012; 3–9. Available at: http://www.sciencedirect.com/science/article/pii/B9781845696658500019
12. Siracusa V, Lotti N. Biobased plastics for food packaging. *Reference Module in Food Science*. Elsevier; 2018. Available at: http://dx.doi.org/10.1016/B978-0-08-100596-5.22413-X.
13. Fortunati E, Mazzaglia A, Balestra GM. Sustainable control strategies for plant protection and food packaging sectors by natural substances and novel nanotechnological approaches Fortunati. *Journal of the Science of Food and Agriculture.* 2019; **99**(3): 986–1000.

14. Abral H, Ariksa J, Mahardika M, Handayani D, Aminah I, Sandrawati N, et al. Transparent and antimicrobial cellulose film from ginger nanofiber. *Food Hydrocoll.*. 2020;
15. Abral H, Basri A, Muhammad F, Fernando Y, Hafizulhaq F, Mahardika M, et al. A simple method for improving the properties of the sago starch films prepared by using ultrasonication treatment. *Food Hydrocoll.*. 2019;
16. Abral H, Ariksa J, Mahardika M, Handayani D, Aminah I, Sandrawati N, et al. Highly transparent and antimicrobial PVA based bionanocomposites reinforced by ginger nanofiber. *Polym. Test.* 2019;
17. Kuruvila ME, Sudhakaran N, Madhu V. Starch-based plastic : strength, degradation in soil and water absorbance. *International Journal of Advanced Scientific Research and Management.* 2018; **3**(11): 155–8.
18. Sapuan SM, Ilyas RA. *Characterization of sugar palm Nanocellulose and its potential for reinforcement with a starch-based composite.* Sugar Palm Biofibers, Biopolym Biocomposites. 2018: 189–220.
19. Sirviö JA, Visanko M, Ukkola J, Liimatainen H. Effect of plasticizers on the mechanical and thermomechanical properties of cellulose-based biocomposite films. *Ind. Crop. Prod.* 2018; **122** (June): 513–21.
20. Cazón P, Vázquez M, Velazquez G. Cellulose-glycerol-polyvinyl alcohol composite films for food packaging: evaluation of water adsorption, mechanical properties, light-barrier properties and transparency. *Carbohydr. Polym.* 2018; **195** (April): 432–43. Available at: https://doi.org/10.1016/j.carbpol .2018.04.120.
21. Qian S, Zhang H, Yao W, Sheng K. Effects of bamboo cellulose nanowhisker content on the morphology, crystallization, mechanical, and thermal properties of PLA matrix biocomposites. *Compos. Part B Eng.* 2018; **133**: 203–9.
22. Chaitanya S, Singh I. Ecofriendly treatment of aloe vera fibers for PLA based green composites. *Int. J. Precis. Eng. Manuf. – Green. Technol.* 2018; **5**(1): 143–50.
23. Brundage MP, Bernstein WZ, Hoffenson S, Chang Q, Nishi H, Kliks T, et al. Analyzing environmental sustainability methods for use earlier in the product lifecycle. Vol. **187**, *J. Clean. Prod.* 2018. p. 877–92.
24. Grant T. Life cycle assessment (LCA) and degradable polymers. *ACS Symp. Ser.* 2012;
25. Bohlmann GM. Biodegradable packaging life-cycle assessment. *Environ. Prog.* 2004;
26. de Koeijer B, Wever R, Henseler J. Realizing product-packaging combinations in circular systems: shaping the research agenda. *Packag. Technol. Sci.* 2017 [cited 2018 Mar. 21]; **30**(8): 443–60. Available at: https://liu.diva-portal.org/smash/get/diva2:1130181/FULLTEXT01.pdf.
27. Ingrao C, Lo Giudice A, Bacenetti J, Mousavi Khaneghah A, Sant'Ana de SA , Rana R, et al. Foamy polystyrene trays for fresh-meat packaging: life-cycle inventory data collection and environmental impact assessment. *Food Res. Int.* 2015; 76: 418–26. Available at: http://dx.doi.org/10.1016/j.foodres .2015.07.028.
28. Accorsi R, Versari L, Manzini R. Glass vs. plastic: life cycle assessment of extra-virgin olive oil bottles across global supply chains. *Sustainability.* 2015
29. Bernstad Saraiva A, Pacheco EBAV, Gomes GM, Visconte LLY, Bernardo CA, Simões CL, et al. Comparative lifecycle assessment of mango packaging made from a fiber-composite and from cardboard polyethylene/naturmaterial. *J. Clean. Prod.* 2016; **139**: 1168–80. Available at: http://dx.doi.org/ 10.1016/j.jclepro.2016.08.135.
30. Günkaya Z, Banar M. An environmental comparison of biocomposite film based on orange peel-derived pectin jelly-corn starch and LDPE film: LCA and biodegradability. *Int. J. Life Cycle Assess.* 2016; **21**(4): 465–75. Available at: http://link.springer.com/10.1007/s11367-016-1042-8.
31. Humbert S, Rossi V, Margni M, Jolliet O, Loerincik Y. Life cycle assessment of two baby food packaging alternatives: glass jars vs. plastic pots. *Int. J. Life Cycle Assess.* 2009; **14**(2): 95–106.
32. Madival S, Auras R, Singh SP, Narayan R. Assessment of the environmental profile of PLA, PET and PS clamshell containers using LCA methodology. *J. Clean. Prod.* 2009; **17** (13): 1183–94. Available at: https://linkinghub.elsevier.com/retrieve/pii/S0959652609000936.

33. Nikoliae S, Kiss F, Mladenoviae V, Bukurov M, Stankoviae J. Corn-based Polylactide vs. PET bottles – cradle-to-gate LCA and implications. *Mater. Plast.* 2015; **52**(4): 517–21.
34. Mansor M., Salit M., Zainudin E., Aziz N., Ariff H. (2015) Life cycle assessment of natural fiber polymer composites. In: Hakeem K., Jawaid M. Y., Alothman O. (eds), *Agricultural Biomass Based Potential Materials*. Springer, Cham.
35. La Rosa AD, Cicala G. LCA of fibre-reinforced composites. In: *Handbook of Life Cycle Assessment (LCA) of Textiles and Clothing*. Elsevier; 2015. p. 301–23. Available at: https://linkinghub.elsevier.com/retrieve/pii/B9780081001691000149.
36. Ingrao C, Gigli M, Siracusa V. An attributional life cycle assessment application experience to highlight environmental hotspots in the production of foamy polylactic acid trays for fresh-food packaging usage. *J. Clean. Prod.* 2017; **150**: 93–103. Available at: http://dx.doi.org/10.1016/j.jclepro.2017.03.007.
37. Venkatachalam V, Spierling S, Horn R, Endres HJ. LCA and Eco-design: Consequential and Attributional Approaches for Bio-based Plastics. *Procedia CIRP*. 2018; **69** (May): 579–84. doi: 10.1016/j.procir.2017.11.086.
38. Alvarenga RA, Dewulf J, De Meester S, Wathelet A, Villers J, Thommeret R, et al. Life cycle assessment of bioethanol-based PVC. *Biofuels Bioprod. Biorefin.* 2013; **7**(4): 396–405. Available at: http://doi.wiley.com/10.1002/bbb.1398.
39. Parameswaranpillai J, Vijayan D. Life cycle assessment (LCA) of epoxy-based materials. *Micro and Nanostructured Epoxy/Rubber Blends*. Wiley-VCH Verlag GmbH & Co.; 2014.421–432. p.
40. Lorite GS, Rocha JM, Miilumäki N, Saavalainen P, Selkälä T, Morales-Cid G, et al. Evaluation of physicochemical/microbial properties and life cycle assessment (LCA) of PLA-based nanocomposite active packaging. *LWT-Food Sci. Technol.* 2017; **75**: 305–15. Available at: http://dx.doi.org/10.1016/j.lwt.2016.09.004.
41. Khoo HH, Isoni V, Sharratt PN. LCI data selection criteria for a multidisciplinary research team: LCA applied to solvents and chemicals. *Sustain. Prod. Consum.* 2018; **16** (July): 68–87. Available at: https://doi.org/10.1016/j.spc.2018.06.002.
42. Gbededo MA, Liyanage K, Garza-Reyes JA. Towards a Life Cycle Sustainability Analysis: A systematic review of approaches to sustainable manufacturing. *J. Clean. Prod.* 2018; **184**: 1002–15. Available at: https://doi.org/10.1016/j.jclepro.2018.02.310.
43. Ferreira Saraiva LE, Naponucena L de OM, da Silva Santos V, Silva RPD, de Souza CO, Evelyn Gomes L.S.I, et al. Development and application of edible film of active potato starch to extend mini panettone shelf life. *LWT-Food Sci. Technol.* 2016.
44. Ingrao C, Tricase C, Cholewa-Wójcik A, Kawecka A, Rana R, Siracusa V. Polylactic acid trays for fresh-food packaging: a carbon footprint assessment. *Sci. Total Environ.* 2015; **537**: 385–98. Available at: http://dx.doi.org/10.1016/j.scitotenv.2015.08.023.
45. Korol J, Burchart-Korol D, Pichlak M. Expansion of environmental impact assessment for eco-efficiency evaluation of biocomposites for industrial application. *J. Clean. Prod.* 2016; **113**: 144–52. Available at: http://dx.doi.org/10.1016/j.jclepro.2015.11.101.
46. La Rosa AD, Cicala G. LCA of fibre-reinforced composites. In: *Handbook of Life Cycle Assessment (LCA) of Textiles and Clothing*. 2015. p. 301–23.
47. Suwanmanee U, Leejarkpai T, Mungcharoen T. Assessment the environmental impacts of polylactic acid/starch and polyethylene terephthalate boxes using life cycle assessment methodology: cradle to waste treatments. *J. Biobased Mater. Bioenergy* **7**, 259–266, 2013. Available at: http://openurl.ingenta.com/content/xref?genre=article&issn=1556-6560&volume=7&issue=2&spage=259.
48. An XH, Liu P, Meng Q, Su CG, Zhao S. Research on life cycle environmental property of biomass biodegradable packaging material. *Appl. Mech. Mater.* 2014.

49. Bernstad Saraiva A, Pacheco EBAV, Gomes GM, Visconte LLY, Bernardo CA, Simões CL, et al. Comparative lifecycle assessment of mango packaging made from a polyethylene/natural fiber-composite and from cardboard material *J. Clean. Prod.* 2016.
50. Ilyas RA, Sapuan SM, Atikah MSN, Asyraf MRM, Ayu Rafiqah S, Aisyah HA, Mohd Nurazzi N, Norrahim MNF. Effect of hydrolysis time on the morphological, physical, chemical, and thermal behavior of sugar palm nanocrystalline cellulose (Arenga pinnata (Wurmb.) Merr). *Textile Research Journal.* 2020. In print. doi: 10.1177/0040517520932393.

23

Reuse and Recycle of Biobased Packaging Products

R.A. Ilyas[1,2], S.M. Sapuan[3,4], F.A. Sabaruddin[2], M.S.N. Atikah[5], Rushdan Ibrahim[6], M.R.M. Asyraf[7], M.R.M. Huzaifah[4], S.O.A. SaifulAzry[4], and Z.M.A. Ainun[8]

[1] *School of Chemical and Energy, Faculty of Engineering, Universiti Teknologi Malaysia,81310 UTM Johor Bahru, Johor, Malaysia*

[2] *Centre for Advanced Composite Materials, Universiti Teknologi Malaysia,81310 UTM Johor Bahru, Johor, Malaysia*

[3] *Advanced Engineering Materials and Composites Research Centre (AEMC), Department of Mechanical and Manufacturing Engineering, Faculty of Engineering, Universiti Putra Malaysia (UPM), Serdang, Selangor, 43400, Malaysia*

[4] *Laboratory of Biocomposite Technology, Institute of Tropical Forest and Forest Products (INTROP), Universiti Putra Malaysia (UPM), Serdang, Selangor, 43400, Malaysia*

[5] *Department of Chemical and Environmental Engineering, Universiti Putra Malaysia (UPM), Serdang, Selangor, 43400, Malaysia*

[6] *Pulp and Paper Branch, Forest Research Institute Malaysia, Kepong, Selangor, 52109, Malaysia*

[7] *Department of Aerospace Engineering, Faculty of Engineering, Universiti Putra Malaysia (UPM), Serdang, Selangor, 43400, Malaysia*

[8] *Program of Pulp and Paper & Pollution Control, Laboratory of Biopolymer and Derivatives, Institute of Tropical Forestry and Forest Products, Universiti Putra Malaysia (UPM), Serdang, Selangor, 43400, Malaysia*

23.1 Introduction

Nowadays, our concern toward the environment is increasing due to the massive amount of pollution in the world [1–9]. Many nations are struggling to develop their countries without taking into consideration the effect of pollution, causing an increase in the numbers

Bio-Based Packaging: Material, Environmental and Economic Aspects, First Edition. Edited by S.M. Sapuan and R.A. Ilyas

Figure 23.1 Garbage on Laysan Island in the Hawaiian Island.

of factories and the application of harmful chemicals and materials in the production process [10].

In the twentieth century, people have used plastic bags for daily needs, mainly for shopping purposes, because they are cheap and convenient [11]. However, the excessive use of plastic bags has contributed a lot of damage to our Mother Earth [12–25]. The hazardous chemical contents in plastic can lead to negative impacts on the environment as the disposal of nonbiodegradable plastic waste can affect the environment, especially toward humans, animals, agriculture, and marine life. For humans, its application may lead to cancerous diseases due to carcinogenic substances. Meanwhile, the effect on marine life includes eventual death due to blockage of digestive tracts of marine animals as they confuse the bags for food and are drowned by plastic bag entanglement [10, 26, 27]. The significant impact of plastic bags toward agriculture includes a decline in soil fertility and nitrogen fixation, including a reduction in crop yield.

Plastic bags are nonbiodegradable and can take up to 300–1000 years in the soil to photodegrade, which means they will break down to small toxics bits that are extremely harmful toward the surrounding environment [10, 26, 28]. By the time it takes for them to photodegrade, our planet will continuously be contaminated, in which most of the toxic chemical will end up in the soil and sea. Moreover, low education and awareness by consumers on the way to reuse and recycle plastics bags have made the situation worse. Figure 23.1 shows a situation on Laysan Island in the Hawaiian Islands that is full of garbage. If this situation keeps occurring, it will harm the environment and marine life.

Recently, concern about climate issues and environmental impact toward the environment has led to extensive research to reduce the effect of nonbiodegradable plastics [29, 30]. Researchers have suggested substituting plastic bags with bio-based materials that can be used for packaging purposes. Bio-based material is made up of living things or is derived from plants and other renewable sources, including agricultural, marine, or forestry materials [17, 31–38]. The characteristic of bio-based plastic bags is less harmful, especially to

Table 23.1 Examples of polymers of each category.

Category	Examples
Category 1 (Natural polymers)	• Starch • Cellulose • Lignin • Gluten
Category 2 (Polymers from bio-based building blocks)	Polylactic acid (PLA) Polypeptides
Category 3 (Polymers from microorganisms)	• Polyhydroxyalkonoates (PHA)

Figure 23.2 *Examples of products that use bio-based materials for packaging.*

soil and the environment. Even though bio-based plastics still contain some toxic ingredients, the alternative of replacing conventional plastic bags with biodegradable plastic bags is most welcome [10, 39].

The application of alternative plastic bags should be less harmful to the environment, especially agriculture. Intensification of public awareness regarding climate change and environmental impacts has led biodegradable plastic manufacturers to respond to international pressure. The decreasing application of conventional plastics has increased the potential of a new bio-based plastics generation for a much broader global usage [10].

Bio-based polymers can be divided into three parts based on their sources and production [40]. The first category is natural polymer, which is directly extracted from natural polymers and biomass, the second category is polymers that are made of bio-based building blocks, and the third category is polymers that are made from microorganisms. Table 23.1 shows the examples of polymers for each category. Each category has different properties that are suitable for different types of applications.

Bio-based material is widely used for packaging, including for food. Figure 23.2 shows some products that use bio-based material for packaging. These products are widely used for packaging because of their properties, which include biodegradable, biocompatible, flexible, and tough performance. To make sure people know what kind of packaging is used, various labels have been developed. Figure 23.3 shows the labels that were used to recognize bio-based plastic packaging. This effort has been given tremendous support from various agencies to implement 100% bio-based packaging. Therefore, this chapter will draw the attention of researchers toward the reuse and recycling of bio-based packaging products.

NEN bio-based content
The bio-based label by the Netherlands Standardization Institute (NEN) is based on the European standard EN 16785-1, which enables the independent assessment of claims about the bio-based content of products. The certification system applies to basic materials as well as intermediate and finished products.

DIN CERTCO bio-based certification
DIN CERTCO's bio-based certification system is based on standard ASTM 6866, CEN/TS 16137 and/or the ISO 16620. The bio-based carbon content in a product is measured using the ^{14}C method and expressed as a percentage of the total carbon in the product.

OK biobased
OK biobased is a certification system that verifies the material composition of a product. The Vinçotte labelling and certification scheme is based on the ^{14}C method and measures the bio-based carbon content as a percentage of the total carbon contained in the product/material. This method is backed up by the international standards CEN/TS 16137 and ASTM 6866.

BioPreferred label
The United States Department of Agriculture's certified bio-based product label assures the consumer that a product or package contains a verified amount of renewable biological ingredients. The label is linked to the US standard ASTM 6866, which is the US equivalent of CEN/TS 16137:2011 Plastics – Determination of bio based carbon content (see Annex 1).

Seedling
The "Seedling" industrial compostability label is connected to the standard EN 13432/EN 14995 and a certification process managed by the independent certification bodies DIN CERTCO and Vinçotte. European Bioplastics is the trademark owner. Biodegradable products that bear the "Seedling" label fulfil all of the criteria stipulated in EN 13432 in relation to industrial compostability.

DIN-Geprüft industrial compostable and OK compost
Both labels guarantee compliance with EN 13432, the European standard for compostability in industrial composting installations.

OK compost HOME
Vinçotte introduced the OK compost HOME label to certify compostability at ambient temperatures under home composting conditions, i.e. a minimum degradation of 90% in 12 months.

This label does not refer to an international standard, but is instead based on a certification scheme defined by the label provider Vinçotte.

DIN-Geprüft Home Compostable
The internationally recognised "DIN-Geprüft" mark from DIN CERTCO can also be awarded to products, which can be composted in the garden (home composting). Certification is possible on the basis of the French standard NF T51-800 and the Australian standard AS 5810.

Figure 23.3 *Labels for bio-packaging products [41].*

23.2 Waste Management Efficiency for Bioplastics

Bioplastics are good for the future and end-of-life options, including reuse, organic recycling, energy recovery, and mechanical recycling. The good part of the bioplastic volume produced today is that it can be recycled using conventional components, whereby separate recycling streams for some material types exist. In addition, by using compostable plastic and biodegradable products, such as bags for food packaging, can strengthen organic recycling using waste management methods and assist in improving waste management.

The way bioplastics waste is recycled depends on the kind of bioplastic material used and the product, its inherent quantities, and available recovery systems. Biodegradable plastics are frequently known as a possible solution to this problem because they can be decomposed by microorganisms without producing harmful residues or emissions during decomposition. On the other hand, the process of composting is dependent on some conditions. Products that are good for industrial composting are suitable for the conditions in a composting factory, but not for wildlife. Figure 23.4 illustrates the way of discarding bioplastic products and further simplifies understanding of the statement.

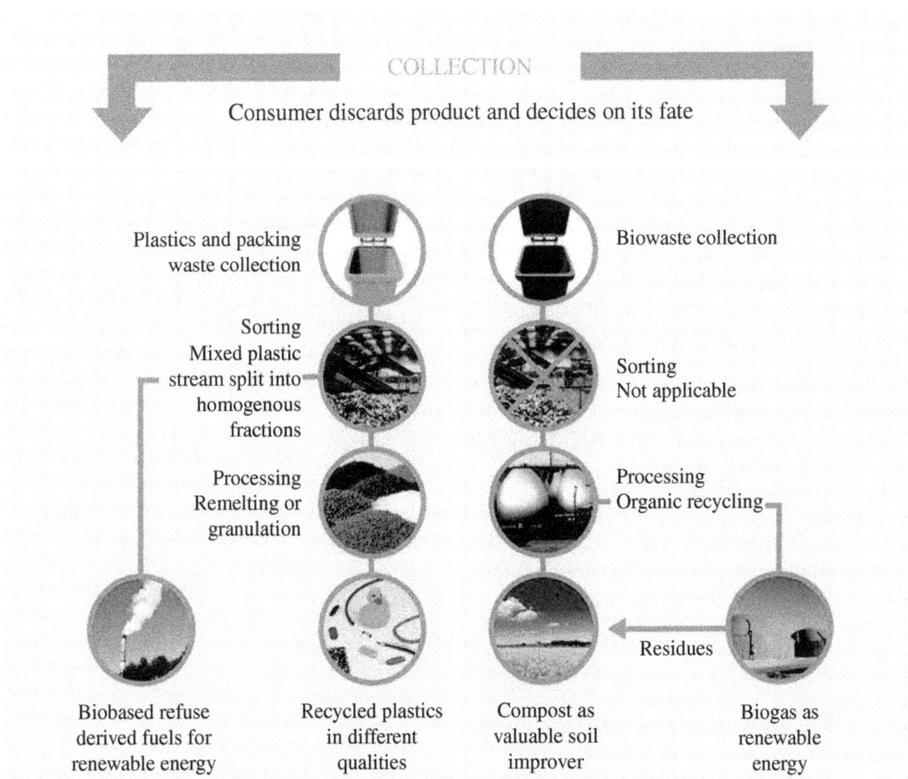

Figure 23.4 Ways of discarding bioplastics product [42].

It is important for the user to be aware of the fact that no matter what kind of packaging is used, it must have appropriate recovery and disposal methods. Landfilling is a problem to resource efficiently. Although it is still one of the prior disposal methods used in most countries, increasing progress of a phase-out is taking place [43].

23.3 Prevention and Reduction

This method of the waste level needs to use materials and manufacturing processes that limit the resource used and raise the functional product performance. Plastics have constantly proved their quality, which is lighter, thinner, and stronger. The law of competing as well as the ecological and market demand economy has to direct bioplastics in raising their resource and performance efficiency. The total number of products available in the market and the speed of the growth rate have testified to the success of bioplastics [42].

23.4 Reuse Bio-Based Products

The reuse of a bio-based product will retain its form as well as its ability to perform the desired task. This basically also requires its chemical composition to remain unchanged. Mechanical recycling might alter the original form and a certain amount of product value due to melting and extruding or other processes of creating a recycled batch [44], but usually the chemical composition is left intact. In chemical recycling, the reformation of materials into original intermediates [45] will remain its several functions, but for polymerization to reform the original polymer it is then compulsory to return the material to be used as a reliable end-product.

Carbon dioxide is created and released during the biodegradation process. However, it is only to the effect of completing the carbon cycle for bio-based products. Figure 23.5 shows the relationship of product recirculation through general examples of reuse, recycle, and feedstock renewal. Reuse is therefore desirable to recycle, whereas the renewal of biomass feedstocks by organic recycling is the least popular mode of recirculation. The best end-of-life option will be determined by the product applications. Reuse and recycling are not viable or are prohibitively costly in these situations, so biodegradation is the most favorable end-of-life choice.

Figure 23.6 explains a few options for reusing and recycling products and their order of preference. Remanufacture can be considered to produce the same end result as closed loop recycling, but is more preferable because it is more direct and maintains the shape of the component without the need to reformulate or mold the material back into its original condition. Other types of reuse are also resource efficiency strategies. Recyclable products should always be designed specifically so as not to cause the feedstocks needed for their manufacture to be depleted because of design flaws.

23.5 Packaging Material Recycling

The recycling of bio-based polymeric composites can only be applied for all thermoplastic polymers and not for elastomers and duro-plastics. There are three terms that illustrate the biopolymer: (i) biodegradation: degraded biologically via enzymatic

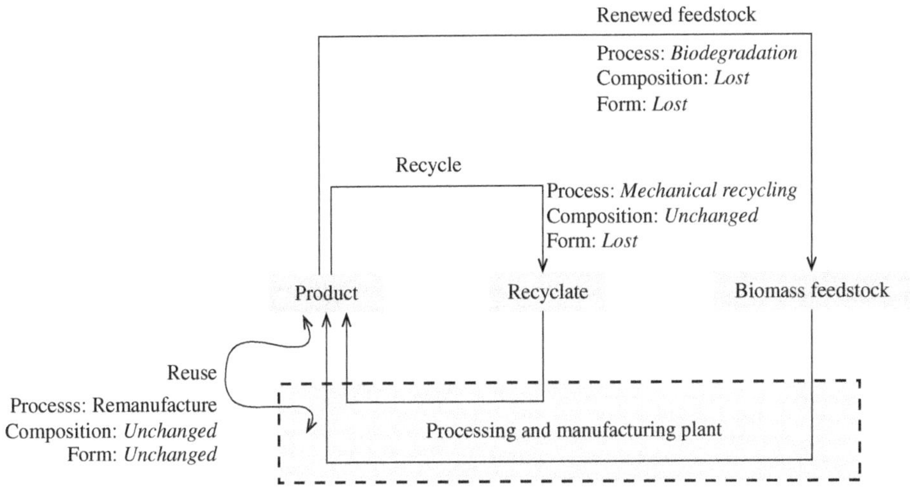

Figure 23.5 *Product recirculation through general examples of reuse, recycle, or feedstock renewal [46].*

reaction, (ii) biodegradability: decomposed through anaerobic and aerobic processes; and (iii) compostability: have the ability to completely decompose via aerobic biodegradation [47]. On the other hand, bioplastics can be from different types of materials that differ in their properties. There are three main groups, including (i) bio-based, nonbiodegradable, such as bio-based polyethylene (PE) or bio-based polyethylene terephthalate (PET), (ii) bio-based and biodegradable materials, such as polylactic acid (PLA), polyhydroxyalkonoates (PHA), or starch blends, and (iii) fossil-based and biodegradable materials [48].

All bio-based materials are applied to produce packaging from bioplastics. They can be processed via all customary plastic processing technologies in which their quality can be easily matched with the traditional products. However, the fate of bioplastics after application needs to be addressed. Therefore, a joint effort between waste management and packaging industries should be effectively worked to ensure appropriate solid waste management [49]. Recycling is one of the best methods that can help to reduce the production of solid waste generated from plastics. In addition, recycling all consumer goods materials can provide many environmental advantages. According to Oever et al. [50], material recycling is defined in the European standards EN13430 and EN16848 as the reprocessing of a used product material into a new product. There are two types of recycling method, namely mechanical recycling and chemical recycling, which are described in the next section.

To increase total recycling and to raise circular reuse, action is required in other parts of the chain. The main factor is to understand which collection system gives the highest result or recovery from household wastes. The feasibility of recycling beverage cartons from multiple collection systems and the method to mix sorting and recycling technologies to some packaging materials need to be understood. Wageningen Food and Biobased Research provides the industry and governments with information that is required for making strategies to raise the rate of recycling [51].

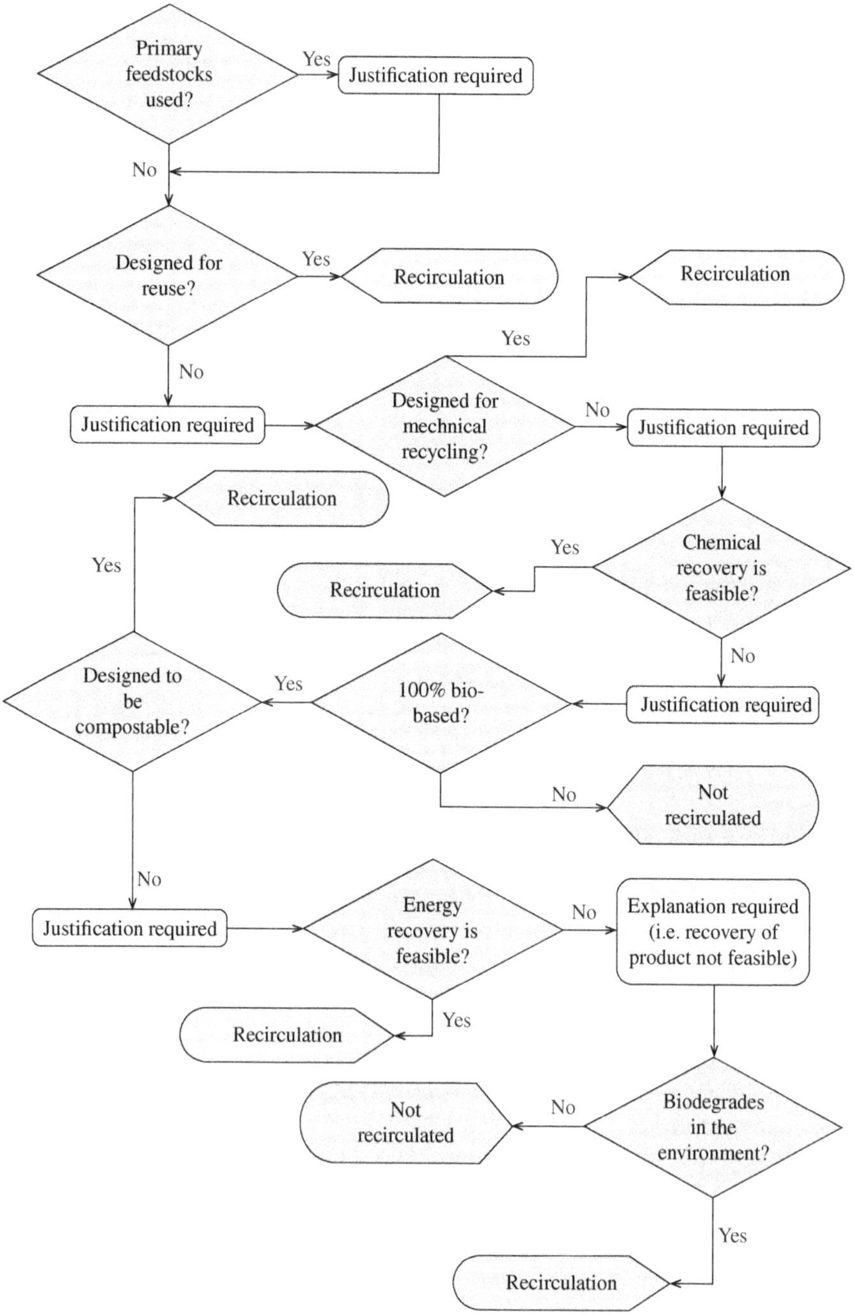

Figure 23.6 *Decision flowchart for bio-based products designed to be recirculated, which applies to each disassembled part [46].*

23.6 Mechanical Recycling Process

Examples of plastic products that are easily recyclable are big hollow containers, big surface films, or construction materials. Recycling of plastic products depends on product design, the cost effectiveness of the process, and material composition. The work is easier for products that are not made from a complex material mix, which can be separated into recyclable materials. The new bio-based types of PET, PE, or, in the future, polypropylene (PP) are physically and chemically similar to their usually used fossil-based material, and thus can be easily thrown away in established recycling plants. New separate streams, for example, for PLA will be shown in the short to medium period. Many tests and research projects for PLA are now in progress in Germany, Belgium, and the USA [43]. Figure 23.7 shows the end-of-life options for bioplastics in a mechanical recyling process.

23.7 Organic Recycling or Composting

By using compostable and biodegradable plastic products like packaging, bags, or cutlery may raise methods for the end-of-life treatment of these products. In addition to mechanical recycling, industrial composting, which is organic recycling or organic recovery, has become a waste management method. This is clearly a benefit when plastic materials are put together with biowaste. If this happens, mechanical recycling is not valid for either the biowaste or plastics. Utilization of decomposed plastics makes multiple wastes suitable for organic recycling.

23.8 Impact of Aging and Recycling on the Quality of Plastic Materials

The most crucial key to affect the recyclability is the different types of plastic that are put together as plastic packaging wastes. Multilayer packages are harder to recycle as compared to packages that contain only one type of plastic material. Through a recycling process,

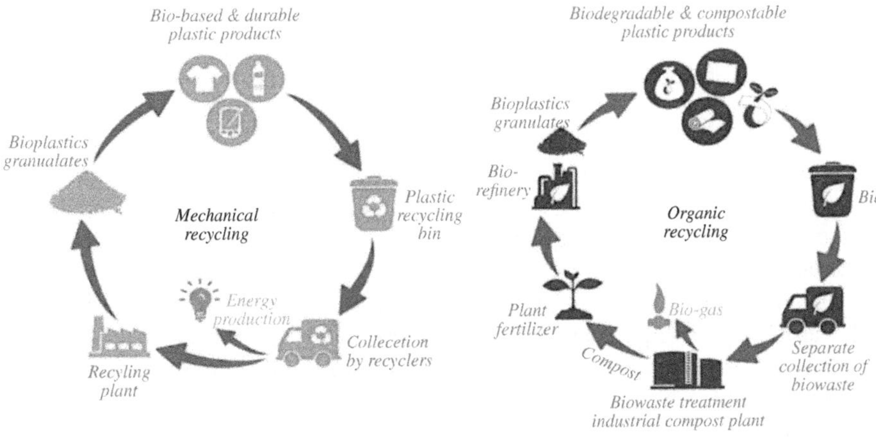

Figure 23.7 The end-of-life options for bioplastics in the mechanical recyling process. Source: Adapted from [42]).

the characteristics of plastic materials can be changed. Aging of different plastics is influenced by different factors so they need to be degraded differently, such as by exposure to sunlight, oxygen, air, water, heat, cold, and microorganisms. Radiation from the sun and thermal degradation have different degrading types, which are thermal, mechanical, radiation, hydrolytic, ultrasonic, chemical, and biological degradation. For example, photodegradation is the most usual for plastics. The important difference of the early steps is the degradation category. In thermal degradation, early yields are from the thermal disassembling of chemical bonds in macromolecules, whereas in photodegradation the yields are from photodissociation of chemical bonds. Plastics are continuously processed during recycling, which can be affected by thermal degradation. Generally, the main aging factor of packaging has only a limited contact with UV irradiation [52].

Other than temperature and radiation, oxygen is a crucial factor in the degradation of plastics. Photo-oxidative degradation is a process that causes many changes in the appearance and characteristics of plastics, such as secondary reactions cross-linking, chain scission, and photo-oxidation. The most common changes in materials are cracking, embrittlement, color changes, deterioration of mechanical properties, and a significantly shortened lifetime.

With continuous re-processing of plastics by affecting them to either thermal or photo aging, or even by combining both, the causes of recycling on the characteristics of plastic materials can be proven. The amount and type of stabilizers in pristine polymers that are used in research is a crucial factor when analyzing the results of artificial aging experiments. Only after 40 repeated extrusions does virgin low-density polyethylene (LDPE) undergo a cross-linking reaction where the yield of continuous extrusions can be processed to be changed. Accompanying the cross-linking reaction is a huge degradation of material. Chain scission is the result of unstable pristine LDPE thermal aging. Cross-linking prevails during photo-aging. No obvious changes happened in high-density polyethylene (HDPE) after 10 cycles of repeated extrusion when making repeated processing of HDPE. HDPE has already undergone chain scission reactions after two consecutive extrusions post-consumer. The yield in thermal aging of pristine HDPE is a rise in crystallinity. After aging in an oven at 100°C for over 10 months, the material maintained good mechanical characteristics. Re-processing of both post-consumer and virgin PP was proven to affect chain scission [52, 53].

23.9 Conclusions

In conclusion, to save the environment it is important to keep supporting the usage of bio-based packaging. This type of material can be utilized in many ways, not only in food packaging but also in the healthcare industry. However, the greatest challenges of applying this material in the industry are low economic feasibility, the after effect of bio-based materials to the environment, and little support from consumers. This lack of support for the recycling and composting program of "green" materials of bio-based plastics is shown by the fact that most of these products still end up in landfills, along with other materials. Therefore, an intense approach to the disposal of these biomaterials is needed. In addition, state and federal regulations need to be revised and updated regularly to avoid improper compostability claims. Extensive scientific and economic research needs to be done to ensure products are made from reasonable bio-materials.

References

1. Syafri E, Sudirman, Mashadi, Yulianti E, Deswita, Asrofi M, et al. Effect of sonication time on the thermal stability, moisture absorption, and biodegradation of water hyacinth (*Eichhornia crassipes*) nanocellulose-filled bengkuang (*Pachyrhizus erosus*) starch biocomposites. *J. Mater. Res. Technol.* 2019; **8**(6): 6223–31. Available at: https://doi.org/10.1016/j.jmrt.2019.10.016.
2. Abral H, Ariksa J, Mahardika M, Handayani D, Aminah I, Sandrawati N, et al. Highly transparent and antimicrobial PVA based bionanocomposites reinforced by ginger nanofiber. *Polym. Test.* 2019; **81** :106186. Available at: https://doi.org/10.1016/j.polymertesting.2019.106186.
3. Abral H, Basri A, Muhammad F, Fernando Y, Hafizulhaq F, Mahardika M, et al. A simple method for improving the properties of the sago starch films prepared by using ultrasonication treatment. *Food Hydrocolloids.* 2019;**93**:276–83. Available at: https://linkinghub.elsevier.com/retrieve/pii/S0268005X1831539X.
4. Abral H, Atmajaya A, Mahardika M, Hafizulhaq F, Kadriadi, Handayani D, et al. Effect of ultrasonication duration of polyvinyl alcohol (PVA) gel on characterizations of PVA film. *J. Mater. Res. Technol.* 2020 : **9**(2): 2477–2486. Available at: https://doi.org/10.1016/j.jmrt.2019.12.078.
5. Abral H, Ariksa J, Mahardika M, Handayani D, Aminah I, Sandrawati N, et al. Transparent and antimicrobial cellulose film from ginger nanofiber. *Food Hydrocolloids* 2020;**98**:105266. Available at: https://doi.org/10.1016/j.foodhyd.2019.105266.
6. Atiqah A, Jawaid M, Sapuan SM, Ishak MR, Ansari MNM, Ilyas RA. Physical and thermal properties of treated sugar palm/glass fibre reinforced thermoplastic polyurethane hybrid composites. *J. Mater. Res. Technol.* 2019 ;**9**(2):1606–1618. Available at: https://doi.org/10.1016/j.jmrt.2019.06.032.
7. Nurazzi NM, Khalina A, Sapuan SM, Ilyas RA, Rafiqah SA, Hanafee ZM. Thermal properties of treated sugar palm yarn/glass fiber reinforced unsaturated polyester hybrid composites. *J. Mater. Res. Technol.* 2019: **8**(5): 3726–3732 Available at: https://linkinghub.elsevier.com/retrieve/pii/S2238785419308312.
8. Aisyah HA, Paridah MT, Sapuan SM, Khalina A, Berkalp OB, Lee SH, et al. Thermal properties of woven kenaf/carbon fibre-reinforced epoxy hybrid composite panels. *Int. J. Polym. Sci.* 2019: 1–8.;Available at: https://www.hindawi.com/journals/ijps/2019/5258621.
9. Asyraf MRM, Ishak MR, Sapuan SM, Yidris N, Ilyas RA. Woods and composites cantilever beam: A comprehensive review of experimental and numerical creep methodologies. *J. Mater. Res. Technol.* 2020 Jan;. Available at: https://doi.org/10.1016/j.jmrt.2020.01.013.
10. Jalil MA, Mian MN, Rahman MK. Using plastic bags and its damaging impact on environment and agriculture: An alternative proposal. *Int. J. Learn. Dev.* 2013;**3**(4):1.
11. Pavani P, Rajeswari TR. Impact of heavy metals on environmental pollution. *J. Chem. Pharm. Sci.* 2014;**94**(3):87–93.
12. Barlaz MA, Ozge Kaplan P, Ranji Ranjithan S, Rynk R. Evaluating environmental impacts of solid waste management alternatives. *BioCycle*. 2003: **44**(10): 52–56.
13. Ilyas RA, Sapuan SM, Ibrahim R, Abral H, Ishak MR, Zainudin ES, et al. Effect of sugar palm nanofibrillated cellulose concentrations on morphological, mechanical and physical properties of biodegradable films based on agro-waste sugar palm (*Arenga pinnata* (Wurmb.) Merr) starch. *J. Mater. Res. Technol.* 2019**8**(5):4819–30. Available at: https://doi.org/10.1016/j.jmrt.2019.08.028.
14. Ilyas RA, Sapuan SM, Ibrahim R, Abral H, Ishak MR, Zainudin ES, et al. Thermal, biodegradability and water barrier properties of bio-nanocomposites based on plasticised sugar palm starch and nanofibrillated celluloses from sugar palm fibres. *J. Biobased Mater. Bioenergy* 2020;**14**:1–13.
15. Ilyas RA, Sapuan SM, Ishak MR, Zainudin ES. Development and characterization of sugar palm nanocrystalline cellulose reinforced sugar palm starch bionanocomposites. *Carbohydr. Polym.* 2018; **202**:186–202. Available at: https://linkinghub.elsevier.com/retrieve/pii/S014486171831066X.

16. Jumaidin R, Saidi ZAS, Ilyas RA, Ahmad MN, Wahid MK, Yaakob MY, et al. Characteristics of cogon grass fibre reinforced thermoplastic cassava starch biocomposite: water absorption and physical properties. *J. Adv. Res. Fluid Mech. Therm. Sci.* ,. 2019;**62**(1):43–52.
17. Jumaidin R, Khiruddin MAA, Asyul Sutan Saidi Z, Salit MS, Ilyas RA. Effect of cogon grass fibre on the thermal, mechanical and biodegradation properties of thermoplastic cassava starch biocomposite. *Int. J. Biol. Macromol.* 2019 : Available at: https://doi.org/10.1016/j.ijbiomac.2019.11.011.
18. Azammi AMN, Ilyas RA, Sapuan SM, Ibrahim R, Atikah MSN, Asrofi M, et al. Characterization studies of biopolymeric matrix and cellulose fibres based composites related to functionalized fibre-matrix interface. In: Kheng Lim Goh, M.K. Aswathi, Rangika Thilan De Silva, and Thomas Sabu (Eds.), *Interfaces in Particle and Fibre Reinforced Composites*. 1. London: Elsevier; 2020. 29–93. Available at: https://linkinghub.elsevier.com/retrieve/pii/B9780081026656000030.
19. Halimatul MJ, Sapuan SM, Jawaid M, Ishak MR, Ilyas RA. Effect of sago starch and plasticizer content on the properties of thermoplastic films: mechanical testing and cyclic soaking-drying. *Polimery* 2019;**64**(6):32–41.
20. Sanyang ML, Ilyas RA, Sapuan SM, Jumaidin R. Sugar palm starch-based composites for packaging applications. In: Mohammad Jawaid, Swain, and Sarat Kumar (Eds.),*Bionanocomposites for Packaging Applications*. Cham: Springer International Publishing; 2018. 125–47. Available at: http://link.springer.com/10.1007/978-3-319-67319-6_7.
21. Atikah MSN, Ilyas RA, Sapuan SM, Ishak MR, Zainudin ES, Ibrahim R, et al. Degradation and physical properties of sugar palm starch/sugar palm nanofibrillated cellulose bionanocomposite. *Polimery* 2019;**64**(10):27–36.
22. Ilyas RA, Sapuan SM, Ishak MR, Zainudin ES, Atikah MSN. Characterization of sugar palm nanocellulose and its potential for reinforcement with a starch-based composite. In: S.M. Sapuan, J. Sahari, M.R. Ishak, and M.L. Sanyang (Eds.),*Sugar Palm Biofibers, Biopolymers, and Biocomposites* 1. Boca Raton, FL : CRC Press/Taylor & Francis Group, 2018.: 189–220. Available at: https://www.taylorfrancis.com/books/9780429813542/chapters/10.1201/9780429443923-10.
23. Ilyas RA, Sapuan SM, Atiqah A, Ibrahim R, Abral H, Ishak MR, et al. Sugar palm (*Arenga pinnata* [Wurmb.] Merr) starch films containing sugar palm nanofibrillated cellulose as reinforcement: Water barrier properties. *Polym. Compos.* 2020 ;**41**(2):459–67. Available at: https://onlinelibrary.wiley.com/doi/abs/10.1002/pc.25379.
24. Jumaidin R, Ilyas RA, Saiful M, Hussin F, Mastura MT. Water transport and physical properties of sugarcane bagasse fibre reinforced thermoplastic potato starch biocomposite. *J. Adv. Res. Fluid Mech. Therm. Sci.* 2019;**61**(2):273–81.
25. Ilyas RA, Sapuan SM, Ishak MR, Zainudin ES. Sugar palm nanocrystalline cellulose reinforced sugar palm starch composite: Degradation and water-barrier properties. In: *IOP Conference Series: Materials Science and Engineering*. 2018.
26. The World Counts. Plastic Bag Pollution Facts. The World Counts. 2018.
27. Equinox Center. *Plastic Bag Bans: Analysis of Economic and Environmental Impacts*. Equinox Center. 2013;1–36.
28. Moharam R, Almaqtari M. The impact of plastic bags on the environment: a field survey of the City of Sana'a and the surrounding areas, Yemen. *Int. J. Eng. Res. Rev.* 2014;**2**(4):61–9.
29. Norizan MN, Abdan K, Ilyas RA, Biofibers SP. Effect of fiber orientation and fiber loading on the mechanical and thermal properties of sugar palm yarn fiber reinforced unsaturated polyester resin composites. *Polimery* 2020;**65**(2):34–43.
30. Nurazzi NM, Khalina A, Sapuan SM, Ilyas RA. Mechanical properties of sugar palm yarn/woven glass fiber reinforced unsaturated polyester composites: Effect of fiber loadings and alkaline treatment. *Polimery* 2019;**64**(10):12–22.
31. Ilyas RA, Sapuan SM, Ishak MR. Isolation and characterization of nanocrystalline cellulose from sugar palm fibres (*Arenga pinnata*). *Carbohydr. Polym.* 2018;**181**:1038–51. Available at: http://dx.doi.org/10.1016/j.carbpol.2017.11.045.

32. Ilyas RA, Sapuan SM, Ibrahim R, Atikah MSN, Atiqah A, Ansari MNM, et al. Production, processes and modification of nanocrystalline cellulose from agro-waste: a review. In: Bebrooz Movahedi (Ed.), *Nanocrystalline Materials*. IntechOpen; 2019. p. 3–32. Available at: http://www.intechopen.com/books/trends-in-helicobacter-pylori-infection/floating-drug-delivery-systems-for-eradication-of-helicobacter-pylori-in-treatment-of-peptic-ulcer-d.
33. Ilyas RA, Sapuan SM, Ishak MR, Zainudin ES. Effect of delignification on the physical, thermal, chemical, and structural properties of sugar palm fibre. *BioResources* 2017;**12**(4):8734–54.
34. Ilyas RA, Sapuan SM, Sanyang ML, Ishak MR, Zainudin ES. Nanocrystalline cellulose as reinforcement for polymeric matrix nanocomposites and its potential applications: a review. *Curr. Anal. Chem.* 2018;**14**(3):203–25. Available at: http://www.eurekaselect.com/156148/article
35. Mazani N, Sapuan SM, Sanyang ML, Atiqah A, Ilyas RA. Design and fabrication of a shoe shelf from kenaf fiber reinforced unsaturated polyester composites. In: Hidayah Ariffin, S.M. Sapuan, and Mohd Ali Hassan (Eds.), *Lignocellulose for Future Bioeconomy* Elsevier; 2019. 315–32. Available at: https://linkinghub.elsevier.com/retrieve/pii/B9780128163542000177.
36. Ilyas RA, Sapuan SM, Ibrahim R, Abral H, Ishak MR, Zainudin ES, et al. Sugar palm (*Arenga pinnata* (Wurmb.) Merr) cellulosic fibre hierarchy: a comprehensive approach from macro to nano scale. *J. Mater. Res. Technol.* 2019 **8**(3):2753–66. Available at: https://doi.org/10.1016/j.jmrt.2019.04.011.
37. Ilyas RA, Sapuan SM, Ishak MR, Zainudin ES. Sugar palm nanofibrillated cellulose (*Arenga pinnata* (Wurmb.) Merr): Effect of cycles on their yield, physic-chemical, morphological and thermal behavior. *Int. J. Biol. Macromol.* 2019 ;**123**:379–88. Available at: https://linkinghub.elsevier.com/retrieve/pii/S0141813018331477.
38. Ilyas RA, Sapuan SM. The preparation methods and processing of natural fibre bio-polymer composites. *Curr. Org. Synth.* 2020;**16**(8):1068–70. Available at: http://www.eurekaselect.com/178429/article.
39. Halimatul MJ, Sapuan SM, Jawaid M, Ishak MR, Ilyas RA. Water absorption and water solubility properties of sago starch biopolymer composite films filled with sugar palm particles. *Polimery* 2019;**64**(9):27–35.
40. Shalini R, Singh A. Biobased packaging materials for the food industry. *J. Food Sci. Technol.* 2009;**5**:16–20. Available at: ftp://118.67.228.244/public/food/resource/se-food-update-62-res-05.pdf.
41. AllThings.Bio. How to dispose of bio-based plastics?. 2017. Available at: http://www.allthings.bio/dispose-bio-based-plastics.
42. European Bioplastic. Waste management and recovery options for bioplastics. 2019. Available at: https://www.european-bioplastics.org/bioplastics/waste-management.
43. Johansson C, Bras J, Mondragon I, Nechita P, Plackett D, Šimon P, et al. Renewable fibers and bio-based materials for packaging applications – A review of recent developments. *BioResources* 2012;**7**(2):2506–52.
44. Ignatyev IA, Thielemans W, Vander Beke B. Recycling of polymers: A review. *ChemSusChem* 2014;**7**(6):1579–93.
45. Sasse F, Emig G. Chemical recycling of polymer materials. *Chem. Eng. Technol.* 1998;**21**(10):777–89.
46. Sherwood J, Clark J, Farmer T, Herrero-Davila L, Moity L. Recirculation: A new concept to drive innovation in sustainable product design for bio-based products. *Molecules*. 2016 ;**22**(1):48. Available at: http://www.mdpi.com/1420-3049/22/1/48.
47. Lorber KE, Kreindl G, Erdin E, Sarptas H. Waste management options for biobased polymeric composites. 2015 (May).
48. European Bioplastics e.V. Bioplastics packaging – Combining performance with sustainability. 2015
49. Calabrò PS, Grosso M. Bioplastics and waste management. *Waste Manage. (Oxford)* 2018;**78**:800–1.
50. van den Oever M, Molenveld K, van der Zee M, Bos H. Bio-based and biodegradable plastics: Facts and figures, focus on food packaging in the Netherlands. *Residue Rev.* 2017: 1–65. A report. Available at: http://dx.doi.org/10.18174/408350.
51. Velzen T van. Packaging-material recycling. Wageningen University and Research. 2015.

52. Hahladakis JN, Iacovidou E. Closing the loop on plastic packaging materials: What is quality and how does it affect their circularity? *Sci. Total Environ.* 2018: **630**: 1394–1400
53. Ilyas RA, Sapuan SM, Atikah MSN, Asyraf MRM, Ayu Rafiqah S, Aisyah HA, Mohd Nurazzi N, Norrahim MNF. Effect of hydrolysis time on the morphological, physical, chemical, and thermal behavior of sugar palm nanocrystalline cellulose (*Arenga pinnata* (Wurmb.) Merr). *Textile Research Journal.* 2020. In print. doi: 10.1177/0040517520932393.

24

Socioeconomic Impact of Bio-Based Packaging Bags

M. Chandrasekar[1], T. Senthil Muthu Kumar[2,3], K. Senthilkumar[2,3], S.M. Sapuan[4,5], R.A. Ilyas[6,7], M.R. Ishak[8], R.M. Shahroze[8], and Suchart Siengchin[3]

[1]*School of Aeronautical Sciences, Hindustan Institute of Technology & Science, Chennai, 603103, Tamil Nadu, India*

[2]*Department of Mechanical Engineering, Kalasalingam Academy of Research and Education, Krishnankoil, 626126, Tamil Nadu, India*

[3]*Department of Mechanical and Process Engineering, The Sirindhorn International Thai German Graduate School of Engineering(TGGS), King Mongkut's University of Technology North Bangkok, Bangsue 10800, Bangkok, Thailand*

[4]*Laboratory of Biocomposite Technology, Institute of Tropical Forest and Forest Products (INTROP), Universiti Putra Malaysia (UPM), Serdang, Selangor, 43400, Malaysia*

[5]*Advanced Engineering Materials and Composites Research Centre, Department of Mechanical and Manufacturing Engineering, Faculty of Engineering, Universiti Putra Malaysia (UPM), Serdang, Selangor, 43400, Malaysia*

[6]*School of Chemical and Energy, Faculty of Engineering, Universiti Teknologi Malaysia,81310 UTM Johor Bahru, Johor, Malaysia*

[7]*Centre for Advanced Composite Materials, Universiti Teknologi Malaysia,81310 UTM Johor Bahru, Johor, Malaysia*

[8]*Department of Aerospace Engineering, Uiversiti Putra Malaysia, Serdang 43400, Selangor, Malaysia*

24.1 Introduction

Conventional materials from the synthetic polymers are commonly employed materials in packaging applications. The features of polymer-based packing materials like lower density, durability, mass production, ease in production, abundant availability of raw

Bio-Based Packaging: Material, Environmental and Economic Aspects, First Edition. Edited by S.M. Sapuan and R.A. Ilyas

materials required for production, and drop in cost compared to the conventional petrochemical packaging materials have made the plastics an inevitable material for packaging applications. According to a survey, the demand for petrochemical-based plastic materials in packaging has increased by 39.2% and 42% in Europe and USA, respectively [1].

Despite of their advantages, extensive use of these synthetic polymer materials in the packaging of processed and packed foods around the globe has led to the accumulation of waste. The accumulation of plastic waste has become a serious environmental concern due to their poor biodegradable characteristic and limited recycling ability. Plastic debris affects human health by entering into the food chain. Plastic litter has also polluted our environment and has affected the marine ecosystems [2]. Marine organisms, animals, and birds have been affected, especially due to consumption of food-based packaging materials. Pictures and videos of the animals, birds, or marine animals entangled in plastics and their carcasses with plastic inside their bodies can easily be found on the Internet. All these factors have contributed to the need to develop and implement sustainable and environment friendly bioplastics. Hence, the concept of bioplastic-based packaging has started to gain attention among the global communities. Bioplastic refers to the plastics derived from renewable sources that have a biodegradable characteristic [3]. Bio-packaging materials are seen as a prominent driver for safeguarding the environment and sustainability [4]. Bioplastic packaging materials can be classified into different types, based on the source from which they are produced [5], as shown in Figure 24.1.

24.2 Socioeconomic Factors Influencing the Bioplastic-Based Packaging Materials

The sustainability and economic impact of bio-packaging materials depends on the factors such as consumer perception and acceptance, support by the government through policy making, interest from investors, revenue generation, and availability of these materials in the consumer market, as shown in Figure 24.2. This chapter addresses the socioeconomic impact of the bio-packaging materials in the context of the factors listed above.

24.2.1 Interest from the Investors

Bioplastics have a significant advantage over conventional plastics due to their environmentally friendly nature. Industrial leaders in the conventional plastic manufacturing sector, such as BASF, DuPont, and Braskem, have introduced bioplastic products such as

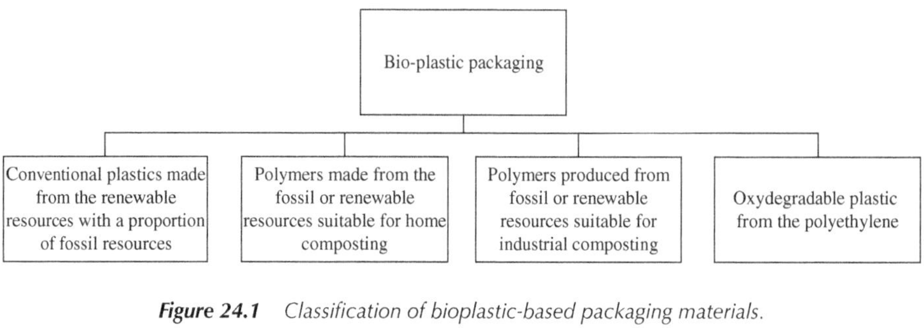

Figure 24.1 Classification of bioplastic-based packaging materials.

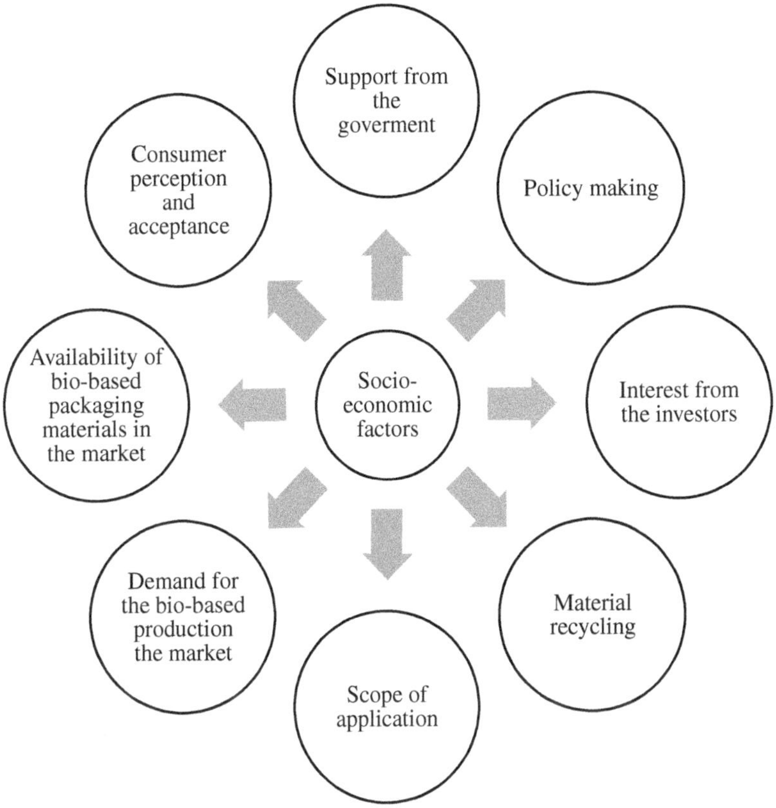

Figure 24.2 *Factors influencing bioplastic-based packaging materials.*

PHA, Biopropanediol (PDO), and Bio-PET [6]. However, the transition from conventional plastics to bioplastics is not an easy task, even for the established firms, due to the following reasons: (i) investment for infrastructure, (ii) production cost involved in the product development, and (iii) understanding the production process.

24.2.1.1 Market Projection on the Production of Bioplastic Materials

Use of bio-based polymer materials in packaging applications has led to the term "bio-based economy" or "bio-economy" [7]. A bio-based economy is a new approach that relates the impact of biologically derived environment friendly materials in terms of income, revenue, and financial aspects of these materials in the consumer market. It also puts the emphasis on economic growth toward the research, development, and policies related to the challenges faced by the communities associated with these materials [8].

Bioplastics produced from biopolymers and other renewable resources have always been outperformed by the conventional plastic, which generates sizable profits to the firms involved in the business. Hence, the production of bioplastics and their use in the market is insignificant compared to the conventional plastics. Table 24.1 shows the market projection of bioplastics from different geographical locations of the world. The

Table 24.1 Production and market projection of the bio-plastics from organizations around the world.

Organization	Market projection
European Climate Change Programme (ECCP)	25 kt in 1998 to 500 kt in 2010 without supportive policies and measures (P&M) and to 1000 kt with P&Ms.
Japanese Biodegradable Plastics Society (BPS)	187 kt by 2010
International Biodegradable Polymers Association and Working Groups (IBAW, Berlin)	2–4 million tonnes until 2020

Table 24.2 Commercially available bio-based polymers, their trade names, manufacturer, and production source [10, 11].

Bio-polymer	Commercial name	Producer and Supplier	Source of production	Price (US$/kg)
Polyhydroxyalkanoate (PHA)	Biopol	Was Monsanto	Direct production of polymer by fermentation or in a crop (usually genetic engineering in both cases)	7–13
	Biomer	Biomer		
Cellulose acetate	Bioceta	Courtaulds	—	3–5
		Mazzucchelli	—	
Starch polymers	National starch Avebe	Eco-FOAM Paragon	Modified natural polymer	3–7
Polylactic acid (PLA)	Nature works PLA	Cargil Dow Polymer	Bio-based monomer (lactic acid) by fermentation, followed by polymerization	2–11
	LACEA	Mitsui	—	
		Hycail		
	Galactic	Galactic		
Polyethylenefuronate (PEF)	Alpha Avantium	Plant bottle	—	—
Polyethylene terephthalate (Bio-PET)	Alpla	Plant bottle	—	—

International Biodegradable Polymers Association and Working Groups (IBAW) has estimated that 2–4 million tonnes of bioplastics will be in use by 2020, though that constitutes only one-in-two hundredth of the overall plastic consumption around the world.

24.2.2 Commercial Producers of Bio-Based Packaging Materials and Scope of Application

Commercialization of bioplastics in packaging applications is a safer and healthy substitute for the conventional petrochemical-based plastic materials [9]. Commonly available completely biodegradable polymers include cellulose polymers, starch polymers, PHAs, and polylactic acid (PLA), as shown in Table 24.2.

Table 24.3 End users of the commercially available completely biodegradable polymers [10–12].

Bio-polymer	Commercial end user	Purpose
Starch polymers	Novamont; Tesco	Nets for fruit
	Novamont; Natura; Albert Heijn; Swiss/German supermarkets	Grocery bags
Polylactic acid (PLA)	McDonalds	Trays and bowls for fast food
	Cristallina/Cargill Dow	Yogurt cup
	Mitsui	Envelope with transparent window, paper bag for bread with transparent window
	Panasonic	Rigid transparent packaging of batteries with removable printed film on back side

Table 24.4 Initiatives from some of the countries on bio-plastics.

Initiative	Country	Description	Reference
Plastic Debris Management Plan 2017–2021	Thailand	• Introduction of eco-packaging design and eco-friendly plastic as substitute for conventional plastic. • Development of the material flow for plastic containers and packaging inventory. • Implement 3R (reduce–reuse–recycle) strategy for managing plastic wastes • Educate stakeholders about alternative materials in the field of plastics.	[13]
Europe 2020 Strategy	Europe	• Creation of jobs in bio-economy • Innovation union and energy efficient Europe	[14]

Table 24.3 shows the use of bioplastic in packaging applications by various end users in the consumer market.

24.2.3 Policy Making and Support from the Government

Production and use of bioplastics have been encouraged by various countries through the new initiatives and change in existing policies to introduce bioplastic in the consumer market, as shown in Table 24.4.

The production of bio-based polymers for packaging applications is beneficial to the society in the following ways:

- Increase in the employment rate through the utilization of manpower for production
- Utilization of agricultural land for the cultivation and harvesting of the starch and cellulose-based polymers
- Subsidy and financial assistance from the government

- Tax reduction
- Development of small-scale industries and medium enterprises through collaborations with established business firms

24.2.4 Consumer Perception and Acceptance by Consumers (According to Countries)

The consumer perception and acceptance of a bioplastic-based packaging material by the public depends on the (i) knowledge about the bioplastic-based material and their re-use, (ii) awareness of the availability of such products in the consumer market, (iii) cost and expenses, and (iv) legal framework in the disposal.

A recent survey in Canada and the US indicates that people prefer bio-based products over the conventional plastics if the bio-based products can fulfill the purpose and have an equivalent price [15]. In a recent study with 89 participants from the European countries such as Czech Republic, Denmark, Germany, Italy, and The Netherlands, it was revealed that the participants had both positive and negative responses to the bio-based products. The study concluded that mixed responses were due to the lack of familiarity with the bio-based products. However, on a positive note, the participants could evaluate the bio-based product with the conventional product in terms of appearance, price, usability, etc. [7]. Paşnicu et al. [8] studied the behavior and perception of the people in Central and Eastern Europe toward the bio-economy. Their results indicate that participants from all the countries had awareness about environmental protection. However, the land allocated for organic crops required for the production of bio-based products varied significantly between the countries.

24.2.5 Challenges for Bioplastics in Packaging Applications

24.2.5.1 Material Performance

Bioplastics made from the polymers obtained from renewable resources have inferior melting strength, poor thermal sensitivity, and lower shear sensitivity than the incumbent plastic materials [16].

24.2.5.2 Recycling

According to Muller et al. [5], there are no governing regulations imposed on the consumers for the disposal of bioplastic waste. This implies that bioplastic wastes are collected along with conventional plastic waste and has to be sorted out further for recycling. Bioplastics and the conventional plastics cannot be distinguished by appearance. Hence, the major limitation is the difficulty in manual separation of bioplastics from conventional plastics. Another reason for not having a separate disposal system for bioplastic is (i) their present use in small volumes compared to the conventional petrochemical-based plastics and (ii) contamination of the post-consumed plastic with residues and impurities. However, a recent research indicates that recycling of bioplastic is possible where the bioplastic is depolymerized to form smaller reusable monomers. In order to recover the bioplastic after use, research activities should be focused toward developing suitable sorting methodologies. Among the existing sorting methodologies in plastic waste separation, only near-infrared

Table 24.5 Modified value chain for sustainability of the bioplastic.

Farmers	Manufacturers and producers	Financiers and bankers
• Cultivate the biomass feedstock that is to be used as a raw material	• Turn the raw materials into commercial bio-plastic • Improvisation of the existing product through research and development • Identify suitable technologies for sorting and recycling of bioplastic • Provide identification labels for bioplastic	• Provide funding for the research activities on bio-plastic. • loans on low interest to the farmers
Universities and research institutions	*Suppliers and retailers*	*Government*
• Initiate joint research activities with industry leaders on inventing new bioplastic materials • Enhance the performance of existing bioplastic materials • Increase the knowledge of bioplastics in the student community through curriculum	• Provide a channel for the supply of bioplastics • Introduce bioplastics to the consumers	• Subsidy and tax deduction to the bioplastic producers • Frame regulations on the reduction of conventional plastic • Ensure the framed regulations are followed • Implement policies to promote the use of bioplastic
Media	*End user*	*Waste treatment companies/facilities*
• Create awareness on the existence of bioplastics in the market • Impart knowledge about safety in the use of bioplastic to the end users	• Use bioplastic over the conventional plastic • Dispose of the waste according to the category it belongs to • Update their knowledge on bioplastics	• Separate the bioplastic wastes based on the identification labels from the manufacturer • Supply bioplastic waste to the manufacturer or the companies associated with recycling

Mistral technology (NIR) was suitable for identifying bioplastic made from PLA [17]. Since the bioplastic-based packaging industry is booming, it is necessary to develop suitable sorting methodologies in the near future [18]. According to Muller et al. [5], a simple label over the bioplastic could help the public to identify them prior to their usage and sort out post-disposal.

24.3 Future Scope

The market share of bio-based polymers is expected to be between 1% and 4% by 2020 and is a comparatively lower value than the market share of the conventional petrochemical-based synthetic polymers [10]. In a recent study, Shahzad Thariq proposed a value chain with responsibilities at each level to increase the production of bioplastic and their use among the consumers [19]. The value chain was modified further at each level to create a sustainable market for the bioplastics, as shown in Table 24.5.

24.4 Conclusion

Bioplastic packaging materials are seen as a viable solution for reducing the environmental impact that is threatening the ecosystem of the world. Major industrial leaders of conventional petrochemical-based plastics have ventured into this booming business considering the environmental conservation and government regulations. According to the literature, socioeconomic factors associated with the knowledge, perception, and acceptance of the bio-based packaging materials among consumers was not common. Government bodies have imposed regulations in restricting the use of synthetic plastics that encourages the use of bioplastics. However, it still lacks in disseminating the information of safety and acceptance of bioplastics among consumers. Based on the recent survey by researchers, consumer behavior toward the usage of bioplastic in food packaging and other applications is positive in countries like Canada, the USA, and some European countries. To conclude, farmers, media, government, manufacturers, financiers/bankers, research institutions, and consumers still have big roles to play and responsibilities to take during the transition from conventional plastic to bio-plastic for a sustainable future.

References

1. Andrady A, Bomgardner M, Southerton D, Fossi C, Holmström A. Plastics in a sustainable society. Mistra Background Paper on Plastics in a Sustainable Society. 2015: 1 – 36.
2. van Crevel R. Bio-based food packaging in sustainable development. 2016 [cited 2020 Dec. 17]. Available at: http://www.fao.org/forestry/45849-%0A023667e93ce5f79f4df3c74688c2067cc.pdf.
3. Van den Oever M, Molenveld K, van der Zee M, Bos H. Bio-based and biodegradable plastics: facts and figures: focus on food packaging in the Netherlands. Wageningen Food and Biobased Research; 2017.
4. Blanc S, Massaglia S, Brun F, Peano C, Mosso A, Giuggioli NR. Use of bio-based plastics in the fruit supply chain: an integrated approach to assess environmental, economic, and social sustainability. *Sustainability.* 2019;**11**(9):2475.
5. Müller G, Hanecker E, Blasius K, Seidemann C, Tempel L, Sadocco P, et al. End-of-life solutions for fibre and bio-based packaging materials in Europe. *Packag Technol Sci.* 2014;**27**(1):1–15.
6. Iles A, Martin AN. Expanding bioplastics production: sustainable business innovation in the chemical industry. *J Clean Prod.* 2013;**45**:38–49.
7. Sijtsema SJ, Onwezen MC, Reinders MJ, Dagevos H, Partanen A, Meeusen M. Consumer perception of bio-based products—an exploratory study in 5 European countries. *NJAS-Wageningen J Life Sci.* 2016;**77**:61–9.
8. Paşnicu D, Ghenţa M, Matei A. Transition to bioeconomy: perceptions and behaviors in central and Eastern Europe. *Amfiteatru Econ.* 2019;**21**(50):9–23.
9. Álvarez-Chávez CR, Edwards S, Moure-Eraso R, Geiser K. Sustainability of bio-based plastics: general comparative analysis and recommendations for improvement. *J Clean Prod.* 2012;**23**(1):47–56.
10. Crank M, Patel M, Marscheider-Weidemann F, Schleich J, Hüsing B, Angerer G. Techno-economic feasibility of large-scale production of bio-based polymers in Europe (PRO-BIP). Final report No. NWS-E--2004-111. Department of Science; 2004:1–256.
11. Okwuejunti E. The role of bio-based polyesters in food and liquid packaging. Master's thesis. School of Energy Systems Mechanical Engineering, Lappeenranta University of Technology, LUT. 2017;1–92.
12. Khan FRM. Business opportunities of ecological packaging: the Biobag® shopping bag. Degree Thesis. Department of Business and Administration, Arcada University of Applied Sciences, Helsinki, Finland. 2017;1–64.
13. Wichai-utcha N, Chavalparit O. 3Rs policy and plastic waste management in Thailand. *J Mater Cycles Waste Manag.* 2019;**21**(1):10–22.

14. Cosma IG. Bio-plastic-between current practices and the challenges of a sustainable future. *Manager.* 2018;(**27**):51–63.
15. Businesswire. Consumer acceptance of bio-based and 'green' household products set to rise; New survey reveals U.S. and Canadian attitudes and recent purchases, Palo Alto, USA. 2011 [cited 2020 Jan. 17]. Available at: http://www.businesswire.com/news/home/20110509005392/en .
16. Shen L, Haufe J, Patel MK. Product overview and market projection of emerging bio-based plastics PRO-BIP 2009. Report of Eur. Polysacch. Netw. Excell. Eur. Bioplastics, Utrecht University. 2009;**243**:1–245.
17. NAPCOR. NAPCOR refutes claims that PLA can be recyled with PET. PET trade organization cites concerns over quality, costs, and overall. Viability of Premise. [cited 2020 Jan. 17]. Available at: http://www.napcor.com/pdf/NAPCOR_PLA.
18. Ilyas RA, Sapuan SM, Atikah MSN, Asyraf MRM, Ayu Rafiqah S, Aisyah HA, Mohd Nurazzi N, Norrahim MNF. Effect of hydrolysis time on the morphological, physical, chemical, and thermal behavior of sugar palm nanocrystalline cellulose (*Arenga pinnata* (Wurmb.) Merr). *Textile Research Journal.* 2020. In print. doi: 10.1177/0040517520932393.
19. Tariq S. Success factors for the adoption of bio based packaging. Master of Science Thesis, KTH logotype, KTH Royal Institute of Technology, Stockholm, Sweden, 2013:1–77.

25

The Assessment of Supply Chains, Business Strategies, and Markets in Biodegradable Food Packaging

K. Norfaryanti[1], Z.M.A. Ainun[2], and S. Zaiton[1,3]

[1] *Institute of Tropical Forestry and Forest Products (INTROP), Universiti Putra, UPM Serdang, Selangor 43400, Malaysia*

[2] *Program of Pulp and Paper & Pollution Control, Laboratory of Biopolymer and Derivatives, Institute of Tropical Forestry and Forest Products, Universiti Putra Malaysia (UPM),Serdang, 43400, Selangor, Malaysia*

[3] *School of Business and Economics, Universiti Putra UPM,Serdang 43400, Selangor, Malaysia*

25.1 The Context of Bio-Packaging

The twenty-first century will be known as the bio-packaging revolution, involving packaging with biopolymers such as protein and starch [1]. The petrochemical polymers consisted of plastics having excellent packaging performance and low cost and were used extensively across the globe until environmental pollution became critical. In addition, the application of plastic packaging posed both technical and logistic issues, although alternatives like classic elimination, incineration, and recycling were practiced [2]. Bio-packaging produced a supply chain of millions of products.

Figure 25.1 illustrates the range of bio-packaging products in the supply chain [3]. The upstream begins with the supplier trading their goods, involving the utilization of resources, disposal, and transportation of the packaging used. At this point, the objective of bio-packaging is to conserve the resources. At the manufacturing point, the traded goods will be packaged, stored, and wrapped before reaching the next point, storing and warehousing. The objective at the manufacturing point is to reduce packaging materials required by packaging, storing, and wrapping activities. Subsequently, the goods are

Bio-Based Packaging: Material, Environmental and Economic Aspects, First Edition. Edited by S.M. Sapuan and R.A. Ilyas

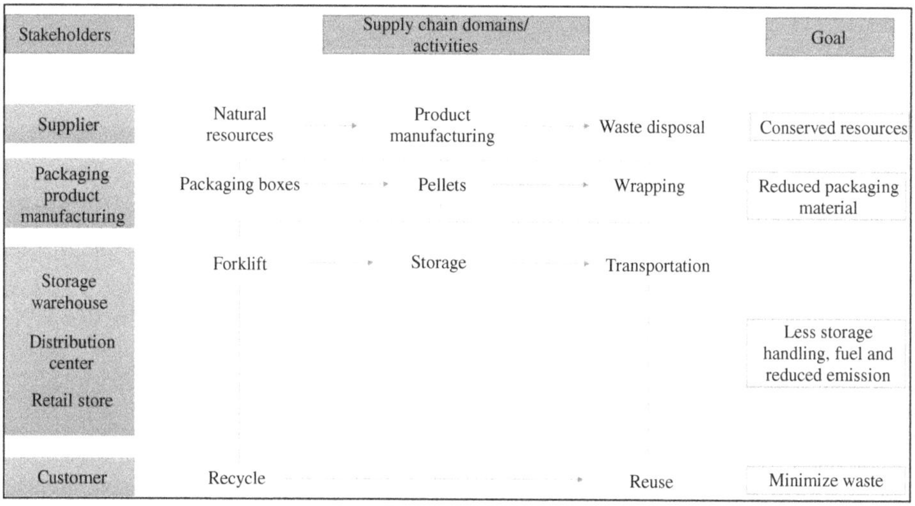

Figure 25.1 *The range of the bio-packaging supply chain.*

transported to various distribution centers and retail stores. Finally, at the customer point, recycling and reusing can minimize the waste from bio-packaging.

Based on the figure above, there is a wide range of bio-packaging that could be covered. This chapter finds that biodegradable food packaging to be the most promising area in pioneering the application of nanotechnology in the food industry [4]. The advancement of research and development of bio-nano composite materials for food applications is expected to grow in the next decade [5]. The main purpose of biodegradable food packaging is to maintain the quality and safety of food products throughout the supply chain and to extend the shelf life. The biodegradable food packaging requires more holistic characteristics than other durable goods because of its safety aspects and the shelf life [6].

There are three main conditions to consider in packaging materials for food: (i) transfer of substances from plastic packaging materials to foods [7], (ii) transfer of some substances from food to packaging that increase the migration of plastic molecules to food [8], and (iii) change of the packaging material properties in terms of lower mechanical properties, to reduce barrier resistance and increase stress cracking. Observing these conditions guide the researchers (i) to search and apply biodegradable packaging and (ii) to produce edible polymers by using food-grade additives that are based on biodegradable materials of natural fibers and substances [2]. In recent bio-packaging, the manufacture of biodegradable films were comprised of (i) synthetic polymers or biopolymer mixtures [9], (ii) microbial polymers, and (iii) agricultural polymers. Consumer demand for safety and health for food products is driving competitiveness among food manufacturers.

25.2 Types of Biodegradable Food Packaging and Its Characteristics

Innovation in food packaging has directed new emerging technologies, specifically bio-packaging. It can be categorized into three types, namely active packaging, smart (intelligent) packaging, and biodegradable packaging, as depicted in Figure 25.2 [10].

Basically, there are five important functions of bio-packaging: containment, protection, processing capacity, presentation of the product, and a packaging content relationship [10].

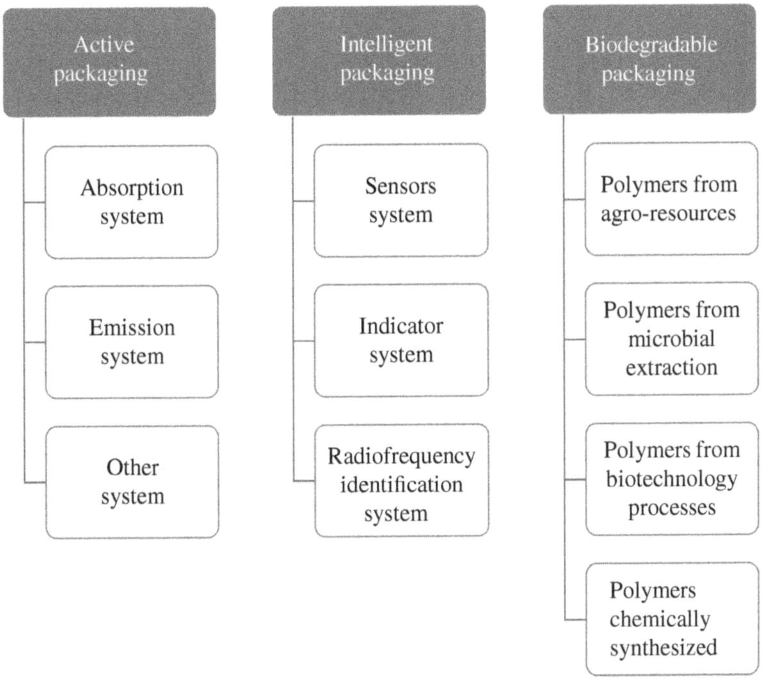

Figure 25.2 Types of biodegradable food packaging structure.

25.2.1 Active Packaging

Active and specific components are incorporated to produce active containers based on the type of interaction, such as absorption, emission, and other systems aimed at maintaining or prolonging the food product quality and shelf life.

Modified atmosphere packaging (MAP) is an effective system that can be applied in reducing the growth of pathogenic microorganisms by promoting oxidative stability, thus prolonging the product shelf life [11]. Several gases can also be used in the preservation of food products [12], such as carbon dioxide [13], oxygen [14], nitrogen [15], carbon monoxide [16], or a combination of one or more of these gases. Antimicrobial packaging is an example of active packaging with the incorporation, addition, or coating of antimicrobial agents to prevent microbial growth [17].

25.2.2 Intelligent Packaging

The incorporation of a system to monitor and communicate can be designed to provide useful information about packaged food. The system is equipped with specified applications with different ways of capturing and displaying information. Basically, there are three intelligent systems, which are sensors, indicators, and radiofrequency identifications systems (RFID) [18, 19]. Intelligent packaging contributes to the improvement of hazard analysis and critical control point (HACCP) and quality analysis and critical control point (QACCP) systems [20]. Intelligent packaging offers promising opportunities that service consumer benefits and convenience [21].

25.2.3 Biodegradable Packaging

Biodegradables are materials that can be degraded and composted by enzymatic action and act as fertilizers [22]. Important materials in biodegradable packaging are biodegradable polymers, which can be classified as (i) polymers from agro-resources, (ii) polymers from microbial extraction, (iii) polymers from biotechnology processes, and (iv) polymers chemically synthesized [22]. The polymers comply with the environmental concerns to a certain extent, such as thermal resistance, barrier, and mechanical properties. Hence, more research is needed to add value, such as the insertion of smart and intelligent components [23]. Samples of biodegradable packaging are (i) polysaccharide-based films, (ii) cellulose-based edible films, (iii) chitosan films, (iv) chitosan coatings, (v) chitosan edible films, (vi) starch-based coatings, (vii) and edible coatings, such as those based on a starch–gelatine blend, antimicrobial coating, pectin films, and alginate and gellan-based coatings [24].

25.3 Biodegradable Food Packaging Supply/Value Chain

The analysis of an industrial value chain is often overlooked in terms of identifying the current activities and their potential impacts in the industry, and how they evolve and could be shaped [25]. It is important to focus and analyze the value chain in order to assess the strategy for the future.

The value chain concept is important in a strategic analysis. It is a tool that has existed for three decades and used to analyze the firm, competitors, performances, and its performance gaps [26, 27]. A value chain is "the series of activities required to produce and deliver a product or service" [27]. The chain is comprised of activities involved in the production from raw materials to the ultimate consumption of the finished product. Layers in a value chain have been described as a sequence comprising suppliers, manufacturers, distributors, and consumers [10].

Figure 25.3 illustrates the three value chains: active and smart nanotechnology supply chains, nanomaterial, and the food packaging value chain. All three chains have framing conditions that consist of stakeholders who governed (regulators, standard agencies, risk agencies, and non-governmental organizations (NGOs)) and provided a financial contribution to the chain (public agencies, investment firms, and venture capitalists). These contributors may differ from one region to another, as well as from one country to another. Active and smart nanotechnology has three main research areas – sensors, enzymes, and RFID – carried out by research communities, such as large material manufacturing firms, universities, and start-ups. On the other hand, a nanomaterial chain has edible films, nanofibers, and nanocomposites as the main research areas, also carried out by large material manufacturing firms, universities, and start-ups. The two nano-led technologies merged into a food packaging value chain at the R&D in packaging and packaging manufacturing. Packaging research and development main research areas involved material science, food contact materials (FCM), and fabrication engineering, which are carried out by the universities, research institutes, and firms. The research activities are then fed to the packaging manufacturers, such as large specialized packaging manufacturers, small and

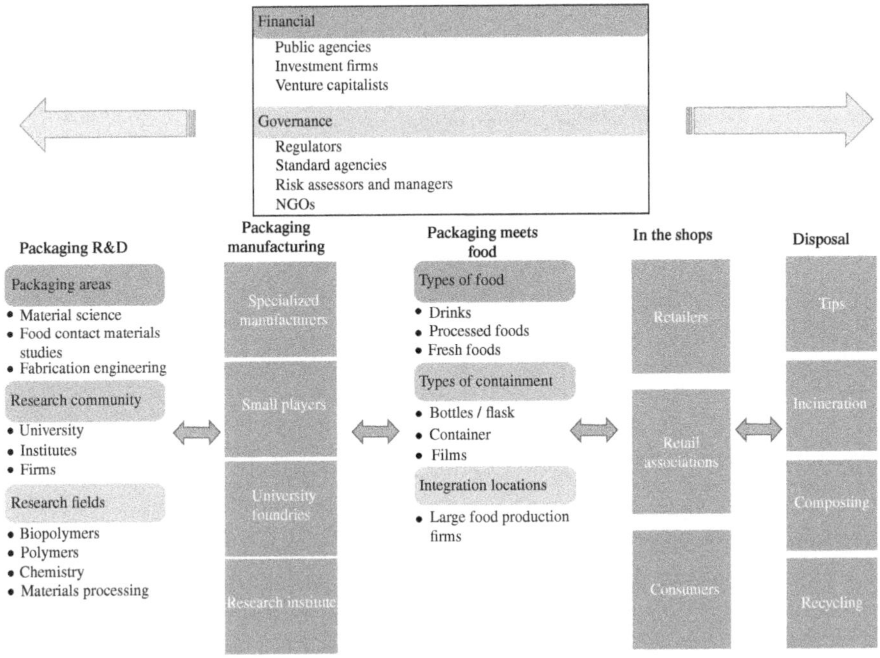

Figure 25.3 *Value chain of the food packaging industry.*

medium sized enterprises (SMEs), universities foundries, and smaller research institutes. The packaging manufacturers provide the various types of food and containments to the large, small, and medium food production firms. The chain continues to the retailers and shops before it ends at the consumers for disposal.

Based on the value chains above, stakeholders involved in research and development (R&D), funding, monitoring, production, marketing, and trading are crucial to ensure the sustainable ecosystems throughout the chains to meet the objective of biodegradable food packaging. This is to illustrate and provide insights into the approaches and actors in foreseeing the potential developments in the food packaging industry [10].

Apart from Figure 25.3, it is insightful to analyze the value chains from the industry players. Mitsui Chemicals Inc. emerged from Mitsui Coal Mine in 1912 and has progressed remarkably since they started coal mining until the 1950s when the business shifted to petrochemicals (see Figure 25.4). Petrochemicals became the driving force behind Japan's chemical industry. In 1997, Mitsui Petrochemicals Industries and Mitsui Toatsu Chemicals Inc. merged and became Mitsui Chemicals Inc. There are four business groups: mobility (elastomers, performance compounds, and performance polymers), health care (materials, nonwoven), food packaging (coating and engineering materials, films, functional sheets, and agricultural chemicals), and basic materials (phenols, PTA, polyethylene terephthalate (PET), industrial chemicals, polyurethane, petrochemicals, feedstocks, and licensing).

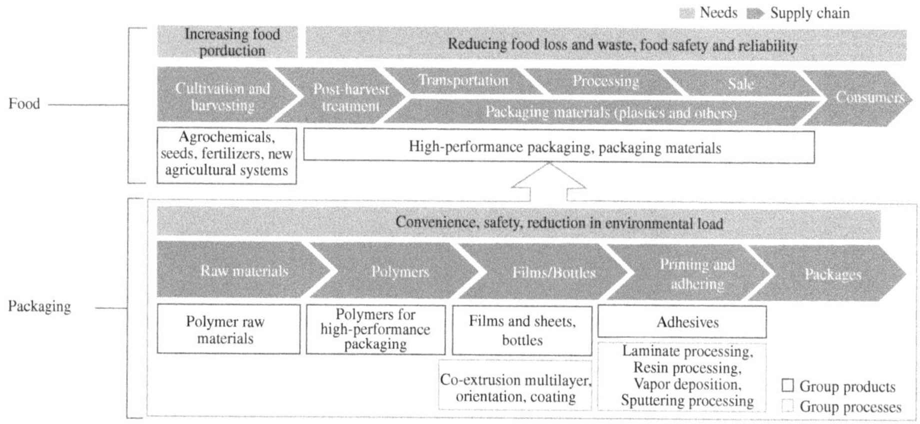

Figure 25.4 *Value chain of the food and packaging business sector for Mitsui Chemicals. Source: Mitsui Chemicals Inc. (2019).*

The above value chain of Mitsui Chemicals Group demonstrated the link between the two sectors of food and packaging as their main products [28]. By addressing individual needs at each stage of the supply chain, new functional value is added through products and services that address the performance and functional needs of end-users. According to the company, the opportunity in the business is highlighted as an expansion in demand for packaging materials due to the development of the food processing industry and improving standards of living in Asia. Currently, they have three main products related to packaging:

1. Coating and engineering materials – polyurethane resins (TAKENATE™, TAKELAC™), packaging adhesives, polyolefin resins, coating resins, and sealants for displays
2. Packaging films – biaxially oriented polypropylene films, cast polypropylene films, and linear low-density polyethylene film (T.U.X™)
3. Industrial films and sheets – semiconductor and electronic components manufacturing process films (ICROS™ TAPE), foam sheets, encapsulant sheets for solar cells (SOLAR ASCE™, SOLAR EVA™)

Based on data in 2017, 42% of the sales made in 2017 was for industrial films and sheets and 35% was for coatings and engineering materials. In their growth strategy, they aimed to continue accelerating the overseas expansion of the food packaging materials business, expand the business scale of industrial films and accelerate the growth of the food packaging business, while simultaneously launching solution businesses aimed at maintaining the quality of food products [28].

25.4 Business Strategies and Market Assessment

According to business databases, there are 16 major global players in the packaging solutions industry [29, 30]. They are involved in a wide range of business sectors of packaging solutions made from biomaterials and petrochemicals. The list is ranked based on their market capital size (US$) for 2019. In some companies that are privately owned and do not have market capital value, the turnover value substitutes for the market capital.

Based on the market capital value, Germany ranked first as the producer of packaging solutions with BASF silicone elastomer (SE) as the largest industry player among all players, with highest market capital value. The USA is the second ranking producer country, with a total of market capital of US$46.64 billion for four corporations, and a turnover of US$14.35 billion for two privately owned corporations. The United Kingdom is the third largest producer for the packaging solutions industry, with US$18.6 billion of market capital. Finland, Ireland, Japan, Brazil, Sweden, and Canada are among the major global industry players for the packaging solutions industry.

They have wide ranges of products but most of them are focusing on packaging technology solutions, producing various types of packaging according to the types, materials, and functional segments. The types differ from flexible to rigid packaging, materials vary from bio-based to petrochemicals, and functional diverge from industrial to home solutions. The list of the products for each industry player is extensive, but Table 25.1 summarizes the variation of products based on the types, materials, and functional segments.

25.4.1 Strategy and Market Projection

The strategies employed by major global industry players were assessed by the information obtained from their annual report, website, and other related reports. According to Table 25.2, it is evident that the global industry players in the packaging solutions have similar strategies, specifically in environmental sustainability agenda. They include reducing CO_2 emissions, forest conservation, reducing waste, using recycled materials, energy saving, innovation in packaging technology, engaging community, and economically gainful enterprises.

Each of the global industry players have their strengths in the market segment geographically, as well as their types of products. Based on the current products, these industry players are significant in European and North American market. As for the Asia market, it is becoming more prominent for certain companies, such as Amcor PLC, Klabin SA, and Tetra Pak International SA.

In order to reach the larger market, it is important to assess the market projection. The production capacities for bio-based and biodegradable plastics reached 1% of total global plastics production in 2015 [47]. Bio-PET, poly(butylenes succinate) (PBS), and poly(lactic acid) (PLA) are expected to expand in the future, while CA and Bio-PA are expected to merge. Hence, it is assumed that by 2020, the shares of both bio-based and biodegradable plastics will rise to 2.55 of fossil plastics production. Market shares of South America, Asia, Europe, and North America on the production of bio-based plastics are 44%, 45.8%, 6.8%, and 3.4%, respectively [48]. Bio-based and fossil-based plastics such as bio-PE and bio-PET can be used in packaging, food service ware, shopping bags, fibers/non-ovens, and agricultural applications [47]. It was expected that the global market for biodegradable polymers increased from GBP 114 million to GBP 206 million in 2010 [49]. Therefore, there is an increase in demand for identifying biodegradable packaging materials and innovative techniques to make degradable plastics [50].

There has been a stronger interest in the growth and application of bioplastics for food packaging in the last century from the food, packaging, and distribution sector [51]. It was

Table 25.1 *The major global industry players for packaging solutions.*

No.	Company name	Market capital (US$) 2019[a]/ turnover[b]	Products/business sectors	Country
1.	BASF SE	69.54 billion	Chemicals, materials, industrial solutions, surface technologies, nutrition and care, agricultural solutions [31]	Germany
2.	International Paper Company	18.18 billion	Industrial packaging, global cellulose fibers, papers [32]	USA
3.	Amcor PLC	17.33 billion	Flexible packaging, specialty cartons, plastic bottles and jars, capsules and closures [33]	USA
4.	Stora Enso Oyj	11.03 billion	Formed fiber, biocomposites, bio-based chemicals, bio-based materials, wood products, intelligent packaging, lignin, paperboard materials, packaging solutions, paper, market pulp, pellets [34]	Finland
5.	Mondi PLC	10.96 billion	Corrugated packaging, flexible packaging, engineered materials, uncoated fine paper [35]	United Kingdom
6.	WestRock Company	10.79 billion	Containerboard, corrugated containers, displays, folding cartons, kraft paper, paperboard, partitions and protective packaging, pulp, recycling [36]	USA
7.	Smurfit Kappa Group PLC	8.755 billion	Packaging, displays, packaging machinery, containerboard, paper and board, recycling [37]	Ireland
8.	DS Smith PLC	7.08 billion	Packaging, recycling, paper, plastics [38]	United Kingdom
9.	Mitsui Chemicals Inc.	5.2 billion	Performance materials, petro and basic chemicals, functional polymeric materials [28]	Japan
10.	Klabin SA	5.13 billion	Forestry, pulp, paper, packaging [39]	Brazil
11.	Rengo Co. Ltd	2.05 billion	Paperboard, corrugated packaging, folding cartons, flexible packaging, heavy duty packaging [40]	Japan
12.	Clearwater Paper Corporation	345.85 million	Tissue and paperboard [41]	USA
13.	Biopak UK Limited	5.87 million	Compostable packaging [42]	United Kingdom
14.	Georgia-Pacific LLC	Na/14.0 billion	Tissue, pulp, packaging, building products, related chemicals [43]	USA
15.	Tetra Pak International SA	Na/334.96 million	End-to-end solutions for food safety, packaging, food processing, services business solution, connecting the food industry [44]	Sweden
16.	NatureWorks LLC	Na/34.5 million	Biopolymers, rigid and flexible packaging, food service ware, health and personal care, durable products, 3D printing films [45]	USA
17.	Kruger Inc.	Na/7.59 million	Tissue, containerboard, packaging, publication papers, energy, specialty papers, biomaterials, recycling, real estate, wines and spirits [46]	Canada

[a]Value obtained from Google Finance.
[b]Value obtained from Crunchbase.

Table 25.2 *Brief strategies and market segments of the major global packaging solutions industry players.*

No.	Company name	Strategy	Market
1.	BASF SE	To combat climate change and global warming, committed to growing production volumes without adding further CO_2 emissions until 2030. We will decouple greenhouse gas emissions from organic growth [31]	Europe, North America, Asia Pacific
2.	International Paper Company	We will continue to lead the world in responsible forest stewardship to ensure healthy and productive forest ecosystems for generations to come. We have teamed up with partners, from small family farmers to global environmental organizations, to sustain forestlands through forest management, conservation, and restoration [32]	North America, Europe, Middle East, Africa
3.	Amcor PLC	Develop all our packaging to be recyclable or reusable by 2025. Significantly increase Amcor's use of recycled materials in our packaging. Work with others to drive consistently greater worldwide recycling of packaging [33]	North America, Europe, Asia
4.	Stora Enso Oyj	Focuses on biodiversity – 100% comes from sustainable sources (FSC and PEFC). Stora Enso Code – ability to meet customer and consumer demand for renewable solutions. Sustainable Development Goals – responsible consumption and production, climate action, and life on land [34]	Europe, China
5.	Mondi PLC	Drive performance along the value chain, invest in assets with cost advantage, inspire our people and grow responsibly, and partner with customers for innovation [35]	Europe, South Africa
6.	WestRock Company	People – investing in our workforce and our communities. Planet – acting as responsible stewards of our environment. Performance – delivering strong financial results and innovative packaging solutions [36]	North America, Europe, South America
7.	Smurfit Kappa Group PLC	Customers – we are committed to delivering innovative customer-focused, sustainable, and cost-efficient packaging and logistics solutions. Environment – to protect it and constantly improve our performance, with reduced emissions and limiting our footprint and impact on the environment [37]	Latin America
8.	DS Smith PLC	The leading supplier of sustainable packaging solutions. We have the scale to invest in innovation, which we then roll out across Europe, using our network of Impact and PackRight Centres [38]	Western Europe and in Central and Eastern Europe

Table 25.2 *(continued)*

No.	Company name	Strategy	Market
9.	Mitsui Chemicals Inc.	To secure its sustainable growth and development in harmony with society. To this end, we are advancing triple bottom line management, which incorporates economic, social, and environmental measures, engaging in operations from a long-term perspective, and promoting measures that are designed to help resolve social issues [28]	Asia, China, USA
10.	Klabin SA	Unique forestry company with a responsible management that is committed to biodiversity. We work in collaboration with our clients and suppliers, always guided by innovation and the constant improvement of our products and processes [39]	Latin America, Africa, Middle East, Asia
11.	Rengo Co. Ltd	Carrying out business with consideration to conserving the global environment as indispensable to the group's sustainable development, and thus endeavors to continually undertake environmental conservation activities [40]	Japan, Asia
12.	Clearwater Paper Corporation	Consider and value the current and future impacts of obtaining our raw materials and operating our facilities. We will evaluate and manage the long-term impacts of our decisions as we meet our customers' requirements for tissue, pulp, and paperboard products [41]	Not available
13.	Biopac UK Limited	Leading supplier of sustainable disposable packaging for food and beverages [42]	Central and Northern Europe
14.	Georgia-Pacific LLC	Innovating sustainable solutions, using resources efficiently, and reducing waste [43]	Na
15.	Tetra Pak International SA	We help you to raise industry standards through sustainable innovation. We have extended our packaging range to include more products with increased renewable content and reduced environmental impact [44]	Europe and Central Asia, North, Central and South America, South East Asia and Oceania
16.	NatureWorks LLC	We continue to work with converters, brand owners, and retailers around the world to help introduce naturally advanced products that span multiple industries and categories [45]	North America
17.	Kruger Inc.	We are deeply committed to the environment and to maintaining meaningful relationships with the people and communities we engage with [46]	Not available

estimated that with growth and demand increasing annually, global production of packaging materials is exceeding 180 million tons per year [52]. The food packaging sector progressively uses films and coatings made from biodegradable materials. G&S Business Communications presented their survey results of "Brand Owner and Consumer Insights on the Sustainability of Flexible Packaging" at the 2018 Flexible Packaging Association Fall Executive Conference in Chicago.

Evidently, to realize a change comes with a cost. For industry players, cost is becoming the most crucial part of the consideration to produce any products. The survey revealed three attributes that are most important to the food packaging industry players: costs (51%), physical protection to the goods (44%), and enhanced brand image (36%) [53]. Effective physical protection attributes emerged from intensive and comprehensive research works but require a large monetary investment by various stakeholders. It is evident that the investment made to path a change has a low return in the short term.

In the FPA study, 86% of consumers reported they "care about sustainability in general," and 79% preferred "products that are in sustainable packaging," but only 27% reported that they "always or often actively seek out products in sustainable packaging," while only 21% are "willing and often pay more for products in sustainable packaging" [53].

The survey results showed that new packaging technology (57%) is expected to drive the change in the packaging industry in the next five years. The second driver is bio-based materials and structures (38%) and biodegradable packaging (38%).

Another important finding of the FPA study is the opportunity to move forward in the market of food packaging. Respondents aged between 18 and 34 (36%) are more eager to consider the full life-cycle impacts of a package and they also actively seek out products with sustainable packaging more often than other age groups and are also willing to pay more for it. The other age groups comprised 35–44 (29%), 45–54 (22%), and 55 and above (13%).

The study indicated a greater opportunity to effectively communicate the value of full life-cycle impacts of flexible packaging with the Millennials, especially as they drive packaging trends in the future through increased buying power and influence.

Other drivers that create the path for food packaging innovation include extending food preservation, urbanization of the population, a stringent waste disposal policy and practice, recyclability and biodegradation, and crude oil price instability [10].

25.4.2 Biodegradable Food Packaging Trends

The main role of food packaging is to protect food products from external influences and harm, to contain food, and to provide customers with information on ingredients and nutrition [54]. Food packaging aims to produce nutrition economically to meet the demands of sector and customer wishes, maintain food security, and minimize the economic impacts of food packaging. The main purpose of food packaging must remain the safe, wholesome, and high quality of food. Food packaging enables customers, at their comfort, to appreciate nutrition as they wish. For instance, through portability and various parts, food packages may be designed to tailor individual lifestyles. Various evolving techniques are being examined given the prospect of incorporation into the fresh, smart food storage systems that satisfy the needs and demands of distinct players across the food production chain [55].

The environmental impact of packaging waste can be minimized through careful selection of materials according to the guidelines of EPO and the review of the environmental impact of packaging [56].

According to European Bioplastics, Asia was the main production for bio-based plastics in 2016 with a 43% market share, followed by Europe (27%), North America (23%), and South America (6%). Hence, Greener Package (2014) expected continual growth of the production of bio-based plastics at a successful rate in Asia, especially Thailand, India, and China [57]. Advanced technologies such as RFID and nanomaterials help the companies with processing, packaging, and manufacturing schemes [58]. Application of nanotechnology in food packaging extends the use of edible and biodegradable films that reduces waste packaging of processed foodstuffs by extending their shelf life to promote the conservation of fresh foodstuffs [59, 60]. The implementation of nanotechnology, it is believed, would extend the use of foodstuffs and biodegradables by generally increasing their mechanical, thermal, and obstacle characteristics, even in a very small context [61, 62].

25.5 Conclusion

Plastics pollution is a global issue concerning many stakeholders. It then sparked the evolution of bio-packaging. Food packaging has many functions and the technologies are evolving at a fast pace to meet the packaging purposes. The value and supply chains of the industry are crucial components in providing insights into the evolution. Apart from the value and supply chains analysis, this chapter has analyzed the products, strategies, and markets of the global industry players. The analysis allowed insights into the macro view of the food packaging industry and the prospects in the future. Strategies of the industry players are aligned with market projections of biodegradable food packaging. By analyzing the global industry players' products, strategies and markets, researchers would be able to chart their research path to help develop the future industry. As a conclusion, the biodegradable food packaging is a fast pace evolving industry that requires high commitment from various stakeholders to play an effective role in making an impactful change for the world.

Acknowledgments

The authors thank Mitsui Chemicals Inc. for permission to use Figure 25.4. They also thank Amira Nabila for assistance in constructing Figure 25.3. The authors acknowledged the support given by the Institute of Tropical Forestry and Forest Products (INTROP), Universiti Putra Malaysia under Higher Institution Centre of Excellence (HICoE), and the Ministry of Education research grant.

References

1. Masuelli, Martin A., ed. *Biopackaging*. CRC Press, 2017.
2. Callegarin, Francesca, et al. Lipids and biopackaging. *J. Am. Oil Chem. Soc.* **74**.10 (1997): 1183–1192.
3. Blanck, T. *Does your Packaging Survive your Supply Chain?* Chainanalytics, CSCP APICS (2015).
4. Chaudhry, Q., Scotter, M., Blackburn, J. Application and implications of nanotechnologies for the food sector. In: *Food Addit. Contam.* **25** (2008), 241–258.

5. Sorrentino, A., Gorrasi, G. and Vittoria, V. (2007) Potential perspectives of bio nanocomposites for food packaging applications, *Trends Food Sci. Technol.*, **18**(2), pp. 84–95. doi: 10.1016/j.tifs.2006.09.004.
6. Rhim, J. W., Park, H. M. and Ha, C. S. (2013) Bio-nanocomposites for food packaging applications, *Prog. Polym. Sci.* Elsevier Ltd, **38** (10–11), pp. 1629–1652. doi: 10.1016/j.progpolymsci.2013.05.008.
7. Mannheim, Chaim H., and Nehama P. Interaction between packaging materials and foods. *Packag. Technol. Sci.,* **3** (3) (1990), 127–132.
8. Johansson, F. Food packaging polymers as barriers against aroma vapours and oxygen in fat or humid environments. *Foods and Packaging Materials Chemical Interactions*, edited by P. Ackermann, M. Jägerstad, and T. Ohlsson, The Royal Society of Chemistry, Cambridge, (1995), pp. 114–117.
9. Gontard, N., and S. Guilbert. Biopackaging technology and properties of edible and/or biodegradable material of agricultural origin, in *Food Packaging and Preservation*, edited by M. Mathlouthi, Blackie Academic and Professional, (1994), pp. 159–182.
10. Masuelli, Martin A. & Zanon, M. 2017. State of the art. *Biopackaging Applications*, 125–47. CRC Press.
11. Arvanitoyannis, Ioannis S., and Alexandros Ch Stratakos. Application of modified atmosphere packaging and active/smart technologies to red meat and poultry: a review. *Food Bioprocess Technol.* **5**.5 (2012): 1423–1446.
12. Sahoo, J., and A. S. R. Anjanejulu. Modified atmosphere packaging of muscle foods: technology, shelf-life and safety aspects. *Indan Food Ind.* **14** (1995): 28–28.
13. Narasimha Rao, D., and N. M. Sachindra. Modified atmosphere and vacuum packaging of meat and poultry products. *Food Rev. Intl.* **18**.4 (2002): 263–293.
14. McMillin, Kenneth W. Where is MAP going? A review and future potential of modified atmosphere packaging for meat. *Meat Sci.* **80**.1 (2008): 43–65.
15. Thippareddi, Harshavardhan, and Randall K. Phebus. *Modified Atmosphere Packaging (MAP): Microbial Control and Quality*. FACTS National Pork Board **4667** (2002): 1–8.
16. Cornforth, D. and Hunt, M. *Low-Oxygen Packaging of Fresh Meat with Carbon Monoxide–Meat Quality, Microbiology and Safety*. In: AMSA white paper series No. **2**. (2008). The American Meat Science Association.
17. Appendini, Paola, and Joseph H. Hotchkiss. Review of antimicrobial food packaging. *Innovative Food Sci. Emerg. Technol.* **3**.2 (2002): 113–126.
18. Kerry, J. P., M. N. O'Grady, and S. A. Hogan. Past, current and potential utilisation of active and intelligent packaging systems for meat and muscle-based products: a review. *Meat Sci.* **74**.1 (2006): 113–130.
19. Vanderroost, Mike, et al. Intelligent food packaging: the next generation. *Trends Food Sci. Technol.* **39**.1 (2014): 47–62.
20. Heising, Jenneke K., et al. Monitoring the quality of perishable foods: opportunities for intelligent packaging. *Crit. Rev. Food Sci. Nutr.* **54**.5 (2014): 645–654.
21. Biji, K. B., et al. Smart packaging systems for food applications: a review. *J. Food Sci. Technol.* **52**.10 (2015): 6125–6135.
22. Trinetta, Valentina. Biodegradable Packaging. *Ref. Module Food Sci.* (2016): 2–4.
23. Siracusa, Valentina, et al. Biodegradable polymers for food packaging: a review. *Trends Food Sci. Technol.* **19**.12 (2008): 634–643.
24. Cazón, Patricia, et al. Polysaccharide-based films and coatings for food packaging: a review. *Food Hydrocoll.* **68** (2017): 136–148.
25. Robinson, Douglas K.R., Value chains as a linking-pin framework for exploring governance and innovation in nano-involved sectors: Illustrated for nanotechnologies and the food packaging sector. *Eur. J. Law Technol.*, Vol. **2**, No. 3, 2011
26. Peppard, Joe; Rylander, Anne. 2006. From value chain to value network: insights for Mobile operators. *Eur. Manag. J.* **24** (2–3),2006, 128–141.
27. Porter, Michael E. 2001. Strategy and the Internet. Harvard Business Review. March 2001, 1–19.

28. Mitsui Chemicals Inc. (2019). Food an Packaging Business Sector. Available at:https://jp.mitsuichemicals.com/en/service/packaging/index.htm.
29. Mordor Intelligence (2019). Biodegradable packaging market – growth, trends, forecast (2019–2014). Available at: https://www.mordorintelligence.com/industry-reports/gobal-biodegradable-packaging-solutions-market-industry.
30. Business Wire (2019). Global biodegradable packaging market to rise by 14.3% over the period 2019–2025. Available at: http://ResearchAndMarkets.com. Accessed on 10th December 2019 at https://www.businesswire.com/news/home/20190814005415/en/Global-Biodegradable-Packaging-Market-Rise-14.3-Period
31. BASF (2019). Packaging and print. Available at: https://www.basf.com/my/en.html.
32. International Paper (2019). Businesses – Industrial packaging, global cellulose fibers and papers. Available at: http://www.internationalpaper.com.
33. Amcor PLC (2019). Markets, products and services. Available at: https://www.amcor.com.
34. Stora Enso Oyj (2019). Our renewable products. Available at: https://www.storaenso.com.
35. Mondi PLC (2019). Products and solutions. Available at: https://www.mondigroup.com/en/home.
36. WestRock Company (2019). Markets and products. Available at: https://www.westrock.com.
37. Smurfit Kappa Group PLC (2019). Products and services. Available at: https://www.smurfitkappa.com.
38. DS Smith PLC (2019). Packaging. Available at: https://www.dssmith.com.
39. Klabin SA (2019). Businesses and products. Available at: https://klabin.com.br/en.
40. Rengo Co. Ltd (2019). Products. Available at: www.rengo.co.jp/english.
41. Clearwater Paper Corporation (2019). Paperboard products. Available at: http://www.clearwaterpaper.com.
42. BioPak UK Limited (2019). Products. Available at: https://www.biopak.com/uk.
43. Georgia-Pacific LLC (2019). Products. Available at: www.gp.com.
44. Tetra Pak International SA (2019). Sustainability. Available at: https://www.tetrapak.com.
45. NatureWorks LLC (2019). What is Ingeo? Available at: www.natureworksllc.com.
46. Kruger Inc (2019). Our sectors. Available at: www.kruger.com.
47. Van den Oever, M., Molenveld, K., van der Zee, M., & Bos, H. (2017). Bio-based and biodegradable plastics – Facts and figures: Focus on food packaging in the Netherlands. *Environ. Sci.* **2**, 907–914.
48. Blanc, S., Massaglia, S., Brun, F., Peano, C., Mosso, A., & Giuggioli, N. R. (2019). Use of bio-based plastics in the fruit supply chain: An integrated approach to assess environmental, economic and social sustainability. *Sustainability*, **11**, 2475. doi: 10.3390/su11092475
49. Auras, R., Harte, B., & Selke, S. (2004). Effect of water on the oxygen barrier properties of poly (ethylene terephthalate) and polylactide films. *J. Appl. Polym. Sci.*, **92**(3), 1790–1803.
50. Mahalik, N. P., & Nambiar, A. N. (2010). Trends in food packaging and manufacturing systems and technology. *Trends Food Sci. Technol.*, **21**(3), 117–128.
51. Kuorwel, K.K., Cran, J.M., Sonneveld, K., Milt, Z.J., & Bigger WS. (2011). Antimicrobial activity of biodegrdable polysaccharide and protein-based films containing active agents. *J. Food Sci.*, **76**, 90–106.
52. Tice, P. (2003). Packaging materials: 4. Polyethylene for food packaging applications. International Life Sciences Institute. Available at: https://www.pac.gr/bcm/uploads/4-polyethylene-for-food-packaging-applications.pdf.
53. Mohan A. M. (2018). Brand owner consumer sustainable packaging perceptions diverge. Available at: https://www.greenerpackage.com/green_resources/brand_owner_consumer_sustainable_packaging_perceptions_diverge.
54. Coles, R., McDowell, D., & Kirwan, M. J. (Eds.). (2003). *Food Packaging Technology* (Vol. **5**). CRC Press.
55. Vanderroost, M., Ragaert, P., Devlieghere, F., & De Meulenaer, B. (2014). Intelligent food packaging: the next generation. *Trends Food Sci. Technol.*, **39**(1), 47–62.

56. Marsh, K., & Bugusu, B. (2007). Food packaging – Roles, materials, and environmental issues. *J. Food Sci.*, **72**(3), R39-R55.
57. Greener Package (2014). Data predict significant growth for bioplastics to 2018. Available at: https://www.greenerpackage.com/bioplastics/data_predict_significant_growth_bioplastics_2018.
58. Nambiar A. N. (2009). Mass customization: Where do we go from here? In: *Proceedings of the World Congress on Engineering*. London United Kingdom.
59. European Bioplastics (2016). Bioplastics market data 2016. Available at: http://docs.europeanbioplastics.org/publications/EUBP_Bioplastics_market_data_report_2016.pdf.
60. Malhotra B. Keshwani A. & Kharkwal H. (2015). Antimicrobial food packaging: potential and pitfalls. *Front. Microbiol.* **6** 611
61. Malathi, A. N., Santhosh, K. S., & Udaykumar, N. (2014). Recent trends of biodegradable polymer: biodegradable films for food packaging and application of nanotechnology in biodegradable food packaging. *Curr. Trends Technol. Sci.*, **3**(2), 73–7
62. Ilyas RA, Sapuan SM, Atikah MSN, Asyraf MRM, Ayu Rafiqah S, Aisyah HA, Mohd Nurazzi N, Norrahim MNF. Effect of hydrolysis time on the morphological, physical, chemical, and thermal behavior of sugar palm nanocrystalline cellulose (*Arenga pinnata* (Wurmb.) Merr). *Textile Research Journal*. 2020. In print. doi: 10.1177/0040517520932393.

26

The Market for Bio-Based Packaging: Consumers' Perceptions and Preferences Regarding Bio-Based Packaging

Carsten Herbes
Institute for International Research on Sustainable Management and Renewable Energy (ISR), Nuertingen-Geislingen University, Neckarsteige 6-10, 72622, Nuertingen, Germany

26.1 Introduction: The Need for Bio-Based Packaging

Packaging fulfills a number of important functions: it protects the packaged goods, facilitates storage and transport, and also serves to communicate with potential customers, especially in the case of consumer goods [1, 2]. Up to one third of the utility a consumer derives from a product has been reported to come from packaging [3], which illustrates the significant role it plays in marketing.

On the other hand, packaging introduces two categories of serious environmental problems: it consumes scarce and finite fossil resources, particularly oil and gas, and it adds to the increasing burden of waste on the planet.

The ubiquitous plastic used in packaging is often the focus of environmental concerns, since plastic production consumes fossil resources both as a feedstock and as a fuel. Over the past decades, production of plastics has risen faster than that of any other bulk material as demand has boomed, largely driven by packaging applications, which account for 36% of plastics demand. Packaging applications are expected to drive future growth in the plastics industry as well. The chemical industry, which produces plastics, uses circa. 12

Bio-Based Packaging: Material, Environmental and Economic Aspects, First Edition. Edited by S.M. Sapuan and R.A. Ilyas

million barrels of oil products, 105 billion cubic meters of natural gas, and 90 million tons of coal per day to produce 960 million tons of primary products of which 44% are plastic, fiber, and rubber. The share of feedstock for the chemical industry in total oil demand was estimated at 12% in 2017 and is expected to rise to 16% by 2050. The industry's share in total demand increases to 18% when the energy used for production is included [4]. The reader might conclude that this rather modest share of total demand minimizes the impact of plastics in packaging on future resource consumption. However, while the plastics share of total oil demand might seem less important than that used for transport or power generation, it is more likely than any other to contribute to future growth in oil demand.

The second category of serious environment problems introduced by plastic packaging is the waste it produces. Plastic waste accounts for a large part of waste worldwide and its volume continues to grow. A recent study suggests that industry experts rate the waste problem caused by plastic packaging as much more important than its demand share of finite resources [5].

In the European Union, one third of all waste is from packaging [6] and only 67% of it was recycled in 2016 [7]. Plastics and glass made up 19% each. While a smaller share than the 41% from paper and cardboard, plastic packaging waste makes a more detrimental impact on the environment [8].

Roughly 40% of the plastics production in the European Union goes into the packaging industry [9], which accounts for nearly 60% of all plastic waste [10]. Globally, 40% of plastic packaging waste is landfilled, 14% is incinerated, and a devastating 32% leak into our waterways; only 14% is recycled [11].

In the United States, circa 50% of packaging waste in municipal solid waste is recycled, 10% is incinerated with energy recovery, and 40% is landfilled. However, recycling rates differ by material. While 75% of paper and cardboard packaging is recycled, the recycling rate for plastic packaging stands at only 13% [12].

Both consumers as well as the industry and policymakers are aware of the problems stemming from packaging waste. In a recent survey of experts in the packaging industry, 93% of the respondents indicated that they believed that the environmental concerns regarding plastic packaging will continue to worsen despite the fact that today they already pose a high risk [5].

Although the problems posed by packaging waste are often seen as a driver of the market for bio-based packaging, this view can be misleading. Unless bio-based materials are biodegradable or easy to recycle, they do not help reduce waste. Facts drive home this sad truth: more than half of the bio-based plastics produced today are not biodegradable [13].

The remainder of this chapter is organized as follows: Section 26.2 presents an overview of bio-based packaging, clarifying for readers the important difference between *bio-based* and *bio-degradable*. Section 26.3 explores consumer broad perception of bio-based plastics, since these materials are an important part of the bio-based packaging material portfolio. Section 26.4 addresses more specific consumer perceptions of bio-based packaging, while Section 26.5 looks at how consumers identify environmentally friendly packaging. Section 26.6 considers industry perspectives, while Section 26.7 concludes with a consideration of problems posed by bio-based packaging materials and potential solutions.

26.2 Bio-Based Packaging: An Overview

The twin problems posed by packaging have prompted responses at the political level, such as the EU Waste Framework Directive, which gives priority to prevention ("reduce"), followed by reuse and recycling – the well-known "3 R's" that have become important principles for the circular economy [14, 15]. Following in this direction is the plastics strategy published by the European Commission [10]. It advocates not only measures related to recycling, but also those that promote the use of renewable resources, especially bio-based materials.

However, the political sector is not alone in turning with interest toward bio-based materials. The plastics industry has started to replace materials derived from oil or natural gas with materials from renewable resources, such as corn starch. Bio-plastics in particular have garnered widespread attention both in the industry as well as in academia [16].

In these discussions, it is vital to distinguish *bio-based* materials from *biodegradable* materials since both often tend to get confused in the discourse on so-called bio-plastics. Figure 26.1 illustrates that describing a material as *bio-based* means broadly it has been produced from renewable resources, i.e. biomass, such as lignin or cellulose or polylactic acids (PLA). Others would define the attribute more narrowly as meaning "materials derived from primarily annually renewable sources" [18], which would largely exclude paper and cardboard.

Defining *biodegradable* is less straightforward as considerable differences exist both in the scientific community and among practitioners as to what qualifies as such. The difficulty stems in large part from the fact that the term specifies neither the time frame of the biodegradation process nor the degree to which the material decomposes. Moreover, there are different types of biodegradation, both aerobic and anaerobic, in which materials are decomposed through biological, especially enzymatic, processes. Ideally, a material described as biodegradable decomposes ultimately into water, carbon dioxide, and minerals. However, as this depends on the material, different definitions and criteria have been used to differentiate biodegradable from non-biodegradable materials [19].

The two attributes, *bio-based* and *biodegradable* are not interchangeable. There are bio-based materials that are biodegradable but other bio-based materials that are not. The fact is that only 43% of the global production capacity for bio-based plastics produce materials that are biodegradable [13]. On the other hand, there are also materials from fossil resources that are biodegradable [16, 20]. Figure 26.1 shows the distribution of packaging materials based on the resources used for production and their biodegradability.

Bio-based packaging materials can be further categorized by their chemical structure and their production methods, as shown in Figure 26.2. Bio-based plastics are often divided into drop-ins, i.e. materials that have the same chemical structure as their fossil equivalents and bio-based materials with a new chemical structure. Another categorization scheme looks at the production process and distinguishes materials using plant biomass from materials derived from animals, direct bacterial production, or production through fungi.

In Figure 26.3, applications for bio-based plastics are illustrated. It can clearly be seen that packaging is the leading application in the market for bio-based plastics. Rigid packaging is strongly dominated by polyethylene terephthalate (PET) and polyethylene (PE), two materials that are bio-based, but *not* biodegradable. In flexible packaging, starch blends

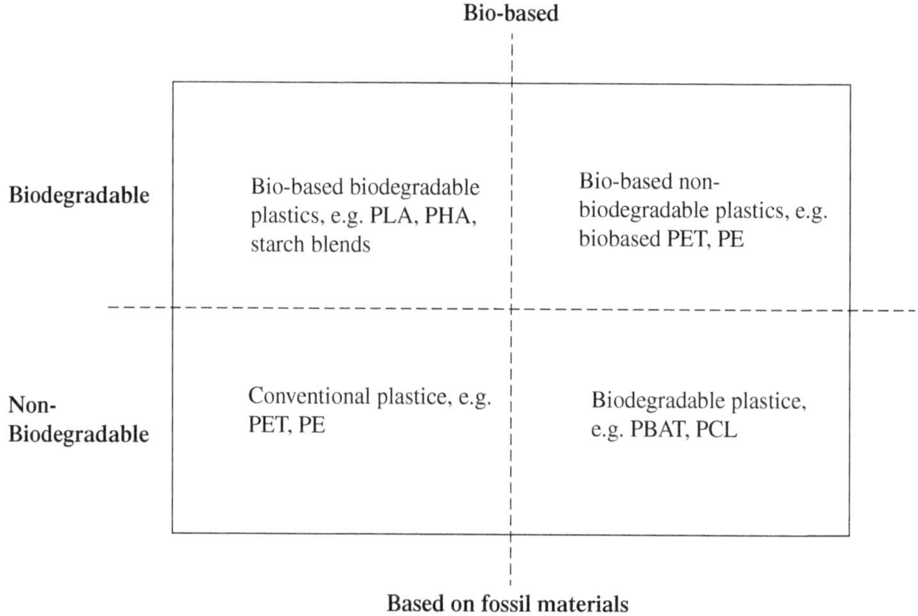

Figure 26.1 Categorization of packaging materials based on the resources used for production and on biodegradability [16, 17]. PLA: polylactic acids, PHA: polyhydroxyalkanoates, PET: polyethylene terephthalate; PE: polyethylene, PBAT: polybutylene adipate terephthalate; PCL: polycaprolactone.

are the foremost material [13]. Overall, the production capacities for bio-based plastics are expected to grow from 2.1 million tonnes in 2018 to 2.6 million tonnes in 2023 [13], numbers which suggest that these materials will likely continue in the near future as niche solutions rather than mainstream applications.

A final remark about bio-based materials is needed to avoid misconceptions. While widely considered promising, bio-based materials cannot be assumed to automatically confer environmental benefits. A recent review study estimated the annual CO_2 equivalent savings that could result from replacing two thirds of the global plastics production by bio-based plastics; the estimates were stunning, ranging from 241 to 316 million tonnes [21]. Impressive numbers like these, however, mask the context-specific question of whether a bio-based material is environmentally superior to its fossil equivalent. The question depends on a number of factors and thus has to be answered specifically for every case [7]. Moreover, first generation bio-plastics from, say, corn introduce another issue, namely competition with food production [22], which points toward using non-food resources [20] and also changing patterns of land use [23].

26.3 Consumer Perception of Bio-Based Plastics

Consumers are clearly aware of the problems stemming from using conventional plastics [24] and they often strongly connect it with the problem of plastics in the ocean and would like to reduce its use [25]. Consequently, they recognize positive effects of

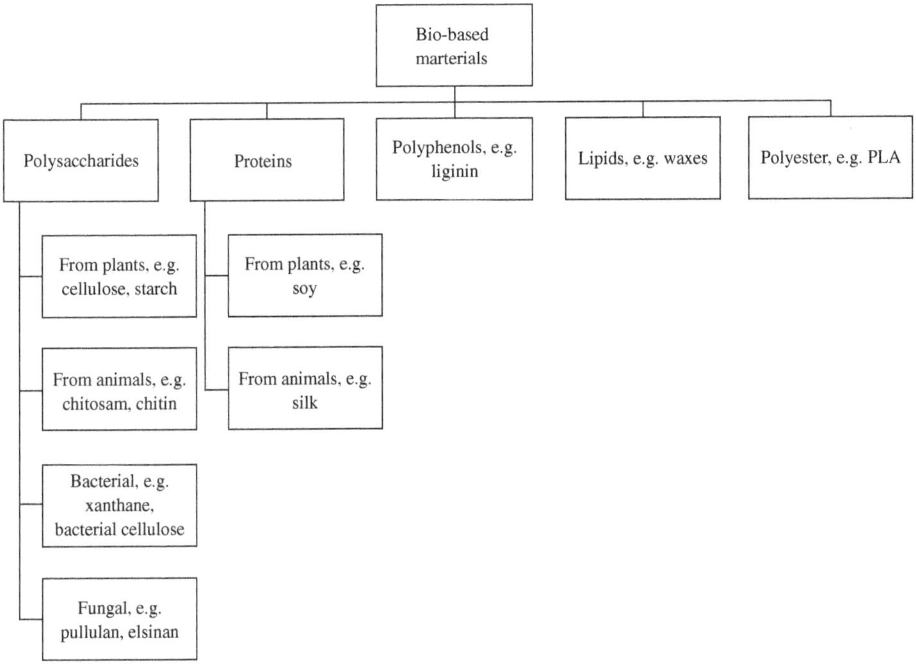

Figure 26.2 *Categorization of bio-based materials [17].*

bio-plastics. Participants in a study from the Netherlands, for example, stated positive environmental and economic effects associated with bio-plastics, contributing to a greener and hence more satisfying lifestyle [26]. Such plastics are used in applications that range from toys to packaging, with innovations emerging at a fast pace.

However, bio-plastics also raise concerns. First, consumers often have a negative perception of plastics in general. Karana (2012) speaks of the image plastics have of being an "environmental criminal" [27]. Add to this concerns arising from consumers' lack of familiarity with and knowledge of bio-based plastics [28, 29]. We saw in 26.2 how easy it is to confuse *bio-based* with *biodegradable*, so it is not surprising that some consumers make this association, with the unwelcome consequence being concerns about the durability of products made of bio-based plastics [28].

However, information for consumers is broadly lacking [30]. Linked to this are suspicions about greenwashing and false claims made by the industry. To trust a product whose very name insinuates environmental wrongdoing, consumers need a neutral institution that can certify the positive environmental effects of bioplastics [26]. This becomes all the more important as consumers raise concerns about the use of land for the production of the biomass used in producing bio-based plastics [6, 26]. Concerns have also emerged about potential environmental damages from producing the biomass [28], an issue that is also reflected in life cycle analysis (LCA) studies of bio-based plastics, which point out sustainability problems posed by these materials [31]. These concerns are not specific to bio-based plastics; they have also been found in studies on consumer perceptions of other bio-based

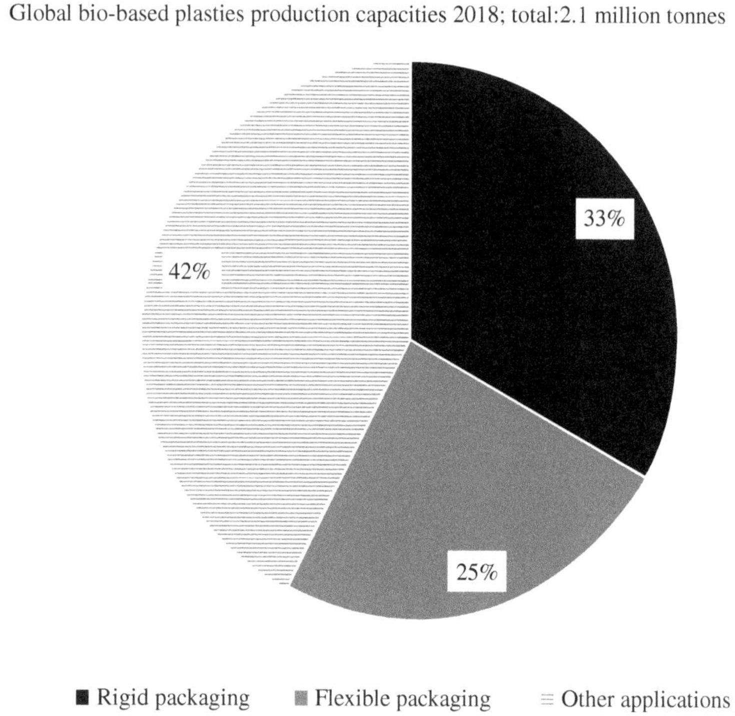

Figure 26.3 *Shares of different applications in the production capacities of bio-based plastics in 2018 [13].*

products [32] and in assessing the use of bio-waste as an environmentally friendly alternative [33] for bio-plastic production.

We can also find expressed concerns about the higher prices and lower quality of bio-based plastics [26, 28], but these simply underscore the broader question of which attributes of bio-based plastic products have the greatest influence on consumer decision making. Scherer et al. (2017) have identified price, geographic origin of the raw materials, share of the bio-based material in the product as well as its environmental effects as being decisive in consumer choice [29]. Kainz (2016) adds labels as an important attribute [28]. Both research teams also demonstrate how the relative importance of attributes differs among consumers depending on attitudes and sociodemographic attributes.

26.4 Consumer Perception of Bio-Based Packaging

While attitudes toward bio-based plastics may influence perceptions of bio-based packaging, bio-based packaging is just one of various environmentally friendly options. Other alternatives such as paper or cardboard have traditionally been used for packaging in large quantities. What consumers value in the end is not the fact that packaging is or is not bio-based, but the degree to which its attributes are considered pro-environmental. Therefore, bio-based packaging has to be viewed in the context of other environmentally

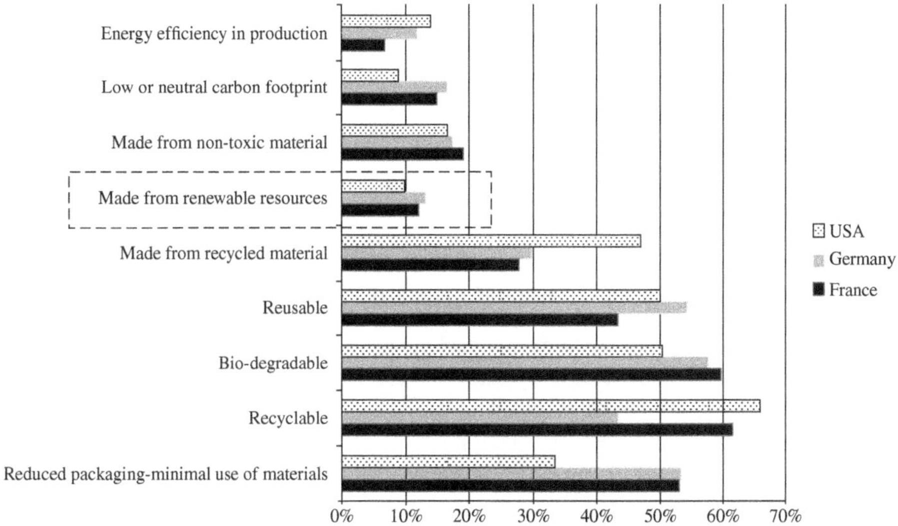

Figure 26.4 Packaging attributes identified as "environmentally friendly" by consumers in a closed question (codings per 100 participants per country) [7].

friendly packaging options, such as packaging from recycled material or reusable glass containers.

Numerous studies have shown that consumers tend to identify the pro-environmental attributes of packaging as those pertaining to its end-of-life characteristics [24, 34–37]. This means that they focus on attributes such as biodegradability, reusability, and recyclability. Figure 26.4 illustrates this tendency to assign greater value to end-of-life attributes than to beginning of life attributes, such as the question of which raw materials are used for producing the packaging. These rank a distant second in the perception of consumers in the US, France, and Germany (Figure 26.4); other stages, such as the production process, transport, or the use phase, are barely considered in assessing environmental friendliness [6].

When looking specifically at rankings of beginning of life attributes, reduced packaging, i.e. the amount of packaging used, features most prominently in the mind of consumers. Attributes related to origin rank lower: "Made from recycled material" is second, followed by "Made from non-toxic materials," and "Made from renewable resources" [6]. This points to a significant problem facing the bio-based packaging industry. When assessing the environmental friendliness of a packaging option, the merits of bio-based materials simply do not yet occupy a prominent position in the consumer mind.

The study by Herbes et al. (2018) also reveals cross-cultural differences in consumer perceptions [6]. In a closed question, German consumers selected "from renewable resources" as an environmentally friendly option more frequently than did their French and US counterparts (see Figure 26.4). In an open question, they also named attributes pertaining to the beginning of life much more often than did respondents from the other two countries. These results may be linked to the strong presence of renewable energies in Germany, which are, in the case of biogas or biofuels, produced from crops.

Among the many options for bio-based packaging materials, biomethane from biogas plants is an innovative possibility [38]. However, it exacerbates concerns that consumers already have. In the study by Herbes et al. (2018), the biomethane-based packaging option was the least popular among all environmentally friendly options, even less popular than other renewable resources that were presented as not biodegradable [6]. Reasons collected from an open question mainly reflected environmental concerns, especially concern that arable land is used to produce the biogas, which is then upgraded to biomethane [6]. This negative perception is not limited to biomethane; general concerns about using arable land for producing bio-based materials have been found in previous studies [20, 22].

26.5 Consumer Identification of Bio-Based Packaging

Now that we have explored what makes packaging environmentally friendly from a consumer perspective, the next crucial question is: How do consumers recognize packaging options that carry these pro-environmental attributes and how do they distinguish them from non-environmentally friendly packaging?

In principle, consumers rely on various cues. These include structural cues such as material, size, and shape; visual cues such as eco-labels, color, or images; informational cues, both numerical and verbal; and sensory cues such as the haptics of the packaging [39]. Past studies have often identified labels as an important cue for consumers [34, 39, 40]. The 2018 study cited in Figure 26.4 found consumers in Germany, France, and the United States all referenced eco-labels on packaging; however, their general lack of understanding of these labels hinders them from being effective [36]. Besides eco-labels, logos, and images [34] as well as the material [36, 41, 42], often signal environmental friendliness. In the 2018 study, French consumers also placed a high importance on the packaging material, notably its color, while consumers in Germany and the United States preferred to read the information on the package and possibly get further information on the Internet [39].

26.6 Industry Perspectives

It is worthwhile to look briefly at what producers and industry experts think about bio-based packaging and what understanding they bring to consumer perceptions. To industry experts, bio-based plastics rank in third place as a solution to the environmental concerns raised by plastic packaging. In the first place, experts rank recycling-focused solutions and in the second, replacing single-use plastic packaging with non-plastic options. Nonetheless, 40% of the brand owners in a recent study said they have specific production goals for the bio-based content of packaging materials [5].

Apparently, industry experts also see a big difference between consumer perception and the facts regarding bio-based packaging: while 74% state that they think bio-based packaging will improve the perception of plastics, only 42% think that it will in reality improve the impact of packaging on the environment [5].

26.7 Conclusion: Problems and Potential Solutions

The European Commission in their Plastics Strategy deplores the small market share of bio-based plastics [10]. However, which problems lie at the heart of this deficiency and which measures could help to increase the share of bio-based packaging are open to debate.

One problem is that the relative environmental performance of bio-based packaging materials is not always clear. In its vision, the Commission focuses on recycling but also mentions renewable alternative feedstocks, albeit with the proviso "... where evidence clearly shows that they are more sustainable compared to the non-renewable alternatives" [10]. This reiterates the point earlier made, namely that bio-based materials are not automatically superior to fossil materials in their environmental performance [31]. Further research performing LCA on various bio-based materials in comparison with fossil materials and taking into account the large variety of materials and use cases could help to improve the decision basis of companies, consumers, and policy makers alike.

Another problem, related to this, is the lack of knowledge held by consumers and their resulting insecurity. Consumers neither understand the relative importance of the different steps in the value chain and nor are they able to evaluate the environmental merits of the materials they are presented in purchasing decisions [24]. Moreover, bio-based packaging materials are often, just like electricity from renewable resources, a credence good, meaning consumers usually cannot experience the attribute "bio-based" before or during use of the material [43]. Like in other areas, simple and easy-to-understand labels could help consumers to evaluate packaging solutions without having to invest much time and effort. However, as other products have witnessed the emergence of a plethora of labels that again tend to confuse consumers [44, 45], an independent state-led label that unifies signals into one consistent visual signposting would be preferable.

In contrast to the marketing of biogas or renewable electricity, producers can for some materials try to make the "bio-based" quality of the material perceptible for consumers. A consumer study showed that it was "visible grains and fibres" [27] that evoked an impression of naturalness among consumers. Product and material designers could use this mechanism deliberately when inventing new products from bio-based materials. However, there is also a risk with this strategy. If producers of conventional materials, e.g. fossil-based plastics, imitate this design and consumers become aware of it, this may further fuel suspicions of "greenwashing" and add to consumer confusion.

Prices significantly influence the diffusion of bio-based plastics and are influenced mostly by high production costs that are in turn driven by a lack of economies of scale and remaining technological challenges [46, 47]. For example, technically demanding processes and high production costs for lignocellulose-based materials [48] are still an obstacle and hinder bioplastics from forestry biomass from being profitable [20]. Lettner et al. (2017) come to the conclusion that while prices for polyhydroxyalkanoates (PHA) and lignin can be expected to drop, leading to further market diffusion, the outlook for other materials like PLA or cashew nut shell liquid (CNSL) is rather uncertain [46].

Finally, regulation would be a strong instrument to increase the bio-based share. Governments around the world have for many years used regulation and incentives to reduce the amounts of single-use plastics [49, 50]. The European Commission published a plastics strategy in 2018, which mentions packaging prominently. Besides regulation on recycling, mandating a specific percentage of bio-based packaging would force a quick stimulus of the market. In the United States, the Energy Policy Act set mandatory volume targets for biofuels as early as 2005, while the European Union first merely set voluntary requirements in 2003 and only introduced mandatory blending requirements in the Renewable Energy Directive of 2009 [51]. These policies have effectively raised the share of biofuels and consequently also altered the acreage shares of different crops [52, 53].

Apart from direct volume targets or blending requirements, policies can also mandate a greenhouse gas (GHG) reduction target or grant subsidies for growing or using biofuels and thus indirectly promote biofuels [54]. Moreover, other policies, such as providing economic incentives, e.g. a plastic charge [55–57] may change consumer behavior. State-led promotion of technological innovation through the support of knowledge development, knowledge diffusion, and others can be part of a policy mix that could drive the future growth of bio-based products and packaging [22].

References

1. Oki Y, Sasaki H. Social and environmental impacts of packaging (LCA and assessment of packaging functions). *Packag. Technol. Sci.* 2000; **13**(2): 45–53.
2. Meherishi L, Narayana SA, Ranjani KS. Sustainable packaging for supply chain management in the circular economy: A review. *J. Clean. Prod.* 2019; **237**: 117582. Available at: http://www.sciencedirect.com/science/article/pii/S0959652619323960.
3. Rokka J, Uusitalo L. Preference for green packaging in consumer product choices – do consumers care? *Int. J. Consum. Stud.* 2008; **32**(5): 516–25.
4. International Energy Agency. The future of petrochemicals: Towards more sustainable plastics and fertilisers; 2018.
5. Gendell A, McTigue Pierce L. 2018 Sustainable packaging study: Comprehensive results of the 2018 Sustainable packaging study conducted by Packaging Digest in partnership with the Sustainable Packaging Coalition. 2018. Available at: https://sustainablepackaging.org/wp-content/uploads/woocommerce_uploads/2018/12/2018-Sustainable-Packaging-Study.pdf.
6. Herbes C, Beuthner C, Ramme I. Consumer attitudes towards biobased packaging – a cross-cultural comparative study. *J. Clean. Prod.* 2018; **194**: 203–18.
7. Eurostat. Packaging waste statistics; 2019. Available at: https://ec.europa.eu/eurostat/statistics-explained/index.php/Packaging_waste_statistics#Waste_generation_by_packaging_material.
8. Leal Filho W, Saari U, Fedoruk M, Iital A, Moora H, Klöga M et al. An overview of the problems posed by plastic products and the role of extended producer responsibility in Europe. *J. Clean. Prod.* 2019; **214**: 550–8. Available at: http://www.redi-bw.de/db/ebsco.php/search.ebscohost.com/login.aspx%3fdirect%3dtrue%26db%3dbuh%26AN%3d134323325%26lang%3dde%26site%3dehost-live.
9. Plastics Europe. Plastics - the facts 2018: An analysis of European plastics production, demand and waste data; 2018. Available at: https://www.plasticseurope.org/application/files/6315/4510/9658/Plastics_the_facts_2018_AF_web.pdf.
10. European Commission. A European strategy for plastics in a circular economy; 2018. Available at: https://ec.europa.eu/environment/circular-economy/pdf/plastics-strategy-brochure.pdf.
11. UNEP. Single-use plastics: A roadmap for sustainability; 2018. Available at: https://wedocs.unep.org/bitstream/handle/20.500.11822/25496/singleUsePlastic_sustainability.pdf?isAllowed=y&sequence=1.
12. United States Environmental Protection Agency. Facts and figures about materials, waste and recycling: Containers and packaging: Product-specific data; 2018. Available at: https://www.-about-materials-waste-and-recycling/containers-and-packaging-product-specific-data.
13. European Bioplastics. Bioplastics market data 2018: Global production capacities of bioplastics 2018–2023. Available at: https://www.european-bioplastics.org/wp-content/uploads/2016/02/Report_Bioplastics-Market-Data_2018.pdf.
14. European Commission. Directive 2008/98/EC on Waste (Waste Framework Directive); 2019. Available at: https://ec.europa.eu/environment/waste/framework.

15. Goyal S, Esposito M, Kapoor A. Circular economy business models in developing economies: Lessons from India on reduce, recycle, and reuse paradigms. *Thunderbird Int. Bus. Rev.* 2018; **60**(5): 729–40. Available at: http://www.redi-bw.de/db/ebsco.php/search.ebscohost.com/login.aspx%3fdirect%3dtrue%26db%3dbuh%26AN%3d131095675%26lang%3dde%26site%3dehost-live.
16. Ballesteros LF, Michelin M, Vicente AA, Teixeira JA, Cerqueira MÂ. Introduction. In: Ballesteros LF, Michelin M, Vicente AA, Teixeira JA, Cerqueira MÂ, editors. *Lignocellulosic Materials and Their Use in Bio-based Packaging*. Cham: Springer International Publishing; 2018. p. 1–11 Available at: https://doi.org/10.1007/978-3-319-92940-8_1.
17. Pandit P, Nadathur GT, Maiti S, Regubalan B. Functionality and Properties of Bio-based Materials. In: Ahmed S, editor. *Bio-based Materials for Food Packaging: Green and Sustainable Advanced Packaging Materials*. Singapore: Springer Singapore; 2018. p. 81–103 Available at: https://doi.org/10.1007/978-981-13-1909-9_4.
18. Robertson G. State-of-the-art biobased food packaging materials. In: Chiellini E, editor. *Environmentally Compatible Food Packaging: Woodhead Publishing Series in Food Science, Technology and Nutrition*: Woodhead Publishing; 2008. p. 3–28 Available at: http://www.sciencedirect.com/science/article/pii/B9781845691943500015.
19. van der Zee M. Biodegradability of polymers – mechanisms and evaluation methods. In: Bastioli C, editor. *Handbook of Biodegradable Polymers*. Shawbury: Rapra Technology Limited; 2005. p. 1–31.
20. Brodin M, Vallejos M, Opedal MT, Area MC, Chinga-Carrasco G. Lignocellulosics as sustainable resources for production of bioplastics – a review. *J. Clean. Prod.* 2017; **162**: 646–64. Available at: http://www.redi-bw.de/db/ebsco.php/search.ebscohost.com/login.aspx%3fdirect%3dtrue%26db%3dbuh%26AN%3d124142676%26lang%3dde%26site%3dehost-live.
21. Spierling S, Knüpffer E, Behnsen H, Mudersbach M, Krieg H, Springer S et al. Bio-based plastics – a review of environmental, social and economic impact assessments. *J. Clean. Prod.* 2018; **185**: 476–91. Available at: http://www.redi-bw.de/db/ebsco.php/search.ebscohost.com/login.aspx%3fdirect%3dtrue%26db%3dbuh%26AN%3d128742401%26lang%3dde%26site%3dehost-live.
22. Imbert E, Ladu L, Morone P, Quitzow R. Comparing policy strategies for a transition to a bioeconomy in Europe: the case of Italy and Germany. *Energy Res. Soc. Sci.* 2017; **33**: 70–81.
23. Razza F, Degli Innocenti F, Dobon A, Aliaga C, Sanchez C, Hortal M. Environmental profile of a bio-based and biodegradable foamed packaging prototype in comparison with the current benchmark. *J. Clean. Prod.* 2015; **102**: 493–500.
24. Heidbreder LM, Bablok I, Drews S, Menzel C. Tackling the plastic problem: A review on perceptions, behaviors, and interventions. *Sci. Total Environ.* 2019; **668**: 1077–93. Available at: http://www.sciencedirect.com/science/article/pii/S0048969719309519.
25. Dilkes-Hoffman LS, Pratt S, Laycock B, Ashworth P, Lant PA. *Public attitudes towards plastics. Resources, Conservation & Recycling* 2019; **147**: 227–35. Available at: http://www.sciencedirect.com/science/article/pii/S0921344919302101.
26. Lynch DHJ, Klaassen P, Broerse JEW. Unraveling dutch citizens' perceptions on the bio-based economy: the case of bioplastics, bio-jetfuels and small-scale bio-refineries. *Ind. Crop. Prod.* 2017; **106**: 130–7. Available at: http://www.sciencedirect.com/science/article/pii/S0926669016307208.
27. Karana E. Characterization of 'natural' and 'high-quality' materials to improve perception of bio-plastics. *J. Clean. Prod.* 2012; **37**: 316–25. Available at: http://www.redi-bw.de/db/ebsco.php/search.ebscohost.com/login.aspx%3fdirect%3dtrue%26db%3dbuh%26AN%3d79481512%26lang%3dde%26site%3dehost-live.
28. Kainz U. Consumers' willingness to pay for durable biobased plastic products: Findings from an experimental auction [Doctoral dissertation]. Munich: Technische Universität München; 2016.
29. Scherer C, Emberger-Klein A, Menrad K. Biogenic product alternatives for children: consumer preferences for a set of sand toys made of bio-based plastic. *Sustainable Prod. Consum.* 2017; **10** 1–14. Available at: http://www.sciencedirect.com/science/article/pii/S2352550916300240.

30. Iles A, Martin AN. Expanding bioplastics production: sustainable business innovation in the chemical industry. *J. Clean. Prod.* 2013; **45**: 38–49. Available at: http://www.sciencedirect.com/science/article/pii/S0959652612002314.
31. Álvarez-Chávez CR, Edwards S, Moure-Eraso R, Geiser K. Sustainability of bio-based plastics: general comparative analysis and recommendations for improvement. *J. Clean. Prod.* 2012; **23**(1): 47–56. Available at: http://www.sciencedirect.com/science/article/pii/S095965261100374X.
32. Sijtsema SJ, Onwezen MC, Reinders MJ, Dagevos H, Partanen A, Meeusen M. Consumer perception of bio-based products – An exploratory study in 5 European countries. *NJAS-Wageningen J. Life Sci.* 2016; **77**: 61–9. Available at: http://www.sciencedirect.com/science/article/pii/S1573521416300070.
33. Morone P, Tartiu VE, Falcone P. Assessing the potential of biowaste for bioplastics production through social network analysis. *J. Clean. Prod.* 2015; **90**: 43–54. Available at: http://www.sciencedirect.com/science/article/pii/S0959652614012761.
34. Scott L, Vigar-Ellis D. Consumer understanding, perceptions and behaviours with regard to environmentally friendly packaging in a developing nation. *Int. J. Consum. Stud.* 2014; **38**(6): 642–9.
35. Jerzyk E. Design and communication of ecological content on sustainable packaging in young Consumers' opinions. *J. Food Prod. Mark.* 2016; **22**(6): 707–16.
36. Boesen S, Bey N, Niero M. Environmental sustainability of liquid food packaging: is there a gap between Danish consumers' perception and learnings from life cycle assessment? *J. Clean. Prod.* 2019; **210**: 1193–206.
37. Klaiman K, Ortega DL, Garnache C. Consumer preferences and demand for packaging material and recyclability. *Resour. Conserv. Recycl.* 2016; **115**: 1–8.
38. Herbes C. Marketing of biomethane. In: Herbes C, Friege C, editors. *Marketing Renewable Energy: Concepts, Business Models and Cases*. Cham, CH: Springer; 2017. p. 151–72.
39. Herbes C, Beuthner C, Ramme I. How green is your packaging – A comparative international study of cues consumers use to recognize environmentally friendly packaging. *Int. J. Consum. Stud.* 2020; **44**: 258–271.
40. Magnier L, Crié D. Communicating packaging eco-friendliness: An exploration of consumers' perceptions of eco-designed packaging. *Intl. J. Retail & Distrib. Mgt.* 2015; **43** (4/5): 350–66.
41. Lindh H, Olsson A, Williams H. Consumer perceptions of food packaging: contributing to or counteracting environmentally sustainable development? *Packag. Technol. Sci.* 2016; **29**(1): 3–23.
42. Steenis ND, van Herpen E, van der Lans IA, Ligthart TN, van Trijp HC. Consumer response to packaging design: The role of packaging materials and graphics in sustainability perceptions and product evaluations. *J. Clean. Prod.* 2017; **162**: 286–98.
43. Kainz U, Zapilko M, Decker T, Menrad, Klaus. Consumer-relevant information about bioplastics. In: Geldermann J, Schumann M, editors. *First International Conference on Resource Efficiency in Interorganizational Networks-ResEff* 2013. Göttingen: Universitätsverlag Göttingen; p. 391–402.
44. Moon S-J, Costello JP, Koo D-M. The impact of consumer confusion from eco-labels on negative WOM, distrust, and dissatisfaction. *Int. J. Advert.* 2017; **36**(2): 246–71. Available at: http://www.redi-bw.de/db/ebsco.php/search.ebscohost.com/login.aspx%3fdirect%3dtrue%26db%3dbuh%26AN%3d121063493%26lang%3dde%26site%3dehost-live.
45. Brécard D. Consumer confusion over the profusion of eco-labels: lessons from a double differentiation model. *Resour. Energy Econ.* 2014; **37**: 64–84. Available at: http://www.redi-bw.de/db/ebsco.php/search.ebscohost.com/login.aspx%3fdirect%3dtrue%26db%3dbuh%26AN%3d97106401%26lang%3dde%26site%3dehost-live.
46. Lettner M, Schöggl J-P, Stern T. Factors influencing the market diffusion of bio-based plastics: results of four comparative scenario analyses. *J. Clean. Prod.* 2017; **157**: 289–98. Available at: http://www.sciencedirect.com/science/article/pii/S095965261730793X.
47. Babu RP, O'Connor K, Seeram R. Current Progress on bio-based polymers and their future trends. *Prog. Biomater.* 2013; **2**(1): 8.Available at: https://doi.org/10.1186/2194-0517-2-8.

48. Ballesteros LF, Michelin M, Vicente AA, Teixeira JA, Cerqueira MÂ. Conclusion and Future Trends. In: Ballesteros LF, Michelin M, Vicente AA, Teixeira JA, Cerqueira MÂ, editors. *Lignocellulosic Materials and Their Use in Bio-based Packaging*. Cham: Springer International Publishing; 2018. p. 95–7 Available at: https://doi.org/10.1007/978-3-319-92940-8_7.
49. Xanthos D, Walker TR. International policies to reduce plastic marine pollution from single-use plastics (plastic bags and microbeads): A review. *Mar. Pollut. Bull.* 2017; **118**(1): 17–26. Available at: http://www.sciencedirect.com/science/article/pii/S0025326X17301650.
50. Schnurr REJ, Alboiu V, Chaudhary M, Corbett RA, Quanz ME, Sankar K et al. Reducing marine pollution from single-use plastics (SUPs): a review. *Mar. Pollut. Bull.* 2018; **137**: 157–71. Available at: http://www.sciencedirect.com/science/article/pii/S0025326X18307033.
51. Skogstad G, Wilder M. Strangers at the gate: the role of multidimensional ideas, policy anomalies and institutional gatekeepers in biofuel policy developments in the USA and European Union. *Policy. Sci.* 2019; **52**(3): 343–66. Available at: https://doi.org/10.1007/s11077-019-09351-5.
52. Gardebroek C, Reimer JJ, Baller L. The impact of biofuel policies on crop acreages in Germany and France. *J. Agric. Econ.* 2017; **68**(3): 839–60.
53. Miljkovic D, Ripplinger D, Shaik S. Impact of biofuel policies on the use of land and energy in U.S. agriculture. *J. Policy Model* 2016; **38**(6): 1089–98.
54. Sorda G, Banse M, Kemfert C. An overview of biofuel policies across the world. *Energy Policy* 2010; **38** (11): 6977–88.
55. Thomas GO, Poortinga W, Sautkina E. The welsh single-use carrier bag charge and behavioural spillover. *J. Environ. Psychol.* 2016; **47**: 126–35. Available at: http://www.sciencedirect.com/science/article/pii/S0272494416300536.
56. Dikgang J, Leiman A, Visser M. Analysis of the plastic-bag levy in South Africa. *Resour. Conserv. Recycl.* 2012; **66**: 59–65. Available at: http://www.sciencedirect.com/science/article/pii/S0921344912001073.
57. Ilyas RA, Sapuan SM, Atikah MSN, Asyraf MRM, Ayu Rafiqah S, Aisyah HA, Mohd Nurazzi N, Norrahim MNF. Effect of hydrolysis time on the morphological, physical, chemical, and thermal behavior of sugar palm nanocrystalline cellulose (*Arenga pinnata* (Wurmb.) Merr). *Textile Research Journal.* 2020. In print. doi: 10.1177/0040517520932393.

27

Regulations for Food Packaging Materials

R.A Ilyas[1,2], S.M Sapuan[3,4], L.N. Megashah[5], Rushdan. Ibrahim[6], M.S.N. Atikah[7], Z.M.A. Ainun[8], Min Min Aung[9,10,11], S.O.A. SaifulAzry[2], and C.H. Lee[4,12]

[1] *School of Chemical and Energy, Faculty of Engineering, Universiti Teknologi Malaysia,81310 UTM Johor Bahru, Johor, Malaysia*

[2] *Centre for Advanced Composite Materials, Universiti Teknologi Malaysia,81310 UTM Johor Bahru, Johor, Malaysia*

[3] *Advanced Engineering Materials and Composites Research Centre (AEMC), Department of Mechanical and Manufacturing Engineering, Faculty of Engineering, Faculty of Engineering, Faculty of Engineering, Universiti Putra Malaysia (UPM), Serdang, Selangor, 43400, Malaysia*

[4] *Laboratory of Biocomposite Technology, Institute of Tropical Forest and Forest Products (INTROP), Universiti Putra Malaysia (UPM), Serdang, Selangor, 43400, Malaysia*

[5] *Department of Bioprocess Technology, Faculty of Biotechnology and Biomolecular Sciences, Universiti Putra Malaysia (UPM), UPM Serdang, Selangor, 43400, Malaysia*

[6] *Pulp and Paper Branch, Forest Research Institute Malaysia, Kepong, Selangor, 52109, Malaysia*

[7] *Department of Chemical and Environmental Engineering, Universiti Putra Malaysia, UPM Serdang, Selangor, 43400, Malaysia*

[8] *Program of Pulp and Paper & Pollution Control, Laboratory of Biopolymer and Derivatives, Institute of Tropical Forestry and Forest Product (INTROP), Universiti Putra Malaysia, Serdang, Selangor, 43400, Malaysia*

[9] *Higher Education Centre of Excellence (HiCoE), Institute of Tropical Forestry and Forest Products, University Putra Malaysia (UPM), Serdang, Selangor, 43400, Malaysia*

[10] *Department of Chemistry, Faculty of Science, Universiti Putra Malaysia (UPM), Serdang, Selangor, 43400, Malaysia*

[11] *Chemistry Division, Centre of Foundation Studies for Agricultural Science, University Putra Malaysia (UPM), Serdang, Selangor, 43400, Malaysia*

[12] *Faculty of Engineering, Universiti Putra Malaysia (UPM), Serdang, Selangor, 43400, Malaysia*

Bio-Based Packaging: Material, Environmental and Economic Aspects, First Edition. Edited by S.M. Sapuan and R.A. Ilyas

27.1 Introduction

Food packaging materials usually consist of various materials, such as paper and boards, glass, and aluminum, which are commonly available from plastics. Selecting food packaging materials is important to protect food from contamination, spoilage, pest attacks, and damage during transportation, storage, and retail sales. The design and materials used in food packages and containers play an important role in improving shelf life and ensuring safety, maintenance, and performance [1]. Plastics have major advantages as compared to many other types of materials due to their durability, good barrier properties, lightness, and cost effectiveness. It has been realized since the 1950s that the use of plastic materials has increased to twentyfold in the world populations and global market. The development in polymerization of organic molecules has brought a large quantity of plastics with various properties that can be widely used. However, most plastic packaging design for foods are for single use and will be thrown away. Therefore, this chapter attempts to review the world's legislations with regard to food packaging materials and waste packaging. Regulations on food packaging do not only cover food contacts, but also environmental and management concerns.

Food packaging has been used for the transport and storage of food since ancient times. While natural materials were initially used and are still being used, people have begun to develop packaging materials that meet their needs. It has been known for a long time that ceramics, paper, metal, glass, and natural resins are used in packaging. The development of food packaging materials only exploded in the last century. The main way to fulfill this need is not only through packaging materials but also through equipment and utensils in contact with food. Industrial food preparation, longer distribution lines, long shelf-life demand, and increased consumption of packaged small portions require the specific properties of packaging materials.

In addition, the appearance and advertising of packaged food, which should persuade consumers to buy the food, has an effect on the packaging material composition. Furthermore, environmental aspects have a growing impact. Material recycling has become a growing industry since natural sources are not infinite. An improvement in biodegradable packaging is required in order to control waste problems. However, all these latest changes in food contact materials (FCM) that are driven by environmental influences should not endanger food safety.

An increasing concern has become obvious with the growing use of packaging materials. Responsible authorities have begun making food contamination threat inventories. Protection of consumers is the driving force of any action taken by the authorities. Therefore, in one way or another, the migration of substances from packaging materials is restricted. Most countries have constructed their own regulatory system and, as a result, there are different methods used throughout the world. Notwithstanding that fact, several countries have recognized the regulations of other countries and accepted the safety of packaging material when complying with such recognized regulations.

Many regulatory systems use conservative methods to determine the exposure and optimum levels of migration. It may be prescribed to be extracted and migrated into a food stimulant or in real food. Several countries require restrictions on packaging systems, while others depend on substance migration. Whichever system is used, the authorities will always have an enforcement mechanism. The industry has done a lot of work and produced assets on a voluntary or commentary basis in order to illustrate compliance with appropriate rules concerning their materials.

Food packaging materials are capable of maintaining the quality of the food as well as safety of the food from other contaminants [2–8]. Food packaging materials act as a protective barrier against penetration of water, oxygen, and other things that can affect the quality of food. A different packaging material had been used for years with various physical and chemical properties. There is an effort to retain different types of packaging materials to improve the effectiveness in maintaining the quality and safety of food. Characteristic properties required for the food packaging material are shown in Figure 27.1.

Most food packaging used nowadays are made of plastic, which is the latest packaging material. Various plastic substances were produced in the 1800s, but were not ready to be used as packaging. It was not until the invention of cellophane in 1908 that the foundations of plastic packaging were laid. In 1946, a spray bottle named "Stopette" was introduced. A few decades later, PET (*polyethylene terephthalate*) plastic bottles were introduced in 1977 and then became the standard container for plastic bottles.

Some researchers have found new technologies to replace traditional food packaging materials, namely nanostructured materials [10–17]. Nanostructured materials are attractive in food packaging, as their enhanced functional properties, such as mechanical strength and barrier properties, have antimicrobial and antioxidant activities that maintain quality and extend shelf life in various food applications [17–27]. Several nanostructured materials (i.e. silicate nanoclay, nanoparticles of metallic oxide, and nanocrystal cellulose) were used as functional properties in food packaging systems. However, there is still a controversy over safety concerns of nanostructured materials [28].

Governments, academia, and private sectors in the United States (US), Europe, Australia, Asia, New Zealand, and other countries have generally initiated the regulation on the use of nanotechnology. So far, the US has not regulated any definition of nanomaterial,

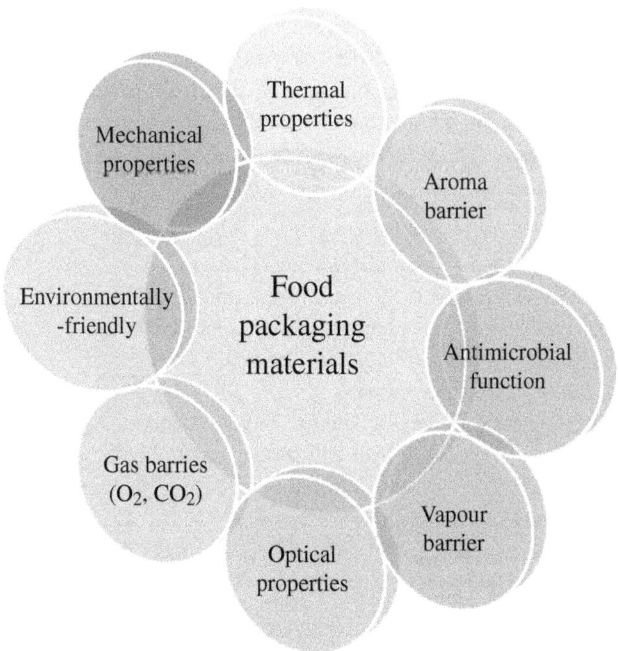

Figure 27.1 *Characteristic properties of food packaging material [9].*

nanotechnology, nanoscale, or any other related terms. According to the Food and Drug Administration (FDA), nanomaterials in regulated products are not mandatory [28]. Therefore, the aim of this present work is to provide insights regarding regulation aspects and global market of packaging for food industry applications in Europe, Australia, Asia, Africa, North America, and South America.

27.2 Asia

Since the 1950s, Asian countries have provided a general rule of law with several regulations and guidelines in addressing packaging/container materials, especially for food packaging. The regulations were developed as a result of environmental and health problems related to toxic, hazardous, and accumulation of plastic packaging materials. The regulation has been created by expert groups, including nutritionists, dieticians, physicians, and food scientists. In this section, the regulations of food packaging/container materials in several Asian countries, including Malaysia, Japan, China, and Thailand are discussed.

27.2.1 Malaysia

Recently, the Ministry of Health has shown a great deal of concern about single-use plastics that are widely applied in consumer products, such as drinking bottles, carrying bags, plastic wrappers, and food packaging. Single-use plastics refer to plastics that are designed to be discarded after use and are often widely identified as being used for food packages. For far too long, the State Government has drawn up a regulation aimed at providing a policy for all stakeholders, including local industries and communities. The law on food containers and packaging materials in Malaysia is all in accordance with the corresponding legislation of the Malaysia's Food Act 1983 and the Food Regulations of 1985 [29]. According to Claudio [30], the risk of chronic contaminant ingestion from food packaging is difficult to estimate because it is well known that the chemicals hidden in packaging/containers could possibly migrate into foods. The following are regulations for food packaging, as described in Section VI, No. 27 to 36A:

Regulation 27 *Forbidden use of harmful packages*
The legislation bans the use of food container/packaging which yields or could yield to its contents of any toxic, hazardous, or tainting substance or lead to food deterioration.

Regulation 28 *Safety of packages for food*
The legislation prohibits the use of food containers/packages that are capable of imparting lead, antimony, arsenic, cadmium, or any other toxic substance to any food that is prepared, packaged, stored, delivered, or exposed in it or that is not acid-resistant unless the package, appliance, container, or vessel meets the test described in the 13th Schedule as follows:

A. *Test for food container/packaging for storage*
   The surface of the ware to be tested shall be washed with detergent and rinsed with clean water. The remaining water shall be removed from the washed ware by rinsing it with a leaching solution, consisting of 4% (v/v) acetic acid in water and left for 24 hours at room temperature. The leachate must not contain higher than 0.2 ppm of antimony, arsenic, cadmium, or more than 2 ppm of lead in the container/packaging for storage.
B. *Test for food container/packaging for cooking*
   The package is heated to 120°C and filled to two-thirds of its active volume with boiling leaching solution of 4% (v/v) of acetic acid in water and is gently cooked for 2 hours. The package is to remain at room temperature for 22 hours. The food container leaching solution shall contain less than 0.7 ppm of individual antimony, arsenic, cadmium, and a minimum lead of 7 ppm

Regulation 29 *Forbidden use of polyvinyl chloride (PVC) containing excess vinyl chloride monomer*
The regulation prohibits the use of food containers/packaging made from PVC that contains vinyl chloride monomer of greater than 1 mg/kg.

Regulation 30 *Foods packaged in containers of PVC must not contain excess vinyl chloride monomer*
The regulation prohibits the sale of food in packages that contain more than 0.05 mg/kg of vinyl chloride monomer.

Regulation 31 *The use of packages for non-food products is prohibited*
The regulation prohibits the use of food packaging/containers for use in any non-food product.

Regulation 32 *Recycling of packages prohibited*
It is also illegal to use a recycling box for certain items, such as sugar, flour, and edible oil as well as containers used for alcoholic beverages or shandy.

Regulation 33 *Packages for similar products that can be recycled*
The legislation also restricts the use of packages for other products that have been previously used.

Regulation 33A. *Other food packages that can be recycled for alcoholic beverages, shandy, vegetables, and fruit*
However, certain food products with the same purposes, such as alcoholic beverages and shandy or vegetables and fruit are permitted to be recycled in this regulation.

Regulation 34 *Presumption as to the use of any package*
For Regulations [32] and [33], whereby the specific food should be labeled or marked.

Regulation 35 *Prohibited use of damaged package*
A damaged package/container, which includes leakage, corrosion, and perforation, are also strictly forbidden.

Regulation 36 *Not to put toys, coins, etc., in foods*
The regulation also prohibits the inclusion of other non-food products, such as toys, coins, or any other in the sale of such food or packages.

Minimize the use of iron powder
The regulation provides a list of substances that may be in the sachet of reduced iron powder, such as calcium chloride, calcium hydroxide, iron oxide, etc. The package shall be marked with the words "OXYGEN ABSORBER" or any term with similar meaning, followed by the words "DO NOT EAT CONTENTS" and "CONTAINS IRON POWDER."

However, the regulation is more concerned about consumer health rather than issues of waste management of this food packaging/container. Environmental issues related to plastic waste have become a major issue in Malaysia, which ranked eighth among the top countries that mismanaged plastic waste disposal. A recent official report by the Ministry of Energy, Science, Technology, Environment, and Climate Change (MESTECC) has emphasized in Malaysia Plans to encourage the plastic industry to replace a single-use plastic with eco-friendly materials [31]. The roadmap, which aims toward zero single-use plastics for a cleaner and healthier environment by 2030, was implemented in October 2018.

In Phase 1 of the roadmap (2018–2021), nationwide charges will be put on the use of plastic bags at a minimum of RM0.20 and also the use of plastic straws in drinking water will be banned. People are encouraged to bring their own food containers or to provide biodegradable and compostable plastic packaging materials or products based on biomass to fixed premises. In the Ipsos Malaysia survey, shown in Figure 27.2, people were asked about plastic bags, plastic straws, food packaging, and other recycled materials. This survey showed that plastic bags were the most used item, which accumulated by about 79% in the past three months, followed by plastic bottles, straws, food packaging/containers, etc. It also revealed that 37% to 47% of Malaysians were aware of the reusability of food packaging and containers.

Meanwhile, in Phase 2 of the roadmap (2022–2025), the State Government strongly insists that biodegradable and compostable products for food packaging, food containers, cutleries, etc., should be used. They are therefore providing funding for research and development (R&D), an incentive for manufacturing, and any activities aimed to prevent plastic pollution. In conjunction with biodegradable materials, such as polyhydroxyalkonates (PHA) and other biodegradable commodities, more tests will also be established to provide a reduction of plastic uptake. Therefore, the final part of this roadmap was set out in Phase 3 (2026–2030) to increase the amount of sustainable packaging manufactured for local consumption. It is foreseen that the ideal solution for overcoming the environmental problem is related to product chains, including health protection and food safety promotion.

27.2.2 Japan

Japan is one of the most advanced countries in Asia due to its fast technological growth. This factor also affects the productivity of its food packaging industry. The efficiency of their system leads to high demands from consumers throughout the world. Despite the growing demands, Japan still emphasizes the quality of food products produced by its industry.

Gap in awareness as to which plastic items are recyclable

MALAYSIA

	Plastic items used in past three months	Awareness that item can be recycled	Average time used before disposal
Plastic bags	79%	58%	3.5%
Plastic bottles	74%	68%	68%
Plastic straws	65%	42%	2.9%
Food packaging	52%	37%	1.9%
Plastic food containers	51%	47%	5.4%
Plastic cups	47%	52%	5.3%
Cotton buds	43%	20%	1.5%
Polystyrene containers	41%	30%	1.7%
Plastic cutleries	36%	36%	2.9%
Polystyrene cups	32%	31%	1.3%
Disposable coffee cup	26%	31%	1.1%
Plastic films	18%	27%	1.4%

Based on findings from interviewing 3000+ Malaysians nationwide in October 2018 – June 2019

Ipsos

Figure 27.2 *Malaysia's awareness regarding. Recyclable plastic items. Source: From Ref. 32.*

As part of its marketing plan, the Japanese government has considered alternative food packaging to gain trust from its customers.

In Japan, there are two national laws for food safety, which are the Food Sanitation Law and the Food Safety Basic Law. They were enacted under The Constitution of Japan Article 25 [1] that pledges: "All people are entitled to maintain the minimum standards of healthy and cultured living." The Food Sanitation Law was enforced in 1947 and regulated the safety of all food, appliances, food additives, and non-drug packages/containers, while the Pharmaceutical Affairs Law was introduced as the regulation on food safety management.

In 1959, the Ministry of Health and Welfare established food, food additives, and other material specifications and standards, known as Notification No. 370 in Japan. After hearing the statement of the Pharmaceutical Affairs and Food Sanitation Council, the Minister of Health (MOH), Labor, and Welfare may establish specifications and standards of equipment or packages/containers used in business or on sale or materials of equipment or packages/containers that do not conform to the requirements.

27.2.3 China

According to Chinese regulations, the definition of food containers and packaging materials is paper, wood, bamboo, metal, porcelain, chemical fiber, plastic, enamel, rubber, and food-containing glass products as well as food-contacting paint products, including machines, pipes, conveyor belts, containers, appliances, and tableware, etc. The China Food Hygiene Law (CFHL) is the primary law governing food container and packaging material regulation in China. The CFHL was approved at the 16th meeting of the 8th Standing Committee of the National People's Congress. It entered into force on 30 October 1995. The

Ministry of Health is responsible for establishing, maintaining, and revising hygiene-related standards for food, food seasoning, and food packaging materials according to the CFHL. Figure 27.2 shows the current framework of the food packaging material standard in China.

It is also responsible for the establishment and approval of related rules and regulations on food hygiene. Food packaging materials must also comply with the general regulations of the Food Hygiene Act, the food hygiene rules and regulations, and any relevant standards for compulsory hygiene control. For various food packaging materials, there are a total of 40 hygiene standards. These compulsory standards cover all commonly used plastics, including polyvinyl chloride (PVC), polyethylene (PE), polypropylene (PP), polystyrene (PS), polyethylene terephthalate (PET), and acrylonitrile butadiene styrene (ABS). If it is necessary to use a new resin finishing plastic articles in food packaging materials, the approval of the MOH is required. Some examples of hygiene control standards (GB) are, for plastics, GB 9691-88 for PE resin hygienic standard of food packaging and GB 9687-88 for PE finished products.

Basically, there are several major regulations on packaging materials in food safety laws, such as:

- Article 20
  This law outlines a ceiling for the usage of health harmful materials in food products, such as pathogens, pesticide residues, heavy metals, and many more.
- Article 36
  This mentions the responsibility of food manufacturers to examine the license of manufacturing and the product's certificate from the suppliers of the raw products. Products without any certificate need to proceed with inspection before they can be used to ensure that the quality of the raw material complies with the requirements of safety standards.
- Article 38
  Food manufacturers should conduct quality inspection on all products before releasing the goods into the market.
- Article 44
  Documents on the safety assessment of any new raw materials used for food-related products should be prepared for the health administrative agency of the State Council.
- Article 62
  All food products released into the international market must meet the requirements of the national food safety standard.
- Article 63
  This law is upgraded from Article 44 whereby the food products that will be exported to another country comply with this article.

27.2.4 India

In India, the Food Safety and Standards Authority of India (FSSAI) has renewed the food packaging regulations. This is to improve food safety and to avoid any contamination that may happen. FSSAI replaces all the requirements in the Food Safety and Standards Regulations 2011, especially in packaging materials as well as labeling. This new guideline states that the materials of recycled plastic are banned from the usage of packing, storing, or carrying food products. In addition, the materials from newspaper and similar materials are also banned due to the carcinogenic effect. The main objective of the new regulations is to prevent food contents from exposure to chemical, physical, and microbiological contamination, particularly when the food is packed. India also reviews the standard inks that are used on food packaging [33]. India has also taken a big step in order to save the environment.

27.3 Europe

Legislations at the national and societal level in Europe will and always will continue to coexist. Different directives and laws were drawn up a long time ago for the community level, while the European Union (EU) Framework Regulation regulates FCMs at the union level, including packaging, machinery, and kitchen ware. A total of 17 individual FCMs were allowed by the Framework Regulation for specific requirements. Six such specific requirements are included in Table 27.1.

There are two methods used to control the use of food contact products of recycled paper. Plastic, which is depolymerized into monomers or oligomers, must satisfy the same demands as virgin plastics under EC 10/2011. When the plastic is mechanically recycled and transformed into pellets, the European Food and Safety Authority (EFSA) will implement the regulation. It is controlled under Directive 2007/42/EC for regenerated cellulose films. The directive includes a favorable list of drugs that can be used in the production process. It is most likely that the printed surfaces will not come into touch with food items. If the cellulose is to come in touch with food, a written statement must be included.

Taking into account active materials and articles in both Regulation 1935/2004/EC and Regulation 450/2009/EC; "materials and articles intended to increase the shelf-life or maintain or improve the condition of packaged food" are allowed. They are intended to deliberately incorporate parts that will release or absorb substances into or from packaged food

Table 27.1 *General regulation on food contact materials (FCM).*

General regulation on PCM	
Regulation EC 1935/2004 (on materials and articles intended to come in contact with, foot)	
Regulation EC 2023/2006 (on Good Manufacturing Practices)	
Specific materials	
Ceramics	Directive 84/500/EEC
Epoxy reims	Regulation EC 1895/2005
Regenerated cellulose film	Directive 2007/42/EC
Recycled plastics material	Regulation EC 282/2008
Active and intelligent packaging	Regulation EC 450/2009
Plastics	Regulation EU 10/2011
Specific regulations	
Regulation EU 321/2011 (restricting the use of bisphenol A in polycarbonate infant feeding bottles)	
Regulation EU 284/2011 (import procedures for polyamide and melamine plastic kitchenware from China and Hong Kong)	
Regulation EC 1895/2005 (restricting the use of certain epoxy resins)	
Directive 93/11/EEC (regulating the release of N-nitrosamines and N-nitrosatable substance from rubber teats and soothers)	

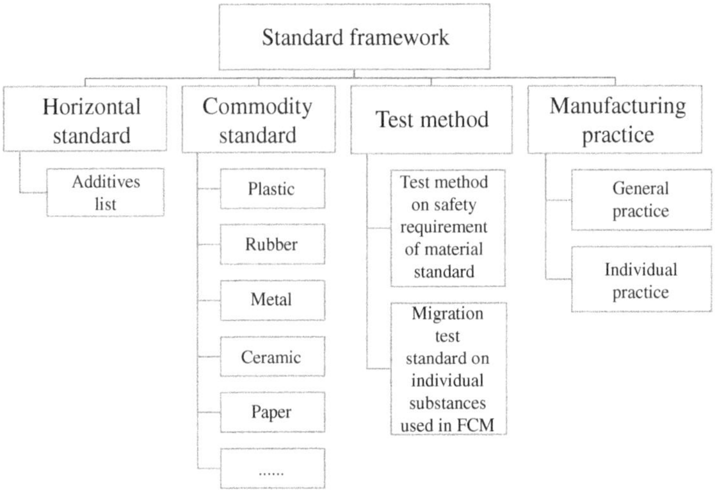

Figure 27.3 *Current framework of food packaging material standard in China.*

or the food environment [34]. Intelligent fabrics and papers, on the other hand, mean: "equipment and papers monitoring the situation of stored goods or service environment."

In Europe, regulations on such packaging contents continue to evolve (see Figure 27.3). Initially, all European laws on food contact originated from individual member states and were implemented by them. With the creation of the EU, member countries are elected to harmonize the laws in order to create a single industry and solve problems as well as overcome trade obstacles. To date, the EU law on food products has shielded customer safety by ensuring that no substance in touch with foods will motivate a response that could alter the structure or organoleptic characteristics of these foods (i.e. flavor, shape, texture, or perhaps odor). In order to enable packaging to benefit from technological development, Regulation 1935/2004/EC repeals these laws. This is essential in the EU because all packaging products (including individuals who are advised to add drugs to meals) are subjected to any or all FCM necessities along with general migration boundaries (OMLs) and particular migration restrictions (SMLs). In 2007, Heckman critically remarked on the EU strategy to find a brand new ranking between A&I packaging and the rest of food-contact products as a test for manufacturing purposes and not to complete a regulatory gap. It is clear that the use of supporting methods, such as unique branding, government schooling, or marketing can accomplish this objective within the traditional strategy. Moreover, trying to add regulatory problems where there is no demand for government concern is of dubious value and can be prevented as a consequence in giving governments the right to enforce additional duties in fields where there is no need.

It follows that active packaging aims to extend food shelf life and therefore maintains or perhaps improves its quality, whereas the purpose of intelligent packaging is to indicate and observe food freshness [35]. Many different types of effective and smart products and papers are available [36]. For example, the substances responsible for active or intelligent performance will be contained in a small paper bag during a separate container or the substances can be incorporated directly into the packaging material. Therefore, the crucial

Table 27.2 Examples of active packaging applications for use within the food industry.

Absorbing/scavenging properties	Oxygen, carbon dioxide, moisture, ethylene, flavors, taints. UV light
Releasing/emitting properties	Ethanol carbon dioxide, antioxidants, preservatives, sulfur dioxide, flavors, pesticides
Removing properties	Catalyzing food component removal: lactose, cholesterol
Temperature control	Insulating materials. Self-heating and self-cooling packaging, microwave susceptors and modifiers, temperature-sensitive packaging
Microbial and quality control	UV and surface-treated packaging materials

goal here is to produce purposeful products that adopt active material in their framework that behave or are safe in a regulated way. In addition, this benefits packagers by simplifying the processing and eliminating the possibility of inadvertently consuming a sachet, and thus increases customer security. In addition, effective and smart materials and papers also consist of one or many parts or parts of different products, such as fabrics, paper, and board, or coatings and varnishes.

Considering active packaging involves additives or "freshness enhancers" that will engage in an overwhelming number of packaging applications, thereby improving the primary packaging system conservation performance. Active packaging involves additives that are capable of scavenging or transporting oxygen, carbon dioxide, ethylene, humidity, and/or odor and flavor; release of oxygen, carbonic acid gas, humidity, alcohol, sorbates, antioxidants, and/or various preservatives and antimicrobials; and/or humidity control (see Table 27.2). The broad range of effective storage systems has particular applications for personal food goods that may well extend the shelf life. There are several trials pertaining either to the implementation of effective storage to water trade (Table 27.3) [37–39] or to effective storage techniques (Table 27.4) [36, 40].

The regulation framework authorizes the use of active and intelligent packaging in Regulation 1935/2004/EC – provided that the packaging can be demonstrated to improve the safety, quality, and shelf life of packaged foods [41]. In Article 1, it is stated that the aim of the legislation is to ensure a human health security and to safeguard customer rights, so that the regulation applies to all products and papers. The Food Contact Materials Regulation (EC) No. 1935/2004 also included effective and smart packaging (see Figures 27.4 and 27.5). According to the legislation, only drugs defined as food additives or food flavorings can be released into the environment. Another EC 450/2009 directive sets out extra security criteria for effective and smart packaging. For example, an item sold by businesses using effective and smart packaging must be followed at the retail level by a statement of compliance including customer data.

Regulation 450/2009/EC has recently incorporated the general requirements laid down in Regulation 1935/2004/EC for the safe use of active and intelligent packaging. The new Regulation also lays down specific requirements for the marketing of active and intelligent materials and articles intended for food contact. It is mentioned that the substances responsible for the active and intelligent functions can be either contained in separate containers (e.g. small sachets of oxygen absorbers) or incorporated directly into the packaging material (e.g. films absorbing oxygen).

Table 27.3 *Applications of active packaging technologies.*

Type of application	Foods
Oxygen scavengers	Ground coffee, tea, roasted nuts, potato chips, chocolate, fat powdered milk, powdered drinks, bread, tortillas, pizza, pizza crust, refrigerated fresh pasta, fruit tones, cakes, cookies, beer, deli meats, smoked and cured meats, fish, cheese
Carbon dioxide absorbers	Ground coffee
Carbon dioxide emitters	Meat, fish
Moisture absorbers	Dry and dehydrated products, meat, poultry, fish
Ethylene scavengers	Kiwifruit, banana, avocados, persimmons
Ethanol emitters	Bread, cakes, fish
Antimicrobial releasing films	Dry apricots
Antioxidant releasing films	Cereals
Flavor absorbing films	Navel orange juice
Flavor releasing films	Ground coffee
Color containing films	Surimi
Antifogging films	Some fresh fruit and vegetable packages
Antisticking films	Soft candies, cheese slices
Light absorbers	Pizza, milk
Time–temperature indicators	Microwaveable pancake syrup, refrigerated pasta, deli items
Gas permeable/breathable films	Ready-to-eat salads
Microwave susceptors	Ready-to-eat meals

Table 27.4 *Examples of some currently known active packaging systems.*

Type of active packaging system	Substances used and mode of action
Oxygen scavengers	Enzymatic systems (glucose oxidase-glucose. Alcohol oxidase-ethanol vapor) Chemical systems (powdered iron oxide, catechol, ferrous carbonate, iron-sulfur, sulfite salt-copper sulfate, photosensitive dye oxidation, ascorbic acid oxidation, catalytic conversion of oxygen by platinum catalyst)
Carbon dioxide absorbing/emitting	Iron powder-calcium hydroxide, ferrous carbonate-metal halide
Moisture absorbing	Silica gel, propylene glycol, polyvinyl alcohol, diatomaceous earth
Ethylene absorbing	Activated charcoal, silica gel-potassium permanganate, Kieselguhr, bentonite, Fuller's earth, silicon dioxide powder, powdered Oya stone, zeolite, ozone
Ethanol emitting	Encapsulated ethanol
Antimicrobial releasing	Sorbates, benzoates, propionates, ethanol, ozone, peroxide, sulfur dioxide, antibiotics, silver-zeolite, quaternary ammonium salts
Antioxidant releasing	BHA, BUT, TBIIQ, ascorbic acid, tocopherol
Flavor absorbing	Baking soda, active charcoal
Flavor releasing	Many food flavors
Color containing	Various food colors
Antilogging and antisticking	Biaxially oriented vinylon. Compression rolled oriented HOPE
Light absorbing/regulating	UV blocking agents, hydroxybenzophenone
Monitoring	Time-temperature indicators
Temperature controlling	Nonwoven microperforated plastic
Gas permeable/breathable	Surface treated, perforated, or microporous films
Microwave susceptors	Metallized thermoplastics
Insect repellant	Low toxicity fumigants (pyrethrins, permetluin)

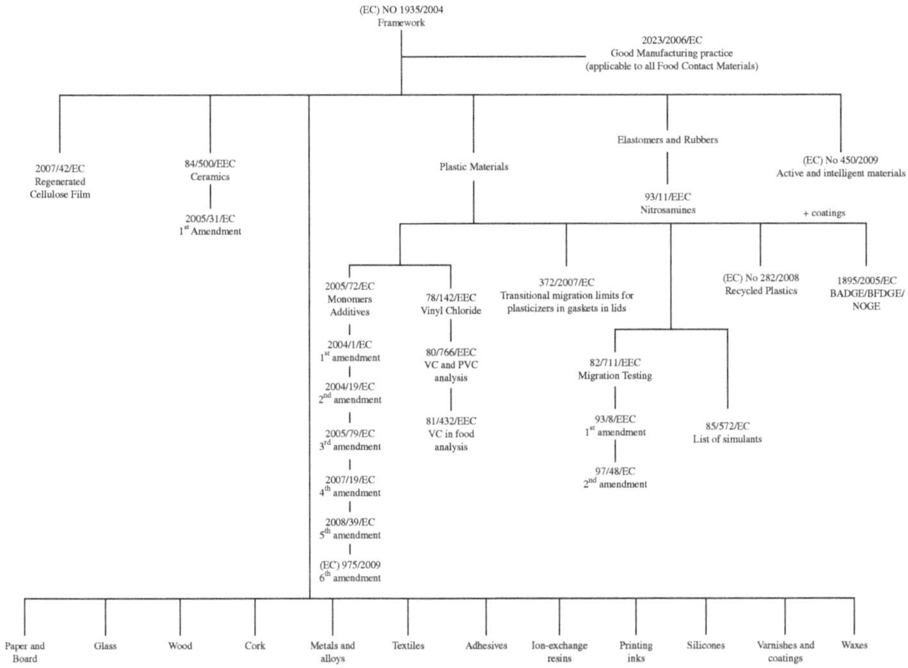

Figure 27.4 *EU food-contact materials legislations. Source: Ref. 42.*

As a conclusion for this section, the important purpose of food packaging is required to give priority to the safety, wholesomeness, and quality of food. Therefore, the problems caused by packaging waste on the environment can be reduced by taking care of the materials selected, Environmental Protection Agency (EPA) guidelines followed, and the expectations of packaging checked in terms of environmental impact. The industry, government, and consumers have to step up their knowledge efforts to promote continuous improvement and an understanding of the functional characteristics of packaging.

27.4 North America and South America

This section explains the way countries in North America and South America, especially the United States, regulates food packaging. The laws and regulations governing food packaging are often neglected by the public, but the truth is, it has a wide recognition with food laws in the legalization. Food packaging serves a range of functions that are not limited to the containment of food and beverages, but also have a significant impact on marketing new products that are able to determine a food company's success in the large and enormous competitive food industries. They provide physical protection and slowing product deterioration to offer safety and confidence to consumers and convey information of the particular contained food or beverages. More progressive food packages have included susceptors to help the browning of foods cooked in microwave ovens, highlights that connected

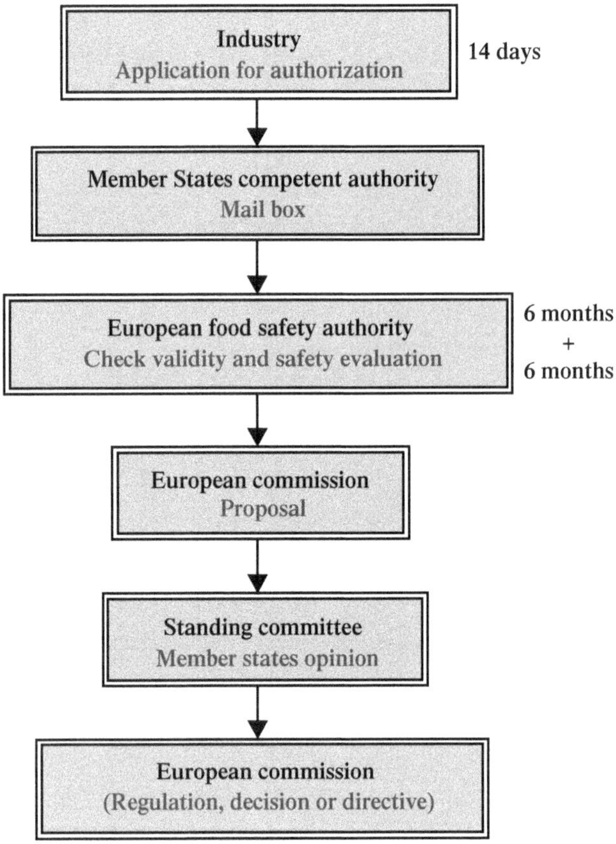

Figure 27.5 *Authorization procedure as defined by Reg. 1935/2004 EC. Source: Ref. 41.*

organically or chemically with their substance to amplify the expiry life, and the capacity to screen the condition of the food.

There are many types of packaging when it comes to food packaging that has a direct or indirect contact with the food or beverage. Figure 27.6 shows the percentages of food packaging materials that are widely used in the market. They come in various forms based on the requirements and criteria, depending on the marketing needs or supply chain. For example, they showcase the brand identity and food information that are needed by consumers. Material that has a direct contact with food is known as a "food contact material" and has strict laws and regulations to guarantee the health and welfare of the consumer.

For certain food packaging, the contact material decides its name. For instance, a plastic container is made of plastic and has direct contact with the food content. The materials in contact with the foodstuff for a glass jar are glass and include a cover coated with metal. On the other hand, beverages containing aluminum cannot have direct contact with the beverage, yet it is the coating on the inner layer of the can that has direct contact. Another example are a few kinds of paper that can likewise be coated with a layer of plastic for packaging or food wrapping (e.g. with a greaseproof coating). The range of FCMs is not

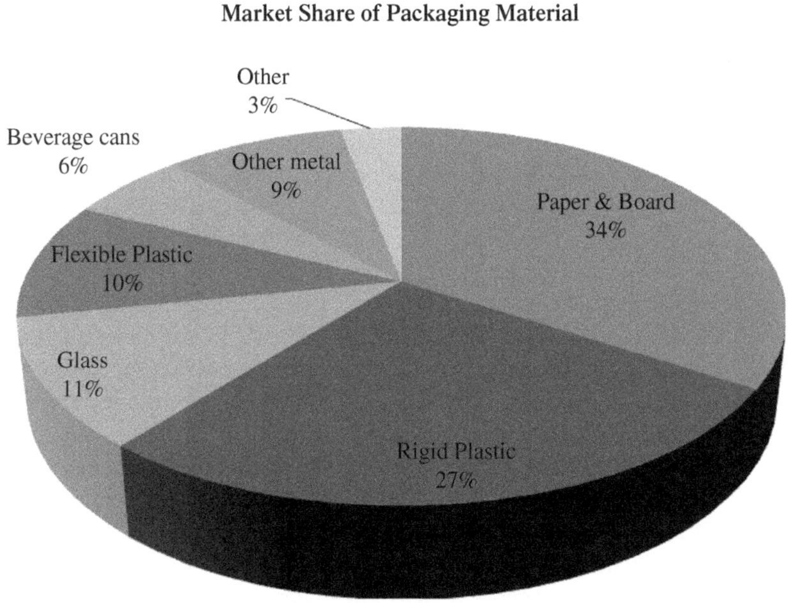

Figure 27.6 Percentages of food packaging materials that are widely used in the market [43].

restricted to food and refreshment packaging, but to any other material that has contact with the food – either during storage, handling and filling, or utilization, for example, cooking utensils used in cooking or preparing food in the kitchen or packaging processes [43].

US. regulation 21CFR174 stated that any FCM should not discharge synthetic or chemical compounds into the food at an amount that can affect human health [44]. Therefore, it is important to understand the chemical composition of the material that is used as food packaging so that health impacts that might be caused by the food packaging through the process of migration can be studied. Migration happens when these compounds in the food packaging material diffuse and mix into food, and thus are harmful to human health and body. Some of the primary packaging materials that can be found in the market include, but are not limited to, ceramics, glass, metal, paper and board, plastics, printing inks, wax, wood, et al. [44].

27.4.1 History of Formal Food Packaging Regulation in the US

With the introduction of the Food Additives Amendment of 1958, the formal regulation of food packaging in the US was drafted. A new section was added in the amendment to the Federal Food, Drug, and Cosmetic Act (FFDCA). This section defined "food additive" as "any substance the intended use of which results or may reasonably be expected to result, directly or indirectly, in its becoming a component or otherwise affecting the characteristics of any food. . . ." A substance or material can only gain exception that is categorized under this category if that substance is "generally recognized as safe (GRAS)" or otherwise it is an exclusion from the Act.

This definition is meant for a particular feature of the US food packaging regulatory scheme that is not common to different jurisdictions, particularly that all food contact substances that will fairly be expected to diffuse or move into food are regulated as food additives – almost like the manner when one substance is added into the food like a sugar substitute or/and emulsifier, and is regulated. Significantly, under Section 409 of the Act, a food additive that is not the subject of a regulation or otherwise cleared by the FDA for its intended use is taken into account to adulterate food. No showing of the potential for harm is necessary under such circumstances to ascertain adulteration.

Three exceptions were listed under the Act. The first is commonly referred to as the "prior sanctioned" substance exception. Prior to 1958, some food manufacturers of food packaging materials would consult with the FDA or the US Department of Agriculture (USDA) beforehand to ensure whether a particular material to package food was subject to any objection if used or vice versa. This was often done to assure customers of the safety of the product when it is used with food. The subjects are known as "prior sanctioned" substances if there are no objections approved and they were exempted from the food additive requirements for further clearance or authorization by the FDA.

During the 1950s, there were continuous and repetitive reports and accusations claiming that chemicals added in food were causing serious diseases that were harmful toward human safety and health, including cancer. A series of hearings were conducted on the subject of food additives due to this. These hearings raised the suspicion that food additives might be the cause of diseases and cancer. However, during the hearings, there was no information reported to prove that food materials that have direct contact with food were the potential source of any public health issue. Moreover, hearings to propose a legislation to require the FDA to give pre-clearance of food additives were held in 1956, whereby the proposed lists included the materials used to prepare, produce, manufacture, process, pack, treat, provide logistic and transport, or hold food. Even though there was no information and proofs, indicating that packaging or processing materials could spoil the quality of food or cause harm to human health and safety, they fell under the regulatory range of the substances that have been prescribed as substances that are directly added into the food and food additives.

Although the US has no comprehensive federal legislation regulating the packaging industry, the federal government has enforced its authority to regulate food, drug, and cosmetic packaging to ensure consumer safety and confidence with their purchases.

27.4.2 US Food Packaging Regulations

The U.S. FDA controls the packaging and labeling of food. The aim of these controls is to upgrade the safety of food conveyed throughout the United States and keep customers educated about the food they consume. The FDA classifies any packaging that comes into direct contact with food as a "food contact substance." It is the obligation of the Office of Food Additive Safety (OFAS) and the Centre for Food Safety and Applied Nutrition (CFSAN) to guarantee the security of these food contact substances. This implies that packaging materials like papers, coatings, plastics, food colorants, and adhesives must be directed and considered secure for utilization. The EPA requires antimicrobial innovation to be built into plastic and materials that are utilized in food packaging to avoid the development of microbes, mold, mildew, parasites, discoloration, and odor. The FDA rules, moreover, require that package labeling incorporate expiration or "best if used by" dates, nutrition guidelines, taking care of informational arrangements and the packaging company's contact data. Allergen labeling has moreover been a necessity since 2006. The Food Allergen Labeling and Consumer Act not only requires revelation in case the item contains potential allergens like peanuts, fish, milk, eggs, shellfish, soy beans, or wheat, but must give notice if the item was made or bundled in a facility utilizing any of the previously mentioned allergens.

27.4.3 Environmental Impact of Materials Used in Food Packaging

The FDA is responsible for protecting public health by ensuring the safety, efficacy, and security of human and veterinary drugs, biological products, and medical devices, and by ensuring the safety of United State of America (USA) nation's food supply, cosmetics, and products that emit radiation. In addition, the FDA's principal concern is the security of the packaging material's contact with food, whereby the National Environmental Policy Act (NEPA) has required the agency's assessment of the packaging materials environmental affect as well. All food-packaging clearances require an Environmental Assessment (EA), taken after the Finding of No Significant Impact (FONSI) and clearance by the FDA. The FDA basically audits the natural effect that stems from the packaging material's utilization and transfer. To be specific, they require information on the recycling effect of the material. Companies attempting to bundle materials in unused holders must give a method of reasoning as to why broadly recycled packaging materials like rigid single utilization of food holders or PET – the sort of plastic utilized by most soda bottlers – canno t be used.

27.4.4 Rigid Plastic Containers

The state of California's Inflexible Plastic Packaging Container (RPPC) Law was ordered to constrain the sum of disposed plastic waste filling in the California landfills and advance the utilization of recycled plastic. RPPC are characterized as holders that are fully comprised of plastic with the special case of caps, covers, and names. The holders must keep up with their shape not withstanding that the item they are holding has a capacity of at least eight liquid ounces and a most extreme capacity of five gallons, and must be competent in reopening/closing. The company title that shows up on the holder will be assigned by the state as the item producer. Oregon and Wisconsin have also ordered comparative unbending plastic laws. The statutes of all three states offer a few alternatives to producers, whereby the holders must be recyclable or reusable, or must comprise of already reused materials.

27.4.5 Regulations

The food industry needs to follow an integrated approach to ensure the safety of food from FCM to consumers. Table 27.5 shows a summary of various country treatments across the continents of North America and South America. For this to take place, information and knowledge need to be shared across all channels to the supply chain.

Table 27.5 A summary of various country treatments.

Country	Authority responsible	Type of regulation for FCM
United States, US	US Food and Drug Administration	Main regulatory body. Modified positive list system with premarketing clearance. Federal Food, Drug, and Cosmetic Act (FFDCA, Title 21 United States Code 348). 21 CFR170.3 defines the term FCM. • FCM fall into – positive lists (indirect additives) (21CFR, 174–179), GRAS (21CFR, 182–n 170.39).
	USFDA Science Board, USFDA Food Advisory Committee Center for Food Safety and Applied Nutrition (CFSAN)	Advisory scientific body. "Indirect Additive" Database – consist of regulated indirect food additives
Countries having non-specific/few regulations for FCM		
North America and Canada	Canada's Food Directorate Surveil-lance – Canadian Food Inspection Agency & Bureau of Chemical Safety in Health	Responsible for food safety from FCM • Positive list system • Letter of no objection (LONO) on voluntary basis • Controlled separately under Division 23, Part B of the Food and Drug Act and Regulations.
South American countries	Common market of the south MERCOSUR (Argentina, Brazil, Paraguay, Uruguay, Venezuela)	Harmonized Regulations. • Positive list approach based on EU's regulation 2011 and USFDA. • Legal confirmation for a specific substance to be taken from concerned authorities. • Final FCM/package needs to have premarket approval/registration. • If FCM not present in positive list, they must be petitioned for inclusion in MERCOSUR positive lists to national sanitary authorities. • GMC 26/03 – defines FCM.
E.g. Argentina		Ministry of Health – regulatory agency CONAL (National Food Commission) – advisory scientific body.
Brazil		Ministry of Health – through its regulatory agency – ANVISA – National Agency of Sanitary Surveillance. In Argentina and Brazil FCM are defined and regulated independently.

Table 27.5 (continued)

Country	Authority responsible	Type of regulation for FCM
Mexico		Ministry of Health (SSA) – through COFEPRIS. • Have no specific regulations on FCM and no specific regulations for their use. • Have regulations on packaging materials. • Have no specific regulations on FCM. • Only limited to few food packaging standards in their legislation for F&B. For e.g. Some standards on Ecuador, Bolivia, and Peru food packaging material regulations.
Other South American countries – Chile, Bolivia, Ecuador, and Peru		

27.4.6 The US Exposure Approach to FCM Legislation

When discussing the control of FCM in the US in the context of global FCM control systems, it is very important to recognize the difference in approach from that prevailing in the EU. Meanwhile, superficially, in terms of toxicology assessment and analysis, there are similarities, whereby the two systems have fundamental differences and represent almost polar opposites in terms of regulatory systems.

The EU takes a subjective toxicological approach to assess each substance and place it on a positive list. The US take an exposure approach to assess the use of material and its contribution to the diet, basing its decisions on the degree of toxicology testing that is required, if any. It is essentially based upon the Paracelsusian principle "the dose makes the poison." Therefore, a series of exemptions to the regulations are either legislated for or are followed through custom and practice, whereby the exposure levels are anticipated to be low or minimal.

The other fundamental difference is the respective responsibilities of the packaging manufacturer and the regulatory authority. In the US, it is up to the manufacturer to interpret the regulations and decide whether approval is needed or not. The FDA considers it unnecessary and even inadvisable – that it should be consulted – seeing this as the role of the manufacturer.

27.4.7 The Regulatory Enforcement Process in the United States

The FDCA sets up the US FDA as the body in charge of the guidelines on FCMs. This work is done at the CFSAN by the OFAS. Food additive substance status is a pivotal piece of the assessment and implementation process at the Department of Agriculture's Food Safety and Inspection Service (FSIS), which administers meat and poultry plants and items.

The Department of the Treasury's Bureau of Alcohol, Tobacco, and Firearms surveys add substance status in the clearance of packaging materials [45].

27.4.8 A Practical Approach to the US Food Contact Materials Regulatory Regime

Due to the cost of toxicological testing and regulatory compliance, it is important that manufacturers adopt a structured approach to their decisions regarding the pre-clearance of FCMs [46]. The first step is to check if the substance and use are already pre-cleared under a published FAP (Food Additive Petition) regulation, FCN (Food Contact Notification), or are subject to a pre-1958 prior sanction. The next step is to establish that migration can potentially occur as, by definition, if there is no migration, it is not a food additive; hence it is not subject to regulation.

Exemptions play a large part in the US system. Therefore, it must be established if the substance is regarded as safe (GRAS), below the threshold of regulation (TOR), a housewares, or if it falls under the basis polymer doctrine, all of which may exempt it. The OFAS will assist in establishing this if requested.

Once a manufacturer determines that pre-clearance is necessary for the marketing of their material, there are two protocols available. In the vast majority of cases, they will assemble the required data, submit this to the FDA via the FCN system, and, assuming that there are no objections from the FDA, the notifiers will be allowed to market their material after 120 days. The notification is posted on the FDA FCN website and applies to the notifier only for the use/s as stipulated. Alternatively (and infrequently since the introduction of the FCN process), an FAP may be applied for. This includes a similar data to the FCN. However, the difference is that, once approved, the petition is published in Title 21 of the Code of Federal Regulations (21 CFR) as the regulation that gives approval to all manufacturers. This process from application to publication can take from two to four years.

More noteworthy, consideration has been paid to food packaging over the previous few years, with the portion of that consideration within the frame of expanded investigation concerning the security of FCMs. A few of the concerns stem from progress in innovation and the requirement for the FDA to survey a few of the choices it has made earlier to those advances, but it is additionally due to assaults by the environmental group and other self-defined "public interest" organizations on the security of food packaging materials. One case is perfluorinated substances.

A request to delist three perfluoroalkyl ethyl compounds utilized in water and oil repellent coatings for paper and paperboard to bundle fluid and greasy food was submitted to the FDA in March 2015. In investigating the request, FDA seemed undecided about the dietary presentation from food contact utilization of the substances. In expansion, there were no toxicological considerations appearing to be regenerative or of a formative poisonous quality for the three substances, but as of late accessible information did appear of a poisonous quality of some of the organized substances. As a result, the FDA concluded that there was not a sensible certainty of no harm for the food contact utilization of the three substances and revised the food-added substance regulations in January 2016 to not allow their utilization. Interestingly, the three substances had not been utilized within the US since 2011; the people who submitted the appeal demonstrated that they were concerned that the food packaging made with these substances was being exported by other countries.

The utilization of food contact substances can be prohibited in personal states or neighborhood purviews. In March 2018, the senator of Washington State marked a charge, prohibiting food packaging that contains perfluoroalkyl and polyfluoroalkyl substances. The ban will become compelling in 2022 unless the state Department of Ecology recognizes a more secure elective at that point. In San Francisco, a ban on the deal of food benefit products that contain fluorinated chemicals took effect in 2019. It is unnecessary to mention that food packaging materials are also subject to the caution necessities of California's Proposition 65 (Safe Drinking Water and Toxic Enforcement Act of 1986).

As food packaging proceeds to gain more consideration, particularly with respect to the security of individual components, the food and food packaging industries will have to be kept side-by-side of the most recent science and the way it fits in with the statutory and administrative necessities outlined to guarantee the security of packaging materials for the utilization of food.

27.5 Australia and Africa

27.5.1 Regulations for Food Packaging Materials in Australia

There is a substantial global trade in packaged food products and part of this market is in Australia/New Zealand, whereby the food packaging value is far-reaching. It allows the transportation of food, prevents microbial contamination, increases shelf life, and transmits information while providing consumer convenience. However, there are a lot of chemicals involved in food. Increased demands for recycled packaging materials and the potential for chemical migration from unknown complex mixtures of recycled and reused packaging materials is also present.

The safety of food supply in relation to the risk of Chemical Migration from Packaging into Food (CMPF) depends on whether the industry in Australia and New Zealand is aware of and voluntarily complies with the US and/or EU regulations and/or other codes of practice (CoP), guidelines, or self-imposed safety requirements that are in line with the requirements of the Food Act.

Resources lost to waste are growing at an unsustainable rate in Australia, which are many times higher than the population growth. In 2010–2011, Australia's municipal waste stood at 628.3 kg per capita, which is 18.5% above the Organisation for Economic Co-operation and Development (OECD) average of 530.2 kg per capita. In addition, some of these trends are driven by the Australian demographics, such as aging populations, increasing single and two-person households, and increasing health issues; some are driven by regulations, such as food standards and standards for therapeutic goods; some are driven by market competition, such as consumer appeal and online retail; and some are driven by voluntary consumer protection measures [47].

In Australia, the National Notification and Assessment Scheme for Industrial Chemicals (NIC-NAS) had developed a working definition of industrial nanomaterials and guidance on testing nanomaterial health effects. In addition, the Food Standards Australia New Zealand (FSANZ) has assessed the capacity of Australia and New Zealand's food regulatory framework on human health risks as a result of nanotechnology under existing legislation [29].

27.5.2 Reducing Environmental Harm in the Natural Environment

The problems behind this scenario are complex: littering is generally linked to consumer behavior in a particular situation; less harmful packaging materials from an environmental point of view may jeopardize the health of the product or consumer, and in the case of takeaway meals, the number of companies, their products and locations, and the number of customers and their consumption. The complexity, however, does not deny the reality. Table 27.6 shows the Australian Packaging Covenant Strategic Plan 2017–2022. In what is considered a global environmental crisis, it is estimated that in every square kilometer of the ocean, there are 18 000 pieces of plastic floating, and thus the marine life of Australia is dramatically affected [47].

27.6 Regulation for Food Packaging Materials in Africa

South Africa's food system is a major contributor to economic growth and food security. However, the food industry is challenged by risks to food safety, such as foodborne diseases, food fraud, and a general lack of effective regulation enforcement. The Government's inability to regulate the food sector effectively is a contributing factor in the increase of risks in food safety. There were 978 cases of laboratory listeriosis with 674 confirmed diseases in South Africa between January 2017 and March 2018 [48]. About 30% of these patients died, according to the WHO [48]. This represents a fatality rate that is higher than the European Union recorded in 2015 [49].

Food contamination from natural toxins, particularly mycotoxins, is a clear and serious problem for the world's agriculture and food industries. One of the most important mycotoxins in cereal commodities is deoxynivalenol (DON or Vomitoxin), group B trichothecene produced by *Fusarium graminearum*, *Fusarium culmorum*, and *Fusarium crookwellense*. DON could mainly lead to vomiting, digestive disorders, oxidative damage, and toxicity to reproduction [50].

27.6.1 Foods Based on Cereals and Wheat Production

The risk of DON exposure to infants (0–4 years) and young children (5–12 years) of certain major mycotoxins was assessed due to in-store maize grains. Infants and children living in the savannah zone in northern Guinea had a high risk of exposure to DON with TDI values of 63 and 25 times higher than the recommended 1 μg/kg bw/day TDI. Children and young children in Nigeria who consumed contaminated maize with mycotoxin were vulnerable to adverse health effects [50].

Wang et al. [50] suggested that exclusive breastfeeding should be encouraged in the country, particularly during the first six months of life; additional food ingredients should be diversified, including other cereals, such as sorghum, millet, and cereals that were less susceptible to mycotoxin contamination.

27.6.2 Beers

DON exposure varied from one country to another among different agro-ecological zones. In some African countries with a relatively developed industry, the raw material DON

Table 27.6 Australian Packaging Covenant Strategic Plan 2017–2022 [47].

Pillars		Key objectives	Strategic priorities and initiatives	Key performance indicators
1. Resource Efficiency	A/sustainable Design: a life cycle approach	Increase the adoption of sustainable design of packaging by: • embedding packaging sustainability in business strategy and processes, • enabling informed packaging design decisions through eco-design assessment; and • ensuring Packaging Sustainability Guidelines adapt and remain relevant to industry and environmental needs.	Development of tools and resources to support and encourage industry performance: 1. guide design, through the continuous evolution of Packaging Sustainability Guidelines, eco-design tools, and education and information; and 2. encourage high performance, through awards and recognition schemes.	By 2020, 90% of signatories have a long-term sustainability strategy in place that: • incorporates the APC Packaging Sustainability Guidelines, • is integrated into business processes; and • has clear and measurable targets. By 2022, 60% of signatories have achieved or exceeded sector specific renewable and recycled content targets based on 2017 levels.
	B/circular economies: waste to resources	Find supply chain solutions to recover packaging and create sustainable closed loop economies.	Bring industry together to identify the barriers to the recovery and reuse of waste packaging, develop strategies to address and build innovation across industry sectors and the supply chain.	By 2022, APCO will have developed proven viable approaches to remove 50% of current problem packaging types or materials from the waste stream. By 2022, 90% of signatories will have actively participated in closed loop collaboration.

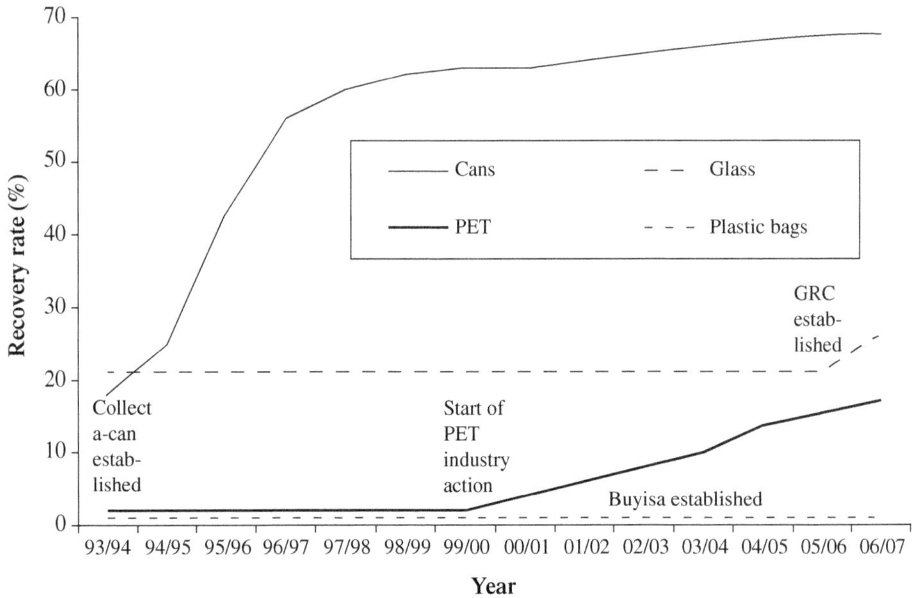

Figure 27.7 *Over time, recovery rates in SA for packaging of cans, glass, PET, and plastic bags [52].*

contamination is lower than in others. The risks of infant and child exposure, however, are not all unacceptable. Although many African countries have begun to develop strategies to reduce the incidence of mycotoxins in food, some African countries are still lacking in food prevention and control of mycotoxins [50].

As an example, palmyra palm leaves are used to weave boxes where cooked food is transported and small banana leaf bags are used to contain coffee beans, which in some parts of Africa are a traditional gift [51].

27.6.3 Food Packaging; Reuse, Reduce, and Recycle

Increased consumption and lack of the culture of recycling in South Africa (SA) resulted in an ever-increasing volume of used packaging, reaching landfill sites. This not only uses increasingly scarce and valuable landfill airspace; it also represents a loss of potentially valuable materials, such as steel, glass, and plastics, which could otherwise have been re-processed into new products, thus displacing the need to exploit virgin raw materials [52]. Figure 27.7 shows the recovery rates in SA for the packaging of cans, glass, PET, and plastic bags.

If the only goal was to increase recycling, that would not be a problem. However, there is a growing demand for packaging with the ever-increasing purchasing power and consumerism in South Africa. Therefore, unless the recovery rates increase sufficiently to absorb the increasing amount of packaging being produced, the amount of packaging reaching the landfill sites will continue to increase in the absence of reduction efforts. This suggests the need for a combination of both regulations (and at least mutually agreed targets) to reduce packaging quantities as well as industry initiatives focused on recycling efforts. This requires both political will in the face of probable opposition to regulations and increased awareness-raising efforts to encourage source separation [52, 53].

Generally speaking, any FCM should not release chemical substances into the food in quantities that could affect human health. Therefore, it is crucial to understand the chemical composition of the packaging material and the levels at which these compounds can be partitioned into foodstuffs – a process known as migration – in order to assess the health impacts associated with food packaging.

The primary purpose of food packaging must continue to preserve food safety, health, and quality. The environmental impact of packaging waste can be minimized by prudently selecting materials, following EPA guidelines, and reviewing the environmental impact expectations of packaging. Knowledgeable attempts by the industry, government, and consumers will encourage continuous improvement, and recognize the functional qualities of packaging will prevent much of the well-intentioned but ill-advised solutions that do not appropriately account for both pre-consumer and post-consumer packaging factors.

27.7 Conclusion

All in all, the growth of packaging products that is mainly from plastics has raised environmental health concerns and packaging legislation worldwide. In the comprehensive regulations, some are focused on certain substances used in food packaging and the reusability/recycling of plastic packaging to reduced disposal issues. In general, the methods provided by the primary agency in regard to food packages are particularly concerning the effect on consumer health and environmental protection. To date, there have been many efforts toward improving the recycling volume of packages and containers by increasing the recycled grades or replacing them with biodegradable materials. This way could be effective in improving the environment health while maintaining the performance of the plastic industry.

References

1. Sanyang ML, Ilyas RA, Sapuan SM, Jumaidin R. Sugar palm starch-based composites for packaging applications. In: *Jawaid, Mohammad, Swain, Sarat Kumar (Eds.), Bionanocomposites for Packaging Applications*. Cham: Springer International Publishing; 2018. p. 125–47. Available at: http://link.springer.com/10.1007/978-3-319-67319-6_7.
2. Ilyas RA, Sapuan SM, Ibrahim R, Abral H, Ishak MR, Zainudin ES, et al. Effect of sugar palm nanofibrillated cellulose concentrations on morphological, mechanical and physical properties of biodegradable films based on agro-waste sugar palm (*Arenga pinnata* (Wurmb.) Merr) starch. *J. Mater. Res. Technol.* 2019; **8**(5): 4819–30. Available at: https://doi.org/10.1016/j.jmrt.2019.08.028.
3. Ilyas RA, Sapuan SM, Ishak MR, Zainudin ES. Water transport properties of bio-nanocomposites reinforced by sugar palm (*Arenga pinnata*) nanofibrillated cellulose. *J. Adv. Res. Fluid Mech. Therm. Sci. J.* 2018; **51**(2): 234–46.
4. Syafri E, Sudirman, Mashadi, Yulianti E, Deswita, Asrofi M, et al. Effect of sonication time on the thermal stability, moisture absorption, and biodegradation of water hyacinth (*Eichhornia crassipes*) nanocellulose-filled bengkuang (*Pachyrhizus erosus*) starch biocomposites. *J. Mater. Res. Technol.* 2019; **8**(6): 6223–31. Available at: https://doi.org/10.1016/j.jmrt.2019.10.016.

5. Halimatul MJ, Sapuan SM, Jawaid M, Ishak MR, Ilyas RA. Effect of sago starch and plasticizer content on the properties of thermoplastic films: mechanical testing and cyclic soaking-drying. *Polimery.* 2019; **64**(6): 32–41.
6. Ilyas RA, Sapuan SM, Ishak MR, Zainudin ES, Atikah MSN, Huzaifah MRM. Water barrier properties of biodegradable films reinforced with nanocellulose for food packaging application : A review. In: *6th Postgraduate Seminar on Natural Fiber Reinforced Polymer Composites 2018.* Institute of Tropical Forestry and Forest Product (INTROP), Universiti Putra Malaysia, Serdang, Selangor; 2018. p. 55–9.
7. Atikah MSN, Ilyas RA, Sapuan SM, Ishak MR, Zainudin ES, Ibrahim R, et al. Degradation and physical properties of sugar palm starch/sugar palm nanofibrillated cellulose bionanocomposite. *Polimery.* 2019; **64** (10): 27–36.
8. Ilyas RA, Sapuan SM, Ishak MR, Zainudin ES, Atikah MSN., 2018 Characterization of sugar palm Nanocellulose and its potential for reinforcement with a starch-based composite. In: S.M. Sapuan, J. Sabari, M.R. Ishak, M.L. Sanyang (Eds.), *Sugar Palm Biofibers, Biopolymers, and Biocomposites.* 1. I Boca Raton, FL: CRC Press/Taylor & Francis Group, CRC Press p. 189–220. Available at: https://www.taylorfrancis.com/books/9780429813542/chapters/10.1201/9780429443923-10.
9. Rhim J, Park H, Ha C. Bio-nanocomposites for food packaging applications. *Prog. Polym. Sci.* 2013; **38** (10–11): 1629–52. Available at: https://linkinghub.elsevier.com/retrieve/pii/S007967001300049X.
10. Ilyas RA, Sapuan SM, Atiqah A, Ibrahim R, Abral H, Ishak MR, et al. Sugar palm (*Arenga pinnata* [Wurmb.] Merr) starch films containing sugar palm nanofibrillated cellulose as reinforcement: water barrier properties. *Polym. Compos.* 2020; **41**(2): 459–67. Available at: https://onlinelibrary.wiley.com/doi/abs/10.1002/pc.25379.
11. Jumaidin R, Ilyas RA, Saiful M, Hussin F, Mastura MT. Water transport and physical properties of sugarcane bagasse fibre reinforced thermoplastic potato starch biocomposite. *J. Adv. Res. Fluid Mech. Therm. Sci.* 2019; **61**(2): 273–81.
12. Ilyas RA, Sapuan SM, Ishak MR, Zainudin ES. Sugar palm nanocrystalline cellulose reinforced sugar palm starch composite: Degradation and water-barrier properties. In: *IOP Conference Series: Materials Science and Engineering.* 2018.
13. Ilyas RA, Sapuan SM, Ishak MR, Zainudin ES. Development and characterization of sugar palm nanocrystalline cellulose reinforced sugar palm starch bionanocomposites. *Carbohydr. Polym.* 2018; **202**: 186–202. Available at: https://linkinghub.elsevier.com/retrieve/pii/S014486171831066X.
14. Ilyas RA, Sapuan SM, Ibrahim R, Abral H, Ishak MR, Zainudin ES, et al. Thermal, biodegradability and water barrier properties of bio-Nanocomposites based on plasticised sugar palm starch and Nanofibrillated celluloses from sugar palm fibres. *J. Biobaased Mater. Bioenergy* 2020; **14**: 1–13.
15. Jumaidin R, Saidi ZAS, Ilyas RA, Ahmad MN, Wahid MK, Yaakob MY, et al. Characteristics of cogon grass fibre reinforced thermoplastic cassava starch biocomposite: water absorption and physical properties. *J. Adv. Res. Fluid Mech. Therm. Sci.* 2019; **62**(1): 43–52.
16. Halimatul MJ, Sapuan SM, Jawaid M, Ishak MR, Ilyas RA. Water absorption and water solubility properties of sago starch biopolymer composite films filled with sugar palm particles. *Polimery.* 2019; **64**(9): 27–35.
17. Jumaidin R, Khiruddin MAA, Asyul Sutan Saidi Z, Salit MS, Ilyas RA. Effect of cogon grass fibre on the thermal, mechanical and biodegradation properties of thermoplastic cassava starch biocomposite. *Int. J. Biol. Macromol.* 2019; Available at: https://doi.org/10.1016/j.ijbiomac.2019.11.011.
18. Ilyas RA, Sapuan SM, Ishak MR. Isolation and characterization of nanocrystalline cellulose from sugar palm fibres (*Arenga pinnata*). *Carbohydr. Polym.* 2018; **181**: 1038–51. Available at: http://dx.doi.org/10.1016/j.carbpol.2017.11.045.
19. Ilyas RA, Sapuan SM, Ibrahim R, Atikah MSN, Atiqah A, Ansari MNM, et al. Production, processes and modification of nanocrystalline cellulose from agro-waste: a review. In: Bebrooz Morabedi (Ed.), *Nanocrystalline Materials.* IntechOpen; 2019. p. 3–32. Available at: http://www.intechopen.com/books/trends-in-helicobacter-pylori-infection/floating-drug-delivery-systems-for-eradication-of-helicobacter-pylori-in-treatment-of-peptic-ulcer-d.

20. Azammi AMN, Ilyas RA, Sapuan SM, Ibrahim R, Atikah MSN, Asrofi M, et al. Characterization studies of biopolymeric matrix and cellulose fibres based composites related to functionalized fibre-matrix interface. In: *Interfaces in Particle and Fibre Reinforced Composites*. 1. London: Elsevier; 2020. p. 29–93. Available at: https://linkinghub.elsevier.com/retrieve/pii/B9780081026656000030.
21. Abral H, Ariksa J, Mahardika M, Handayani D, Aminah I, Sandrawati N, et al. Highly transparent and antimicrobial PVA based bionanocomposites reinforced by ginger nanofiber. *Polym. Test.* 2019; (October): 106186. doi: 10.1016/j.polymertesting.2019.106186.
22. Ilyas RA, Sapuan SM, Ishak MR, Zainudin ES. Effect of delignification on the physical, thermal, chemical, and structural properties of sugar palm fibre. *Bioresources* 2017; **12**(4): 8734–54.
23. Abral H, Basri A, Muhammad F, Fernando Y, Hafizulhaq F, Mahardika M, et al. A simple method for improving the properties of the sago starch films prepared by using ultrasonication treatment. *Food Hydrocoll.* 2019; **93**: 276–83. Available at: https://linkinghub.elsevier.com/retrieve/pii/S0268005X1831539X.
24. Ilyas RA, Sapuan SM, Sanyang ML, Ishak MR, Zainudin ES. Nanocrystalline cellulose as reinforcement for polymeric matrix nanocomposites and its potential applications: a review. *Curr. Anal. Chem.* 2018; **14**(3): 203–25. Available at: http://www.eurekaselect.com/156148/article.
25. Abral H, Ariksa J, Mahardika M, Handayani D, Aminah I, Sandrawati N, et al. Transparent and antimicrobial cellulose film from ginger nanofiber. *Food Hydrocoll.* 2020; **98** (August 2019): 105266. Available at: https://doi.org/10.1016/j.foodhyd.2019.105266.
26. Ilyas RA, Sapuan SM, Ibrahim R, Abral H, Ishak MR, Zainudin ES, et al. Sugar palm (*Arenga pinnata* (Wurmb.) Merr) cellulosic fibre hierarchy: a comprehensive approach from macro to nano scale. *J. Mater. Res. Technol.* 2019; **8**3): 2753–66. Available at: https://doi.org/10.1016/j.jmrt.2019.04.011.
27. Ilyas RA, Sapuan SM, Ishak MR, Zainudin ES. Sugar palm nanofibrillated cellulose (*Arenga pinnata* (Wurmb.) Merr): effect of cycles on their yield, physic-chemical, morphological and thermal behavior. *Int. J. Biol. Macromol.* 2019; **123**: 379–88. Available at: https://linkinghub.elsevier.com/retrieve/pii/S0141813018331477.
28. Sothornvit R. Nanostructured materials for food packaging systems: new functional properties. *Curr. Opin. Food Sci.* 2019; **25**: 82–7. Available at: https://doi.org/10.1016/j.cofs.2019.03.001.
29. Li C, Tangpitayakul S. Southeast Asia food contact. In: Rinus Rijk, Rob Veraart (Eds.), *Global Legislation for Food Packaging Materials*. Wiley; 2010. p. 345–54.
30. Claudio L. Our food: packaging & public health. Vol. **120**, *Environ. Health Perspect.* 2012. p. A232–7.
31. MESTECC. Malaysia's roadmap towards zero single-use plastics 2018–2030. 2018.
32. Ipsos. The last straw: Discouraging single-use plastic. Environment. 2019.
33. India announces new food packaging regulations. Available at: http://PackagingLaw.com. 2019.
34. Yildirim S, Röcker B, Pettersen MK, Nilsen-Nygaard J, Ayhan Z, Rutkaite R, et al. Active packaging applications for food. *Compr. Rev. Food Sci. Food Saf.* 2018; **17**(1): 165–99.
35. Han JH, Ho CHL, Rodrigues ET. Intelligent packaging. In: Jung H. Han, *Innovations in Food Packaging*. Elsevier Ltd, 2005. p. 138-155.
36. Ozdemir M, Floros JD. Active food packaging technologies. *Crit. Rev. Food Sci. Nutr.* 2004;**44**(3); 185–193. doi: 10.1080/10408690490441578.
37. Kerry JP, O'Grady MN, Hogan SA. Past, current and potential utilisation of active and intelligent packaging systems for meat and muscle-based products: A review. *Meat Sci.* 2006; **74**(1):113–130. doi: 10.1016/j.meatsci.2006.04.024.
38. Kruijf DD, Beest V V., Rijk R, Sipiläinen-Malm T, Losada PP, Meulenaer DD. Active and intelligent packaging: applications and regulatory aspects. *Food Addit. Contam.* 2002;
39. Vermeiren L, Devlieghere F, Van Beest M, De Kruijf N, Debevere J. Developments in the active packaging of foods. *Trends Food Sci. Technol.* 1999 Mar; **10**(3): 77–86. doi: 10.1016/s0924-2244(99)00032-1.
40. Curcio M, Puoci F, Iemma F, Parisi OI, Cirillo G, Spizzirri UG, et al. Covalent insertion of antioxidant molecules on chitosan by a free radical grafting procedure. *J. Agric. Food Chem.* 2009; **57**(13):5933–5938. doi: 10.1021/jf900778u.

41. Restuccia D, Salomone R, Spizzirri UG, Saija G, Ioppolo G, Parisi OI, et al. Industrial applications. In: Barros-Velázquez J, editor. *Antimicrobial Food Packaging*. San Diego: Elsevier; 2016. p. 215–27. Available at: https://linkinghub.elsevier.com/retrieve/pii/B9780128007235000164.
42. Restuccia D, Spizzirri UG, Parisi OI, Cirillo G, Curcio M, Iemma F, et al. New EU regulation aspects and global market of active and intelligent packaging for food industry applications. *Food Control* 2010; **21** (11): 1425–35. Available at: http://dx.doi.org/10.1016/j.foodcont.2010.04.028.
43. Muncke J. Food packaging materials. *Food Packag. Forum*. 2012. Available at: https://www.foodpackagingforum.org/food-packaging-health/food-packaging-materials.
44. Department of Health and Human Services. *Code of Federal Regulations Title 21*. U.S. Food & Drug Administration. 2019. Available at: https://www.accessdata.fda.gov/scripts/cdrh/cfdocs/cfCFR/CFRSearch.cfm?CFRPart=174&showFR=1.
45. Heckman JH. Food packaging regulation in the United States and the European Union. *Regul. Toxicol. Pharmacol.* 2005; **42**(1): 96–122. Available at: https://linkinghub.elsevier.com/retrieve/pii/S0273230005000243.
46. Baughan JS, Attwood D. Food packaging law in the United States. In: Rijk RVR, editor. *Global legislation for Food Packaging Materials*. Weinheim, Germany: Wiley-VCH; 2010. p. 223–39.
47. Australian Packaging Covenant Organisation. Australian Packaging Covenant Strategic Plan 2017–2022. 2019; 1–16.
48. WHO. Listeria. 2018 [cited 2019 Apr. 5]. Available at: https://www.who.int/mediacentre/factsheets/listeriosis/en.
49. Boatemaa S, Barney M, Drimie S, Harper J, Korsten L, Pereira L. Awakening from the listeriosis crisis: food safety challenges, practices and governance in the food retail sector in South Africa. *Food Control* 2019; **104**: 333–42. Available at: https://linkinghub.elsevier.com/retrieve/pii/S0956713519302178.
50. Wang L, Liao Y, Peng Z, Chen L, Zhang W, Nüssler AK, et al. Food raw materials and food production occurrences of deoxynivalenol in different regions. *Trends Food Sci. Technol.* 2019; **83**: 41–52. Available at: https://linkinghub.elsevier.com/retrieve/pii/S0924224418306447.
51. Ashok B, Obi Reddy K, Feng H, Varada RA. Preparation and properties of cellulose/Thespesia lampas microfiber composite films. *Int. J. Biol. Macromol.* 2019; **127**: 153–8. Available at: https://linkinghub.elsevier.com/retrieve/pii/S0141813018352085.
52. Mohanty CRC. Reduce, reuse and recycle and resource efficiency as the basis for sustainable waste management. 2011. Available at: https://sustainabledevelopment.un.org/content/dsd/csd/csd_pdfs/csd-19/learningcentre/presentations/May9am/1-Learning_Centre_9May_ppt_Mohanty.pdf.
53. Ilyas RA, Sapuan SM, Atikah MSN, Asyraf MRM, Ayu Rafiqah S, Aisyah HA, Mohd Nurazzi N, Norrahim MNF. Effect of Hydrolysis Time on the Morphological, Physical, Chemical, and Thermal Behavior of Sugar Palm Nanocrystalline Cellulose (Arenga Pinnata (Wurmb.) Merr). *Textile Research Journal*. 2020. In print. doi: 10.1177/0040517520932393.

Index

Bio-Based Packaging: Material, Environmental and Economic Aspects, First Edition. Edited by S.M. Sapuan and R.A. Ilyas